Charcoal Analysis

Methodological Approaches, Palaeoecological Results and Wood Uses

Proceedings of the Second International Meeting of Anthracology, Paris, September 2000

Edited by

Stéphanie Thiébault

BAR International Series 1063
2002

Published in 2016 by
BAR Publishing, Oxford

BAR International Series 1063

Charcoal Analysis

ISBN 978 1 84171 443 1

BAR Publishing is the trading name of British Archaeological Reports (Oxford) Ltd. British Archaeological Reports was first incorporated in 1974 to publish the BAR Series, International and British. In 1992 Hadrian Books Ltd became part of the BAR group. This volume was originally published by Archaeopress in conjunction with British Archaeological Reports (Oxford) Ltd / Hadrian Books Ltd, the Series principal publisher, in 2002. This present volume is published by BAR Publishing, 2016.

Printed in England

BAR
PUBLISHING

BAR titles are available from:

BAR Publishing
122 Banbury Rd, Oxford, OX2 7BP, UK
EMAIL info@barpublishing.com
PHONE +44 (0)1865 310431
FAX +44 (0)1865 316916
www.barpublishing.com

Contents

Introduction

The Second International Meeting of Anthracology held in Paris during four days (13th-16th) in september 2000, brought together for the first time since ten years about 120 scholars in order to discuss the progress of a discipline that is still young: Anthracology. During these four days, fifty five papers and fifteen posters were presented. Three main subjects defined in advance were each introduced by an special paper given by a major personality, respectively: L. Chabal (CNRS, Montpellier), J.-L. Vernet (University of Montpelier II), L. Castelletti (Como Italy).

The first subject, entitled "New approaches in Anthracology", concerned the methodology of the discipline. It could further be divided into three themes:
The first of these related to the development of computerise atlases for the anatomical determination of wood. Since the 1980's, the application of the discipline to all the regions of the world and, in particular to the tropical and subtropical areas of Africa and America, has emphasised the need of wood atlases based on modern specimens in order to identify (to the genus, or at least to the family level) the numerous woody taxas growing in these areas.

The second theme, developed by at least four lecturers, concerned the statistical treatment of data.

Finally the third theme presented the first results obtained by the analysis of "off site" charcoal and tried to estimate the significance of this type of material in the sediments.

Two days were dedicated to the second topic: "Anthracology and Palaeoecology" that involved a majority of the participants. Much progresses was noticed in this domain both concerning the chronological (for example, the analysis of Pleistocene sites were presented) and the geographical extension of the studies. Numerous countries, until then little or not studied at all, now provide us with interesting results on the evolution of plant environments. Let us just quote the work of our German colleagues in Africa, as well as those, very important, concerning ancient fires and vegetation history in Guyana, Brazil, Patagonia (Chile) and on the Easter island, the latter having since long been considered a desert.

The third subject: "Anthracology and Palaeoethnobotany". dealt with the uses of wood. This subject, fully developed during the conference, studied from a more ethnobotnatical point of view various uses of wood since millennia, not only for fuel but also for construction works, crafts or religious purposes, such as graves or cremations.

The present volume shows that the research on the relationship between humans and their environment in the past, by the reconstruction of the past vegetation cover, becomes more and more important. This book constitutes both a state of the question and an overview of research the main topics that are likely to be developed in the near future.

On the lines of the conference, this volume is organised in three different sections. The first deals with progress in methodological issues such as statistics (Piqué and Barcelo, Scheel-Ybert), taphonomy and processing of data (Dufraisse). It also exposes the first results of new applications of charcoal analysis such as the study of ancient fires (Berger and Thiébault, Galop *et al.*, Bazile *et al.*) and quantitative eco-anatomy (Terral).

The second part focuses on the palaeoecology, *i.e.* on the relationship between Man and plants during the upper Pleistocene (Damblon and Haesaerts, Maspero, Ntinou and Kotjabopoulou, Fletcher), and the Holocene (from Heinz to Scheel-Ybert) in various parts of the world. The papers of Delhon and Solari *et al.* show the benefit of the multidisciplinary environmental approach when applied to an archaeological site.

The importance of charcoal analysis, notably by the study of ancient charcoal burning sites, in the reconstruction of the history of is stressed by five papers (from Marguerie to Bonhôte *et al.*). Further, the articles of Gaudin and Montanari *et al.* show the contribution of charcoal studies in the ecological and geographic modelling of past landscapes.

The last part of the volume presents, from an ethnobotanical point of view, studies on (Guiot) different uses of wood: as fuel (Théry-Parisot) or for constructions and wooden artefacts (from Durand to Miejac).

Aknowledgments

It is a pleasure for me to express my gratitude to the various institutions who largely contributed to the success of the Second International Meeting of Anthracology: the French National Centre for Scientific Research (Centre National de la Recherche Scientifique), the French Foreign Ministry (Ministère des Affaires Etrangères), the Ministry of Education and Research (Ministère de l'Education et de la Recherche) the University of Paris X-Nanterre, theCNRS's Mixed Research Unit 7041 (UMR 7041, Archéologies et Sciences de l'Antiquité), the Regional delegation of Ile de France West and North (Délégation Régionale Ile de France ouest et nord,) and the Museum of Popular Art and Traditions (Musée des Arts et Traditions populaires). I would also like to thank the persons who dedicated their time and efforts to this meeting: Jean-Marie Pernaud who is the co-organizer, Claire Delhon, Laurent Fabre, Henri-George Naton, Hugues Pessin and Margareta Tengberg who took care of practical matters during the conference. Finally, I would like to address my thanks Sander van der Leeuw for encouraging me during the preparation of this volume as well as Laura Loth and particularly Margareta Tengberg for reading through the manuscripts and all the reviewers.

Stéphanie Thiébault
UMR 7041 CNRS, Archéologies et Sciences de l'Antiquité, 21 allée de l'Université, 92023 Nanterre cedex, France.

Preface

When opening the Second International Congress of Anthracology, it is particularly pleasant for me to thank the organisation committee that has accomplished a thankless and considerable amount of work to organise these days. Stéphanie Thiébault and Jean-Marie Pernaud have much insisted that I make this introduction. It is a redoubtable honour as your intention is to veneer me as the predecessor. By doing this, you over-estimate my merits as the history of sciences shows that the period we live in and the people we meet are of primordial importance. In fact, at the end of the 1960s, the archaeologists and prehistorians who from that point collected charcoal in a systematic way, in order to obtain 14C datings, suddenly discovered that this material could also be identified (something that had already been done by the Abbot Breuil in 1903!). Moreover, this was a period when not much was known, neither about the prehistoric environment, nor about the ecology of the Quaternary period, especially in the Mediterranean domain. I thus owe my "first steps" in anthracology, at the Botanical Institute of Montpellier, to Armand Pons and Sébastien Santa.

In this period, every identification brought up new information and it was then enough to ask the right questions and to progress step by step:
- What is the difference in signification between charcoal fragments dispersed in the archaeological levels and those found in concentrations?
- Should we count or weigh the charcoal?

At the same time, it was of the outmost importance to accumulate chrono-ecological sequences, illustrated in anthracological diagrams, and to convince students, who were sometimes kamikazes and always stubborn, that charcoal analysis can lead to anything...and here we are in the year 2000.

Since then, the anthracological "tool" continues to improve. Today, we distinguish between "archaeo-anthracology", focusing on archaeological charcoal; "pedo-anthracology", studying charcoal from natural soils and "geo-anthracology", concerning natural contexts others than soils.

The initial research, concerning charcoal from prehistoric and historic contexts remains fundamental, especially when it comes to studying the relationship between humans and their environment. This type of research is moreover favoured by the existence of a privileged relationship between anthracologists and archaeologists. Another domain, more spatially oriented and independent of archaeological sites, has appeared little by little, focusing on the role of natural fires or fires caused by humans for agricultural or pastoral reasons. For the ecologist, these phenomena are major elements in the dynamics of ecosystems.

In fact, many complex situations cannot be understood without combining archaeological, geopedological, geomorphological and ecological approaches. In this way, the anthracologist can approach the concept of landscape, defined as the result of co-evolutionary processes involving both man and the surrounding ecosystems.

Anthracology is a living discipline and its practice, following the tradition of ecological and archaeological fieldwork, has allowed undeniable progress concerning quantitative aspects, as well as sampling methods and treatment of results. Its position, between natural sciences and human sciences, is indisputably an advantage but requires an awareness of the principles of pluridisciplinarity. It is thus clear that the anthracological discipline should play an important role for any contemporary scholar studying environmental changes.

According to his academic training, the anthracologist is either a specialised archaeologist or a specialised ecologist. The ecologist-anthracologist will without doubt be better prepared for treating issues concerning ecosystems, their dynamics and their climatic or anthropogenic evolution. The archaeologist-anthracologist will adopt a radically different approach, where the study of various aspects of human society will lead him to consider natural and anthropogenic environments.

Nevertheless, these specialists cannot really advance in their research if the bases of the discipline are not taken into account. Who will be concerned with the methodology and especially with anthracological computer applications? Who will work on taxonomical issues? Who will create databases? For these reasons, we have to think about how to employ new types of researchers, different from the traditional scholars or university teachers, for example technical personnel, who do not have the constraint of publishing.

In the past ten years, since the first congress held in Montpellier, the anthracological discipline has evolved, much due to scientific and technical progress. The change of perspective, from the isolated site to a larger man-made space, has become one of the major objectives of archaeology. Simultaneously, microscope technology (optical and electronical) has improved, as have computer techniques and the systematic treatment of images. ^{14}C dating by AMS now allows the anthracologist to fulfil one of his oldest dreams: from one charcoal fragment, whatever its size, we will be able to obtain both an identification and a date. The research using stable isotopes should be continued, as well as the measuring of ^{13}C rates directly from charcoal fragments, in order to study the dryness-humidity ratio.

During these last ten years, through numerous publications and Ph.D. theses, the anthracological research has contributed to our knowledge of past environments, the relationship between fires and vegetation communities and, more generally, the human impact on the environment and the formation of landscapes in many different regions on earth: Alaska, South Africa, South America, Australia, Europe, Nigeria, Sudan, Morocco, Syria, Indonesia, etc.

Many articles have been published in major or international reviews, even though none of them is specialised in the discipline. The most important of these are Vegetation History and Archaeobotany, Review of Palaeobotany and Palynology, Quaternary Science Review, The Holocene, Quaternary Research, Journal of Archaeological Science and the Revue d'Archéométrie. To this we have to add the numerous papers published in monographs on specific sites and which represent a natural contribution of natural sciences to archaeology.

The papers of the congress are divided into three different themes. Posters are equally presented.

The first theme concerns methodological issues, including taphonomy, post-depositional processes, computerised identification keys, comparative and quantitative eco-anatomy, statistical anthracology and dating methods.

The second theme, bringing together the majority of the papers, focuses on palaeoecological issues and can be divided into three sections according to the studied contexts:
- traditional sites, allowing the elaboration of chronological and palaeoecological sequences;
- natural sites and sites affected by fire;
- modern coalmine sites.
Various regions of the world are studied: the loess areas of Eurasia, Germany, France (mountains as well as plains), Spain, Italy, Croatia, Greece, Turkey, Burkina Faso, Namibia, Egypt, the Near East, Chilean Patagonia, tropical America and the Easter islands.

The third theme is more ethnobotanical in nature and treats subjects such as the use of wood in different geographical contexts (Arctic, Polynesia, Egypt), traditional construction methods in rural areas and even funerary rituals and mining.

When enlarging its perspectives beyond the traditional fields of archaeology and ecology, anthracological research can contribute in a substantial manner to discussions on biodiversity, especially when it comes to ecosystems, their formation and their exploitation by humans.

In this respect, the understanding of the present requires knowledge of the past. Knowledge of past events - those that occurred during the last million years and especially since the last Interglacial/Glacial periods, from 120 000 BP until today - is necessary in order to apprehend the impact of climatic and anthropogenic factors on the ecosystems, at different time scales that can span over a century as well as a millennium.

Several major research projects have already been undertaken within these parameters in France, such as "Evolution of Hominids" (Evolution des Hominidés), "Palaeoenvironments, life and society" (Paléoenvironnements, vie et société), "Interdisciplinary project on environmental research" (the former PIREN, Programme Interdisciplinaire de Recherche sur l'Environnement), the project of "Global change and biological crisis" (Changement global et crises biologiques), the Late Glacial/Holocene project on the evolution of fauna.

At present, one can hope that future research will be oriented towards a few major issues, most of which are already addressed in this congress:
- Increasing our knowledge of historical botany and vegetation history. The obtained data will allow a better understanding of the evolution of biodiversity, its dynamics and the formation of landscapes. This type of research concerns cave sequences, which, despite their discontinuity, represent important contributions and complete the information obtained from palynological studies carried out in peatbogs or in lacustrine contexts.
- Analysing periods of crisis and their consequences, in order to ascertain the factors required for maintaining biodiversity. Among the environmental crises, fire events are a good indicator of aridity if they can be related to specific phenomena such as the ENSO-El Nino movements, known on a global scale. It is necessary, however, to distinguish between natural fires and fires caused by herding communities. This distinction might seem simple in theory, but is much more difficult to do once in the field.
- Studying the role of human impact on the environment during early (Neolithic) and more recent (protohistorical and historical) periods that resulted in the formation of present ecosystems. Anthracological research has yielded and still yields enormous amounts of results - for example for the Mediterranean area - concerning the privileged domain of "archaeo-anthracology". Recently, "pedo-anthracological" and "geo-anthracological" studies contribute greatly to the progress of the discipline, due to the spatialisation of numerous archaeological studies and also to the influence of ecologists, who are interested in the dynamics of ecosystems.
- Finally, the creation of models, a procedure already applied to some projects, will allow a global approach, in particular concerning the exploitation of resources by prehistoric and traditional societies.

In conclusion, one can hope that besides the more frequent meetings already planned, a web-site for permanent exchanges will be created. It could be called "GRAPE" (Groupe de Recherche en Anthracologie et Paléoécologie).

Jean-Louis Vernet
Professor at the University of Montpellier II (Sciences et Techniques du Languedoc)
Directeur d'Etudes à l'Ecole Pratique des Hautes Etudes
163, rue Augustre broussonet, 34095 Montpellier cedex, France.

Firewood management and vegetation changes: a statistical analysis of charcoal remains from Holocene sites in the north-east Iberian Peninsula

Raquel Piqué and Joan Anton Barceló

Divisió de Prehistòria, Universitat Autònoma de Barcelona, 08193 Bellaterra, Spain.
E-mail: Raquel.pique@uab.es
E-mail: jbarcelo@seneca.uab.es

Abstract

The purpose of the current study is to present some statistical applications on charcoal analysis from the prehistory of Catalonia in order to resolve specific questions concerning firewood management and vegetation changes in the region. We show some results obtained through the application of correspondence analysis to anthracological data published for Holocene sites from the north-east of the Iberian Peninsula. It is our belief that a wide regional database allows us to evaluate taxonomical diversity and its relation to the chronological, social and geographical factors that caused this variety. Our objective is to establish a representative pattern of use and management of firewood. Applying different statistical tests on charcoal taken from the archaeological site at Mas Castellar (Pontòs, Catalonia) allows us to represent the pattern and evaluate the causes of the observed variability. Statistical tests on the data have allowed us to determine the importance of the deposit process formation.

Key-words: *Prehistory, Catalonia, statistical applications, charcoal, firewood managment, vegetation changes.*

Résumé

Cet article présente quelques exemples d'applications statistiques effectuées sur les résultats des analyses de charbons de bois issus d'occupations préhistoriques en Catalogne, dans le but de répondre aux questions concernant la gestion du combustible et les changements de végétation. Les résultats sont issus de l'analyse des correspondances effectuée sur les données publiées de gisements holocène situés au nord-est de la péninsule ibérique. Cette importante base de données permet d'évaluer la diversité taxonomique et sa relation avec les facteurs chronologiques, sociaux et géographiques à l'origine de cette variété. Le but est aussi de montrer la représentativité du modèle d'utilisation et de gestion du combustible. L'application de plusieurs tests statistiques, sur des charbons issus du site de Mas Castellar (Pontos, Catalogne), permet de présenter un modèle et d'évaluer les causes de la variabilité observée. Les tests statistiques démontrent l'importance des modes de formations des dépôts.

Mots-clés : *Préhistoire, Catalogne, statistiques, charbons de bois, gestion du combustible, changement de végétation.*

Introduction

The goal of this paper is to present a hypothesis concerning the causes of variability in charcoal assemblages. This is done by using statistical tests in an effort to understand palaeoenvironmental variation and to explain ancient practices of forest exploitation for obtaining firewood.

We contend that social and economic factors are crucial in the quantitative composition of charcoal assemblages (Piqué, 1999a; Piqué, 1999b). Archaeological charcoal results from, above all, human activity in the region. Charcoal remains are the result of diverse work processes performed by social agents in order to obtain various kinds of social benefits. It follows, therefore, that these work processes are not the result of random activity. Labour is always regulated, organised, and socially determined. Although the environment influences the species that can be used, it does not determine their specific use. We believe that the appearance of charcoal fragments at a site is primarily related to the amount of firewood that was burned and that this amount was not exclusively determined by environmental supply.

Differences between archaeological collection methods and the nature of archaeological sites affect the reliability of any analysis. If sample size and taphonomic processes are not taken into account, the study of data is biased, and the true sources of variability are hidden by spurious variation due to the archaeological processing.

Another factor is that fragmentation of charcoal remains depends on diverse uncontrolled factors: the initial biomass, the morphology and the properties of the wood, the process of pyrolisis, the work of cleaning and maintenance of burning areas, processes after deposition, such as trampling, and finally, the archaeological excavation in question (Smart and Hoffman, 1988; Chabal, 1997). Without a doubt, the process of charcoal fragmentation is one of the least controlled of all the factors that take part in the formation of charcoal assemblages. At the same time, it is the most important factor, given that it is directly related to the lesser and greater appearance of remains.

Considering the above factors, one controversial question is: how do we quantify the remains and for what purpose? The question must be raised whether the quantification of charcoal remains gives us information about the landscape and its transformation, or conversely, whether this only complicates our understanding of landscapes in the past even further. On one hand, the quantification of charcoal remains gives us information about the intensity of use of primary material. On the other hand, we still must take into account the fact that the number of fragments available at the site depends on their state of conservation. The causes of the variability in the charcoal assemblages must relate as much to depositional as to post-depositional factors, among which we include the actions of the archaeologists themselves.

Interpretation of archaeological charcoal assemblages does not depend exclusively on numerical analysis of remains. If quantification goals are not clear, we cannot obtain clear information from data. Numerical analysis should be well connected to prior hypotheses. Before analysing remains numerically , the assumptions on which our study is based must be specified.

Presence/absence data are more likely to lead us closer to the conditioning environmental factors, because the variables are more easily controlled than in frequency analysis. Furthermore, this strategy allows the elimination of distortions caused by differences of criterion and sample collection methods. Frequency analysis provides much more detailed information; however, there are more variables than in qualitative analysis. When the archaeologist is interested in isolating one specific source of variation (notably, ecological conditions) it is more useful to begin with a more controlled data set than with one where a hypothesis on variation sources cannot be made before analysis. This analytical decision does not mean that frequency analysis is inaccurate or overly prone to error. We simply suggest that both analyses should be used to analyse different factors responsible for the formation process of the archaeological record. Quantitative analysis and related techniques can be used to define variation related to the social management of resources, whereas qualitative analysis is more useful for detecting taxonomical variation due to ecological conditions.

This study aims at interpreting the variability in both charcoal analysis records and particularly at attempting to answer the following questions:
- Why are some species absent from charcoal assemblages?
- Why are some species better represented than others in a stratigraphical unit or phase?

Palaeoecological interpretation of charcoal analysis data from a Holocene site in the north-eastern Iberian Peninsula

First, we used an instrumental methodology that permits us to detect clear changes in the use of firewood throughout the sequence of Holocene deposits in the north east of the Iberian Peninsula. Second, we analysed the possible causes of the variability observed in these deposits.

The work was conducted using Holocene site data published by various authors: Bazile-Robert, 1980; Boquer, 1995; Bosch, J. *et al.* 1996; Bosch, A. *et al.* 1998; Burjachs and Ros, 1992; CRPS, 1985; García Argüelles, 1982; Guilaine *et al.* 1982; Jordà *et al.* 1992; Mir and Freixas, 1993; Piqué, 1996a, 1996b, 1997, 1998, 1999c; Ros, 1992, 1993, 1994, 1997, 1998.

One of the problems in comparing charcoal assemblages is the different level of taxonomical classification (species, genus, family) that we use, either due to different criteria used by researchers, due to classification difficulties stemming from anatomical similarity, or to the fact that samples are not in an optimum state. This leads to taxa proliferation that makes comparison more difficult. In some cases, we grouped species of the same genus together, especially when the ecological requirements of the species are not in opposition in order to mitigate the discrepancies mentioned above. This is the case of the genus *Acer, Prunus, Fraxinus, Ulmus, Sambucus* and *Viburnum* that are represented by diverse species according to the authors of the present analysis. Pomoideae and *Pinus* type *sylvestris*, *nigra* and *uncinata* have also been grouped together due to anatomical similarity between taxa.

The data were processed using multidimensional classification methods, specifically the analysis of binary correspondences developed by M. Jambu (Jambu, 1995; Baxter, 1994) with the aim of determining whether or not the relationship between absent and present species is related to chronology, altitude, and geographical position of sites. We have used the software package STAT-LAB, from SLP, Inc. because it allows the use of binary input (Comby and Sagot, 1994).

In accordance with our initial hypothesis, we employ presence/absence data of species in various deposits in the north-east of the Iberian Peninsula (fig. 1). The most notable feature of the results provided by the analysis of binary correspondences is the low contribution of the axes in explaining the variability of the assemblages. The first six factorial axes explain only 46.6% of the variance. This fact should be interpreted in terms of the high diversity between assemblages.

Although the correlation is rather low, the species tend to be associated with a gradient of bioclimatic conditions coherent with their present day ecological requirements. This is demonstrated in the diagram of axes 1 and 2 (fig. 2). Only 23.08% of total variance can be explained in ecological terms. Taxonomical variation between sites is also generated by other factors, which are related to the specific social conditions of each site. The plotting of axes 3 and 4, explaining more than 12% of total variance, and where we have not detected any regular pattern related to differential landscape features, shows this.

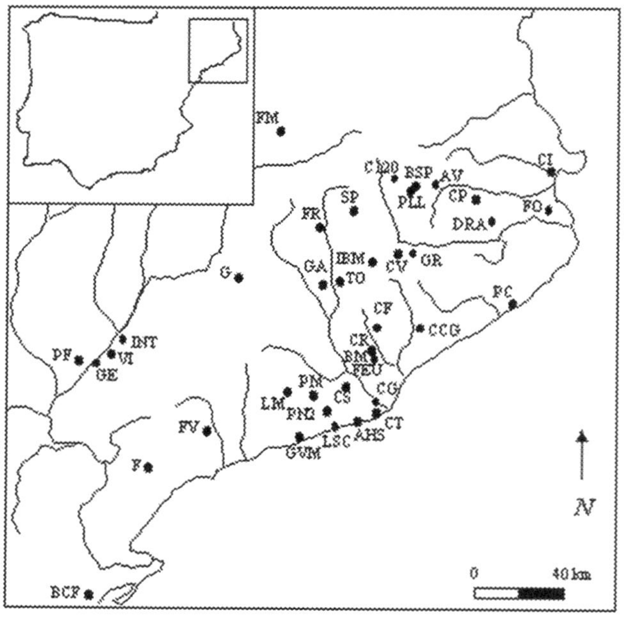

Figure 1:Location of archaeological sites in the northeast of the Iberian Peninsula. (AHS), Bauma del Gai (GA), Bauma Guilanyà (G), Bauma del Serrat del Pont (BSP), Bòbila Madurell (BM), Ca n'Isach (CI), Camí de Cal Grau (CCG), Can Feu (CF), Can Guardiola (CG), Can Roqueta (CR), Can Sadurní (CS), Can Tintorer (CT), Cingle Vermell (CV), Cova 120 (C120), Cova de l'Avellaner (CA), Cova del Frare (CF), Cova del Toll (TO), Cova d'en Pau (CP), Feixa del Moro (FM), Filador (F), Font del Ros (FR), Genó (GE), Grioteres (GR), Institut Batxillerat Manlleu (IBM), Intervenció 46-47 (INT). La Draga (DRA), La Fonollera (FO). La Massana (LM), La Solana (LSC), Les Guàrdies (GVM). Plansallosa (PLL), Pou Nou 2 (PN2), Puig Castellet (PC), Pujolet Moja (PM), Punta Farisa (PF), Sota palou (SP), Vilot (V).

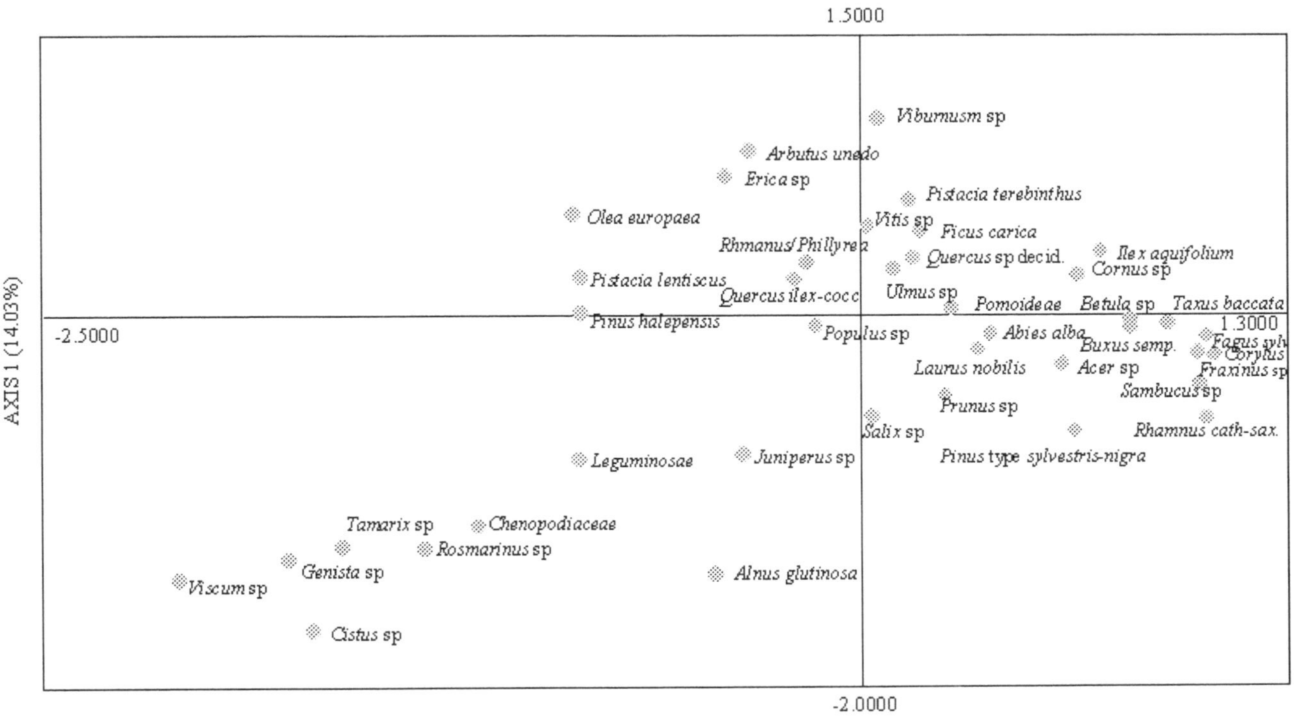

Figure 2: Binary Correspondence Analysis of archaeological sites in the north-east of the Iberian Peninsula; diagram of species in axes 1 and 2.

Axis 1 shows the highest amount of total variance (14.03%). Here, the supra-oromediterranean species are situated at the positive pole with respect to the thermo-mesomediterranean ones, which are positioned at the negative pole. On axis 2 (9.05%) *Alnus Glutinosa, Cistus* sp., *Juniperus* sp., and *Tamarix* sp., appear associated on the negative pole. *Arbutus unedo, Erica* sp., *Olea europea*, *Quercus* sp. deciduous, *Rhamnus-Phillyrea*, and *Viburnum* sp. all fall within the opposite group, and appear at the positive extreme of axis 2.

With reference to the variability among deposits (fig. 3), it should be noted that the Mesolithic period deposits, independent of their geographical location, have a very similar

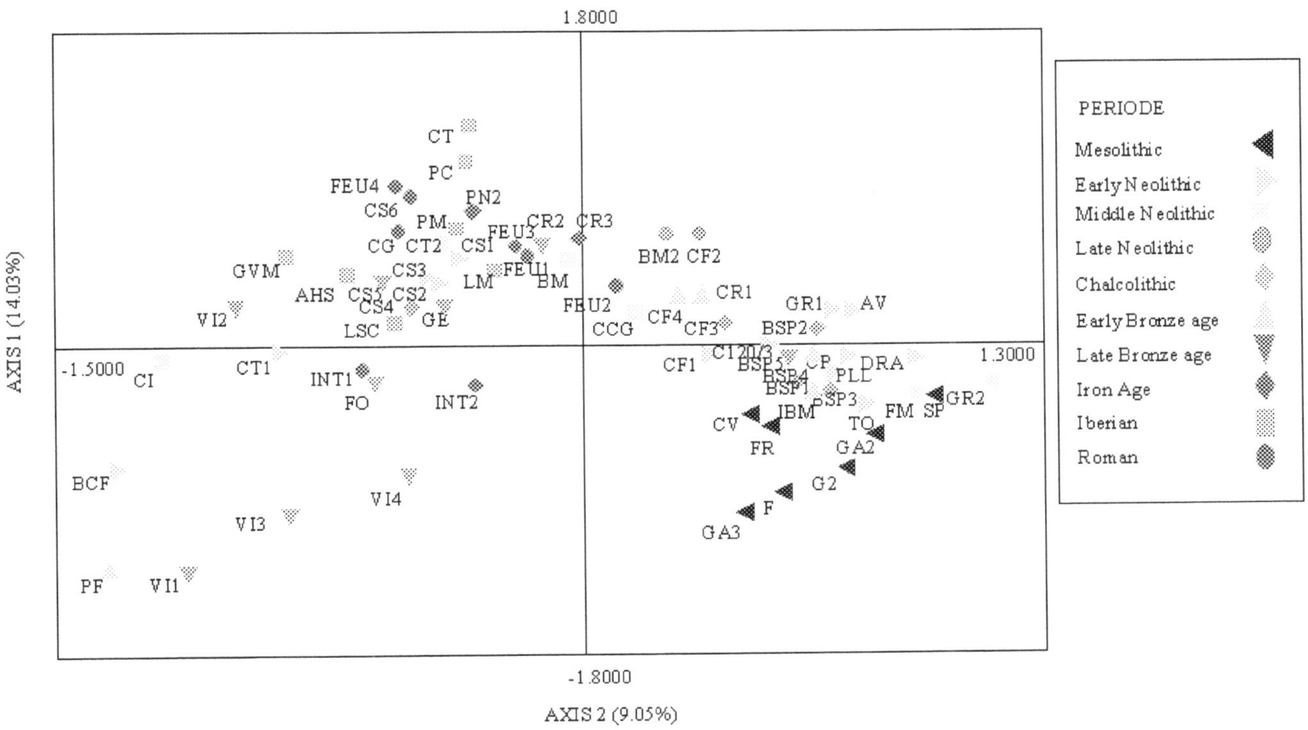

Figure 3: Binary Correspondence Analysis of archaeological sites in the north-east of the Iberian Peninsula; diagram of sites in axes 1 and 2.

taxonomic composition. The same is true for Iberian and Roman deposits, while the Neolithic, Chalcolithic and Bronze Age deposits do not show a regular pattern. The observed similarity during the Mesolithic period could be explained by the homogeneous nature of the landscape in the north-east of the peninsula. There is a clear predominance of supra-oromediterranean vegetation. Another explanation could be strategies of resource exploitation closely conditioned by environmental supply. Nevertheless, the cause of the similarity among the Iberian and Roman deposits would be different, as they are related to specialised forest exploitation practices, which are characteristic of complex societies. The taxonomical similarities among these later deposits are not so much a result of environmental supply as of growing energy necessities. This change in energy needs could have led to a search for production methods that maximised efficiency.

Beginning in the Neolithic period, deposits situated in north Catalonian counties contrast dramatically with those in southern, interior and coastal counties. The North is currently dominated by supramediterranean vegetation, while in southern, interior and coastal counties, termomediterranean vegetation figures predominately. Figure 3 shows the northern deposits situated on the positive pole, axis 1, and the southern deposits on the negative pole, axis 2. The deposits situated in the intermediate zone between humid and dry Catalonia are positioned in the central part of the diagram. It is interesting to note the opposition between deposits situated in the interior, with a more continental climate, and those localised on the coast. Other researchers have already documented differentiation of landscape vegetation in varying bioclimatic regions of the north east of the peninsula by using charcoal analysis and palynological data (Ros and Burjachs, 1992; Ros, 1992; Riera and Esteban, 1994).

In conclusion, we suggest that the taxonomical composition of the assemblages studied reflects the broad features of the variations in the supply of timber resources produced by transformations of the landscape encountered and caused by climatic phenomena of the Holocene period.

It is also possible to detect differences in the landscape caused by latitude and altitude variations, a phenomenon that develops in the north eastern area of the peninsula beginning in the Neolithic period and is well documented in the pollen analyses of the zone. The particular strategies in taking advantage of timber resources developed by specific societies could be the reason for the distribution of Iberian and Roman deposits that seem to indicate a similar pattern in the exploitation of these resources.

Analysis of principal components of quantitative variables shows similar results. The total variance explained by the first six components is almost identical to the qualitative analysis sample (46.29%). Nevertheless, the general layout is much less regular: the first factor of the quantitative analysis only explains 10% of the variance, as opposed to the 14% explained by the quantitative analysis. This axis only allows us to distinguish one homogenous deposit group characterised by the high proportion of *Cistus* sp., *Pistacia lentiscus, Rosmarinus officinalis, Alnus glutinosa* and *Pinus halepensis*. No other homogenous assemblage can be identified. The second axis (8% variance) encompasses those sites with high values of the following taxa: *Arbutus unedo, Erica* sp., *Quercus ilex-coccifera*.

This pattern of site variability suggests the influence of other factor apart from environmental variability. We have seen that the absence/presence pattern of variability can be explained partially with reference to ecological factors, but the causes of the differences in the frequencies of each taxon go much beyond ecological diversity. Many different social processes, including the research activity itself, are responsible for the variations detected in the number of fragments between sites and periods.

Charcoal analysis and forest exploitation techniques

In the first part of this paper, we have shown how the taxonomic composition of the assemblages studied reflects the broad features of transformations experienced in the landscape over the passage of time. Nevertheless, we consider that the exploitation of new resources does not always indicate a supply change in these environmental resources, but could be due to the appearance of new needs in energy production (related to population increase and artisan activities) and the development of other methods of resolving these needs (such as the intensification of gathering or the specialisation in fuel production).

The results obtained from the study of archaeological charcoal remains at the Mas Castellar site in Pontós (Girona, Spain) are presented here. Four phases of occupation in a time scale between the fifth and second centuries BC (Pons, 1997) have been identified. First, the object of the present

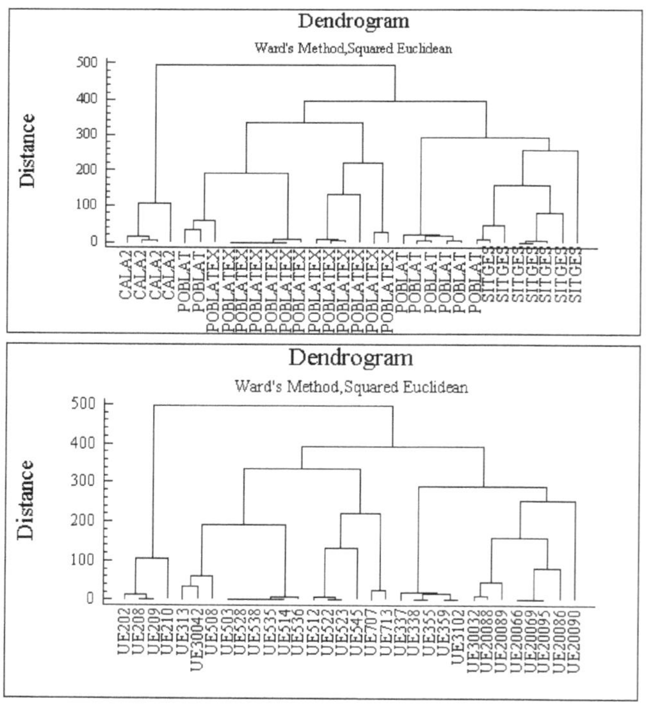

Figure 4: Euclidean distances between stratigraphic units for phase 3. Mas Castellar, Pontòs (Girona, Spain).

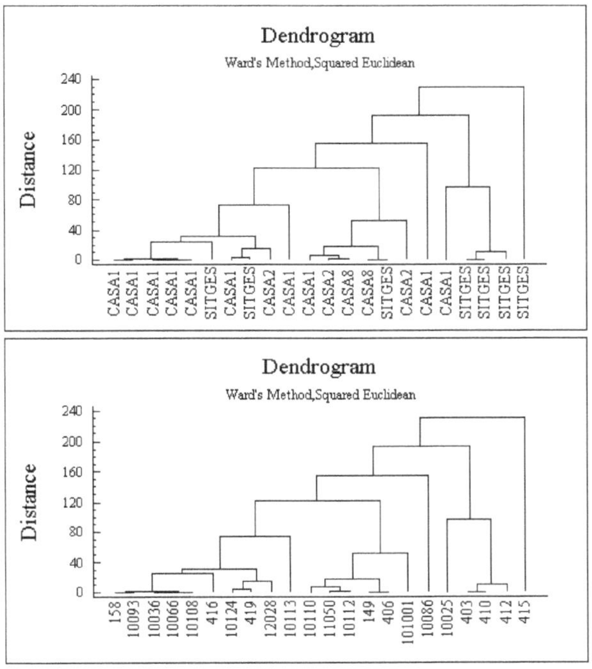

Figure 5: Euclidean distances between stratigraphic units for phase 5b. Mas Castellar, Pontòs (Girona, Spain).

analysis was to study the pattern of provisioning and use of fuel in each of the documented phases and also to evaluate any possible differences and to analyse the causes. The stratigraphic units studied permit us to document a high heterogeneity in the processes of formation and function. For this reason the first question was to know up to what point the observed variability between chronological phases and stratigraphic units was caused by particular depositional and post-depositional formation processes of the compared units.

Second, we raised the question whether once the processes of site formation were eliminated as a cause of variability, we could establish changes in the forest exploitation practices over the passage of time.

The stratigraphic units in phases 3 and 5b proved to best demonstrate the heterogeneity of the site formation processes. In order to analyse the significance of the observed differences between various units and their possible causes, we calculated the Euclidean distances between them for each one phase. We have used the Ward method for this calculation, using algorithms in the SYSTAT 5.2 software package. The statistical test was applied, taking into account the percentages of the species in the stratigraphic units with a number of remains above 10. The application of this test allows us to check the statistical proximity of the stratigraphic units according to their taxonomical composition, with the majority grouping themselves according to the sectors they come from. These groupings also coincide with specific formation processes in each sector. So in the resulting dendrogram (fig. 4) for phase 3, archaeological features such as the function of the area studied and the process of incorporation of charcoal in the deposits have without doubt intervened decisively in the taxonomic composition of the assemblages. The cereal processing pit and indications of incendiary fires in Unit Cala2, and the village outside the walls, called poblatex on the diagram, or deposit of waste in the rubbish dumps (Unit abocador 38) and the silos (called in the graphic sitges) are good examples of this.

The resulting dendrogram for phase 5b reflects, on the other hand, that the majority of the stratigraphic units of the rural establishment have a very close taxonomical composition, regardless of the formation process causing the assemblage. Nevertheless, some of the stratigraphic units from silo 101 are exceptions to this rule, and others reflect perhaps other specific uses of timber (for example post holes and a ritual deposit which was identified in silo 101).

Phases 4 and 5 were represented by stratigraphic units that were configured by formation processes similar to the strata of silos reused as rubbish dumps. No significant differences between stratigraphic units can be noted in their taxonomical composition.

Given that formation processes of the units studied are one of the causes of the observed variability in charcoal assemblages, it is necessary to eliminate any possible source of distortion before looking for differences or similarities in the pattern of firewood exploitation through time. For this reason we have eliminated those stratigraphic units not related to the provisioning of firewood (ritual areas, incendiary fires) when processing the total data of each phase.

With the goal of analysing the differences or similarities between the 4 phases of Mas Castellar, we have applied a correspondence analysis to the resulting contingency table from these data, using the algorithms in SPSS. In the graphic representation of the first two factors of this analysis (which explain 82% of the variability of the assemblage), we can observe a clear differentiation between phases (fig. 5). The curved or horseshoe pattern of the graph is a consequence of the Guttman effect, and it is natural to correspondence analysis in the presence of seriation structures. According to Baxter (1994), when phases at either end of the spectrum have nothing in common, they are "forced" to be approximately equidistant on the plot, because the second axis appears approximately to be a quadratic function of the first. Our study shows that as a consequence, the oldest levels are notably different from the more recent ones.

The oldest period of the site, phase 3, distinguishes itself clearly from the others due to a significantly greater presence of *Erica* sp., and of other shrub-like species, which are rare in the later phases. In the more recent periods (phases 4, 5a and 5b) evergreen *Quercus* sp. (type *ilex-coccifera*) and deciduous *Quercus* are predominant. The differences among these later phases are centred on the frequency of appearance of deciduous *Quercus* (less represented in phase 5a). Thus, what is observed in this ordering of the phases, according to their taxonomic composition, is a product of a change in the exploitation pattern of the timber resources. This pattern is oriented towards species of the shrub strata of the forest (*Erica* sp.) during phase 3 and towards tree strata of the forest formations in phases 4, 5a and 5b (*Quercus ilex-coccifera* and deciduous *Quercus* sp.).

This change in forest management may be a product of a change in overall production strategy of timber resources

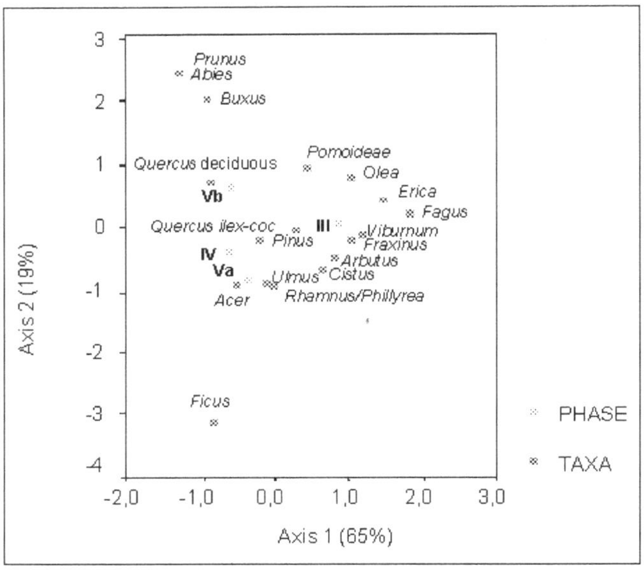

Figure 6: Simple Correspondence Factorial Analysis between the 4 phases of Mas Castellar, Pontòs (Girona, Spain).

among the phases studied. We should not forget that between the oldest phase of Mas Castellar and the later ones, there was a transformation in the types of settlement, which changed from a large, walled village to a small, rural establishment. This could have caused a decrease in the quantity of production activities during the more recent phases and, for this reason, less pressure on the local environment. These conditions make it unnecessary to resort to rather inefficient firewood (like shrubs), which appear in very small quantities in the archaeological record.

The application of different statistical tests on charcoal data from Mas Castellar allows us to interpret the pattern of use and management of fuel for each of the individualized episodes. Archaeological charcoal fragments are the result of diverse work processes which a society undertakes to obtain the various types of goods it requires to satisfy its social needs (Piqué, 1999b). The social dimension of the charcoal remains needs to be taken into account in the interpretations made of them. The results of the present study have permitted us to verify that positional factors in the formation of strata have influenced much of the charcoal distribution within the space. In conclusion, we consider that an analysis of the formation process of the studied assemblage is an essential step to be taken before we can understand how firewood was used.

References

BAXTER M. 1994 - *Exploratory Multivariate Analysis in Archaeology*. Edinburgh: Edinburgh University Press.

BAZILE-ROBERT E. 1980 - Les groupements à *Amygdalus* et *Prunus* de la fin du Tardiglaciaire et du début du Postglaciaire en Méditerranée Nord-occidentale. *Geobios*, 13 fasc.5, p. 777-781.

BÓQUER S. 1995 - *El jaciment de l'institut de batxillerat Antoni Pous*. Barcelona: Dept. de Cultura de la Generalitat de Catalunya.

BOSCH A. *et al.* 1998 - El poblat neolític de Plansallosa. *L'explotació del territori dels primers agricultors-ramaders de l'Alta Garrotxa*, Girona: Museu Comarcal de la Garrotxa.

BOSCH J. *et al.* 1996 - El Barranc d'en Fabra: un assentament neolític a l'aire lliure. *Tribuna d'Arqueologia* 1994-1995. p. 51-62.

BURJACHS F., ROS M.T. 1992 - *Paleoambient de l'Epoca Neolítica en el NE de la Península Ibèrica*. IX Colloqui Internacional d'Arqueologia de Puigcerdà. Estat de la Investigació sobre el Neolític a Catalunya. Puigcerdà i Andorra 1991.

C.R.P.E.S. 1985 - *Sota Palou, Campdevànol. Un centre d'Intervención Prehistòrica postglaciar a l'aire lliure*. Sèrie Monogràfica nº 5. Girona: Diputació de Girona.

CHABAL L. 1997 - Forêts et sociétés en Languedoc (Néolithique final, Antiquité tardive). L'anthracologie, méthode et paléoécologie. *Documents d'Archéologie Française*, 63. Paris: Editions de la Maison des Sciences de l'Homme.

COMBY D., SAGOT A. 1994 - *STAT-LAB by SLP. Le logiciel d'exploration de données*. Guide de l'Utilisateur. Ivry : Statistiques-Logiciels-Pédagogie.

GARCIA ARGÜELLES P. 1990 - Estrato 4 de Filador (Priorat, Tarragona): un ejemplo de transición Epipaleolítico-Neolítico en el Sur de Catalunya, *Saguntum* 23, p. 61-76.

GUILAINE J. *et al.* 1982 - Prehistoric Human Adaptations in Catalonia (Spain). *Journal of Field Archaeology*, Vol. 9, p. 407-416.

JAMBU M. 1995 - *STAT-Lab. Logiciel d'Analyse des Données*. Paris : Slp-Statistiques.

JORDÀ J.F., MORA R., PIQUÈ R. 1992 - La seqüència litoestratigràfica y arqueològica del yacimiento de la Font del Ros (Berga, Barcelona). *Cuaternario y Geomorfologia*, vol. 6, p. 21-30.

MIR A., FREIXAS A. 1993 - La Font Voltada, un yacimiento de finales del Paleolítico Superior en Montbrión de la Marca (La Conca de Barberà, Tarragona). *Cypsela* X, p. 13-21.

PIQUÉ R. 1996a - Aproximación a l'entorn vegetal durant el Paleolític y Mesolític a la vessant sud de los Pre-Pirineus a partir de los carbons vegetals Actes X Colloqui Internacional d'Arqueologia de Puigcerdà. Cultures y Medis. *De la Prehistòria a l'Edat Mitjana. Puigcerda y Oceja* 1994.

PIQUÉ R. 1996b - La gestió dels recursos forestals a La Draga (Banyoles, Pla de l'Estany), *Rubricatum* 1, p. 57-63.

PIQUÉ R. 1997 - Anàlisi Antracològica. In: MARTÍ, M. Excavacions arqueològiques a la ronda Sud de Granollers. *Excavacions Arqueològiques a Catalunya*, 14. Barcelona: Departament de Cultura de la Generalitat de Vatalunya. p. 229-231.

PIQUÉ R. 1998 - Dinàmica paleoambiental entre el 80.000- 5.000 BP al nord-est de la Península Ibèrica: les dades antracològiques, *Revista d'Arqueologia de Ponent*, 8, p. 7-17.

PIQUÉ R. 1999a - Quantification in archaeobotany: charcoal analysis and fire-wood management" New Techniques for old times. Computer applications and quantitative methods in archaeological research *BAR*. p. 189-200.

PIQUÉ R. 1999b - *Producción y uso del combustible vegetal: una evaluación arqueológica*. Treballs d'Etnoarqueologia, 3. Madrid: Consejo Superior de Investigaciones Científicas and Universitat Autònoma de Barcelona.

PIQUÉ R., BARCELÓ J. A., NOGUERA M. 1999c - *Transformaciones del paisaje durante el Pleistoceno y el Holoceno en el Nordeste de la Península Ibérica: una evaluación a partir del análisis antracológico* Actas del 3° Congresso de Arqueologia Peninsular 1999. Porto.

PONS E. 1997 - Estructures, objectes i fets culturals en el jaciment protohistòric de Mas Castellar-Pontós. *Quaderns de Prehistòria i Arqueologia de Castelló*, 18 p.

RIERA S., ESTEBAN A. 1994 - Vegetation history and human activity during the last 6000 years on the central Catalan coast (northeastern Iberian Peninsula), *Vegetation History and Archaeobotany*, 3, p. 7-23.

ROS M.T. 1988 - L'aplicació de l'anàlisi antracològica a l'arqueologia catalana. *Cota Zero* n°4, p. 51-60.

ROS M.T. 1992 - Les apports de l'Anthracologie à l'étude du paléoenvironnement végétal en Catalogne (Espagne). Les Charbons de Bois, les Anciens Ecosystèmes et le Rôle de l'Homme, Colloque de Montpellier, 1991. *Bulletin de la Société Botanique de France*, 139, Actualites Botaniques (2/3/4), p. 483-493.

ROS M.T. 1993 - Análisis antracológico del yacimiento del Bronze Medio de Punta Farisa (Fraga, Huesca). *Estudios de la Antigüedad*, 5, p. 45-47.

ROS M.T. 1994 - Anàlisi Antracològica in: ALCALDE G. *et al. Procés d'ocupació de la Bauma del Serrat del Pont a partir del 1450 AC*. Girona: Museu Comarcal de la Garrotxa.

ROS M.T. 1997 - Anàlisi Antracològica in: ALCALDE G. *et al. Procés d'ocupació de la Bauma del Serrat del Pont entre 2900 i el 1450 cal AC*. Girona: Museu Comarcal de la Garrotxa.

ROS M.T. 1998 - Anàlisis Antracológica in: MAYA J.L., CUESTA F., LOPEZ CACHERO J. Genó. Un poblado del Bronce Final en el Bajo Segre (Lleida). Barcelona: *Publicacions Universitat de Barcelona*. p. 187-190.

SMART T.L., HOFFMAN E.S. 1988 - Environmental interpretation of archaeological charcoal in: HASTORF C.A., V.S. POPPER *Current Paleoethnobotany: Analytical Methods and Cultural Interpretations of Archaeological Plant Remains*. Chicago:The University of Chicago Press. p. 167-205.

Evaluation of sample reliability in extant and fossil assemblages

Rita Scheel-Ybert

CNRS-UMR 5059, Paléoenvironnements, Anthracologie et Action de l'Homme,
Institut de Botanique, 163, rue Auguste Broussonnet, 34090 Montpellier, France.
E-mail: scheel@crit.univ-montp2.fr

Abstract

In palaeoenvironmental studies, evaluation of samples reliability is usually conducted by analysing saturation curves, but Gini-Lorenz concentration curves provide a very interesting complementary method. Gini-Lorenz index was first employed in phytosociological studies on European extant herbaceous formations and in anthracological studies in the Mediterranean region. Analysis of 68 phytosociological surveys of various extant Brazilian plant associations attempted to verify whether or not this approach is also valid for studies in the inter-tropical region. We propose that Gini-Lorenz indices give a measure of vegetation diversity, *i.e.* the higher the index, the lower is diversity in the analysed sample. It can be employed for the estimation of sampling efficiency, since small samples present elevated diversity indices that are "too high", because the vegetation structure is not well represented. These results were applied to the analysis of anthracological samples from Sambaqui do Forte (Rio de Janeiro State, Brazil). Analysis of saturation and Gini-Lorenz curves suggest that a minimum of 200 or 300 charcoal fragments per sample is necessary to give a representative image of the past vegetation.

Key-words: *sampling efficiency, palaeoenvironment, plant diversity, ecology.*

Résumé

La construction de courbes de Gini-Lorenz constitue une méthode complémentaire à celle des courbes de saturation dans la définition de l'effectif minimal de l'échantillon à analyser pour des reconstitutions paléoenvironnementales. L'utilisation de cet indice a été introduite dans le cadre d'études phytosociologiques sur des formations herbacées actuelles d'Europe et d'études anthracologiques en région méditerranéenne. Afin de confirmer la validité de cette approche en milieu tropical, nous avons analysé 68 relevés phytosociologiques actuels de diverses formations végétales brésiliennes. Nous avons vérifié que l'indice de Gini-Lorenz fournit une mesure de diversité de la végétation : plus fort est l'indice, plus basse est la diversité de l'ensemble analysé. Il peut être utilisé pour évaluer la suffisance de l'échantillonnage, car un échantillon insuffisant présente en général une diversité trop élevée, parce que la structure de la végétation n'est pas représentée de façon adéquate. Ce résultat a été appliqué à l'analyse du matériel anthracologique du Sambaqui do Forte (État de Rio de Janeiro, Brésil). Pour chaque niveau de ce site, des courbes de saturation et des courbes de concentration de Gini-Lorenz ont été construites. Nous en avons conclu qu'un minimum de 200 à 300 fragments de charbons de bois par échantillon est suffisant pour obtenir une image représentative de la végétation.

Mots-clés : *suffisance de l'échantillonnage, paléoenvironnement, diversité végétale, écologie.*

Introduction

Reliable palaeoenvironmental reconstructions depend on the number of elements analysed (charcoal pieces, pollen grains, diatoms etc.). The minimal sample is currently defined by the construction of saturation curves, but the study of Gini-Lorenz concentration curves may provide an interesting complementary approach.

This approach was first applied to palaeoenvironment in an anthracological study from the Mediterranean region (Chabal, 1991, 1997), where Gini-Lorenz concentration index evaluated for archaeological charcoal samples is frequently of 20:80. This author maintains that under Mediterranean or temperate climatic conditions this index does not vary much in the nature, nor in the anthracological spectra. Otherwise, it would reflect ecological disturbances. L. Chabal recommends that the optimal sample number should be defined by the stabilization of Gini-Lorenz concentration index for curves constructed for an increasing number of charcoal pieces. An optimal sample effective of 300 to 400 charcoal pieces, per stratigraphic unity, might assure the statistical validity of the anthracological study.

Considering this information, we tried to answer some questions concerning the tropical vegetation:
1 - is this approach valid for woody plant formations from tropical regions, *i.e.* can Gini-Lorenz concentration indices be applied to these formations?
2 - if so, are the "normal" indices also of 20:80?
3 - is a stabilization of the Gini-Lorenz concentration index verified in the analysis of ancient samples, and can this stabilisation give an estimate of the optimal sample effective?
4 - what does the Gini-Lorenz concentration index represent?

Gini-Lorenz curves: theoretical approach

French researchers introduced the application of Gini-Lorenz curves to vegetation studies in floristical and phytosociological analysis of extant herbaceous formations (Poissonet, 1968; Jacquard *et al.* 1968; Poissonet and Poissonet, 1969). These curves represent the frequency of each species in relation to the floristic richness of the sample (Daget and Poissonet, 1971; Chabal, 1997). Cumulative relative frequencies (in percentages) are plotted in ordinates, and species rank percentages (*i.e.* cumulative percentage contribution, in relation to the total number of species, of each species) in abscissas (fig. 1). The species are classified in decreasing order of relative frequencies. We used IVI (index of importance value) as a second ranking criterion. Gini-Lorenz index is defined by the intersection of the curve with the second diagonal of the plot.

Considering the equilibrium constants in the floristical composition of plant communities, Poissonnet (1968) noticed a relation between species frequency and specific contribution, which he associated with a concentration law of 20:80, or "Gini-Lorenz law". The 20:80 index means that 20 % of the species from a community comprise 80 % of the individuals observed. Poissonnet (1979) assumes that a concentration index of 20:80 is characteristic of a steady plant community. When submitting an herbaceous phytocoenosis to chemical disturbance, the author observed that a new equilibrium state, different from the previous one, was attained some time after. In this second state, proportions between species and their relative importance in the community varied, but the concentration index remained the same. However, it has never been demonstrated whether unstable plant communities present concentration indices other than 20:80.

species rankink	Ni	IVI	$F = \frac{Ni \cdot 100}{\Sigma Ni}$	F cumulate	$\frac{rank. 100}{Nsp}$
1	300	24,66	12,55 %	12,55 %	2 %
2	252	19,29	10,54 %	23,09 %	4 %
3	249	25,93	10,41 %	33,50 %	6 %
4	227	18,12	9,49 %	42,99 %	8 %
5	140	15,55	5,86 %	48,85 %	10 %
...	...	...	...	...	...
49	1	0,24	0,04 %	99,96 %	98 %
50	1	0,16	0,04 %	100,00 %	100 %

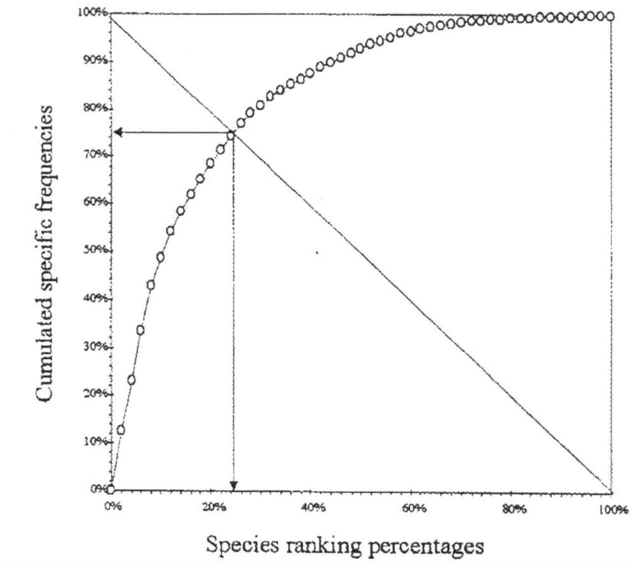

Figure 1: Construction of Gini-Lorenz curves (Total number of individuals = 2391; Number of species = 50. Data taken from Trindade, 1991).

The specific frequencies obtained from twelve different methods of phytocoenosis analysis of the same grassland community indicate that, provided reliable sampling is carried out, any method of phytosociological study leads to symmetrical Gini-Lorenz concentration curves characterized by an index of 20:80 (Poissonnet and Poissonnet, 1969; Ogereau and Poissonnet, 1988).

Extant tropical plant formations

In order to verify if this approach can be applied under tropical conditions, Gini-Lorenz curves of 68 phytosociological surveys were analyzed (fig. 2). The best represented vegetation types are semi-deciduous forest, "restinga" (Brazilian characteristic coastal vegetation), "cerrado" (Brazilian savanna) and riparian forest.

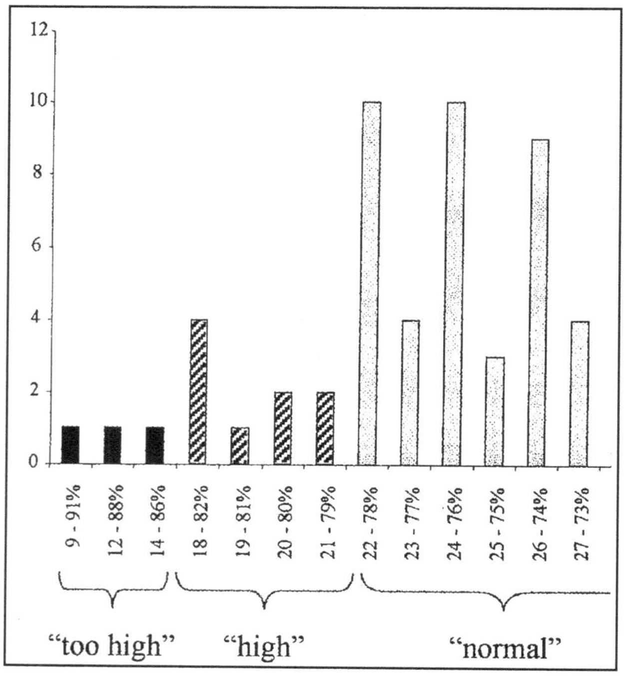

Figure 2: Distribution of Gini-Lorenz indices in 68 phytosociological surveys from the extant vegetation.

These curves show Gini-Lorenz indices varying from 40:60 to 9:91. 20:80 indices are not very frequent (fig. 3). Most of the analysed surveys present indices between 28:72 and 22:78 (*c.* 25:75); modal classes all within this interval, and these indices are considered as "normal". Indices between 21:79 and 18:82 (*c.* 20:80) are considered as "high". Indices higher than 15:85 are considered "too high" and those lower than 29:71, "too low".

In order to explain the variation in the concentration indices, each study was analysed following information given by the authors (tabl. 1). Gini-Lorenz indices are not characteristic of particular formations, or of sampling methods or inclusion criteria. The more usual sampling methods are plots, transects, quadrats and points. The more frequent inclusion criteria are diameter at breast height (DAP 2,5 to 10 cm), diameter at the base (DAB 2 to 10) and minimum height (Hmin 50 cm to 2 m). Frequently, each station is divided into various components (shrubby-arboreal formation in opposition to herbaceous, trees of different height classes etc.).

The number of individuals sampled varies from 80 to 4718. All surveys presenting Gini-Lorenz indices that were "too low" correspond to sampling of less than 500 individuals. There is no direct relation between individual numbers and sampling area, for it depends also on sampling methods and vegetation characteristics. Sampling area varies from 2,6 ha to less than 0,008 ha. Surveys associated with larger areas always present indices higher than 28:72, but in the case of stations with very reduced area (< 0,4 ha) "too low" concentration indices are much more frequent (tabl. 1).

We believe that concentration indices "too low" are associated with insufficient sampling, where the vegetation structure is not adequately represented.

However, "too high" Gini-Lorenz indices are associated with a high number of individuals (> 1500), therefore with a good sampling. For two of these surveys (indices 9:91 and 12:88), it is assumed that vegetation disturbances in the past led to the dominance of a limited number of species (Costa, 1992). In the other survey (index 14:86), the strong dominance of a few species is explained by the selectivity that the environment of this riparian forest, submitted to periodic flooding, imposes on the different taxa (Silva *et al.* 1992). In all these cases, Shannon diversity indices are abnormally low in relation to what should be expected for tropical forests.

Our results indicate that Gini-Lorenz indices give a measure of plant diversity. The higher the index, the lower the diversity in the formation analysed.

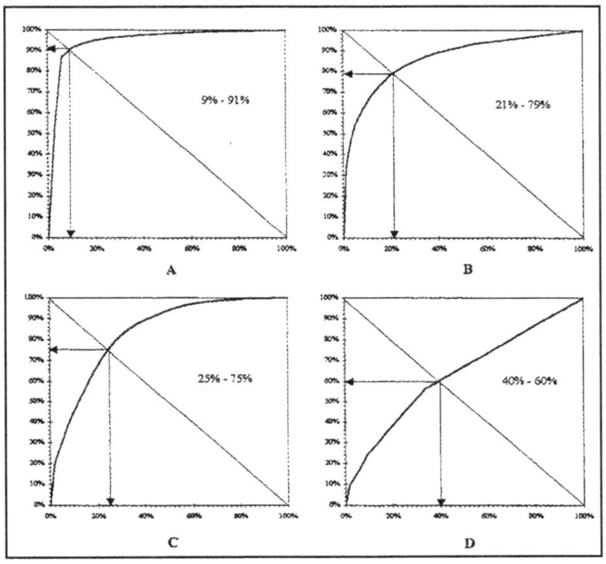

Figure 3: Examples of Gini-Lorenz curves from the extant vegetation. A – G.-L. index "too high" (Costa, 1992); B – G.-L. index "high" (Sá, 1993); C – G.-L. index "normal" (Silberbauer-Gottsberger and Eiten, 1987); D – G.-L. index "too low" (Peixoto and Gentry, 1990).

Diversity, taken in its ecological sense, depends on the specific richness (number of taxa in a community) and on the frequency of each taxon in the community. A plant community has low diversity if it has few species and if their abundance is uneven. On the contrary, diversity is high if the community has many species and if their abundance is fairly even. For an equal number of species, diversity will be higher in the station where frequency distribution is most even.

Low plant diversity in the temperate zone explains why Gini-Lorenz indices are higher than those generally obtained for tropical formations. Higher plant diversity in the Tropics leads to indices lower than 20:80 in the great majority of the analysed surveys (tabl. 1, fig. 3).

Conditions favoring the establishment of a secondary succession and vegetation disturbances also induce an increase in Gini-Lorenz concentration indices, *i.e.* lower diversity. Initial stages of secondary succession are usually associated with low diversity (Kurtz, 1994). One good example comes from Matthes (1992), who studied the dynamics of secondary succession in a semi-deciduous forest after fire. Phytosociological surveys were made in different plots, three and six years after the disturbance. Three years after the fire, Gini-Lorenz concentration indices in plots without that did not burn or were submitted to moderate fire levels are near 25:75 "normal", while plots submitted to intense fire show indices near 20:80, indicating a lower diversity than in less disturbed plots (fig. 4). Six years after the fire, there is a drop in Gini-Lorenz indices in most of the plots, suggesting the installation of new taxa, and therefore a higher diversity. Concentration indices are near 25:75 in all the studied plots, although it is of 20:80 in the survey including the complete data.

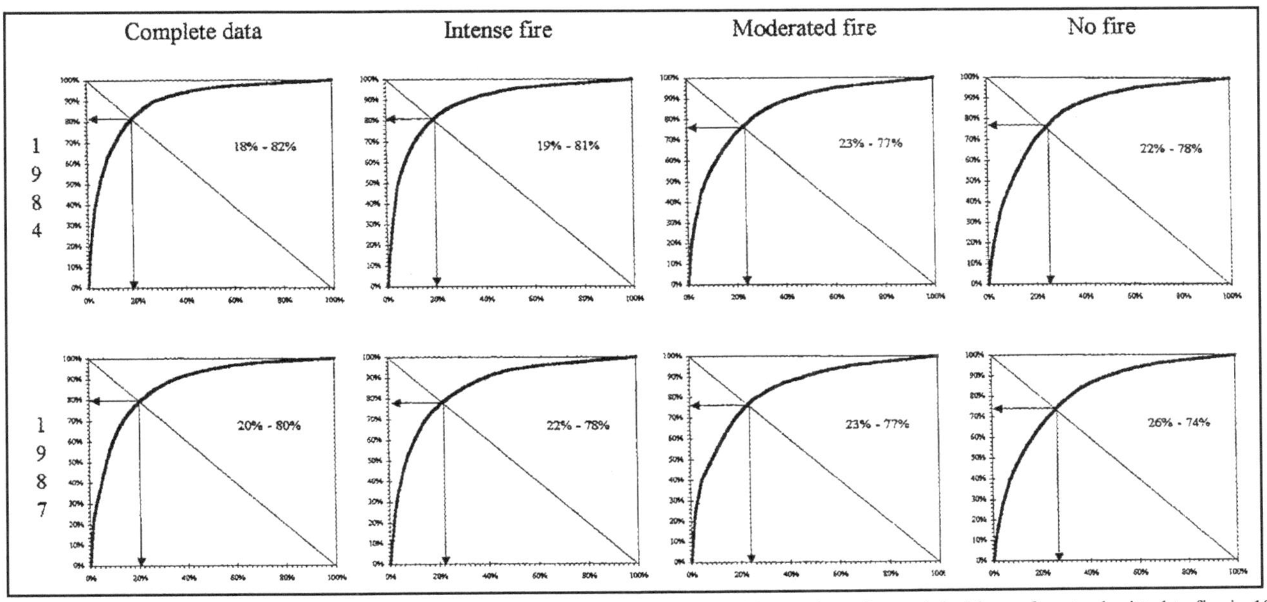

Figure 4: Gini-Lorenz curves obtained from a secondary succession dynamic study in different plots of a semi deciduous forest submitted to fire in 1981. Mata de Santa Genebra, Campinas, São Paulo State (Matthes, 1992).

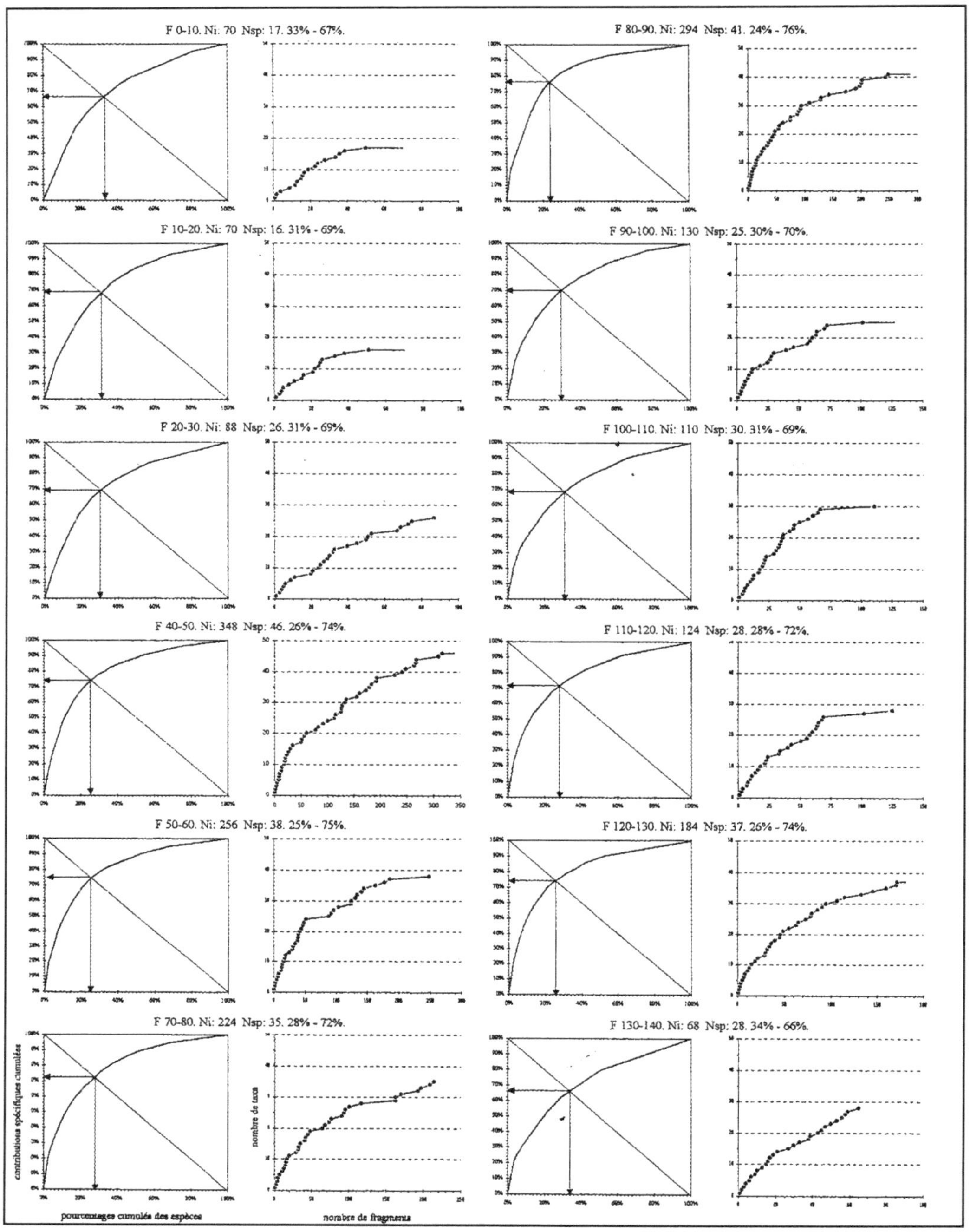

Figure 5a: Gini-Lorenz and saturation curves from Sambaqui do Forte (Cabo Frio, Rio de Janeiro State) archaeological samples (Scheel-Ybert, 1998).

Ancient samples

Study of Gini-Lorenz indices was applied to the evaluation of sampling reliability in the anthracological study of the Sambaqui do Forte, on the southeastern Brazilian coast (Scheel-Ybert, 1998, 2000). Saturation and Gini-Lorenz concentration curves were constructed for each sampling level (fig. 5a and b). Gini-Lorenz curves were constructed for an increasing number of fragments for all levels presenting more than 200 charcoal pieces (fig. 6).

A clear elevation of Gini-Lorenz indices is evidenced by the increase of the number of analysed charcoal pieces:
- for the first 25-50 charcoal pieces, the curves are very asymmetrical and always present very low concentration indices (generally lower than 35:65);
- for 100-150 pieces, most of the curves present indices near 30:70;
- for 200 pieces, 60 % of the curves present indices of *c.* 25:75;
- from the six samples of more than 300 pieces, only one present a "too low" concentration index;
- from the four samples of more than 400 identifiable charcoal pieces, three present concentration indices of *c.* 25:75, and one "high" concentration index (21:79);

Above 200 analysed elements on, most of the samples present concentration indices similar to those normally found

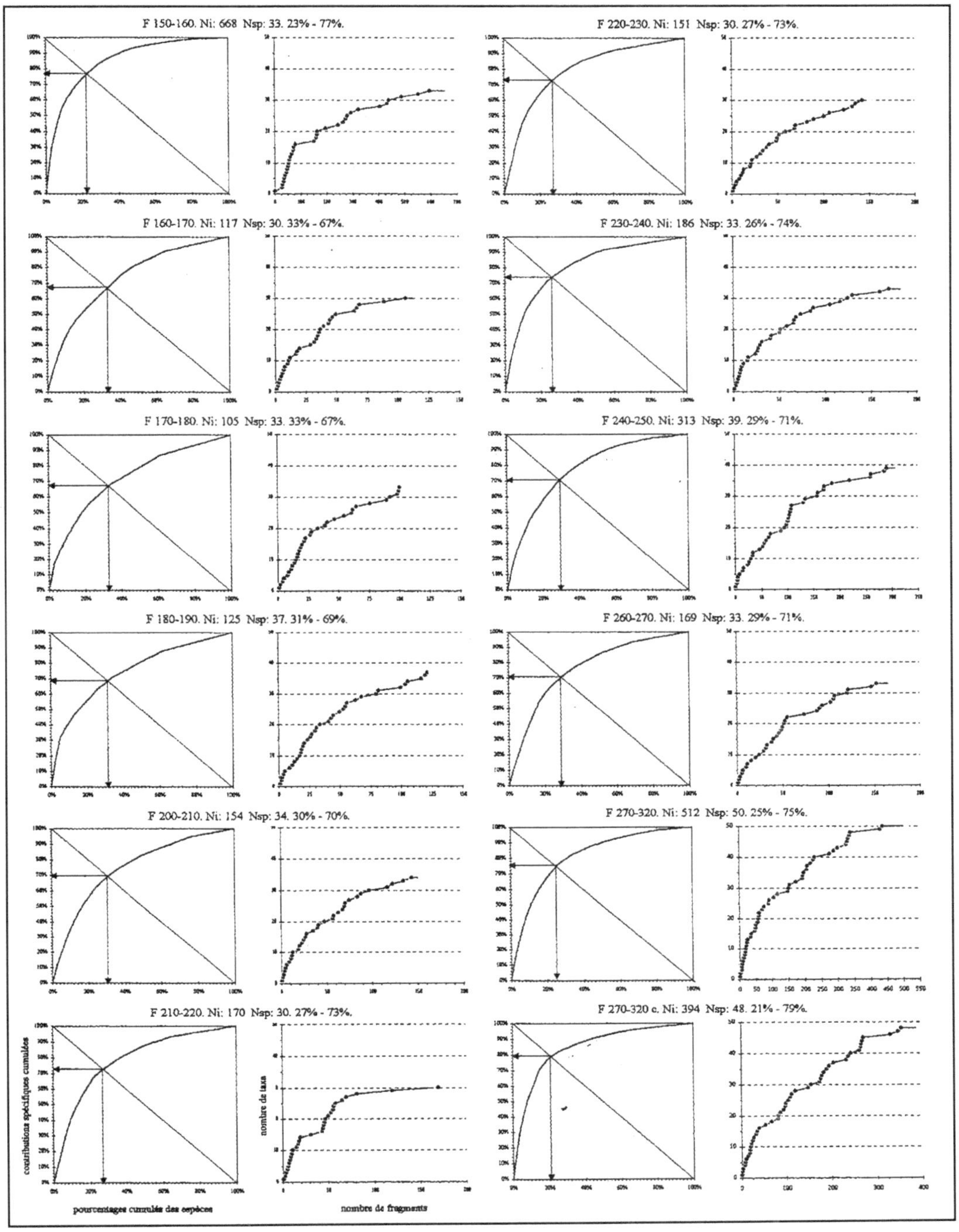

Figure 5b: Gini-Lorenz and saturation curves from Sambaqui do Forte (Cabo Frio, Rio de Janeiro State) archaeological samples (Scheel-Ybert, 1998).

for the extant vegetation. Samples of less than 150 charcoal pieces present almost always "too low" concentration indices, indicating an insufficient sampling.

Most of the saturation curves constructed for this sambaqui samples are not stabilised, as is frequently the case in phytosociological (Cavassan *et al*. 1984; Sá, 1993; Kurtz, 1994; Pereira and Araujo, 1995 etc.) or palynological (Ledru, 1991) studies from tropical regions. However, all the curves constructed for samples of more than 200 elements may be considered as acceptable.

These results agree with those obtained by the analysis of Gini-Lorenz curves, and indicate that a minimum of 200 or 300 elements per sample may give a representative image of the surrounding vegetation. Lower numbers may sometimes give correct results, but these samples are less reliable and their study should be limited to qualitative considerations.

However, the analysis of a higher number of charcoal pieces (> 400) is recommended, in order to reinforce the value of the palaeoecological interpretations proposed.

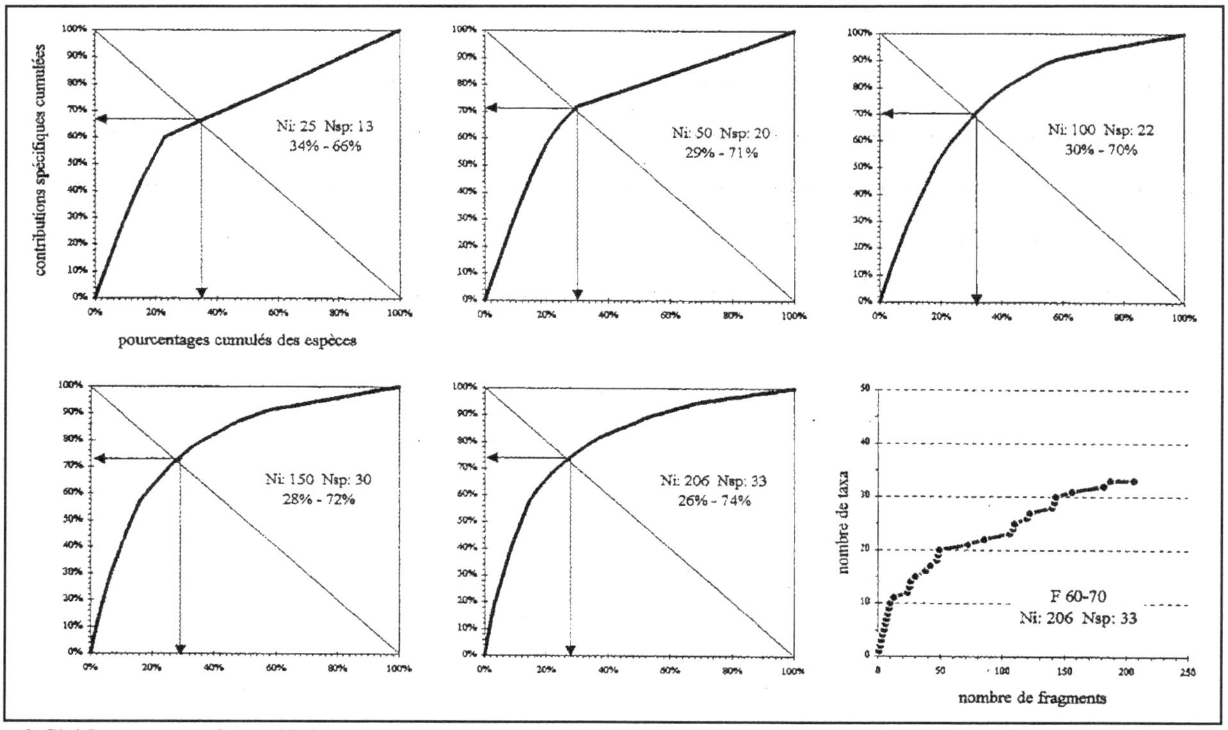

Figure 6: Gini-Lorenz curves for the 25, 50, 100, 150 and 206 first analyzed fragments at level 60-70 cm from the Sambaqui do Forte and saturation curve for the total sample.

Conclusions

Gini-Lorenz concentration indices give a measure of plant diversity. The higher the concentration index, the lower is the diversity in a plant formation. This index might be useful in characterising the successive state of plant formations, therefore allowing its use in ecosystems modelisation. It can be used to evaluate sampling reliability in extant and fossil assemblages, because small samples generally present an excessively high diversity, since the vegetation structure is not well represented.

Brazilian plant formations are characterised by mean concentration indices of 25:75. Indices near 20:80 are generally associated with secondary succession processes. Indices higher that 15:85 are related either to vegetation disturbances or to very specialized plant associations, in which environmental factors favor the dominance of a few well adapted species, and therefore lower diversity. Indices lower than 30:70 are due to insufficient sampling.

Saturation and Gini-Lorenz concentration curves constructed for archaeological samples showed that the analysis of a minimum of 200 or 300 charcoal pieces per sample give a representative image of the surrounding vegetation.

This approach, consisting of the application of a single method of analysis to extant vegetation and to ancient assemblages, is particularly well adapted to multidisciplinary studies. It can be applied to several palaeoenvironmental sciences, such as anthracology, palynology, diatoms or ostracodes studies.

ACKNOWLEDGEMENTS

This work was accomplished during a Ph.D. thesis at the University of Montpellier, France, while holding a doctoral scholarship from the CAPES (MEC, Brazil). Thanks are due to Maurice Roux, Paule Ogereau, Philippe Daget and Jacques Poissonnet for discussions and helpful comments during its accomplishment.

References

CASTRO A.A.J.F. 1987 - *Florística e fitossociologia de um cerrado marginal brasileiro, Parque Estadual de Vaçununga, Santa Rita do Passa Quatro*, SP. M.Sc. Thesis. Inst. Biologia, UNICAMP, Campinas. 231 p.

CATHARINO E.L.M. 1989 - *Estudos fisionômico-florísticos e fitossociológicos em matas residuais secundárias no município de Piracicaba*, SP. M.Sc. Thesis. Inst. Biologia, UNICAMP, Campinas. 181 p.

CAVASSAN O., CESAR O., MARTINS F.R. 1984 - Fitossociologia da vegetação arbórea da Reserva Estadual de Bauru, Estado de São Paulo. *Revista Brasileira de Botânica* 7 (2), p. 91-106.

CESAR O., PAGANO S.N., LEITÃO FILHO H.F., MONTEIRO R., SILVA O.A., MARINIS G., SHEPHERD G.J. 1988 - Estrutura fitossociológica do estrato arbóreo de uma área de vegetação de cerrado no município de Corumbataí (Estado de São Paulo). *Naturalia* 13, p. 91-101.

CESAR O., LEITÃO FILHO H.F. 1990 - Estudo fitossociológico de mata mesófila semidecídua na Fazenda Barreiro Rico, município de Anhembi, SP. *Revista Brasileira de Biologia* 50 (2), p. 443-452.

index	Ni	Nsp	H'	area	inclusion criteria	sampling method	disturbance	vegetation	author
9 : 91	2816	69	2,01	0,76 ha	CAP 5	plots	(ancient?)	Semideciduous forest (H < 5 m)	Costa 1992
12 : 88	3757	103	2,44	0,76 ha	CAP 5	plots	(ancient?)	Semideciduous forest	Costa 1992
14 : 86	1577	42	2,1	1 ha	DAP 5	plots	no	Riparian forest	Silva *et al.* 1992
18 : 82	372	41	2,08	?	?	plots	?	Trans. semideciduous/*cerradão*	Simonetti 1993
18 : 82	375	36	2,3	0,925 ha	DAP 10	transects	?	Riparian forest	Heringer & Paula 1989
18 : 82	700	68	2,7	0,76 ha	CAP 5	plots	(ancient?)	Semideciduous f. (5m<H<10m)	Costa 1992
20 : 80	612	21	1,9	?	?	plots	?	Riparian forest	Simonetti 1993
21 : 79	640	91	3,13	3200 m	Hmax 2 m	points	tractor (6 years)	*Restinga* forest (herb./shrubby)	Sá 1993
21 : 79	3172	103	3,54	0,5 ha	DAB 3	plots	no	*Cerradão*	Toledo Filho *et al.* 1989
18 : 82	2331	152	3,72	0,105 ha			fire 1981	Total data (1984)	
20 : 80	1995	129	3,72	0,105 ha			fire 1981	Total data (1987)	
19 : 81	1016	92	3,24	0,0475 ha			intense fire	Semideciduous forest (1984)	
22 : 78	823	82	3,36	0,0475 ha	Hmin 0,5 m	plots	intense fire	Semideciduous forest (1987)	Matthes 1992
23 : 77	629	73	3,3	0,0275 ha			moderated fire	Semideciduous forest (1984)	
23 : 77	553	68	3,25	0,0275 ha			moderated fire	Semideciduous forest (1987)	
22 : 78	681	82	3,58	0,03 ha			no fire	Semideciduous forest (1984)	
24 : 76	619	75	3,59	0,03 ha			no fire	Semideciduous forest (1987)	
22 : 78	134	8	1,12	?	?	plots	?	*Tucum* formation	Simonetti 1993
22 : 78	225	23	2,15	?	?	plots	?	Riparian forest	Simonetti 1993
22 : 78	530	71	3,12	0,4 ha	DAP 5	plots	fire 1981	Semideciduous forest	Catharino 1989
22 : 78	867	31	2,44	0,1 ha	CAP 5	plots	?	*Restinga* forest (Area 2)	Sugiyama 1993
22 : 78	1200	113	3,56	300 pts	DAP 3	quadrats	?	Semideciduous forest	Cesar & Leitão Fo. 1990
22 : 78	1256	56	3,09	0,27 ha	CAP 8	plots	?	*Restinga* forest (Area 1)	Sugiyama 1993
22 : 78	1888	70	3,23	1 ha	DAB 3	transects	no	*Cerrado sensu stricto*	Oliveira Filho *et al.* 1989
22 : 78	4718	83	3,4	0,6 ha	DAB 3	plots	anthropic	*Cerrado sensu stricto*	Castro 1987
23 : 77	271	36	2,44	?	?	plots	?	Trans. riparian /semideciduous	Simonetti 1993
23 : 77	1450	30	2,43	0,014 ha	all	plots	?	*Restinga* (sandy vegetation)	Pereira & Araujo1995
24 : 76	211	20	2,1	1 ha	Hmin 2 m	plots	anthropic	*Cerrado*	Rizzini 1975
24 : 76	429	66	3,15	107 pts	Hmin 1 m	quadrats	fire (5 years)	Subtropical forest (shrubby)	Vasconcellos *et al.* 1992
24 : 76	553	102	3,77	0,2 ha	DAB 2,5	plots	tractor (6 years)	*Restinga* forest	Sá 1993
24 : 76	818	95	3,41	1 ha	PA10 <8	points	frequent fire	*Cerrado* (herb./shrubby)	Vincent *et al.* 1992
24 : 76	819	46	2,89	0,4 ha	PA10 8	plots	frequent fire	*Cerrado*	Vincent *et al.* 1992
24 : 76	1042	31	2,64	2,6 ha	DAP 10	transects	?	Riparian forest	Heringer & Paula 1989
24 : 76	1510	91	3,18	0,56 ha	DAP 5	plots	?	*Restinga* forest (total data)	Silva 1990
24 : 76	1541	88	3,7	0,5 ha	CAP 15	plots	no	*Restinga* forest	Fabris 1995
24 : 76	3400	101	3,64	0,375 ha	DAB 3	plots	no	*Cerradão*	Cesar *et al.* 1988
25 : 75	213	32	2,52	?	?	plots	?	Riparian forest	Simonetti 1993
25 : 75	2391	50	3,13	1,65 ha	CAP 15	plots	anthropic	*Restinga* forest	Trindade 1991
25 : 75	4197	54	3,14	1 ha	DAB 10	plots	no	*Cerrado sensu stricto*	Silberbauer & Eiten 1987
26 : 74	199	34	2,84	50 pts	Hmin 1,3 m	quadrats	anthropic	Riparian forest	Mencacci *et al.* 1992
26 : 74	235	70	2,77	1 ha	Hmin 2 m	plots	anthropic	*Cerrado*	Rizzini 1975
26 : 74	288	41	2,87	?	?	plots	?	*Cerrado*	Simonetti 1993
26 : 74	406	49	3,25	?	?	plots	?	*Cerradão*	Simonetti 1993
26 : 74	416	29	2,69	?	DAP 5	plots	?	*Restinga* forest (transition 2)	Silva 1990
26 : 74	590	137	4,2	2,5 ha	DAP 5	quadrats	no	Atlantic forest	Kurtz 1994
26 : 74	640	123	4,07	160 pts	DAP 10	quadrats	no	Atlantic forest	Silva & Leitão Fo. 1982
26 : 74	838	91	3,6	107 pts	Hmin 1 m	quadrats	fire (5 years)	Subtropical forest (total data)	Vasconcellos *et al.* 1992
26 : 74	1883	155	4,27	301 pts	Hmin 1,3 m	quadrats	anthropic	Semideciduous forest	Pagano *et al.* 1987
27 : 73	241	53	2,88	0,76 ha	CAP 5	plots	(ancient?)	Semideciduous forest (H >10 m)	Costa 1992
27 : 73	325	43	3,14	?	DAP 5	plots	?	*Restinga* forest (transition 1)	Silva 1990
27 : 73	344	47	3,16	0,72 ha	DAP 10	quadrats	no	Riparian forest	Gibbs & Leitão Fo. 1978
27 : 73	1481	124	4,21	1 ha	DAP 5	plots	no	Riparian forest	Silva *et al.* 1995
28 : 72	285	37	2,99	?	DAP 5	plots	?	*Restinga* forest (flooded)	Silva 1990
28 : 72	318	40	3,13	?	?	plots	?	Trans. *cerradão*/semideciduous	Simonetti 1993
28 : 72	360	85	3,85	90 pts	DAP 4,77	quadrats	anthropic	Urban semideciduous forest	Nastri *et al.* 1992
28 : 72	484	24	2,55	?	DAP 5	plots	?	*Restinga* forest (sandy)	Silva 1990
28 : 72	516	61	3,48	129 pts	DAP 10	quadrats	ancient fire	Semideciduous forest	Cavassan *et al.* 1984
28 : 72	582	113	4,06	?	?	"relascopy"	possible	Atlantic forest	Negreiros *et al.* 1995
28 : 72	800	114	4,1	200 pts	DAP 5	point-quadrat	?	*Restinga* forest	Sá & Araujo *unpublished*
28 : 72	1588	122	4,19	?	Hmin 1,3 m	quadrats	anthropic, fire	*Cerrado*	Pagano *et al.* 1989
29 : 71	305	29	2,85	0,36 ha	Hmin 2 m	plots	anthropic, fire	*Cerrado*	Rizzini 1975
29 : 71	409	60	3,49	107 pts	DAP 10	quadrats	fire (5 years)	Subtropical forest (arboreal)	Vasconcellos *et al.* 1992
30 : 70	187	26	2,75	0,25 ha	DAP 10	plots	?	*Cerrado sensu stricto*	Guarim Neto *et al.* 1994
30 : 70	266	50	3,44	?	?	plots	?	Trans. semideciduous/riparian	Simonetti 1993
31 : 69	280	90	4,11	0,2 ha	DAP 5	plots	anthropic, fire	Atlantic forest	Guedes 1988
34 : 66	80	30	4,14	< 0,008 ha	DAB 2	centered quadrat	?	Riparian forest	Meguro *et al.* 1996
36 : 64	80	32	4,61	< 0,008 ha	DAB 2	"	?	Altitude forest	Meguro *et al.* 1996
40 : 60	99	65	4	0,1 ha	DAP 10	transects	no	*Tabuleiro* forest	Peixoto & Gentry 1990

Table 1: Gini-Lorenz concentration indices of the extant vegetation phytosociological surveys and information on the number of individuals analyzed (Ni), number of species (Nsp), Shannon diversity index (H'), sampling area, inclusion criteria, sampling methods, vegetation disturbance and concerned plant association. Dates in parenthesis inform the year of the study. Number of years in parenthesis concern period of time since the last disturbance. Abbreviations - CAP: circumference at breast height; DAP: diameter at breast height; DAB: diameter at the base; PA10: perimeter at 10-cm height; Hmax: maximum height; Hmin: minimum height.

CHABAL L. 1991 - *L'Homme et l'évolution de la végétation méditerranéenne; des âges des Métaux à la Période Romaine : recherches anthracologiques théoriques, appliquées principalement à des sites du Bas-Languedoc*. Ph.D. Thesis. USTL. Montpellier. 435 p.

CHABAL L. 1997 - Forêts et sociétés en Languedoc (Néolithique final, Antiquité tardive). L'anthracologie, méthode et paléoécologie. *Documents d'Archéologie Française* 63, 188 p.

COSTA L.G.S. 1992 - *Estrutura e dinâmica de trecho de mata mesófila semidecídua, na Estação Ecológica de Ibicatu, Piracicaba*, SP. M.Sc. Thesis. Inst. Biociências, USP, São Paulo. 187 p.

DAGET P., POISSONET J. 1971 - Une méthode d'analyse phytologique des prairies. Critères d'application. *Annales Agronomiques* 22 (1), p. 5-41.

FABRIS L.C. 1995 - *Composição florística e fitossociologia de uma faixa de floresta arenosa litorânea do Parque Estadual de Setiba, município de Guarapari*, ES. M.Sc. Thesis, Inst. Biociências, UNESP, Rio Claro. 195 p.

GIBBS P.E., LEITÃO FILHO H.F. 1978 - Floristic composition of an area of gallery forest near Mogi-Guaçu, state of São Paulo, S. E. Brazil. *Revista Brasileira de Botânica* 1, p. 151-156.

GODRON M. 1984 - *Écologie de la végétation terrestre*. Paris, ed. Masson. Collection Abrégés. 196 p.

GUARIM NETO G., GUARIM V.L.M.S., PRANCE G.T. 1994 - Structure and floristic composition of the trees of an area of cerrado near Cuiabá, Mato Grosso, Brazil. *Kew Bulletin* 49 (3), p. 499-509.

GUEDES R. 1988 - Composição florística e estrutura de um trecho de mata perturbada de baixada no município de Magé, Rio de Janeiro. *Arquivos do Jardim Botânico do Rio de Janeiro* 29, p. 155-200.

HERINGER E.P., PAULA J.E. 1989 - Contribuição para o conhecimento ecodendrométrico de matas ripárias da região centro-oeste brasileira. *Acta Botânica Brasílica* 3 (2), p. 33-42.

JACQUARD P., DAGET P., POISSONET J., LAROCHE C. 1968 - The expression of production potential and botanical composition of a dense herbaceous formation. In: *Proceedings of Symposium on hill land productivity*. European Grassland Federation. p. 96-108.

LEDRU M.-P. 1991 - *Étude de la pluie pollinique actuelle des forêts du Brésil central : climat, végétation, application à l'étude de l'évolution paléoclimatique des 30 000 dernières années*. Ph.D. Thesis. MNHN, Institut de Paléontologie Humaine. Paris. 193 p.

KURTZ B. 1994 - *Composição florística e estrutura do componente arbóreo de um trecho de Mata Atlântica na Estação Ecológica Estadual do Paraíso, Cachoeiras de Macacu*, RJ. M.Sc. Thesis. Museu Nacional, UFRJ, Rio de Janeiro. 106 p.

MARTINS F.R. 1993 - *Estrutura de uma floresta mesófila*. 2nd ed. Campinas, edit. UNICAMP. Série Teses. 246 p.

MATTHES L.A.F. 1992 - *Dinâmica da sucessão secundária em mata, após a ocorrência de fogo - Santa Genebra - Campinas, São Paulo*. Ph.D. Thesis. Inst. Biologia, UNICAMP, Campinas. 216 p.

MEGURO M., PIRANI J. R., MELLO-SILVA R., GIULIETTI A.M. 1996 - Caracterização florística e estrutural de matas ripárias e capões de altitude da Serra do Cipó, Minas Gerais. *Boletim de Botânica*, USP, 15, p. 13-29.

MENCACCI P.C., SCHLITTLER F.H.M. 1992 - Fitossociologia da vegetação arbórea da mata ciliar de Ribeirão Claro, município de Rio Claro - SP. In: *Anais do 2° Congresso Nacional sobre Essências Nativas*. p. 245-251.

NASTRI V.D.F., CATHARINO E.L.M., ROSSI L., BARBOSA L.M., PIRRÉ E., BEDINELLI C., ASPERTI L. M., DORTA R.O., COSTA M.P. 1992 - Estudos fitossociológicos em uma área do Instituto de Botânica de São Paulo utilizados em programas de educação ambiental. In: *Anais do 2° Congresso Nacional sobre Essências Nativas*. p. 219-225.

NEGREIROS O.C., CUSTODIO FILHO A., DIAS A.C., FRANCO G.A.D.C., COUTO H.T.Z., VIEIRA M.G.L., MOURA NETTO B.V. 1995 - Análise estrutural de um trecho de floresta pluvial tropical, Parque Estadual de Carlos Botelho, Núcleo Sete Barras (SP - Brasil). *Revista do Instituto Florestal*, S. Paulo, 7 (1), p. 1-33.

OGEREAU P., POISSONET J. 1988 - Relations entre les fréquences des espèces et leurs rangs, dans une phytocénose prairiale. Colloques phytosociologiques, *Phytosociologie et Pastoralisme*, 16, p. 757-768.

OLIVEIRA FILHO A.T., SHEPHERD G.J., MARTINS F.R., STUBBLEBINE W.H. 1989 - Environmental factors affecting physiognomic and floristic variation in an area of cerrado in central Brazil. *Journal of Tropical Ecology* 5, p. 413-431.

PAGANO S.N., LEITÃO FILHO H.F., SHEPHERD G.J. 1987 - Estudo fitossociológico em mata mesófila semidecídua no município de Rio Claro (Estado de São Paulo). *Revista Brasileira de Botânica* 10, p. 49-61.

PAGANO S.N., CESAR O., LEITÃO FILHO H.F. 1989 - Estrutura fitossociológica do estrato arbustivo-arbóreo da vegetação de cerrado da Área de Proteção Ambiental (APA) de Corumbataí, Estado de São Paulo. *Revista Brasileira de Biologia* 49 (1), p. 49-59.

PEIXOTO A.L., GENTRY A. 1990 - Diversidade e composição florística da mata de tabuleiro na Reserva Florestal de Linhares (Espírito Santo, Brasil). *Revista Brasileira de Botânica* 13, p. 19-25.

PEREIRA O.J., ARAÚJO D.S.D. 1995 - Estrutura da vegetação entre moitas da formação aberta de Ericaceae no Parque Estadual de Setiba, ES. *Oecologia Brasiliensis* 1, p. 245-257.

POISSONET J. 1968 - *Recherche sur les lois générales d'équilibre dans la composition floristique des formations herbacées denses. Premiers résultats et hypothèses*. Montpellier, CEPE. 18 p.

POISSONET J. 1979 - *Équilibre et déséquilibre des phytocénoses herbacées*. In: Dixièmes Journées du Grenier de Theix. INRA. p. 461-468.

POISSONET P., POISSONET J. 1969 - *Essai de comparaison de méthodes d'analyses phytosociologiques et agronomiques des formations herbacées denses permanentes*. Montpellier, CNRS-CEPE. Document n° 50. 120 p.

RIZZINI C.T. 1975 - Contribuição ao conhecimento da estrutura do cerrado. *Brasil Florestal* 6 (22), p. 3-15.

SÁ C.F.C. 1993 - *Regeneração de um trecho de Floresta de Restinga na Reserva Ecológica Estadual de Jacarepiá, município de Saquarema*, RJ. Thèse de Mestrado. Museu Nacional, UFRJ, Rio de Janeiro. 168 p.

SÁ C.F.C., ARAUJO D.S.D. unpublished - *Données phytosociologiques, Reserva Ecológica de Jacarepiá*, município de Saquarema, RJ. Research report.

SCHEEL-YBERT R. 1998 - *Stabilité de l'écosystème sur le littoral Sud-Est du Brésil à l'Holocène Supérieur (5500-1400 ans BP). Les pêcheurs-cueilleurs-chasseurs et le milieu végétal : apports de l'anthracologie*. Ph.D. Thesis, USTL, Montpellier. 3 vol. 520 p.

SCHEEL-YBERT R. 2000 - Vegetation stability in the Southeastern Brazilian coastal area from 5500 to 1400 14C yr BP deduced from charcoal analysis. *Review of Palaeobotany and Palynology* 110, p. 111-138.

SILBERBAUER-GOTTSBERGER I., EITEN G. 1987 - A hectare of cerrado. I. General aspects of the trees and thick-stemmed shrubs. *Phyton*, Austria, 27 (1), p. 55-91.

SILVA A.F., LEITÃO FILHO H.F. 1982 - Composição florística e estrutura de um trecho da mata atlântica de encosta no município de Ubatuba (São Paulo, Brasil). *Rev. Brasil. Bot.* 5, p. 43-52.

SILVA F.C., FONSECA E.P., SOARES-SILVA L.H., MULLER C., BIANCHINI E. 1995 - Composição florística e fitossociologia do componente arbóreo das florestas ciliares da bacia do rio Tibagi. 3. Fazenda Bom Sucesso, município de Sapopema, PR. *Acta Bot. Brasil.* 9 (2), p. 289-302.

SILVA S.M. 1990 - *Composição florística e fitossociologia de um trecho de floresta de restinga na Ilha do Mel, município de Paranaguá, PR*. M.Sc. Thesis. Inst. Biologia. UNICAMP, Campinas. 146 p.

SILVA S.M., SILVA F.C., VIEIRA A.O.S., NAKAJIMA J.N., PIMENTA J.A., COLLI S. 1992 - Composição florística e fitossociologia do componente arbóreo das florestas ciliares da bacia do rio Tibagi, Paraná: 2. Várzea do rio Bitumirim, município de Ipiranga, PR. In: *Anais do 2° Congresso Nacional sobre Essências Nativas*. p. 192-198.

SIMONETTI C. 1993 - *Relatório de Impacto Ambiental para a CESP*. Consórcio THEMAG-ENGEA-UMAH. Internal report.

SUGIYAMA M. 1993 - *Estudo de florestas na restinga da Ilha do Cardoso, Cananéia, SP*. M.Sc. Thesis. Inst. Biociências, USP, São Paulo. 115 p.

TOLEDO FILHO D.V., LEITÃO FILHO H.F., SHEPHERD G.J. 1989 - Estrutura fitossociológica da vegetação de cerrado em Mogi-Mirim (SP). *Revista do Instituto Florestal*, São Paulo, 1 (2), p. 1-12.

TRINDADE A. 1991 - *Estudo florístico e fitossociológico do estrato arbustivo-arbóreo de um trecho de floresta arenícola costeira do Parque Estadual das Dunas - Natal (RN)*. M.Sc. Thesis. Depto. Botânica, UFRPE, Recife. 168 p.

VASCONCELLOS J.M.O., DIAS L.L., SILVA C.P., SOBRAL M. 1992 - Fitossociologia de uma área de mata subtropical no Parque Estadual do Turvo - RS. In: *Anais do 2° Congresso Nacional sobre Essências Nativas*. p. 252-259.

VINCENT R.C., MIYAZAKI S.L., GOMES E.P.C., MANTOVANI W. 1992 - Estrutura e composição florística do cerrado de Emas, Pirassununga, SP. In: *Anais 8° do Congresso da SBSP*. p. 139-151.

Charcoal analysis in a lake dwelling site (Chalain 19, Jura, France): a sampling model for neolithic lacustrine contexts

Alexa Dufraisse

CNRS-UMR6565, Laboratoire de Chrono-Ecologie,
16, route de Gray, F- 25000 Besançon, France.
E-mail: Adufraisse@aol.com

Abstract

This study is based on the charcoal analysis of a lake dwelling site, Chalain 19, dated to the final Neolithic. Due to the action of water, the dispersion of charcoal in archaeological layers is different in lacustrine contexts compared to terrestrial contexts. Therefore, this paper suggests a sampling model of charcoal fragments. The excavation of the site revealed two different architectural areas. One is represented by a village, constituted of houses protected by a long semi-circular fence. The other architectural unit consists of an isolated construction, with a raised floor. Our argument is based on the spatial analysis of charcoal fragments from the isolated construction and its garbage dump. Lastly, we will see how the modes of deposit allow sampling by cores.

Key-words: *Chalain, Charcoal analysis, Lacustrine context, Methodology, Sampling, Core.*

Résumé

Cette étude s'appuie sur l'analyse anthracologique de la station 19 datant du Néolithique final et située sur la rive occidentale du lac de Chalain (Jura, France). En contexte lacustre, la dispersion des charbons de bois au sein d'une couche archéologique ne suit pas toujours celle observée en milieu terrestre en raison de l'action de l'eau. C'est pourquoi l'objectif de cette étude est de proposer un modèle d'échantillonnage des charbons de bois puisque leur représentativité et la signification des diagrammes anthracologiques en dépendent fortement La fouille de cette station a révélé l'existence de deux aires architecturales distinctes. L'une est représentée par un village constitué de dix à douze maisons d'habitation protégées par une longue palissade semi-circulaire, auquel il est possible d'accéder par un chemin de planches, long de 70 m. L'autre unité concerne un bâtiment isolé du village, à plancher rehaussé. Afin de comprendre la dispersion des charbons de bois, nous nous appuierons sur une analyse spatiale des dépôts de charbons au sein du bâtiment isolé et de son dépotoir. Enfin, nous verrons que ces modes de dépôts peuvent se prêter à des prélèvements par carottages.

Mots-clés : *Chalain, anthracologie, milieu lacustre, méthodologie, échantillonnage, carottages.*

Introduction

The objective of this paper is to present a model for the sampling of charcoal fragments from a particular archaeological context: lake dwelling sites. This study was done within the framework of a larger project which consists of an archaeological and ecological approach to the vegetation around the lakes of Chalain and Clairvaux (fig. 1) during the middle and recent Neolithic. Chalain 19, situated on the western shore of the lake and dated to the final Neolithic, is part of an area of Neolithic lake dwelling villages constituting a complete chronological sequence for the 31st and 30th centuries BC.

The sampling method for charcoal in terrestrial zones is based essentially on the mode of charcoal dispersion in order to obtain a good statistical representativity of the archaeological layer (Chabal, 1997). However, the mechanisms of deposits are very different in lakeshore zones, both during occupation and after abandonment. Therefore, a very complete sampling was carried out in order to analyse the spatial distribution of charcoal and to determine a sampling model that can be used in similar sites.

The action of water on the organic deposits will first be described. Then, a model of sampling will be presented. Finally, we shall see how the modes of deposit enable the obtaining of samples by coring.

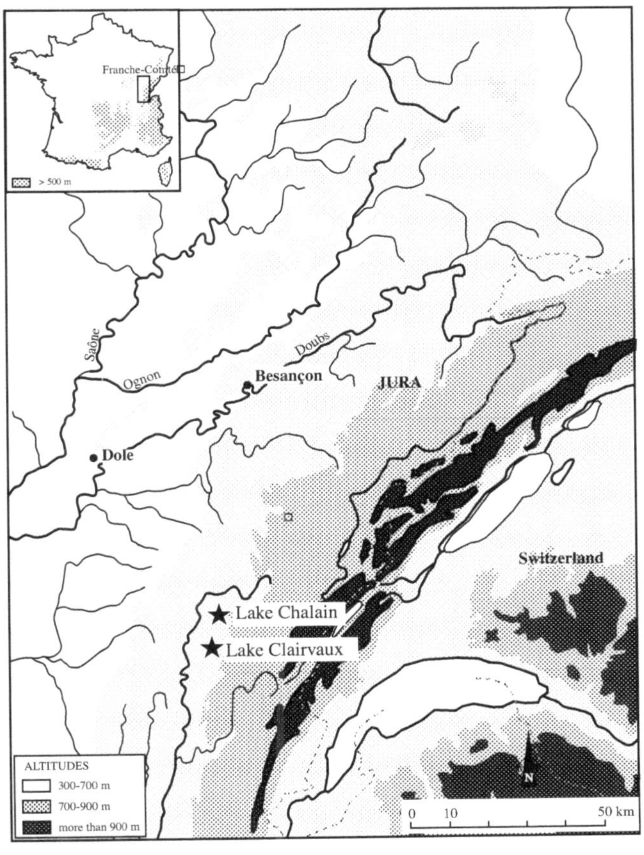

Figure 1: Localisation of the lakes of Chalain and Clairvaux in the Jura.

Archaeological context: action of water on the deposits

In lake dwelling site contexts, water is the main factor affecting the organisation of deposits and thus also of char-

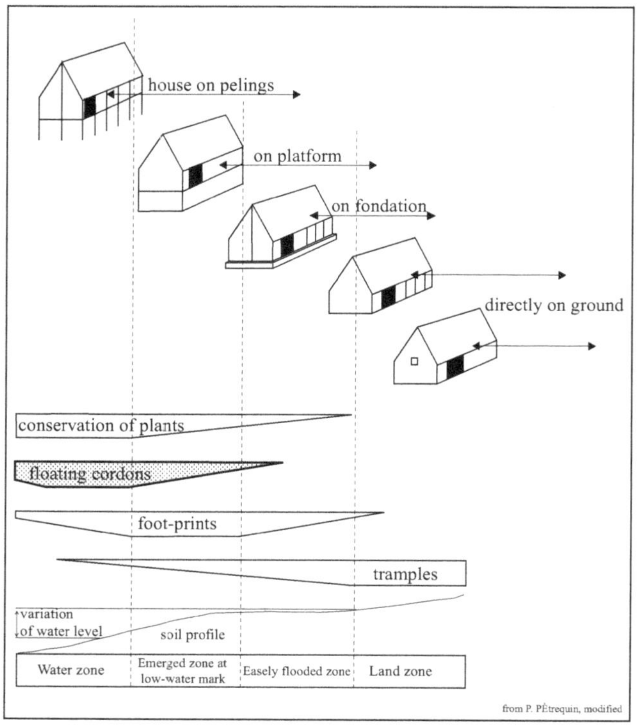

Figure 2: Water levels, architecture and floating cordons.

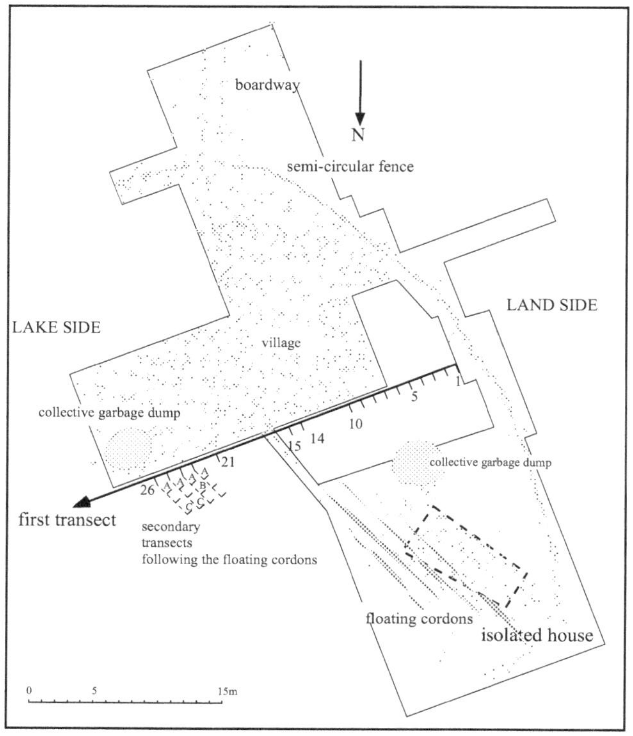

Figure 3: Map of Chalain 19 (old Clairvaux phase) and localisation of floating cordons and transects.

coal dispersion (fig. 2). With respect to an axis perpendicular to the bank, different water levels may be distinguished: a shallow water zone, a zone that emerge at low water levels, an easily flooded zone and a terrestrial zone. According to these zones, the architectural features of houses will also vary, for example the height of the floors above the ground (fig. 2). Different architectural organisations may sometimes be found within the same village (Pétrequin and Pétrequin, 1984).

Sedimentation and erosion

In a lacustrine context, water is a preponderant element in sedimentation and erosion phenomena. In the shallow water zone the action of waves may provoke intense erosion, especially on banks exposed to winds. In the lake-beating zone (lands easily flooded or emerged), plant macroremains float and are carried away by seasonal floods. In the terrestrial zone, natural sedimentation is almost inexistant due to important streams of rainwater, carrying away light, organic material and silt towards depressions (Chalain is situated at 500 m a.s.l., on the first tableland where rainfall attains 1500 mm/year).

Lastly, a decrease in the depth of archaeological layers from the lake towards the terrestrial zone may also be observed.

Floating and accumulation

The floating and accumulation of organic material are two important phenomena in lake-beating and have to be considered in order to understand the deposits. These phenomena are caused by movements of the lake in shallow water or in easily flooded zones. They regularly return organic particles such as charcoal fragments to the open water. These macroremains undergo a sorting, depending on their size and their shape, during the occupation as well as after the abandonment of the village. A selection through floating creates horizons of accumulation or floating cordons that contain fragments of the same size class. Each cordon may include material of any type (bark, charcoal, wood, moss etc.) or shape, but the presence of a main calibrated size material such as moss and twigs is often noted. The width of the horizon seems to depend on the inclination of the bank and the depth of action of the waves. These cordons succeed each other, starting from the lake towards the bank, according to element size and contour lines, independently of the posts that constitute the layout of the village. However, these posts may constitute obstacles to the regular distribution of the cordons.

From these different data, a sampling model for lake-dwelling sites was established, especially for stations situated in shallow water or in emerged and easily flooded zones, such as Chalain 19.

Station 19 of Chalain

The construction and occupation of this village was dated by dendrochronology to between 3200 and 2910 BC, with an interruption at the beginning of the 31st century. Two levels of occupation were identified, the oldest is attributed to Horgen (layer O) and the most recent to the old Clairvaux phase (layer H/K) (Pétrequin and Pétrequin, 2000). Excavation of this station revealed two different architectural areas (fig. 3). The first is a village where the houses were protected by a long semi-circular fence. The other architectural unit consists of an isolated construction, whose function is not yet determined. Garbage was thrown on two large collective dumps, of which one was situated in the village and the other near the isolated house.

Excavation of the layer attributed to the old Clairvaux phase, on which this study is based, revealed the presence of a succession of horizons of accumulation, organised according to the size of the included elements. From the lake towards the bank and the dry-land zone, a first floating cordon composed of small elements, mainly moss and twigs, can be observed. The second cordon consists mainly of twigs, branches (non-burnt) and charcoal fragments. The third is composed of charcoal, branches and horizontal wood pieces. The last cordon is characterised by horizontal wood and patches of bark. These accumulations, due to floating, follow the contour lines and do not always appear clearly during the excavation of the different sectors.

Material and methods

This sampling model is based on a qualitative and quantitative spatial analysis of charcoal fragments found at the site of the isolated house and its garbage dump. These features form a closed unit compared to the rest of the village. As the main objective of this study is to understand the palaeoenvironment, we chose to analyse charcoal as this type of material is better adapted to our purposes than twigs, branches or horizontal wood that was used in architecture. In fact, dispersed charcoal fragments come from domestic fires and represent a longer period of deposit than wooden architectural elements. The archaeological fire layer, mainly composed by carbonized architectural elements, is also clearly distinguishable from the occupation layer.

Sampling methods

Two sampling methods were applied to the isolated house, excluding its garbage dump. The first consisted of the removal of 40 l of sediment from each square situated in the zone of the second and third floating cordon schematically described above, *i.e.* in the zone containing twigs, branches and charcoal. These cordons were observed during the excavation and followed the contour lines.

Two series of samples were also retrieved from outside of the floating cordon, one situated on a perpendicular line and the other on a transverse line in relation to the bank.

The garbage dump forms a small mound, 50-80 cm thick, with the summit situated above the water level. Samples were taken from three squares using a layer subdivision established according to sediment texture and colour.

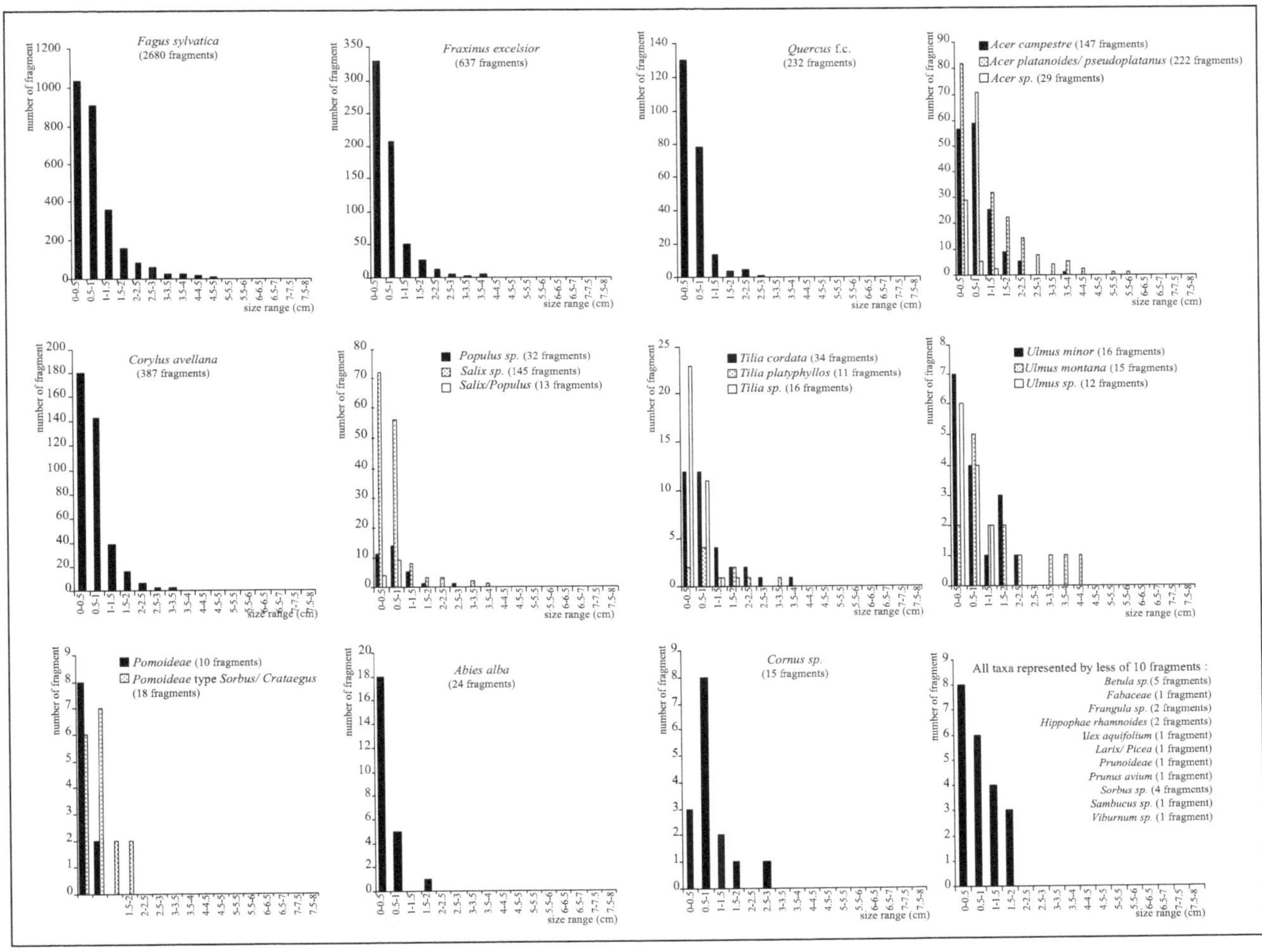

Figure 4: Histograms showing the fragmentation of each taxon from the isolated house (the garbage dump excluded).

Sieving and number of fragments analysed

All the sediment was then sieved through a 2 mm mesh. For practical reasons the relative frequencies of taxa were recorded by counting and not by weighing. For the isolated house the number of identified charcoal fragments was estimated by means of taxonomic curves for each square and not for the total layer (between 164 and 636 fragments/m^2) in order to obtain an optimal number of taxa per square. For the garbage dump, the number of charcoal pieces attains around 200 per sublayer and they are thus susceptible of revealing possible vegetation dynamics. The spatial analysis is based on the identification of about 3625 charcoal fragments from the isolated house, 1245 from the perpendicular and transversal lines and 4652 from the garbage dump.

Fragmentation

In terrestrial archaeological levels, charcoal is subject to fragmentation processes following a statistical pattern of the Poisson distribution type (Chabal, 1997). Charcoalfrom any taxa shows the same fragmentation pattern independently of physical characteristics, especially its density. Due to this fact, results are usually expressed by relative frequencies.

Figure 4 represents the fragmentation state of the charcoal from the isolated house (the garbage dump excluded) and shows that the fragmentation of each taxa follows the same statistical distribution. Water does apparently not sort the species and the results can thus be expressed by relative frequencies. We have also compared the fragmentation in each sampled zone (fig. 5). The statistical distribution is similar from one area to another and looks like the Poisson distribution (small sized fragments are represented by a large majority). Thus, this analysis allows us to confirm that the proportion between species in lacustrine contexts, here shown by relative frequencies, is significant.

A sampling model

To establish a sampling model, spatial analyses of the charcoal from the garbage dump and the floating cordon were undertaken separately.

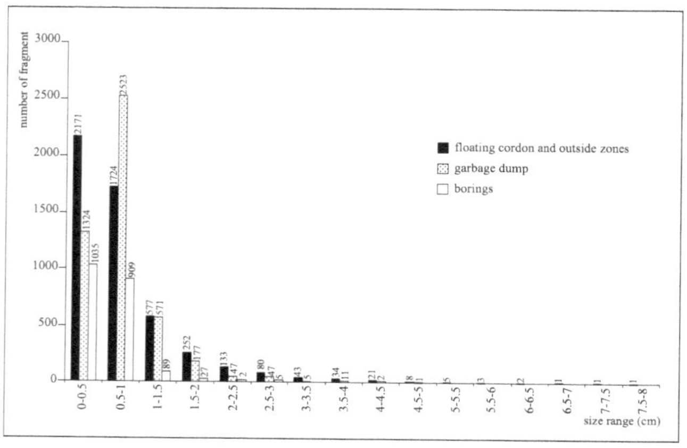

Figure 5: Comparison of the charcoal fragmentation between sampling zones.

Taxa/ Square metre	CQ 122	CQ 124	CR 121	Total
Abies alba	0,3%	-	-	0,1%
Acer campestre	1,9%	1,5%	1,5%	1,7%
Acer platanoides/ pseudoplatanus	5,7%	4,2%	4,7%	5%
Acer sp.	1%	0,3%	1,4%	1%
Angiospermae	0,7%		0,1%	0,3%
Betula sp.	0,1%	0,2%	0,1%	0,1%
Cornus sanguinea	0,6%	0,2%	0,4%	0,4%
Corylaceae	-	-	0,1%	0,02%
Corylus avellana	3%	4,5%	3,3%	3,4%
Ecorce/ Phloéme	1,2%	0,8%	1,9%	1,4%
Fagus sylvatica	62,3%	65,7%	64,5%	63,8%
Frangula sp.	-	0,1%	-	0,02%
Fraxinus excelsior	16,1%	14%	17,4%	16,1%
Ilex aquifolium	-	0,1%	-	0,02%
Pomoideae	0,1%	0,7%	0,3%	0,3%
Populus sp.	0,3%	0,2%	0,2%	0,3%
Prunus avium	-	0,1%	-	0,02%
Quercus f.c.	3,7%	2,9%	1,6%	2,8%
Rhamnus cathartica	0,1%	-	-	0,02%
Salix sp.	1,1%	1,2%	0,5%	0,9%
Salix/Populus	0,4%	0,3%	0,3%	0,3%
Tilia cordata	0,4%	0,5%	0,3%	0,4%
Tilia platyphyllos	0,2%	0,4%	0,2%	0,2%
Tilia sp.	0,6%	0,2%	0,6%	0,5%
Ulmus minor	0,1%	0,1%	-	0,04%
Ulmus montana	0,2%	1,0%	0,4%	0,4%
Ulmus sp.	0,3%	1,1%	0,2%	0,4%
Total	1964	1029	1659	4652
Underterminable	22	10	19	51
Undeterminated		5		5
Minimal number of taxa	18	18	16	22

Table 1: Number of taxa and relative frequencies in the garbage dump

The garbage dump

For several reasons we used the garbage dump as a reference in our study: previous analyses of charcoal from garbage dumps have proved the validity of such structures; the charcoal accumulated over a long period (25 to 35 years) in the dump comes from domestic activities.

Three samples, corresponding to three square meters, were taken. This zone was very rich in charcoal. The number of charcoal fragments attains an average of 932 fragments/l/m^2.

The results from the different square metres were compared (tabl. 1). The total number of taxa is 20 and does not vary from one square metre to another, probably due to the great number of identified fragments/m^2. The specific hierarchy remains the same and relative frequencies reflect a great spatial uniformity: *Fagus sylvatica* is preponderant with approximately 65% and is followed by *Fraxinus excelsior* (16%). Some taxa are present in only one square, for example *Frangula* sp. or *Rhamnus cathartica*.

The floating cordon

The objective of this analysis is to test the statistic significance of charcoal fragments from the floating cordon.

First, an analysis of the spatial richness of charcoal was established in order to understand the charcoal distribution and to emphasise the existence of the floating cordon.

<u>Charcoal dispersal</u> (fig. 6)

All of the identified charcoal fragments situated in the vicinity of the isolated house were included in the spatial analysis. First, the number of charcoal fragments/litre was calculated

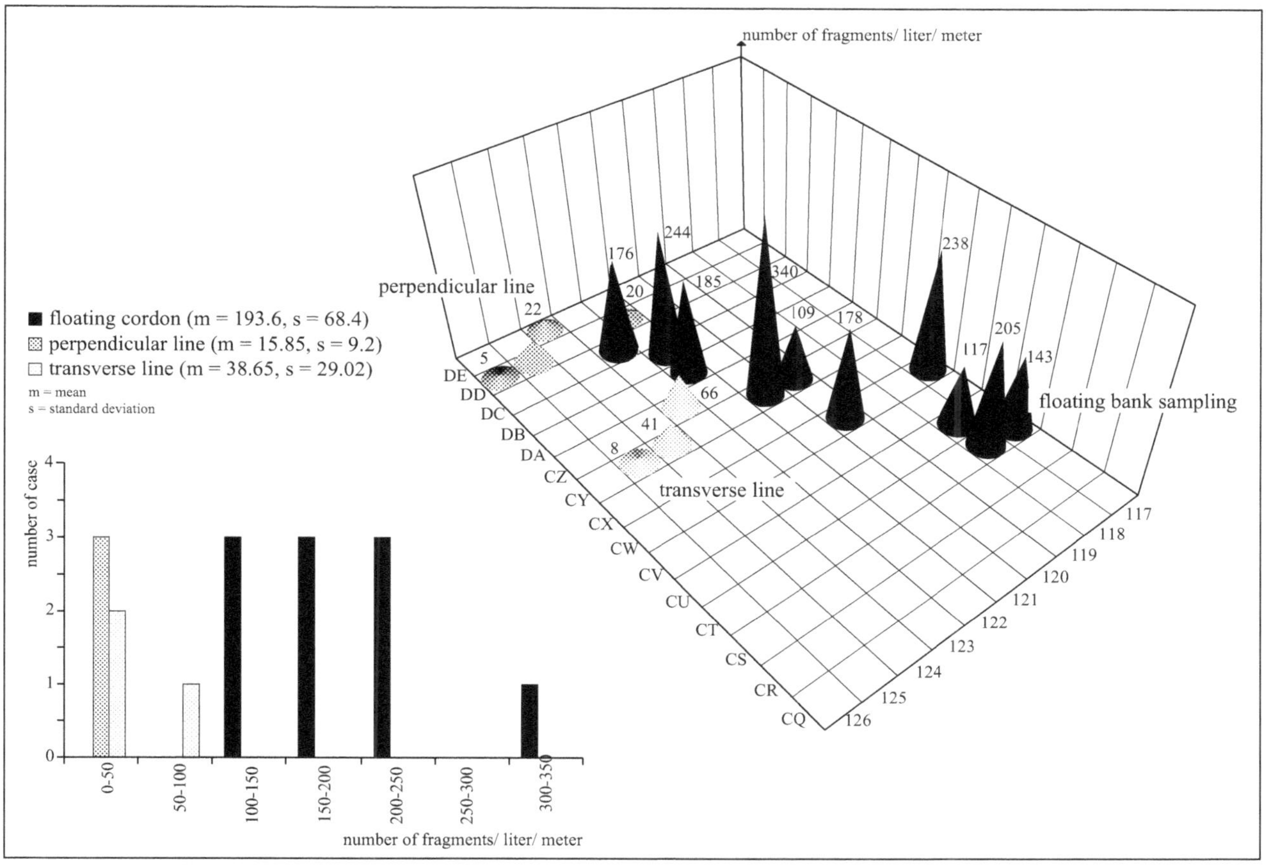

Figure 6: Relative abundance of charcoal per square metre in the isolated house.

for each square meter. All the square meters belonging to the floating cordon (observed during excavation) present values of more than 100 fragments/l/m^2. The average richness of charcoal fragments within the floating cordon is 193.6 charcoal fragments/l/m^2. Samples taken outside the floating cordon are much poorer, with averages of 38.7 fragments/l/m^2 for the transverse line and 15.8 fragments/l/m^2 for the perpendicular line. Still, outside the floating cordon one of the square meters revealed a richness of 408 fragments/l (different taxa). This may be explained by the presence of a non carbonised post that could have caused a local accumulation. The charcoal content of the floating cordon is nine times less than in the garbage dump and these results allow us to distinguish between the square metres that belong to the floating cordon and those who do not. The latter will be taken into account by the next analysis.

Optimal number of fragments (fig. 7)

Even though a great number of taxa had been identified by the charcoal analysis, we wanted to estimate the sufficient and necessary number of fragments to analyse in order to obtain statistically and palaeoecologically representative results. For that purpose, we successively calculated the number of taxa, starting with 30 fragments from every square, then 60, 100, 150 and 200, before including the totality of the fragments. We then compared the minimal number of taxa obtained from a given number of analysed charcoal fragments. The results show that 14 taxa were identified from 312 fragments (30/m^2) while 28 taxa were obtained by the study of 3625 fragments (250/m^2), *i.e.* from a number of fragments ten times higher. Such an effort is however not necessary and the identification of about 300 to 600 charcoal pieces (corresponding in our study to 18 taxa) seems to be sufficient. Some taxa, absent in the garbage dump, are represented in the floating cordon such as *Hippophae rhamnoides, Larix/Picea, Sambucus* sp., *Viburnum* sp. This may be due to a better spatial representation of the samples.

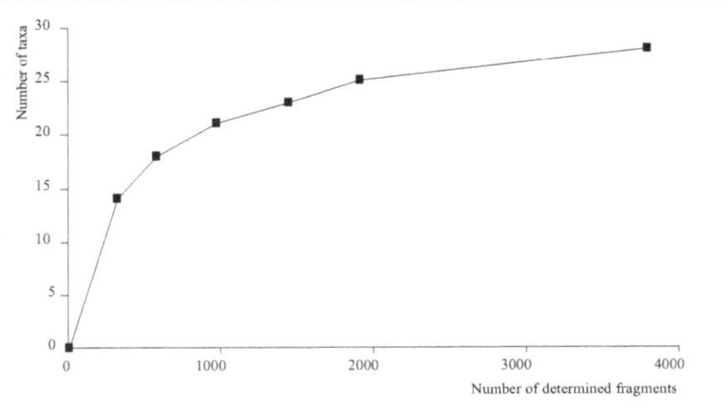

Number of determinated fragments per square meter	Minimal taxa number	Number of total determinated fragments
30	14	312
60	**18**	**574**
100	21	960
150	23	1430
200	25	1895
>250	28	3625

Figure 7: Optimal number of charcoal fragments.

Spatial analysis and the distribution of relative frequencies (fig. 8)

The number of taxa is more important than in the garbage dump, perhaps due to a better spatial representation of the samples. From one square metre to another, the number of taxa varies between 9 (249 analysed fragments) and 18 (380 analysed fragments). The number of taxa varies also with the number of analysed fragments except for two squares (CT 118 and DB 122) where 17 and 18 taxa were identified respectively on the basis of 400 fragments. The relative frequencies of the main taxa are stable from one square metre to another except for CW117, where *Fagus sylvatica* is less represented and *Fraxinus excelsior* more important than in the other square metres. Thus, the floating cordon is characterized by a clear homogeneity in the dispersion of taxa which is probably due to the action of water.

Lastly, the specific hierarchy is the same as in the garbage dump even though *Fagus sylvatica* is slightly more frequent than in the floating cordon. Thus, these results allow us to affirm that the sampling in the floating cordon is statisti cally significant.

Perpendicular and transverse lines: the outer zones (tabl. 2)

In these square metres, the number of taxa is less important than in the garbage dump and the floating cordon. The number of taxa varies from one square metre to another, mainly in proportion to the number of analysed fragments.

The specific hierarchy is the same as in the garbage dump but relative frequencies are slightly different: for example, the relative frequencies of *Fagus sylvatica* vary between 50 and 80%, with an average of 57.6%. This species is thus

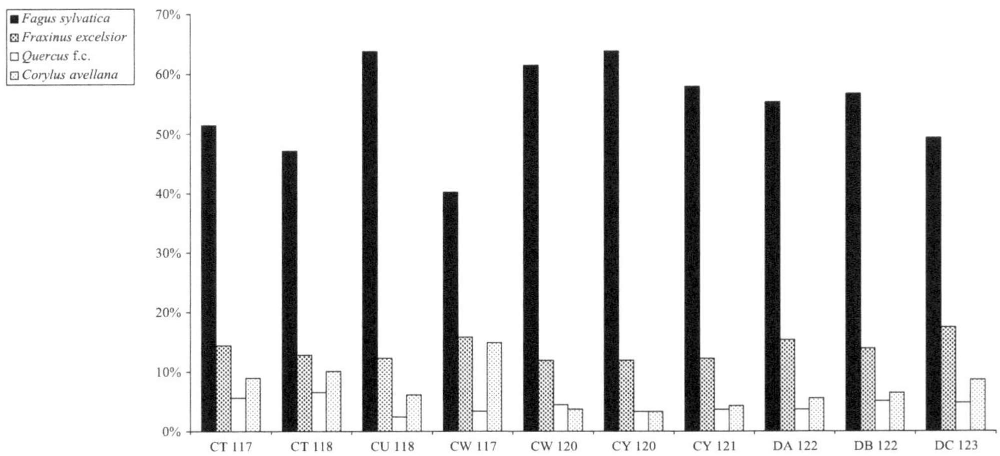

Taxa/ Square metre	CT 117	CT 118	CU 118	CW 117	CW 120	CY 120	CY 121	DA 122	DB 122	DC 123	Total
Abies alba	1,3%	1,9%	-	-	-	-	-	-	-	0,9%	0,6%
Acer campestre	2,1%	1,1%	3,3%	4,3%	1,9%	6%	4,3%	2,1%	4,3%	4,8%	3,4%
Acer platanoides/ pseudoplatanus	4%	5,7%	4,5%	3,3%	4,5%	5,1%	6,7%	2,8%	2,1%	5,1%	4,4%
Acer sp.	0,5%	0,5%	0,4%		0,4%		1,2%	0,9%	0,3%	0,9%	0,6%
Angiospermae	-	-	-	1,0%	-	-	-	-	0,3%		0,1%
Betula sp.	0,2%			0,5%	-	-	-	-	-	0,2%	0,1%
Cornus sanguinea	0,3%	1,4%	-	0,5%	-	0,3%	0,9%	0,3%	0,5%	-	0,4%
Corylus avellana	8,9%	10,1%	6,1%	14,8%	3,7%	3,3%	4,3%	5,5%	6,4%	8,6%	7,3%
Ecorce/ PhloËme	2,4%	2,2%	0,4%	1,9%	7,4%	2,4%	3,1%	4,3%	1,6%	4%	3%
Fabaceae						0,3%	-	-	-	-	0,03%
Fagus sylvatica	51,4%	47,1%	63,7%	40,2%	61,3%	63,7%	57,8%	55,2%	56,5%	49,2%	54,2%
Frangula sp.									0,3%	0,2%	0,1%
Fraxinus excelsior	14,4%	12,8%	12,2%	15,8%	11,9%	11,9%	12,2%	15,3%	13,9%	17,4%	14%
Hippophae rhamnoides	0,2%	0,3%	-	-	-	-	-	-	-	-	0,1%
Ilex aquifolium	-	0,3%	-	-	-	-	-	-	-	-	0,03%
Larix /Picea	-	0,3%	-	-	-	-	-	-	-	-	0,03%
Pomoideae	-	-	-	-	0,4%	0,3%	0,3%	0,9%	0,3%	-	0,2%
Pomoideae type *Sorbus/ Crataegus*	0,8%	0,8%	-	0,5%	0,4%	0,6%	-	0,6%	0,3%	-	0,4%
Populus sp.	0,5%	1,1%	-	2,4%	0,4%	0,6%	-	-	1,3%	1,5%	0,8%
Prunoideae	0,2%	-	-	-	-	-	-	-	-	-	0,03%
Prunus avium	-	0,3%	-	-	-	-	-	-	-	-	0,03%
Quercus f.c.	5,6%	6,5%	2,4%	3,3%	4,5%	3,3%	3,7%	3,7%	5,1%	4,8%	4,5%
Rosaceae	-	-	-	-	0,4%	-	-	-	-	-	0,03%
Salix sp.	4%	3,8%	4,5%	9,1%	1,1%	1,5%	3,7%	4,9%	0,8%	1,5%	3,2%
Salix/Populus	0,2%	1,4%	-	-	-	-	-	-	0,3%	-	0,2%
Sambucus sp.	-	0,3%	-	-	-	-	-	-	-	-	0,03%
Sorbus sp.	-	-	-	-	-	-	-	-	0,3%	0,2%	0,1%
Tilia cordata	1%	-	0,8%	0,5%	0,7%	0,3%	0,9%	0,6%	2,4%	-	0,7%
Tilia platyphyllos	-	-	-	0,5%	0,4%	-	0,3%	-	0,3%	-	0,1%
Tilia sp.	1,4%	0,5%	0,8%	1%	0,4%	-	0,6%	1,8%	0,3%	-	0,7%
Ulmus minor	0,5%	0,3%	-	0,5%	0,4%	0,3%	-	0,6%	0,8%	0,2%	0,4%
Ulmus montana	-	0,8%	0,4%	-	-	0,3%	-		1,6%	0,2%	0,3%
Ulmus sp.	0,2%	0,5%	0,4%	-	-	-	-	0,3%	0,3%	0,4%	0,2%
Viburnum sp.	0,2%	-	-	-	-	-	-				0,03%
Number of fragments	626	367	245	209	269	336	327	326	375	545	3625
Indeterminable	8	12	3	7	7	11	10	24	13	23	118
No identified	2	1	1	2			3			8	17
Minimal number of taxa	17	18	9	15	14	15	11	12	17	14	28

Figure 8: Distribution of the frequencies of the main taxa in the square metres corresponding to the floating cordon.

Taxa/Square metres	CY 124	CY 125	CZ 123	DD 122	DD 125	DD 126	DE 124	Total
Abies alba	0,4%	-	0,5%	0,6%	-	-	0,6%	0,3%
Acer campestre	0,4%	3,2%	5,5%	0,6%	1,1%	-	1,7%	2%
Acer platanoides/ pseudoplatanus	4,6%	3,9%	5%	8,3%	4,4%	3,2%	4,6%	5%
Acer sp.	0,8%	0,6%		1,3%	0,7%	3,2%	0,6%	0,7%
Angiospermae	0,4%	-	-	-	-	-	1,1%	0,2%
Betula sp.	0,8%	-	-	-	-	-	-	0,2%
Corylus avellana	15,9%	12,3%	12,3%	3,2%	4,1%	6,5%	12,6%	10%
Ecorce/ Phloéme	0,8%	1,9%	3,6%	2,6%	0,7%	3,2%	-	1,6%
Fagus sylvatica	49,8%	46,8%	55,0%	64,1%	77,5%	71%	42%	57,6%
Fraxinus excelsior	12,6%	13,6%	7,3%	10,9%	6,6%	9,7%	13,2%	10,3%
Monocotyledonae	-	0,6%	-	-	-	-	-	0,1%
Pomoideae	0,4%	-	0,5%	-	-	-	0,6%	0,2%
Pomoideae type Sorbus/ Crataegus	0,8%	-	-	-	-	-	-	0,2%
Populus sp.	-	0,6%	-	-	-	-	1,7%	0,3%
Quercus f.c.	3,3%	8,4%	5,9%	0,6%	1,1%		17,2%	5,5%
Salix sp.	4,6%	1,3%	0,5%	1,3%	3%	3,2%	2,3%	2,3%
Salix/Populus	-	-	-	3,2%	-	-	0,6%	0,5%
Sorbus sp.	-	-	0,5%	0,6%	-	-	-	0,2%
Tilia cordata	0,8%	0,6%	1,4%	-	0,7%	-	-	0,6%
Tilia platyphyllos	-	1,9%	1,8%	-	-	-	-	0,6%
Tilia sp.	2,5%	1,3%	0,5%	0,6%	-	-	0,6%	0,9%
Ulmus minor	-	1,9%	-	-	-	-	-	0,2%
Ulmus montana	0,4%	0,6%	-	0,6%	-	-	-	0,2%
Ulmus sp.	0,4%	-	-	1,3%	-	-	0,6%	0,3%
Total	239	154	220	156	271	31	174	1245
Indeterminable	9	9	14	4	9	1	4	50
No identified		1						1
Minimal number of taxa	14	13	12	11	8	5	13	20

Table 2: Number of taxa and relative frequencies in outer zones.

less important than in the garbage dump while the frequency of *Corylus avellana* is higher. Thus, a statically representative sampling in these zones is possible but in order to obtain a comparable number of taxa as in the floating cordon, more fragments have to be analysed.

Conclusion

We have seen that different architectural organisations could exist in the same village with respect to the water level (Pétrequin, 1988). There is also a relation between the architecture of the houses, the social and technical functions and the spatial distribution of artefacts and waste. Spatial analysis in Chalain 3 has shown a clear correlation between the distribution of charcoal and the location of fireplaces, underlain by a clay layer. Therefore the charcoal fragments could have resulted from the episodic sweeping of the clay layer (Pétrequin, 1997, p. 588). The charcoal in the garbage dump could result from domestic activities related to the isolated house, such as firewood gathering. According to dendrochronological dates, the accumulation of these charcoal fragments would correspond to a period of 25-30 years.

The very similar results obtained from the garbage dump, the floating cordon and the outer zones allow us to suppose that the charcoal in these three areas resulted from the same type of activities. If the isolated house had a raised floor, a space must have existed between the floor and the ground. The waste could then have been thrown directly on a garbage dump or disposed of through a trapdoor in the floor. In that case some of the waste would have been mixed with the water and trapped under the floor when the house collapsed.

The garbage dump and the floating cordon are rich in taxa that are coherent with current vegetation patterns and thus probably representative from a palaeoecological point of view. The results from old Clairvaux are characterized by a predominant exploitation of *Fagus sylvatica*. The analysis of the others layers shows a vegetal dynamic between *Fraxinus excelsior* and *Fagus sylvatica* between the Horgen and the old Clairvaux (Dufraisse, 2000).

To conclude, at lake dwelling sites the charcoal sampling can be restricted to the floating cordon and does not have to be based on the entire surface of the studied archaeological layer.

These conclusions lead us to consider sampling by cores in the cordons of lake dwelling sites, especially when excavations have not been carried out or when excavated sites have been dated by dendrochronology or ^{14}C. One of the main aims of our research is to obtain a detailed picture of the firewood supply for a chronological sequence spanning from 3200 to 2500 BC. The suggested sampling method appears particularly interesting given the number of non-excavated villages on the shores of the lakes of Chalain and Clairvaux. Such a procedure would equally permit the completion of the chronology.

Core samplings of charcoal

What type of lake stations should be studied?

Two types of stations may lend themselves to core sampling: stations that have already been excavated and precisely dated by dendrochronology, such as Chalain 3. The coring can then be done in parts of the site that were left in place. The second type of stations are listed during coring and are mostly dated by ^{14}C.

Material and methods

Two types of auger can be used. The first one is the Russian auger, mostly used in palynology and with a diameter of 5 cm. It reaches the deepest archaeological layers and can provide a highly legible stratigraphy. The second type is a pedological auger which permits the obtaining of core samples with a diameter of about 10 cm. The larger diameter presents a certain interest but the stratigraphy cannot always be read as precisely as with the Russian auger.

According to the sedimentary dynamics, we propose a sampling protocol that was tested at Chalain 19. A main transect, perpendicular to the bank ("the first transect" of fig. 3) was made in order to localise the floating cordons in every archaeological layer. Along this transect, the drilling of cores every 2, 5 or 10 m was enough and already produced a few samples. In order to complete the first sampling, secondary transects were made either in the most dilated layers or where the number of charcoal fragments was the most important (within the floating cordons) following the contour lines (localised along a line parallel to the bank). Both of these transects provide supplementary information on sedimentary dynamics.

Results (tabl. 3)

Only a palynology auger was used. Three archaeological layers could be sampled: H, K, and O. The first transect was made along a line perpendicular to the bank, with samples taken at every metre. In order to complete the sampling, a second transect, perpendicular to the first and following the

H

Taxa/ Corings	S23	S23A	S24A	S24Abis	S25A	S25C	S26	Total
Abies alba	-	-	-	-	-	-	-	-
Acer campestre	3,6%	-	-	6,3%	4,8%		0,9%	2,8%
Acer platanoides/ pseudoplatanus	4,3%	-	-	6,3%	3,2%	2,9%	2,8%	3,5%
Acer sp.	2,9%	-	-	1,6%	3,2%		2,8%	2,2%
Angiospermae	2,2%	-	-	0,0%	1,6%		5,6%	2,2%
Betula sp.	-	-	-	1,6%		2,9%		0,4%
Carpinus betulus	-	-	-	-	-	-	-	-
Cornus sanguinea	-	-	-	-	-	-	-	-
Corylus avellana	5,8%	6,9%	19,2%	9,4%	12,9%	20%	8,4%	9,8%
Ecorce/ Phloéme	2,9%	3,4%	3,8%	1,6%		2,9%	1,9%	2,2%
Fabaceae	-	-	-	-	-	-	-	-
Fagus sylvatica	29,7%	48,3%	38,5%	34,4%	48,4%	20%	34,6%	34,9%
Fraxinus excelsior	24,6%	20,7%	15,4%	15,6%	17,7%	28,6%	24,3%	21,9%
Pomoideae	1,4%	-	-	-	-	-	0,9%	0,7%
Pomoideae type *Pirus*	-	-	-	-	-	-	-	-
Pomoideae type *Sorbus/ Crataegus*	-	-	3,8%	-		-	-	0,2%
Populus sp.	0,7%	3,4%	-	-	3,2%	-	0,9%	1,1%
Prunus type *avium/padus*	-	-	-	-	-	-	-	-
Quercus f.c.	5,1%	3,4%	0,0%	10,9%	1,6%	2,9%	2,8%	4,3%
Salix sp.	2,9%	6,9%	11,5%	7,8%	1,6%	17,1%	9,3%	6,7%
Salix/Populus	6,5%	-	-	-	-	-	-	2,0%
Taxus baccata	-	-	-	-	-	-	-	-
Tilia cordata	-	3,4%	-	-	1,6%	2,9%	1,9%	1,1%
Tilia platyphyllos	0,7%	-	-	1,6%	-	-	0,9%	0,7%
Tilia sp.	2,9%	-	-	1,6%	-	-	0,9%	1,3%
Ulmus minor	-	-	3,8%	-	-	-	-	0,2%
Ulmus montana	-	3,4%	3,8%	-	-	-	-	0,4%
Ulmus sp.	3,6%	-	-	-	-	-	0,9%	1,3%
Viburnum sp.	-	-	-	1,6%	-	-	-	0,2%
Total	138	29	26	64	62	35	107	461
Indeterminable			1		1		1	3

K

Taxa/ Corings	S23	S23A	S24	S24A	S24Abis	S25	S26A	Total
Abies alba	-	-	-	-	-	-	-	-
Acer campestre	2,1%	5,3%	2,0%	3,5%	-	2,5%	-	2,7%
Acer platanoides/ pseudoplatanus	4,3%	-	4,1%	1,2%	11,8%	2,5%	4,3%	3,0%
Acer sp.	-	2,6%	6,1%	2,4%	5,9%	7,5%	4,3%	3,7%
Angiospermae	2,1%	-	4,1%	1,2%	-	5,0%	4,3%	2,3%
Betula sp.	-	-	-	-	-	-	-	-
Carpinus betulus	-	-	-	-	-	-	-	-
Cornus sanguinea	-	-	-	-	-	-	-	-
Corylus avellana	6,4%	5,3%	-	12,9%	-	2,5%	21,7%	7,4%
Ecorce/ Phloéme	2,1%	5,3%	2,0%		5,9%	10,0%		3,0%
Fabaceae	-	-	-	-	-	-	-	-
Fagus sylvatica	34,0%	55,3%	40,8%	48,2%	47,1%	37,5%	34,8%	43,1%
Fraxinus excelsior	38,3%	13,2%	32,7%	14,1%	5,9%	20,0%	30,4%	22,4%
Pomoideae	-	-	-	-	-	-	-	-
Pomoideae type *Pirus*	-	-	-	-	-	-	-	-
Pomoideae type *Sorbus/ Crataegus*	-	-	-	-	-	-	-	-
Populus sp.	2,1%	-	-	-	-	-	-	0,3%
Prunus type *avium/padus*	-	-	-	1,2%	-	-	-	0,3%
Quercus f.c.	6,4%	2,6%	2,0%	2,4%	17,6%	5,0%	-	4,0%
Salix sp.	-	-	-	5,9%	-	-	-	1,7%
Salix/Populus	-	-	-	-	-	2,5%	-	0,3%
Taxus baccata	-	-	-	-	-	-	-	-
Tilia cordata	-	-	-	-	-	-	-	-
Tilia platyphyllos	-	2,6%	-	1,2%	-	-	-	0,7%
Tilia sp.	-	-	6,1%	4,7%	-	5,0%	-	3,0%
Ulmus minor	-	2,6%	-	-	-	-	-	0,3%
Ulmus montana	-	2,6%	-	-	-	-	-	0,3%
Ulmus sp.	2,1%	-	-	-	5,9%	-	-	0,7%
Viburnum sp.	-	2,6%	-	1,2%	-	-	-	0,7%
Total	47	38	49	85	17	40	23	299
Indeterminable	1			1		2		4

O

Taxa/ Corings	S1	S2	S3	S4	S5	S6	S9	S21	S23	S23A	S24A	S24Abis	S25A	S25C	Total
Abies alba	-	-	-	-	-	-	-	-	-	-	-	-	-	-	-
Acer campestre	-	-	-	25%	-	-	-	-	-	6,1%	-	-	2,4%	-	1,0%
Acer platanoides/ pseudoplatanus	-	-	-			20%	3,8%	3%	-	3,0%	-	-	-	-	1,5%
Acer sp.	-	-	-	-	-	-	-	1%	-	3,0%	-	-	-	-	0,5%
Angiospermae		22,2%	-	-	-	-	7,7%	1%	4,9%	-	-	-	-	-	2,3%
Betula sp.	-	-	-	-	-	-	-	-	-	-	-	-	-	-	-
Carpinus betulus	-	-	-	-	-	-	-	-	-	-	-	-	-	-	-
Cornus sanguinea	-	-	-	-	-	-	-	-	-	-	-	-	-	-	-
Corylus avellana	-	33,3%	-	-	-	-	3,8%	4%	3,7%	3,0%	-	2,9%	2,4%	5,6%	3,8%
Ecorce/ Phloéme	-	-	-	-	-	-	38,5%	46%	6,1%	6,1%	42,3%	54,3%	33,3%	-	27,1%
Fabaceae	-	-	-	-	-	-	-	-	-	-	-	-	-	-	-
Fagus sylvatica	100%	33,3%	50%	25%	75%	60%	3,8%	1%	1,2%	3,0%	-	-	-	-	5,1%
Fraxinus excelsior	-	-	50%	25%	-	-	23,1%	36%	78,0%	63,6%	53,8%	42,9%	57,1%	66,7%	50,1%
Pomoideae	-	-	-	-	-	-	-	2%	-	-	-	-	-	-	0,5%
Pomoideae type *Pirus*	-	-	-	-	-	-	-	-	-	-	-	-	-	-	-
Pomoideae type *Sorbus/ Crataegus*	-	-	-	-	-	-	-	-	-	-	-	-	-	5,6%	0,3%
Populus sp.	-	-	-	-	-	-	-	-	-	-	-	-	-	-	-
Prunus type *avium/padus*	-	-	-	-	-	-	-	-	-	-	-	-	-	-	-
Quercus f.c.	-	11,1%	-	25%	25%	-	-	1%	1,2%	9,1%	-	-	-	11,1%	2,5%
Salix sp.	-	-	-	-	-	-	-	-	-	-	-	-	-	5,6%	0,3%
Salix/Populus	-	-	-	-	-	-	7,7%	-	1,2%	3,0%	-	-	-	-	1,0%
Taxus baccata	-	-	-	-	-	-	-	-	-	-	-	-	-	-	-
Tilia cordata	-	-	-	-	-	20%	-	-	-	-	-	-	4,8%	-	0,8%
Tilia platyphyllos	-	-	-	-	-	-	-	1%	-	-	-	-	-	-	0,3%
Tilia sp.	-	-	-	-	-	-	3,8%	1%	2,4%	-	-	-	-	5,6%	1,3%
Ulmus minor	-	-	-	-	-	-	-	-	-	-	-	-	-	-	-
Ulmus montana	-	-	-	-	-	-	-	2%	-	-	-	-	-	-	0,5%
Ulmus sp.	-	-	-	-	-	-	7,7%	1%	1,2%	-	3,8%	-	-	-	1,3%
Viburnum sp.	-	-	-	-	-	-	-	-	-	-	-	-	-	-	-
Total	1	9	10	4	4	5	26	100	82	33	26	35	42	18	395
Indeterminable					1				1						2

Table 3: Relative frequencies per layer and core.

contour lines, was made where charcoal fragments were concentrated, probably at the location of a cordon.

The amount of charcoal fragments, counted for similar volumes of sediment, was found to be highly variable according to the localisation of the core on the bank. For example, cores S23 and S23A of layer O yielded respectively 100 and 82 charcoal pieces while the other cores contained less than 40 fragments. Consequently, secondary transects can be positioned according to concentrations of charcoal, corresponding to the presence of floating cordons in the studied archaeological layers.

By considering the optimal number of charcoal fragments, it is thus possible to complete the sampling by concentrating on the floating cordons. Sometimes the number of charcoal fragments is nevertheless similar from one core to another, as in the case of layers H and K. The total number of fragments is calculated on the basis of all of the cores and correponds to 300-500 charcoal fragments/layer.

The specific richness is relatively high (12-21taxa according to the layer). Some additional taxa are represented only in previous samples (by square metres) but this is probably due to the higher number of identified fragments. The analysis of an equal number of fragments would probably yield more or less the same number of taxa.

The proportions of the various species are different in layers H and K, compared to layer H/K. Yet, the same cultural period, the old Clairvaux, is concerned in the two cases. These discrepancies do not however question the validity of the results. Indeed, the organisation of the village could have undergone changes through time (Pétrequin, 1997). The main species are the same and a vegetal dynamics between *Fraxinus* and *Fagus* appears again.

Consequently and according to the proportions of the various species, the sampled zone near the bank could correspond to an intermediate stage of the human occupation. As our results show a similar diversity in species and are coherent from a palaeoecological point of view, the chosen method of sampling seem to allow a good statistical and palaeoecological representation. Lastly, let us note that by this procedure it was possible to sample all the archaeological layers in just one day.

Conclusion

In conclusion, the analysis of the forming of deposits has revealed the presence of horizons of accumulation formed by plant macroremains suspended in water. This study confirms the mode of dispersion of charcoal in floating cordons.

Analysis of relative frequencies and specific diversity in several samples gave very homogenous results, due certainly to the action of waves. The relatively high number of taxa identified and their palaeoecological coherence confirm further the representativity of these samples. Consequently, sampling from floating cordons seems sufficient in lake dwelling sites. Sampling from cores, following a precise protocol, was carried out at Chalain 19. One objective of our Ph.D. research is to establish a detailed chronology for a period of 25-50 years, out of a millennium of occupation, and this method will make it possible to analyse key stations.

References

CHABAL L. 1997 - Forêts et sociétés en Languedoc (Néolithique final, Antiquité tardive). L'anthracologie, méthode et paléoécologie. Maison des Sciences de l'Homme, Paris. *DAF* 63, 189 p.

DUFRAISSE A. 2001 - Combustible et espace forestier à Chalain, station 19 (Jura).In : *Equilibres et Ruptures dans les écosystèmes depuis 20 000 ans en Europe de l'Ouest.* Colloque international du 18 au 22 septembre 2000. Besançon.

PETREQUIN P., PETREQUIN A.-M. 1984 - *Habitat lacustre du Bénin. Une approche ethnoarchéologique.* Ed. Recherche sur les Civilisations, mémoire 39, 214 p.

PETREQUIN P. 1997 - La répartition des témoins et le fonctionnement de la cellule domestique. *In* : *Les sites littoraux néolithiques de Clairvaux-les-Lacs et de Chalain III.* Chalain station 3 3200-2900 av. J.-C. Volume 2, p. 583-639

PETREQUIN P., PETREQUIN A.-M. 2000 - *Lac de Chalain. Fontenu (Jura). Chalain 19.* 2ème année de fouille programmée tri-annuelle 2000. Rapport intermédiaire. Besançon, Laboratoire de Chrono-écologie. UMR 6565. CNRS-Université de Franche-Comté, multigraphié.

The study and significance of charcoal as an indicator of ancient fires: an application to the middle Rhone valley (France)

Jean-François Berger[1] and Stéphanie Thiébault[2]
1) CNRS-UMR 6130, CEPAM - 250, rue Albert Einstein, Sophia–Antipolis, F- 06560 Valbonne, France.
E-mail: berger@cra.cnrs.fr
2) CNRS-UMR 7041, Archéologies et Sciences de l'Antiquité- 21, Allée de l'Université, F- 92023 Nanterre cedex, France.
E-mail: s.thiebault@mae.u-paris10.fr

Abstract
Since about fifteen years, the systemic approach has been privileged for palaeoenvironnemental reconstructions in the middle Rhone valley. Various methods of research have been developed. The question of the origin of fires during the Holocene is considered, by focusing on soil science, sedimentology and charcoal analysis. After a review of the literature devoted to Holocene fires in terrestrial pedo-sedimentary formations, the biogeographic and climatic contexts of the study are dealt with. The presentation of the state of our knowledge is followed by a description and discussion of the process and the methods chosen. The processes that modify the state of charcoal particles, from their zone of production until the place where they are discovered (taphonomy), are then described in detail and explained. Finally, two examples are presented in order to illustrate the interest of such studies. The first example relates to a river sequence in the Valdaine, dating from the Bronze Age to the modern period; the second relates to the agricultural cycle and the climate in the Tricastin plain during Antiquity (120 BC-500 AD).

Key-words: *Ancient fires, Holocene, Middle Rhone valley, Palaeo-environmental reconstructions, Systemic approach.*

Résumé
Depuis une quinzaine d'années en moyenne vallée du Rhône, la démarche systémique a été privilégiée. Divers axes méthodologiques ont été développés et notamment la pratique systématique du hors site. La question de l'origine des incendies au cours de l'Holocène est envisagée à partir d'une approche principalement centrée sur la pédologie, la sédimentologie et l'anthracologie. Après une revue des études consacrées aux paléo-incendies holocènes, dont les vestiges sont rencontrés dans les formations pédo-sédimentaires continentales, les contextes biogéographique et climatique de l'étude sont exposés. Ensuite, un état de la question est présenté. La démarche sélectionnée et les méthodes utilisées sont décrites et discutées. Les processus qui modifient l'état des charbons depuis leur zone de production jusqu'à leur lieu de découverte aujourd'hui (taphonomie) sont finement décrits et explicités. Enfin, deux exemples sont présentés, afin d'illustrer l'intérêt de telles études. Le premier concerne une séquence fluviatile de l'Age du Bronze à l'époque moderne en Valdaine, le second les cycles agraires et le climat de la plaine du Tricastin dans l'Antiquité.

Mots-clés : *Paléo-incendies, Holocène, moyenne vallée du Rhône, reconstitution paléoenvironnementale, approche systémique.*

Introduction

The systemic approach, which was introduced more than fifteen years ago in the middle Rhone valley, was selected as the most appropriate for the present study. The area is a transition zone between two climates: Mediterranean and Alpine. This study aims at characterising, as precisely as possible, the history of the Holocene pedosedimentary dynamics, and to evaluate the long-term interactions between society and climate, upstream as well as downstream from selected basin slopes. One of the major features interesting this study is to evaluate why and how changes in the landscape have transformed cultural behaviour, and more precisely the choice of settlement location and the nature of the exploitation of the environment (Berger, 2000).

Carbonised vegetal remains are found in various forms and sizes in recent (Holocene) continental deposits in the form of soot and charcoal that is more or less carbonised and identifiable by charcoal analysis. The sampling of these remains is based on systematic sieving of well-defined stratigraphic units in the field. The exploitation of different pedo-sedimentary contexts (river beds, detritic fans, marshy basins, colluvium, irrigation channels, cultivation terraces...) is carried out within the framework of the systemic analyses of river basins of different sizes, located in the north-western part of the Mediterranean basin (*i.e.* the middle Rhone valley) (fig. 1). Good control of the context of charcoal fossilisation, together with species identification (a practice still very rare outside an archaeological context), give us a glimpse of the diversity of the burnt vegetation, both in time (since the late Glacial), and in space (from the local to the regional scale). It enables us to discuss the climatic or anthropogenic origin of fires in the past.

Two applications illustrate these remarks. The first relates to the charcoal found in the palaeochannels and the palaeosoils of a fluvial sequence, trench n°18 at Portes-en-Valdaine (Drôme), which allows us to propose a scenario for the evolution of the Lanson basin and its agrosystem, from the end of the Bronze Age to modern times. The second application concerns the study of the charcoal particles trapped in ancient agrarian structures excavated in the Tricastin plain (Vaucluse), and its contribution to our knowledge of pyro-technology, agro-pastoral cycles and climate in the past.

Holocene fires

Spontaneous fires are attested since the Mesozoic period by charcoal deposits, mainly in the fluvio-estuarine sedimentary rocks (Glasspool, 2000; Falcon-Lang, 2000). During more recent periods, fire signals are found in the remains of most processes that characterise the continental geo-biosphere. They reflect climatic variations and human impact on the environment. Knowledge of the history of fires is thus important in order to evaluate the interaction between vegetation, climate and man.

Hunter-gatherer and farming activities (such as slash-and-burn methods) used to be preferred over climatic events in order to explain the frequency of fires and their negative effects on the landscape during the European Holocene (Simmons and Tooley, 1981; Jacobi *et al.* 1976; D'Anna and Courtin, 1986; Thinon, 1992; Moore, 2000). The evidence for this hypothesis, however, was very often insufficient and

inadequate. The main arguments used in these discussions on the origins and manifestation of fires were:
-the fact that the fires were contemporaneous with palynological evidence from settlements, revealing changes in the AP/NAP ratio;
- the presence of crop plant remains in the deposits;
- the presence of agricultural tools in the deposits.

The assumptions made are more often than not based on indirect evidence, notably on a behavioural analogy with current "primitive" communities, studied by anthropologists or ethnologists. The use of fire by modern agro-pastoral communities is well documented ethnographically (in Australia, Africa, North America...) and by written records from the last few centuries (Sigaut, 1975).

The systematic palaeoenvironmental studies that have been carried out in the past few years in the middle Rhone valley, however, reveal that evidence for fires antedates the emergence of agriculture in south-western Europe during the 6th millennium cal. BC. Moreover, recent palaeoclimatic studies have improved our understanding of "global change". They have shown that, in sensitive areas, large variations in climate and/or long, marked drought periods were favourable to increases in the frequency of fire (Courty, 1990; Fedoroff and Courty, 1989; Vernet *et al.* 1994; Levine, 1996). Historic data with annual resolution, indicate that natural fires are a cyclical phenomenon, the cycle of which is controlled by climatic dynamics (Clark, 1988, 1996). Data concerning the natural origin of fires are quite well known but rarely mentioned, except in Australia or North America where natural dynamics of fire has been observed continuously since the European colonisation, and where the history of agriculture is younger than in Europe.

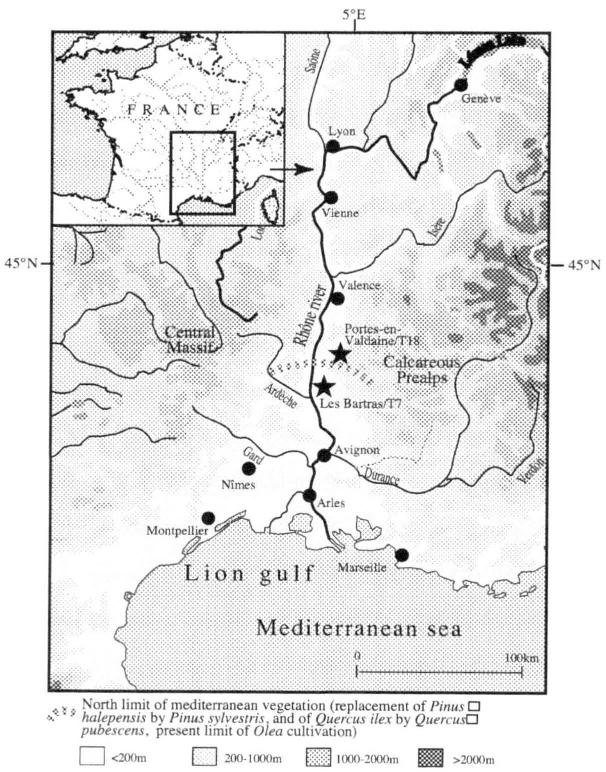

Figure 1: Location of the middle Rhone valley, the studied sequences, and the northern limit of the Mediterranean vegetation according to Ozenda (1994).

In fact, the majority of European researchers are in our opinion too reticent to assume natural cycles of fire until the last part of the Holocene (Thinon, 1992; Simmons and Innes, 1996). In this, they differ from their American, Canadian or Australian colleagues, who have been developing theories about the very ancient origins of cyclical fires for about thirty years (Jones and Chaloner, 1991; Clark, 1982; Laird and Campbell, 2000). The latter consider fire as a process of the ecological regulation, in which human practices play a part, but they do not see fires as an ecological and economic disaster, as is generally assumed in the Mediterranean basin. The perception of fire in these different areas is based on very different cultural traditions.

Recent publications and our own investigations clearly show a pattern of interaction in fire dynamics between the activities of the last hunter-gatherers, the first farmers and the climate over a very long time (Clark, 1988; Bradshaw, 1993; Berger, 1996, 2000). This dialectical relation makes a distinction between fires of climatic and anthropogenic origin difficult. Indeed, the dynamics must be observed over a long time in order to document the existence of cycles, and to decide whether casual links with other processes (the anthropisation and the artificialisation of the environment, the evolution of climate and the vegetation, the erosion of soils) can be proven.

The biogeographic and climatic context of the study

The Mediterranean region forms a transitional bioclimatic area between the temperate and tropical zones, or between the humid and arid zones. The middle Rhone valley is an ecotone that links the northern limit of the Mediterranean climatic zone to the Medio-European and Alpine climatic spheres (fig. 1). Its location makes the vegetation of the middle Rhone valley highly reactive to climatic changes and anthropogenic pressure. The area's climate is of a humid sub-Mediterranean type, marked by highly irregular temperatures and precipitation; its winds form vectors, and its vegetation, associated with the supra-Mediterranean level of vegetation (deciduous oaks, ash, beech and box, Ozenda, 1964), is little sensitive to fire action. The Mediterranean dryness (two months of water deficit in summer) is still marked but the average annual precipitation appears rather abundant (between 700 and 1000 mm). The concentration of precipitation over two seasons (autumn and spring), associated with irregular precipitation (20 to 50 mm), does not, however, allow an optimal exploitation of the water by plants. The reason is that the process of surface streaming drains the major part of the precipitation before its infiltration into soils.

The analysis of the annual and inter-annual climatic averages, during the whole last century, shows that a certain climatic variability characterises the middle Rhone valley.

During the last century, fresher or mountainous and hotter or dryer influences can be observed (Boisson, 1988). The same variations could have occurred during the past for long periods. Recent geomorphological and especially palaeo-pedological data (Berger, 1996, 2000; Bravard,

1997; Provansal *et al.* 2000) confirm the fact that the climatic limit varied during the Holocene according to a south-north gradient. During that period, the middle Rhone valley was alternatively an integral part of each of the three bioclimatic areas mentioned, depending on the evolution of the global climate. Different bioclimatic contexts were present at different times, and had the potential of modifying the structure of the vegetation and the ways in which fires manifested themselves. Because the area is at the interface of three climatic zones, includes a wide range of microclimatic conditions related to various topographical features, and has for a long time been subject to heavy human pressure, its vegetation pattern is the kind of mosaic characteristic of the supra-Mediterranean stratum. (Ozenda, 1964, 1994). The forests consist of oak coppices, dominated by *Quercus* cf. *pubescens*, and pine. Their undergrowth usually contain trees and grasses. Pine forests are the pioneer formation succeeding fire in the area, whereas deciduous oak forests establish themselves only later (Pons and Thinon, 1987; Trabaud, 1989). The transitional aspect of this area is well recorded in the vegetation, which shows for example a clear substitution of *Pinus halepensis* by *Pinus sylvestris* and *Quercus ilex* by *Quercus pubescens* to the north of the Donzère defile, a few kilometres south of the 45th parallel (fig. 1).

State of the question, process and method

State of the question

Ecologists have long studied the consequences of fire on the vegetation and on the functioning of present-day ecosystems (including soils) in southern France (Braun-Blanquet, 1936; Kuhnholtz-Lordat, 1938; Trabaud, 1982, 1992, 1998; Barbero *et al.* 1990), but long-term observations are very scarce. The existing studies show that the consequence of fires (even if they are frequent) is not an irreversible degradation of the indigenous plant communities. In the garrigue or in a pine forest, the evolution of the plant communities after a fire is characterised by the re-establishment of a relative ecological and floristic stability after 10-12 years.

Such studies are, however, based on short term observations (one or two decades) and do not consider the changes that can occur in the long term. The long-term palaeo-ecological record shows different dynamics under the impact of different factors such as climatic change or agro-pastoral activities. Considering the very long term appears to be the only way to investigate the reality of climatic, biological and human cycles that have interacted for 15,000 years in the auto-regulation of the ecosystem. In North America or the United Kingdom, the study of ancient fires concerns mainly the lacustrine edge. The methods used were based on the counting of carbonised microscopic remains in palynological slides, and sometimes also sedimentological slides (Clark, 1988, 1996; Patterson *et al.* 1987; Edwards and Whittington, 2000). The conditions of carriage and fossilisation in lacustrine or palustrine sequences were particularly well studied and quantified by observation of present-day forest fires. Those studies consider the soot particles (whose size is between 3 and 125 μm) which are transported by the atmospheric convection flows generated by fires and winds. To establish a curve that is really representative of fire occurrences, the recording of the event depends directly on the location of the sedimentary basin in relation to the dominant winds during periods of high frequency of fires.

The macroscopic charcoal particles (> 1 mm), first accumulate at the surface of the burnt area and are then either moved by water streaming into the flow system, or integrated *in situ* by pedological processes. Although they are usually not analysed or not considered by palynologists, their size allows identification to the level of the species of the burnt plant associations. Such studies are still very scarce outside archaeology (Jacquiot *et al.* 1973; Thinon, 1992; Vernet *et al.* 1994; Carcaillet and Thinon, 1996; Thiébault *in* Berger, 1996; Delhon, *in* this volume). They are even more rarely done in conjunction with the studies that are necessary to understand the context of their formation and fossilisation, so as to assess the representativity of the analysed samples. The few studies realised with this aim in mind were either carried out by geologists working on coal (Vaughan and Nichols, 1995; Nichols *et al.* 2000) or by soil specialists (Courty, 1982, 1994; Van Vliet-Lanoë, 1988; Berger, 1996).

The systemic dynamics

Charcoal particles are found in a wide range of pedo-sedimentary contexts. In the systemic, multi-scale approach developed by us in the middle Rhone valley (Brochier *et al.* 1991; Berger, 1996, 2000), palaeo-fires are described at different scales of analysis. This enables us to distinguish the local (foothill, bottom of the valley), the regional (flood plains of the Rhone), and the continental (Rhone basin, north-western Mediterranean basin) signals from global signals. The latter are represented by patterns documented in the polar ice caps, in alpine glaciers or at the bottom of the oceans.

Remains of fire, integrated in hydrological and atmospheric cycles, cannot be dissociated from other elements found in their area of sedimentation. Like other mineral components, or certain pollen types, they enable us to characterise the area of production (Berger, 1996; Berger *et al.* forthcoming). And just like pollen, micro-charcoal, phytoliths, seeds, and gastropod shells, they inform us on the composition and the structure of the burnt vegetation (Galop, 1998; Vannière *et al.* 2000; Delhon *et al.* forthcoming; Magnin, 1991).

When the remains of ancient fires are located at the surface of palaeosoils or in agrarian structures (in the foothills, the floodplains etc.) the information is local (pl. 1-7). It is then possible to distinguish the natural or human (agro-pastoral) origins of the fire by:
-studying the agro-pastoral markers microscopically and identifying their impact on the soils (Macphail, 1986; Courty, 1990, 1994; Berger, 1995, 2000);
- studying the modifications of the soils resulting from the hydrological functioning of the fires (Courty, 1982; Berger, 1996);
-studying the traces of soil manuring or carbonised cultivated seeds.
This analytical scale seems to be much more precise when one wishes to discuss the exact localisation, origin and impact of fires (Berger, 2000). But the results are limited by

the discontinuity and the variety of pedo-sedimentary records in the continental area under conditions of geomorphological evolution. This type of study must be repeated in space to be representative of the spatial localisation and evolution of fire systems.

When the palaeo-fire signal is found in a secondary position, this testifies to the transformation of burnt soils, due to erosion, in a process which carries and traps fire indicators in temporary alluvial areas, such as:
- colluvial formations of slopes or slope deposits (combined with eroded and rolled soil aggregates), in drainage ditches filled by surface runoff from adjacent fields (micro-local significance) (pl. 2);
- palustrine or lacustro-palustrine formations (of micro-local to micro-regional significance) (pl. 1);
- alluvial fills of palaeochannels linked to confluences of the Rhone, which vary in size (local or micro-regional signal), in the form of charcoal lenses or beds, whether associated, with eroded and alluvial soil aggregates or not (pl. 1);
- alluvial detritic fans, recording the global history of the water basin (micro-regional to regional signal) (pl. 1);
- deep, early Holocene fluvio-estuarine formations in the Rhone delta, near the city of Arles (Bouches du Rhône) (macro-regional to continental signal) (pl. 1).

The method: sampling and laboratory techniques

Charcoal analysis

Charcoal has been obtained by the sieving of 10 kg of sediment, through a sieve with a 0.5 mm mesh. After drying, the charcoal was sorted under a binocular loupe. The fragments, down to sizes of less than 1 mm, were identified with the help of a reflected-light microscope, after being cut manually. Whenever possible, the three anatomical wood sections (transversal, longitudinal-tangential and longitudinal-radial) were observed and compared with anatomical atlases (Greguss, 1955, 1959; Jacquiot, 1955; Jacquiot *et al.* 1973; Schweingruber, 1978, 1990) and with an extensive reference collection of present-day burnt wood. In the least favourable cases, even charcoal from palaeo-fire layers measuring less than 500μm could be identified under the microscope.

Sediment analysis

The sediment that constitutes the coal/soil matrix was observed on a microscopic scale, using a micromorphological thin section (13 x 6 cm), after induration (Guilloré, 1985). Micromorphological analysis informs us about the elementary mechanisms that are a part of:
- soil modification by fire and the combustion of the vegetation;
- fragmentation, dispersion and integration of charred residues into soils after the passing of a fire;
- transportation and export of charred residues out of the combustion area.

The pedo-climatic and anthropogenic features associated have been identified on the basis of reference work and case studies developed by Courty *et al.* (1989), Courty (1994) and Berger (1996).

Phytoliths were identified directly on thin sections or after extraction and centrifugation in the laboratory of the UMR 7041 in Nanterre (Delhon, *in* this volume).

Dating

The chronology of the fire events is based on a series of absolute dates, made conventionally or using the A.M.S. method, in the laboratories of Gif-sur-Yvette (France), Lyon (France), and Tucson (USA). The events were chronologically correlated whenever the chrono-cultural conditions appeared to be satisfactory.

Taphonomic processes

Wind and water are the primary agents responsible for transporting charcoal from its source to the basins in which it is preserved and sampled. Some exhaustive studies on the production and transport of charcoal by atmospheric convection and wind were published the last few years (Clark, 1988; Thinon, 1992; Patterson *et al.* 1987; Laird and Campbell, 2000), and will not be dealt with here. Transport of charcoals by hydrological processes, from slope runoff to rivers, lakes or oceans, and their post-depositional evolution in soils are less prominent subjects in the literature (Patterson *et al.* 1987; Cwynar, 1978; Thinon, 1992; Berger, 1996). Most of the time, charcoal analysis of soil profiles is less used than pollen analysis (Dimbleby and Speight, 1969; Patterson *et al.* 1987). We will therefore include a short review of water transport, deposition, and post-depositional processes in soils that affect charcoal (fig. 2). It considers the whole of the sedimentary (Vaughan and Nichols, 1995) and pedological (Courty, 1982; Thinon, 1992) cycle subsequent to soil cover erosion after fire. Studies on the "erodibility" of burnt soils show that two factors have an effect on erosion:
- the combustion of part of the organic material which changes the structural stability of the soil (reducing its resistance to erosion);
- the accumulation of hydrophobic substances at the surface which reduces the ratio of infiltration and increases surface flow and consequently erosion (Giovannini *et al.* 1987; Greene *et al.* 1990).

When there is no farming after a fire, its consequences are minimal and the stabilisation occurs very quickly (3-5 years). One can observe a rapid recolonisation and stabilisation of the slopes after a fire that occurred in 1990 in the Massif des Maures (southern France). The surface flows were very diffuse during the first years.

Water transport has consequences for charcoal identification, interpretation and radio-carbon dating which merit our attention.

The effect of water transport on charcoal

Due to their low density (0,3 to 0,6) and their great porosity, charcoal particles can move over long distances. This is no longer the case when, after a stay in a disturbed soil, iron

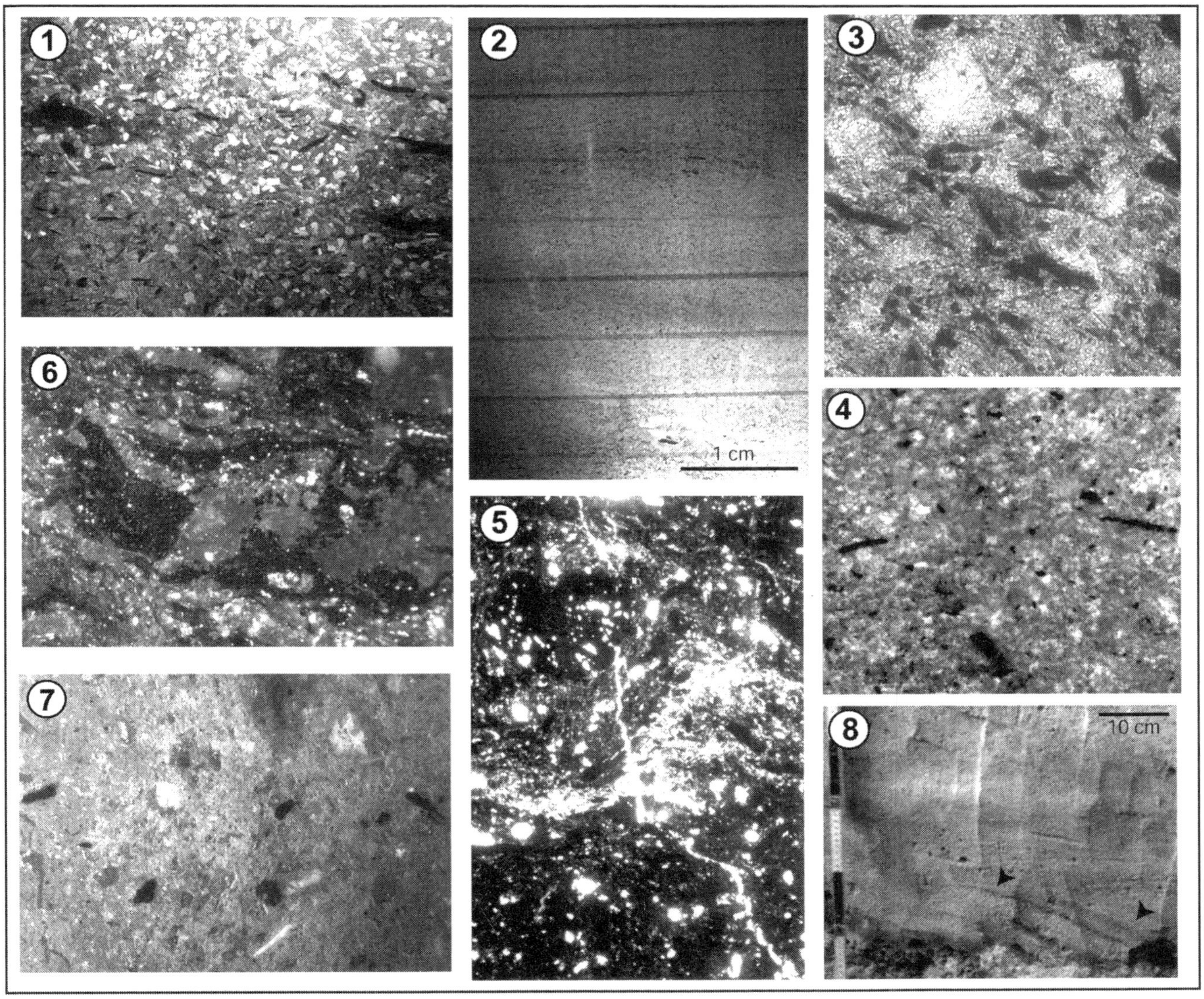

Plate 1: Charcoal in secondary positions in alluvial or paludal contexts:
1-2. charcoal flux in a fluvio-estuarian context of the Rhone delta (Piton core, Arles, Bouches-du-Rhône);
3-4. concentration of small elongate charcoal fragments, probably of Gramineae or Juncaceae type, in an organic-mineral sediment from the alluvial fan of the Roubion river (Drôme), dated to around 7400-7300 BP (natural light, x400);
5. burnt seed and small elongate charcoal fragments from a peat formation in the Roubion river alluvial fan (Drôme), dated to around 7400-7300 BP (natural light, x400);
6. *in situ* burnt fragments of peat in an organic-mineral sediment from a palaeochannel, revealing a drought period favourable to the outbreak of wild (natural?) fires in the middle Rhone valley at the beginning of the Holocene (Bartras site, Vaucluse, incident light, x40);
7. a bed of charcoal in a fluvial soil in the Tricastin floodplain, dated to the early Neolithic, reflecting the fire of a gallery forest (Brassieres site, Vaucluse, incident light, x40);
8. a succession of charcoal deposits in alluvial ridges in a late Antiquity deep channel (1570 BP) marked by a sandy filling (trench 18, Portes-en-Valdaine, Drôme, field photo), resulting from a moderate flow velocity (close to experimental conditions proposed by Nichols *et al.* 2000).

monoxide, manganese inclusions, secondary carbonates or clay deposits make them heavier (fig. 2 and pl. 2-4, 7 and 8).

The size of charcoal particles seems to be a good criterion for determining the distance between the burnt area and the studied deposit. This theory, developed by Clark (1988) or Tolonen (1983), is valid when the charcoal fragments are transported by convective air currents to recording areas, such as lakes or peat-bogs. But it is not really valid in fluvial or colluvial areas where hydrodynamics plays a major role in the sedimentation of charcoal (a moderate flow velocity of 0,35 to 0,55 m s-1 is optimal for sedimentation). The biggest charcoal pieces need more time to be waterlogged, so the duration of their transport in the hydro-system is longer (Nichols *et al.* 2000) (pl. 1-8). Only part of the charcoal is fragmented during transport; the remaining fragments seem to be rolled and remain relatively resistant (Nichols *et al.* 2000).

Charcoal can stay *in situ* and be buried in the ground by sedimentary processes (shrink-swelling deposits) or by bioturbation (pl. 1-7). After a fire, the deposition of charcoal is generally a rapid process (Patterson *et al.* 1987; Bradbury and Waddington, 1973; Thinon, 1992). Transport in the soil is very quick during the first rains after a fire. It stops once the herbaceous and woody plant cover has recovered (Thinon, 1992). The deposit in fluvial contexts and the transport by water have been studied under experimental

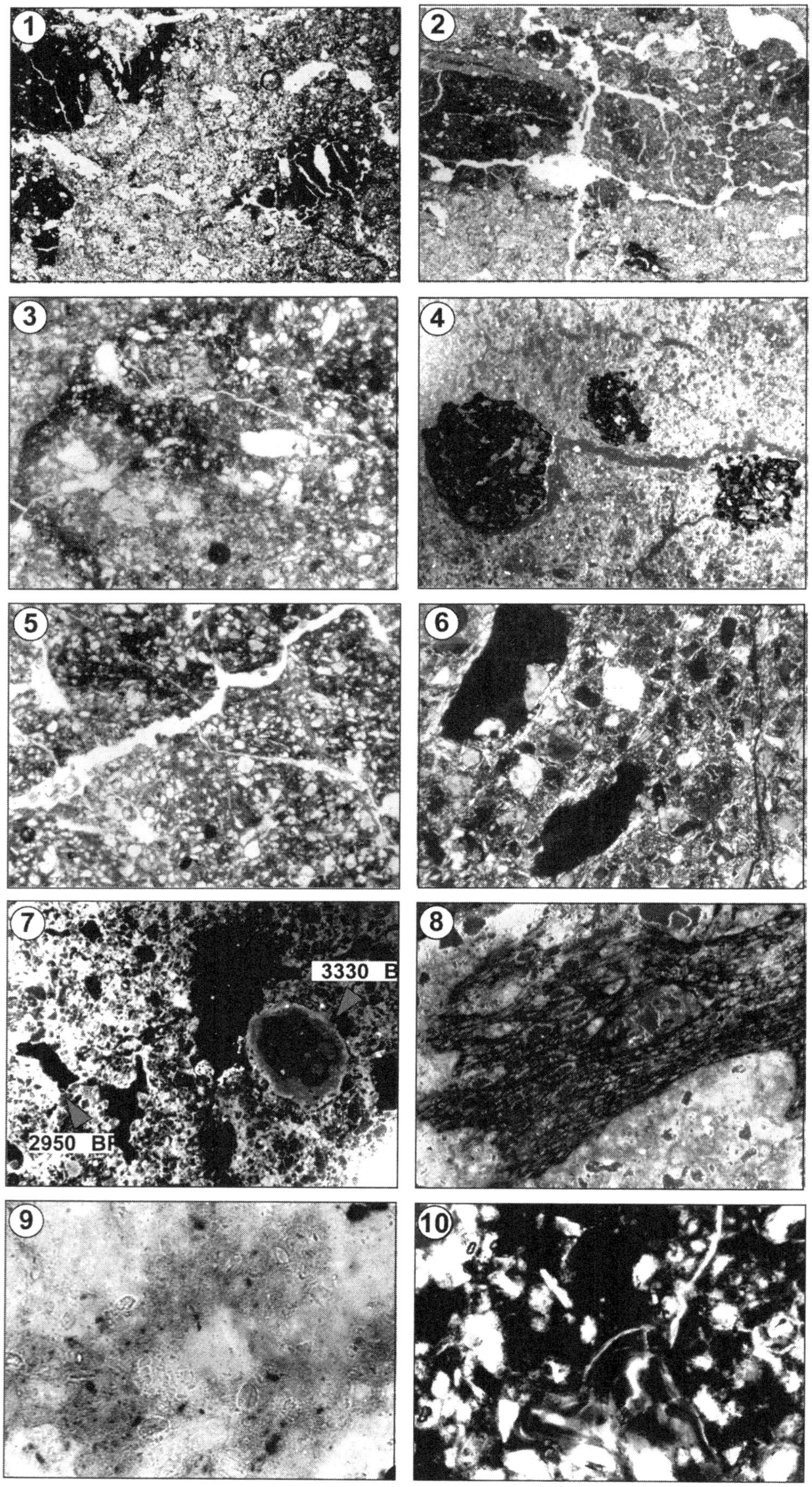

Plate 2: Postdepositional evolution of charcoal in soils:
A. Fragmentation, concentration by earthworm activities and physical pressure on clay sediments:
1. progressive integration of macro-charcoal into the soil matrix by a pedoturbation process. Note the strong bioturbation and the presence, on the right, of burnt aggregates of soil (in dark grey, natural light, x5);
2. burnt surface soil fragments from the Iron Age, with sorted surface crusts formed by the effect of rain-drops on naked soil after fire (Rochefort-en-Valdaine, T61, natural light, x10);
3. traces of a local palaeofire in the bottom of a humid depression (La Touche, T56) at the Early/Middle Bronze Age transition (3370 BP), with evidence of soot being re-integrated into the soil matrix by earth-worm activity. They are concentrated at the periphery of faecal pellets;
4. ferruginous worm pellets in neolithic soils with a concentration of microparticules (charcoal) and the beginning of ferruginisation by siderophilous bacteria (incident light, x40);
5. small charcoal fragments finely integrated into a mid-Holocene black soil matrix of the Valdain basin by vertic processes (see the fissural porosity of a prismatic to polyedric microstructure, T13, LN, x20);
6. detail of photo 5. When the black soil matrix is observed under polarised light, the vertic action of clayey sediments, organising clay domains around grains or soil porosities, is visible (grano- to porostriation type, LP, x100).
B. Ferruginisation process of charcoal, inherited ferruginous charcoal in soils:
7. observation of two "generations" of charcoal (one ferruginised, the other still unaltered) associated in the same stratigraphic unit by a multi-phase evolution of the soil (incident light, x100). The ^{14}C dates of the differents types of charcoal are separated by about of four centuries;
8. detail of ferruginous nodules as observed under incident light. In black, the carbon matter; in light grey, the iron oxides surrounding the charcoal. The hart' nodules are formed around a fragment of *Quercus* charcoal (incident light, x 400).
C. Integration into and coloration of soils by soot:
9. integration of Gramineae micro-charcoal and phytoliths into the matrix of a black paleosoil (8000 BP) (x400);
10. black palaeosoil (8000 BP) with black to dark orange dusty coatings of soot particles (x100).

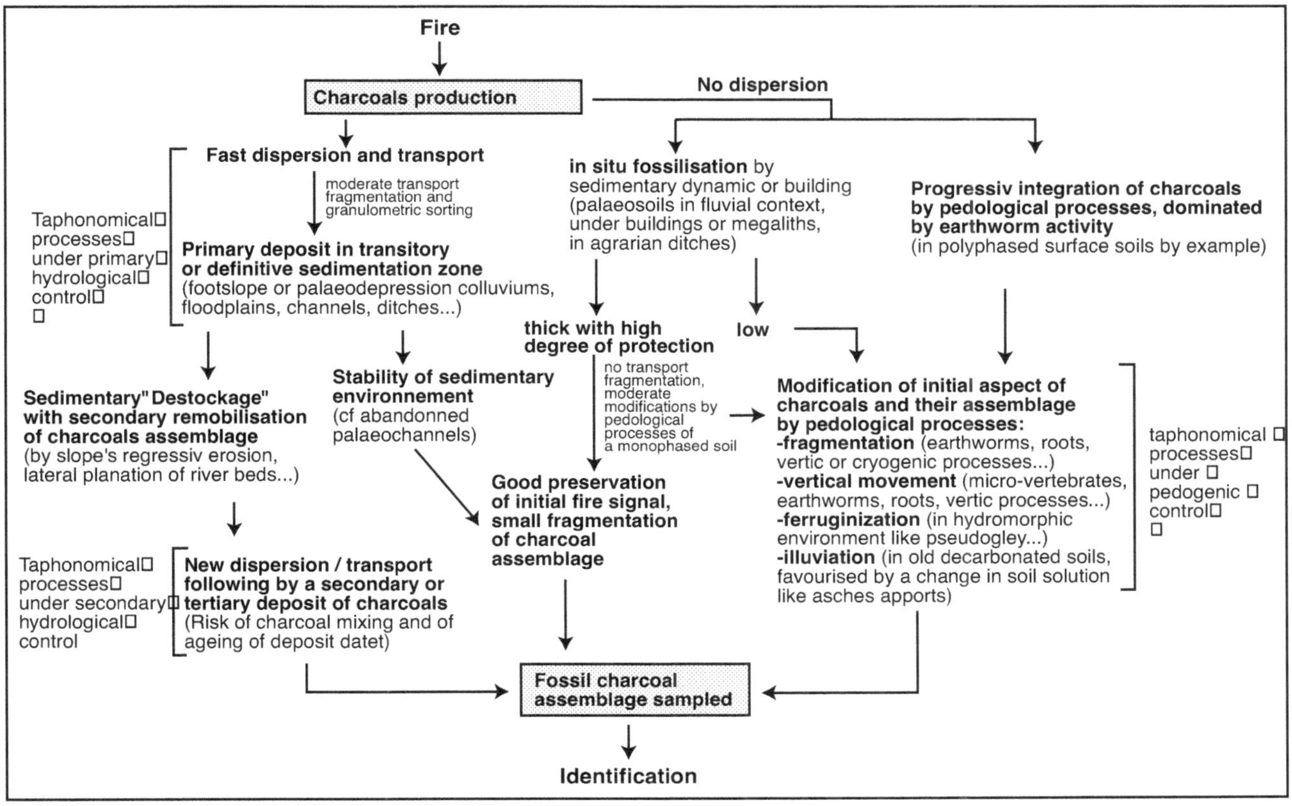

Figure 2: Diagram showing the dispersal, transport, sedimentation and post-depositionnal processes of charcoal assemblages in soils, under hydrological and pedological control (no aeolian processes included).

conditions (Vaughan and Nichols, 1995; Nichols *et al.* 2000).

The effects of pedogenesis on charcoal (fig. 2)

In the soil, lateral and vertical movements of charcoal particles can be provoked by physical processes such as the alternation of freezing and thawing (cryosoils), or by variations in shrinking and swelling in clays (successions of humid and dry phases in vertisolic contexts). The very strong tensions exerted by those mechanical processes provoke major fragmentation of charcoal (pl. 2-1, 5 and 6). In an environment characterised by a vertic behaviour of soils, charcoal is often fragmented into particles of micrometric size and consequently rarely identifiable.

After a fire, the ash residues stimulate biological activities in the soil. The fauna, dominated by earthworms and termites, makes an important contribution to the mobility of charcoal in the soils (Thinon, 1992), to its progressive fragmentation, and to its integration into the fine matrix of the soils (Courty, 1982; Berger, 1996, pl. 2-3 and 4). In temperate areas, experiments show that earthworm activities often modify the organisation and the aspect of the initial charcoal deposit. Earthworms can ingest charcoal pieces of 1.6 mm and fire levels can be buried after the subsidence of their burrows (Thinon, 1992). The effects of root action seem to be minimal for transport, but important for fragmentation, particularly for ring-porous or semi ring-porous wood taxa, or taxa with resin channels (Thinon, 1992).
The speed of deposition and the thickness of a soil after a fire, are important elements for the preservation of the charcoal deposit. The thicker the sediment cover deposited just after a fire, the better the preservation of the deposit, and the lower the risk that new pedo-genesis will disturb the charcoal.

The best situation is, maybe, the instantaneous fossilisation of a burned soil under a building or a tumulus. In Brittany and England, Neolithic or Protohistoric burnt soils were preserved intact in an archaeological context (Gebhardt, 1990; Macphail, 1986). In the middle Rhone valley a fire attributed to the middle Iron Age was very well preserved by a sudden colluvial deposit. Big pieces of charcoal were located close to burnt aggregates and some remains of important earthworm activities just after the fire were also observed (pl. 2- 1 and 2).

On bare surfaces rainfall carries both the micro-fragments of charcoal accumulated at the surface and part of the fine mineral fraction (clays and silts) down into the pores of the soil. They fill the empty spaces in the lower layers (pl. 2-10). This last observation confirms previous results (Slager and Van de Wateringe, 1977; Courty, 1982; Courty *et al.* 1989), which show that the enrichment in K^+, Ca^{++} and Mg^{++} ions due to the advent of ashes could provoke or intensify the illuviation process. Charcoal helps the dispersal of clays and the formation of clayey and dusty deposits, rich in small particles of charcoal.

Soils associated with numerous fires and important bioturbation can thus include a lot of micro-charcoal particles of less than 50 μm (pl. 2-5 and 9). These post-depositional processes can explain the blackish aspect of some palaeosoils or archaeological levels, invoking a gradual integration of various micro-charcoal fragments, soot and humic organic substances into the clayey matrix (Courty and Fedoroff, 1982; Macphail, 1981; Berger, 1996; pl. 2-5 and 9). In the Valdaine area, the colour of some black palaeosoils (dated to around 8000 BP and 5000 BP) can be explained by such ecological and pedological factors (charcoal accumulation, bio-pedo-disturbance, melanisation of the organic matter) (Berger, 1996; Beeching *et al.* in press).

When hydromorphic events impregnate the soil, charcoal in primary positions can be altered, offering a ferruginous or a manganese stained aspect, as a result of bacterial activities. A strong carapace of sesquioxides is built around charcoal or earthworm faeces, which are loaded in soot and micro-charcoal (pl. 2-7 and 8). As a consequence the charred remains are well-protected. Some big pieces are preserved (from 0.5 to 1 cm), but the evolution of the charred remains makes it difficult to distinguish (with the naked eye or under the binocular microscope) the coaly origin of the ferro-manganese nodules and to determine the nature of the ferruginous charcoals. The discrimination can be done on a slide, with a polarising microscope, under incident light (pl. 2-7). Sometimes an appropriate orientation of the charcoal allows an identification (pl. 2-8). Ferruginous forms are often observed in alluvial soils of low plains, in the bottom of humid basins and marls, marked by frequent variations in the water content. They can be related to a localised deterioration of the drainage conditions subsequent to a modification of the functioning of the hydrosystem after a fire. The water-proofing of the soil surface by an intense fire could be at the origin of such coaly ferruginous nodules (Debano *et al.* 1979; Giovannini *et al.* 1987; Scott and Van Wyck, 1990; Greene *et al.* 1990). Generally, such nodules are rarely in a primary position in pedo-sedimentary formations.

In many areas, fires occurred regularly. Two formations of charcoal particles are found in these circumstances: those originating from burning events and pedogenesis (primary charcoal) (pl. 2-7), and those associated with colluvial processes (secondary charcoal).

In such cases the secondary charcoal testifies to the alteration of the fired soil by erosion processes. Such charcoal particles are often rounded in appearance (pl. 2-4 and 7). They are very resistant and can be compared to the ferro-manganese nodules (so-called "plomb de chasse") of the soil scientist. The radiocarbon date of these charcoal particles informs us on the age of the fire and not on that of the erosion, so that the radiocarbon dates obtained from charcoal found in mixed pedo-sedimentary units will be located between the date of the first and the second fire.

At La Touche (T56), in the Valdaine basin, it was possible to distinguish microscopically, prior to dating, between two formations of charcoal, one in a primary and the other in a secondary position. This enabled us to subsequently identify and date separately two types of charcoal located in the same sedimentary unit: one with fresh fractures, the other rolled and ferruginised (pl. 2-7). Their dates differ by several centuries (2950 BP and 3330 BP respectively). The ferruginous type was the oldest of the two. That confirms the hypothesis that these particles were buried for a long time in a hydromorphic soil before a period of erosion, associated with a later burning episode that added a more recent charcoal deposit.

Two geoarchaeological case studies

What follows are the results of two case studies that provide good insight to conditions of deposit and the origin of the charcoal particles. They concern recent Holocene alluvial sequences and irrigation canals excavated in two small areas in the Valdaine and the Tricastin, both part of the middle Rhone valley (T7 and T18, fig. 1).

In the stratigraphic units concerned, the superposition of taxa belonging to different vegetation assemblages is often observed. In the middle Rhone valley such taxa belong to the following associations:
- a deciduous oak forest (a supra-Mediterranean association) with: *Quercus* f.c., *Acer* sp., *Ulmus minor*, Rosaceae, *Buxus* pro parte ...
- a meso-Mediterranean oak forest (which is rare), with sclerophyllous oak, *Quercus ilex*, as well as *Arbutus unedo* and *Olea europaea*;
- a riparian association: *Vitis* pro parte, *Fraxinus* pro parte, *Salix, Populus, Alnus*;
- the taxa associated with human activities: *Buxus* pro parte, Fabaceae, *Juniperus* pro parte and cultivated species such as *Vitis* pro parte and *Juglans regia*;
- a mountain flora or one associated with a cold climate: *Pinus sylvestris, Abies alba, Fagus sylvatica, Taxus baccata, Betula* sp. pro parte.

In the case of studies with a wide range of charcoal particles, these may in principle come from:
- areas with distinct ecological characteristics (such as high spatial diversity), corresponding to a vegetation mosaic, the size of which varies with the size of the basin and the drainage or irrigation systems involved;
- locations where different plant associations succeed each other in time without any sedimentary deposit between the phases (temporal diversity).

In this case, however, the rapidity of sedimentation excludes the latter kind of station (no more than ten years can have elapsed between each studied formation). Particular attention will therefore have to be paid to the systematic recurrence, during certain periods and in particular sedimentary contexts, of mountain species, such as *Fagus sylvatica, Pinus sylvestris* and *Abies alba*.

The evolution of the Lanson basin and its agro-system (Valdaine)

The pedo-sedimentary sequence (of 265 m) cut into the alluvial formations of the recent Holocene (fig. 3-1, 2 and 3) by more than 2 m. In some instances, yellow gravel from the Pleniglacial, and some fragments of middle Holocene palaeosoils (6000 - 4000 BP) are observed (Berger, 1996;

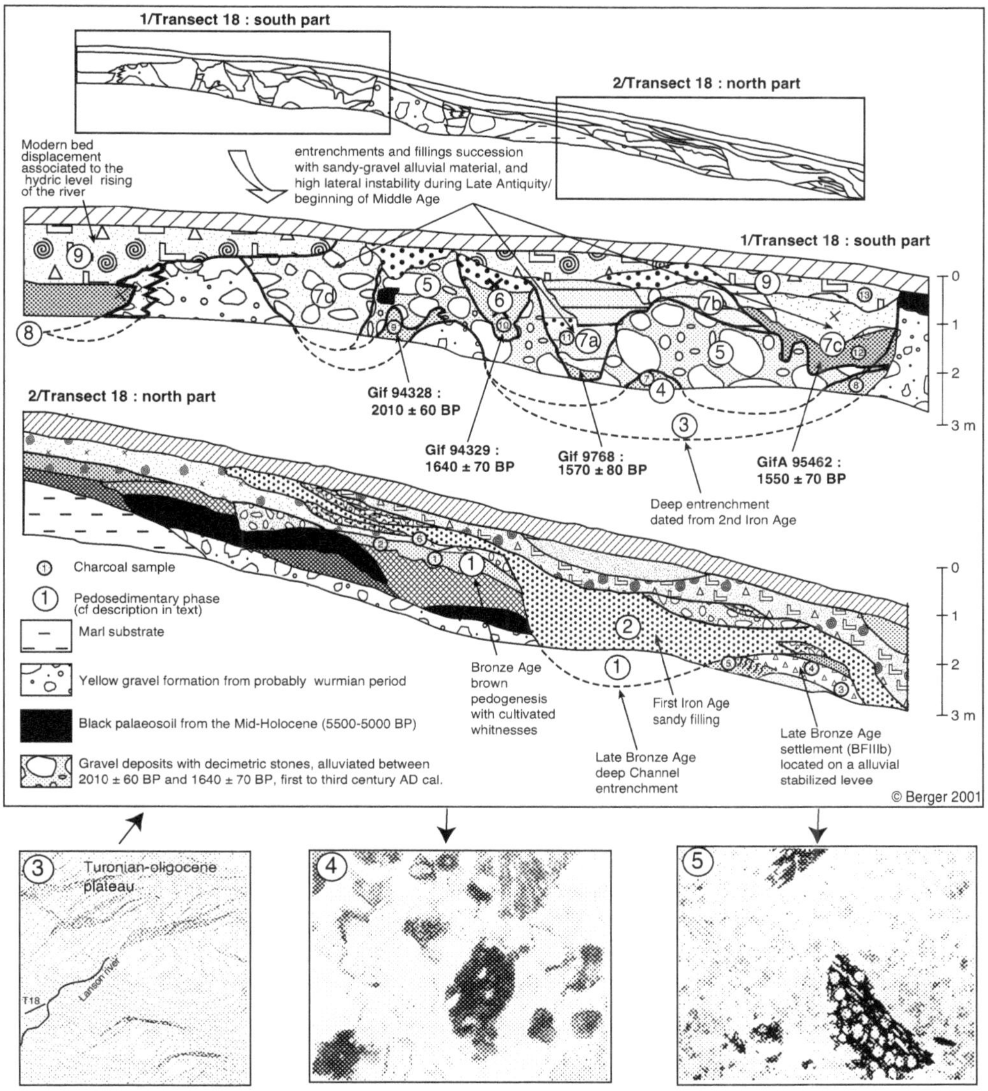

Figure 3:Location, stratigraphy and thin section photos of trench 18 in the river bed of the Lanson river (Portes-en-Valdaine, Drôme):
1. south part of trench 18;
2. north part of trench 18;
3. location of trench 18 in the alluvial plain of the Lanson river, below the southern hills and the plateau of the Valdain Basin;
4. concentration of decarbonated red soil micro-aggregates (terra rossa?), in the Lanson sandy alluviums filling a channel dated to 1570 ± 70 BP (phase 7a, sample 11, LN x 40);
5. concentration of charcoal (oak and pine) in the Lanson sandy alluvial filling of a channel dated to 1570 ± 70 BP (phase 7a, sample 11, LN x 10).

fig. 3- 2). The spatial heterogeneity of the fluvial record in the northern Mediterranean area is well represented in the Lanson sequence. After several metamorphoses of the hydrosystem after about 3000 BP, related to variations in the flow of the river, the bed of the Lanson moved to its present alluvial plain (Berger, 1996). This lateral mobility left behind several palaeochannels and some ancient embankments and palaeosoils, in which numerous pieces of charcoal were sampled. The analyses and the comparison of the results obtained by anthracology and geomorphology allows us to summarise the history of this little basin (cf. fig. 3-1 and 2, table 1) as follows:

1. - Deep incision of the river bed during the Bronze Age, associated with a period of geomorphological stability characterised by the formation of a brownish soil with secondary carbonates in the alluvial plain (fig. 3). An archaeological settlement of late Bronze Age phase 3B (900-750 BC cal.) was located near the active channel on the eastern bank. On the opposite bank a palaeosoil was associated with traces of ploughing and some LB3B potsherds. Charcoal analysis shows that the deciduous oak forest, which recovered during the Bronze Age as a result of the abandonment of this area by humans during the early and middle Bronze Age (Thiébault, 1999; Berger *et al.* 2000), was again heavily exploited (tabl. 1). This exploitation of the deciduous oak forest occurred in piedmont areas. At the end of the period of human occupation (at the transition between the Bronze and the Iron Ages), the presence of ash (*Fraxinus excelsior*), Pomoïdeae and beech (*Fagus sylvatica*), may testify to clearance of the oak forest. Beech, a mountain species, took over from oak (de Beaulieu, 1977; Thiébault, 1988).

2. - During the Iron Age (before 2010 BP ± 60 BP, *i.e.* the second century BC cal.) the channel filled up rapidly, provoking a heightening of the river banks and an overflow of alluvium, which covered the Bronze Age settlement (fig. 3). The watercourse moved towards the north (fig. 3-2).

	900-750BC cal.					750-161 BC	2010 ±60 BP			1640 ±70 BP	1570 ±80 BP	1550 ±70 BP	XIV-XIXe c.
N° Sample	1	2	3	4	5		7	8	9	10	11	12	13
Kind of sediment	cultivated palaesoil		settlement level		pedogeneised bank	overflowing alluvium	channel filling						
Meter	m70	m73	m5 to 15	m12 to 16	m 25	m70/T3	m192	m161	m221	m206	m200	m165	m162
Cultural attribution	**BF3b**				**BF3b/Fer**	**Iron Age**	**Tene final/HE**	**Tene final/HE**	**Tene final/HE**	**Low empire**	**Low Emp/M-A**	**Low Emp/M-A**	**Modern**
Pedosedimentary phases	1	1	1	1	1	2a	2b	2b	2b	3	4	4	5
Species													
Angiosperma										3			
Coniferous								1					
Pinus cf.*sylvestris*						13	7	55			10		3
Fagus sylvatica	1				2		2		3		12	1	1
cf. *Juniperus sp.*									22				
Quercus(deciduous)	4	15	10	25	8				8	5	10	2	
Acer campestre											1		
Betulaceae										1			
Buxus sempervirens										6	1	1	3
Corylus avellana										7		7	
Pomoïdeae					1								
Sorbus/Crataegus										2			
Sorbus sp.				1									
cf.*Quercus ilex*										3			
Fraxinus excelsior					1								
Populus sp.										2		1	
Vitis sp.							1						
Total = 262	5	15	10	26	12	13	10	56	33	29	34	12	7
Unidentified		4									3		

Table 1: Taxa identified by charcoal analysis at Portes en Valdaine, Trench n° 18.

3. - Just before 2010 BP ± 60 BP, during the late La Tène, a new embankment of the river formed at the northern edge of the sequence. This fluvial behaviour occured elsewhere in the region (fig. 3-1).

4. - Around 2010 BP ± 60 BP, the channels filled up with fine sand and silt containing charcoal particles. This phenomenon was associated with fires in the upper basin (fig. 3-1).

5. - Between 2010 BP ± 60 BP and 1640 BP ± 70 BP, the river system became braided. This happened in association with a progradation of the gravel at the bottom of the Lanson, and with extreme lateral instability of the riverbed (displacement of nearly 80 m!) (fig. 3-1). This hydrological crisis developed at the time of the Roman Empire and is well documented in the area (Berger, 1995, 1996). The described processes took place in a changed vegetation pattern, dominated by coniferous forests in which pine (*Pinus sylvestris*) was prevalent, accompanied by beech (tabl. 1). The samples attributed to the recent and final phases of the La Tène and the transition to the Roman period (n° 7 and 8) testify to the local eradication of deciduous oak forests. The species that took the place of the exploited deciduous oak forests, testify to a colder climate. Thus, sample n° 9 shows the disappearance of *Pinus sylvestris* in favour of an association of juniper (*Juniperus* sp.), and, to a lesser extent *Quercus* f. c. and *Fagus*. The identification of these species in a northern Mediterranean bioclimatic context recalls the cycles occurring in the same area at the end of the late Glacial and the beginning of the Holocene (Vernet and Thiébault, 1987; Thiébault, 1988; Heinz and Thiébault, 1998), but in this case the reasons are of course different.

6. - Just before 1640 BP ± 60 BP, in the Late Roman period, the incision of the watercourse was followed by the infilling by a fine silt, rich in charcoal particles (fig. 3-1). The charcoal from this fill shows considerable taxonomic variation (tabl. 1). The species are characteristic of an open woodland, which was recolonised under dry climatic conditions. Deciduous oak is found alongside *Buxus*, evergreen oak (cf. *Quercus ilex*) and hazelnut (*Corylus avellana*). Poplar (*Populus* sp.) characterises the riparian forest. The species identified are essentially pre-forestry species, testifying to the reconquest of the environment by the deciduous oak forest.

7. - A deep incision, occurring before 1570 BP ± 80 BP, is followed by a sandy fill (fig. 3-1, 4 and 5), full of micro-aggregates of Mediterranean red soils (terra rossa from the plateau), and numerous charcoal lenses and deposits, as well as burnt cereal seeds (cf. the lower part of fig. 3). Extreme lateral mobility of the watercourse can be observed until after 1550 BP ± 70 BP, causing the encasing of several channels with heavy fills. The initial recovery of the forest, in which mountain species are still present (*Pinus sylvestris* and beech in particular (tabl. 1) is to be confirmed. The presence of these two taxa could be local, as a consequence of a cooling of the climate, but it could also be explained by the heavy incidence of flooding identified for this period and which must have occurred over large distances.

8. - After the third hydrological crisis the river bed moved to the south, cut into older parts of the alluvial formation and, during the Middle Age *sensu lato* (VI-XIIth centuries AD) formed a new embankment.

9. - During the modern period the floor of the riverbed was raised (it is dated by the presence of grey medieval potsherds and glazed pottery in the alluvium). This phenomenon is associated with heavy deposits of brown and red sands containing charcoal, which are evidence of a new phase of erosion of red soils, in relation to fire (fig. 3 - 1). The charcoal analysis shows an open landscape with pine, beech and boxwood.

These observations corroborate recent regional syntheses on the evolution of the river systems (Bravard *et al.* 1997; Provansal *et al.* 2000; Berger, 2000).

The analysis of charcoal sampled in trench n°18 in Portes-en-Valdaine points, for the first half of the first millennium BC, to a relatively forested landscape dominated by deciduous oak. Significant deforestation can be observed at the end of the late Bronze Age 3B, probably as a consequence of the reclaim of abandoned agricultural land which had become covered by the oak forest. The latter is clearly identifiable in the archaeological and ecological record of the region during the ancient and middle Bronze Age (Berger *et al.* 2000). This forest clearance is characterised in the second half of the millennium by oak being replaced, under cooler climatic conditions, first by pine and then by beech. The climatic cooling of this first phase subsequently intensifies, as is seen from the presence of yew

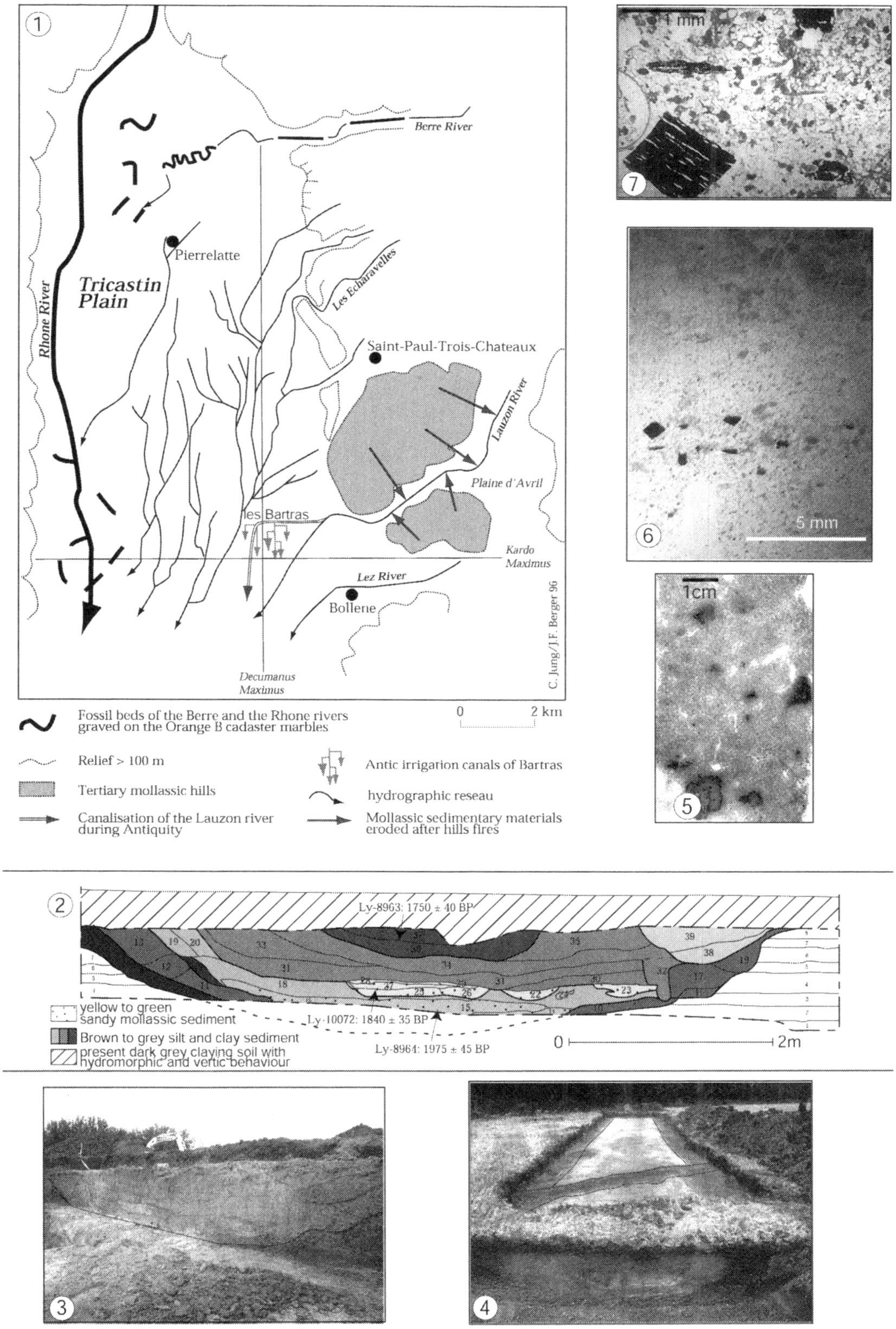

Figure 4: Location, stratigraphic context and photos (field and thin sections) of the hydraulic system of Bartras in the alluvial Tricastin plain (Vaucluse):
1. geographic and hydrologic context of the Bartras hydraulic system;
2. stratigraphy of the main collector of the antique irrigation system;
3. photo of the main collector (trench 7);
4. photo of a secondary irrigation canal with sandy alluvium (in light grey) cut by a modern drainage ditch filled by dark brown clay (in dark grey);
5. photo of a sandy gravel sedimentary feature with small rounded gravel of mollassic origin (thin section);
6. photo of a charcoal bed associated with sandy mollassic features (thin section, natural light);
7. detail of photo 6, with submillimetric, rounded to subangular charcoal fragments (thin section, natural light).

Datation	1avJ.-C.			I-II ap. J.-C.		II-III ap J.-C.						IV-V			
Sample	1	2	3	5	10	8	9	6	7	11	12	13(6)	13(7)	15	16
cf. *Abies*											*				
cf. *Pinus sylvestris*							*								
Fagus sylvatica	*			*	*	*	*	*	*		*		*		
cf. *Acer*							*								
cf. *Buxus sempervirens*						*									
cf. *Crataegus/Sorbus*							*								
cf. *Juniperus sp.*						*			*						
cf. *Pomoïdeae*						*			*	*					
cf. *Sambucus sp.*							*								
cf. *Ulmus sp.*												*	*	*	
Fraxinus sp.						*	*						*		
Prunus sp.						*	*				*				*
Quercus cf. *ilex/ coccifera*				*		*	*				*	*		*	*
Quercus f. c.	*	*	*	*	*	*	*	*		*	*	*	*	*	
Rhamnus/Phillyrea						*									
Feuillu				*	*	*	*			*	*	*		*	*
Gymnosperme		*		*			*				*	*			
	Local origin			Regional origin						L. O	Regional origin				L.O.
Sedimentary phases	1a	1b	1c	3	5b	7	8	8	8	8	8	9a	9a	9c	

Table 2: Taxa identified by charcoal analysis at Bartras, trench n°7 (L. Rousseau, 1996).

(*Taxus baccata*) in a contemporaneous sample from trench n° 19.

The expansion of pine to the detriment of oak, before 2010 BP ± 60 BP, may be associated with two converging factors. On the one hand, a climatic cooling favourable for the descent of pines from the nearby hill and mountain slopes and, on the other hand, an increase in the incidence of fires due to human (agro-pastoral) activities, which also favours pine (a heliophilous and hardy species). Indeed, several present-day environmental and ecological studies show that fire plays an important role in the perturbation of woods by *Pinus sylvestris* in north-western Europe (Bradshaw, 1993; Lowe, 1993; Little *et al.* 1996; Agee, 1998; Lageard, 2000). The analysis of molluscs recovered in alluvium (Magnin *in* Berger, 1996), due to the washing down of debris from the calcareous plateau located above the studied site, show the presence of a dry woodland during this period, and thus support the results of the charcoal studies.

The vegetation history documented by the charcoal analysis of trench n° 18 can be summarised in 5 main steps. The first concerns the exploitation of a deciduous oak forest in the piedmont zone by the inhabitants of the area in the late Bronze 3B period. During the transition between the Bronze and Iron Ages, the oak forest decreases in importance. *Pinus sylvestris* takes its place. After that, we see a vegetation that is dominated by juniper, but in which *Quercus* slowly emerges. The renewed spread of a deciduous oak forest is clearly marked by the presence of pioneer species, but human pressure and a slight climatic degradation, quickly prevent this development. The landscape crisis is highly visible in the sedimentary record of the late Antiquity (phases 3 and 4, between 1640 ± 70 BP and 1550 ± 70 BP, IVth to the beginning of the VIIth century cal. AD), and is due to a conjunction of anthropogenic and climatic factors. During this period, the evolution of the river indicates an increase in rainfall and a quick succession of erosi-

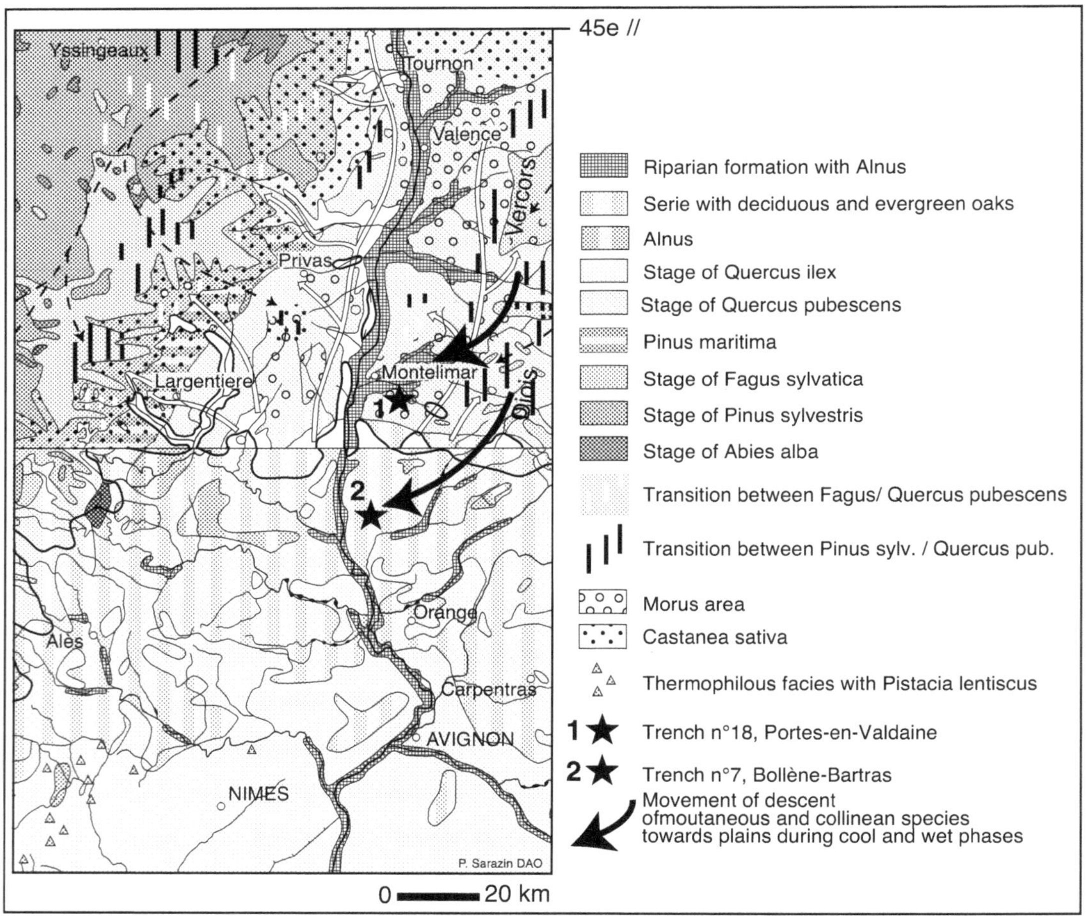

Figure 5: Vegetation dynamics in the middle Rhone valley during the Augustan Age and the Early Roman Empire (first century BC cal. to third century AD cal.) according to anthracological data from off-site analysis. *Pinus sylvestris*, beech and fir series evolve from their present stage in the Diois and Vercors mountains towards the low alluvial plains of the Rhone valley and its adjacent prealpin rivers (adapted from the vegetation map of Avignon and le Puy, CNRS eds).

ve periods at a regional scale (Provansal *et al.* 2000; Berger, 2001). The erosion of higher areas seems to be linked to a hiatus in human occupation and soil management that occurs during the Vth century. The exploitation of the calcareous plateau, intensively deforested by agro-pastoral fires, is responsible for an important phase of erosion of its soil cover (terra rossa). It is indirectly documented by the presence of alluvium in the river bed and at the foothills. The presence of micro-fragments of terra rossa in the alluvium, as well as of oak and pine wood among the charcoal particles, points to anthropogenic fires (fig. 3- 4 and 5, tabl. 1), as does the occurrence of carbonised cereal seeds. The erosion of old soils, remains of the Pleistocene, is to be related to the agricultural exploitation of the plateau and to the development of rural settlement. This is confirmed by the spatial analyses of archaeological remains, which point to an increase in the number of settlements in late Antiquity and in the beginning of the Middle Ages (Berger, 1995; Ode and Odiot, 2001). This change in the agricultural use of the landscape is accompanied by a deterioration of the drainage and irrigation systems of the plain, from 1500 BP (VI/VIIth centuries cal. AD) on (Berger, 1995, 2001). During the modern period, a return to the rhexistasic conditions of the late Antiquity is due to the same (climatic and anthropogenic) processes. It follows a period of relative stability of the north Mediterranean geosystem, in the Middle Ages (Berger and Brochier, in press). On the calcareous plateaux, this new crisis signifies the end of the erosion of the terra rossa of which hardly anything is left today.

The vegetation history of the settlement of Bartras near Bollène (Vaucluse)

The settlement of Bartras near Bollène (Vaucluse), excavated during the construction of the T.G.V. from Lyon to Marseille, illustrates what can be done with the study of a hundred irrigation ditches (fig. 4). It is the first known example of a large-scale irrigation system dating back to Antiquity in Gallia Narbonensis. It is part of the grid of the "Cadastre B" of Orange, which organised the landscape of this region since the beginnings of the Augustan Age (around 30 BC). A radiocarbon date (based on charcoal found in the lower part of the ditches) confirms the historical dates (1975 $\pm$ 45 BP, first century cal. BC). The geoarchaeological study of the sediment from this system allows us to distinguish sedimentation phases common to all of the studied ditches (fig. 4 - 2).

The study of trench n° 7 (fig. 4 - 2 and 3), which is the main irrigation channel, is here taken as an example. The charcoal analyses were done by L. Rousseau (1996). The results show a remarkably close correlation between the nature of the matrix, the origin of the deposit and the identified plant species. The sandy, yellow-green sediments, point to a high hydrological activity (cf. samples 6, 8, 9, 13 (6), 13 (7) and 15; fig. 4-5). This fits well with the plant species identified from this context: *Fagus sylvatica* (samples 8, 9, 6 and 7), *Pinus sylvestris* (sample 9), and cf. *Abies* (sample 12). Moreover, the samples are taxonomically rich (table 2).

The colour and the petrographic composition of some of the sediments (sandy burdigalian molasses with glauconite and foraminifera; fig. 4-3 and 4-5) indicate that these deposits predominantly originated from the hills of Saint-Restitut (300 m), in the upper part of the river basin, less than 5 km away. The canals covering that area date back to Antiquity (fig. 4 - 1). The relief has not changed since then, and the present-day floristic components point to recent forest reconstruction dominated by deciduous and sclerophyllous oak, with here and there *Pinus halepensis* on sunny slopes and *Pinus sylvestris* on northern slopes (fig. 5). The presence of *Fagus sylvatica*, *Pinus sylvestris* and cf. *Abies* in a vegetation characterised by meso- and supra-mediterranean species can be explained by the important regional climatic fluctuations which occurred between the first century BC and the beginning of the third century AD. These fluctuations are clearly attested in the Rhone basin, by numerous geomorphological and palaeopedological studies (Bravard *et al.* 1992; Berger, 1996; Guibal and Teissier, 1998; Provansal *et al.* 2000). During this period, elements of the hill and mountain flora descended to lower altitudes (fig. 5). The present day distribution of pine and beech is located between 10 and 20 kilometres to the east and the north-east of the site, on the first flanks of the Rhone foothills, and fir at an altitude of 1000 m, in the mountains of the Diois and the Vercors (cf. fig. 5).

The grey-beige silty deposits are, on the contrary, characteristic of a low hydrodynamic regime, and they are associated with rather local and hydromorphous charcoal remains, formed after floating down from the surroundings (samples 1, 2, 3, 11 and 16).

These observations allow us to evoke the exploitation of the hilly areas during the Roman period. The difference between the abundance of charcoal in the molassic deposits (fig. 4 - 6 and 7) and the relative scarcity in the local deposits can be explained by changes in the exploitation of the soil. During certain periods of Antiquity, the exploitation of the hills by fire must have had important erosive consequences (notably under humid climatic conditions), because charcoal fragments are discovered in the sands and gravels from the hillslopes (fig. 4 - 5), together with aggregates from brownish soils or rendzinas (fig. 3 - 4 and 5). Similar observations were made for the same (Roman Imperial) period, in channels n° 8, 15, 56 and 10 of the same settlement (Bartras), but also in Girardes, at Lapalud, and at Mondragon Le Duc (located in the Tricastin plain; cf. T. 8 and pit S7), in the "Negades" at Caderousse (US24-25), in the plain of Orange (at the confluence of the Rhone and the Aigues), and also in several samples from trenches located in the Valdaine (Thiébault *in* Berger, 1996; Thiébault and Rousseau *in* Berger and Jung, 1999). They all point to regional changes at the scale of the middle Rhone valley.

Conclusion

The kind of systemic studies that were initiated more than fifteen years ago now open new perspectives for the understanding of the ancient environment. The nature of fires, their origin and their periodicity during the Holocene, are investigated by the means of analyses focused on pedology, sedimentology and anthracology. The middle Rhone valley is, due to its biogeographic position, located at a climatic junction, and the great potential offered by the investigation of the trajectory of the T.G.V. from Lyon to Marseille constituted a privileged occasion for research on the signals left by ancient fires. Even though the precise origin of the fires - natural or anthropogenic - cannot be strictly established, the first results reveal that they were largely anterior to the emergence of the first farmers, 8 000 years ago. Moreover, it appears that the last hunter-gatherers, the first farmers and the climate often influenced the outbreak of fires.

The analysis of charcoal sampled in pedo-sedimentary sequences, although not commonly used, yields reliable and exact information on palaeo-fires.

Before determining the species, it is fundamental to study their taphonomy, their dispersion, and their burial processes in detail. Wind and water are the main agents responsible for the transport of charcoal. Conservation and sedimentation depend on a range of factors (the nature of the sedimentary process, bioturbation, pedogenesis ...). The charcoal discovered in palaeosoils and sediment consists, contrary to that from archaeological settlements, of very small fragments (less than 2 mm) which require an adaptation of the techniques traditionally used by the charcoal analyst. Only river beds (palaeochannels), some palaeosoils, and agricultural structures sometimes provide large size charcoal fragments (1 to 2 cm). The charcoal particles are progressively integrated in the majority of soils by earthworm activities and other biological processes occurring in the layers below those where bio- and pedoturbation occur, or they can be moved into colluvial or alluvial formations by erosion and/or flooding.

The two examples presented here illustrate some of the interesting results of this type of analysis. Trench n° T 18 at Portes en Valdaine provides evidence of a cyclical evolution of the landscape in which crises, desertification, climatic degradation and agricultural periods alternate. Similarly, the study of microscopic charcoal from drainage ditches or irrigation channels constructed during Antiquity allows us to evoke cycles of exploitation of the soil in the plain and in the hills.

The analysis of microscopic charcoal as an indicator of ancient fires is a valuable complement to the more classical kind of charcoal analyses from archaeological excavations. It sometimes both allows a local interpretation and provides subtle evidence of climatic variations and human interventions at the time-scale of the Holocene. Such studies, when combined with pedo-sedimentary data, provide information on both the cultural and the natural cycles which determinate the long-term evolution of landscapes.

AKNOWLEDGEMENTS
This work is partly supported by the archaeological operation of the Mediterranean fast railway (coordinator T. Odiot) and by the ATP of the CNRS "Premiers paysans en moyenne vallée du Rhône", coordinated by A. Beeching and J.-L. Brochier. We wish to thank H. Valladas and M. Fontugne (Laboratoire des Sciences du Climat et de l'Environnement, UMR 1572, Gif-sur-Yvette) for their collaboration and the radiocarbon laboratory. We kindly thank Professor S. E. van der Leeuw for his help in translating this paper and Dr R.I. Macphail for its advices.

References

AGEE J.K. 1998 - Fire and pine ecosystems. In: Richardson D. M. eds., *Ecology and Biogeography of Pinus*. Cambridge University Press, Cambridge, p. 193-208.

BARBERO M., LOIZEL R., QUEZEL P. 1990 - Méditerranée : les incendies et l'évolution des forêts. *La Recherche*, 22, p. 1154.

BEAULIEU, J.-L. (de) 1977 - *Contribution pollenanalytique à l'histoire tardiglaciaire et holocène de la végétation des Alpes méridionales françaises*. Thèse de Doctorat ès Sciences, Marseille. 358 p., 29 fig.

BEECHING A., BERGER J.-F., BROCHIER J. L., FERBER F., HELMER D., SIDI MAAMAR H. in press - *Chasséens : mobiles ou sédentaires ? arguments archéologiques et paleoenvironnementaux*. Actes des 3éme rencontres méridionales de Préhistoire récente, Toulouse, 6 et 7 novembre 1998.

BERGER J.-F. 1995 - Facteurs anthropiques et naturels sur l'évolution des paysages holocènes dans le bassin valdainais (Drôme), Actes des XVe Rencontres Internationales d'Archéologie et d'Histoire d'Antibes, *l'Homme et la dégradation de l'Environnement*, S. E. van der Leeuw ed., APDCA ed., p. 79-114.

BERGER J.-F. 1996 - *Le cadre paléogéographique des occupations du bassin valdainais (Drôme) à l'Holocène,* Thèse de doctorat nouveau régime en Préhistoire, Protohistoire, Ethnologie, Université de Paris I, 325 p.

BERGER J.-F. 2000 - Cycles anthropiques et environnementaux à l'Holocène dans des bassins-versants rhodaniens de rang inférieur (Valdaine et Tricastin, Drôme). In : *Les Temps de l'Environnement,* Actes des Journées du Programme Environnement, Vie et Sociétés PIREVS, Toulouse, Octobre 1997, M. Barrué-Pastor et G. Bertrand Eds, Presse universitaire du Mirail-CNRS, p. 473-500.

BERGER J.-F. 2001 - Evolution des agro-et des hydrosystèmes dans la région médio-rhodanienne, In : *Les campagnes de la Gaule à la fin de l'Antiquité*, colloque AGER IV, Montpellier, 11-14 mars 1998, APDCA eds, Sophia Antipolis, p. 369-403.

BERGER J.-F. 2001 - L'étude géoarchéologique et paléoécologique de fossés bordiers historiques : Démarche, méthodes et apports à l'Histoire agraire rhodanienne. In : La très longue durée, J. Guilaine (dir.), EHESS eds, *Revue d'Etudes Rurales* 153-154, p. 59-90.

BERGER J.-F., JUNG C. 1999 - Developping a methodological approach to the Evolution of mid-Rhodanian Agro-systems during Historical Periods, In: The Archaeology of the Mediterranean Landscape 2: *Environmental Reconstruction in Mediterranean Landscape Archaeology*, K. Walsch, G. Barker, F. Trément, P. Leveau (Eds), Actes du colloque Populus, Aix-en-Provence, octobre 1995, Oxbow books, p. 155-168.

BERGER J.-F., BROCHIER J. L. in press - Les apports de la géoarchéologie à la connaissance des paysages et des climats de l'époque médiévale en moyenne vallée du Rhône. In : *Habitats, nécropoles et paysages dans la moyenne et basse vallée du Rhône du VIIe au XVe siècle, contributions des travaux archéologiques sur le tracé du TGV-Méditerranée à l'étude des sociétés rurales médiévales* (O. Mauffras dir.), DAF grands travaux.

BERGER J.-F., MAGNIN F., THIEBAULT S., VITAL J. 2000 - Emprise et déprise culturelle à l'Age du Bronze en moyenne vallée du Rhône, *Bull. Soc. Preh. Fra.*, t 97, n°1, p. 95-119.

BERGER J.-F., DELHON C., BONTE S., PEYRIC D., TIEBAULT S., BEECHING A., VITAL J. in press - Paléodynamique fluviale, climat, action humaine et évolution des paysages du bassin versant de la Citelle (moyenne vallée du Rhône, Drôme) au cours de l'Atlantique ancien (8000-6000 BP) à partir de l'étude de la séquence alluviale d'Espeluche-Lalo. Actes du colloque PEVS *Paléohydrologie des 15 derniers millénaires*, (J.P. Bravard et M. Magny Dir.), Lyon, 21-23 Juin 2001, CNRS ed.

BOISSON M. 1988 - Le climat de la région de Montélimar, *Recherches Donzéroises*, p. 1-23.

BRADBURY J.P., WADDINGTON J.C.B. 1973 -The impact of European settlement on Shagawa Lake Northeastern Minnesota, USA. *Quaternary Plant Ecology* (Eds. H.J.B. Birks and R.G. West), Blackwell, p. 109-115.

BRADSHAW V. 1993 - Forest reponse to Holocene climatic change: equilibrium or non-equilibrium. In : F. M. Chamber, *Climatic change and human impact on the landscape*. London: Chapman and hall, p. 57-65.

BRAUN-BLANQUET J. 1936 - La forêt d'Yeuse languedocienne (*Quercion ilicis*) : Monographie phytosociologique. SIGMA, communication n°40. *Mémoires de la Société d'étude des sciences naturelles de Nîmes*, 5, Nîmes 1936. 147 p.

BRAVARD J.-P. 1997 - Géoarchéologie des vallées alluviales de Rhône-Alpes depuis le Tardiglaciaire. In : Actes de la table ronde Dynamique du paysage : entretiens de géoarchéologie, Lyon 17 et 18 novembre 1995, (J.-P. BRAVARD et M. PRESTREAU, éd.), *Documents de l'Archéologie de Rhône-Alpes*, 15, p. 129-150.

BRAVARD J.-P., VEROT-BOURRELY A., SALVADOR P. G. 1992 - Le climat d'après les enregistrements sédimentaires fluviatiles étudiés sur les sites archéologiques. *Les Nouvelles de l'Archéologie*, 50, p. 7-13.

BRAVARD J.-P., VEROT-BOURRELY A., FRANC O., avec la collaboration de ARLAUD C. 1997 - Paléodynamique du site fluvial de Lyon depuis le Tardiglaciaire, . In : Actes de la table ronde Dynamique du paysage : entretiens de géoarchéologie, Lyon 17 et 18 novembre 1995, (J.-P. BRAVARD et M. PRESTREAU, éd.), *Documents de l'Archéologie de Rhône-Alpes*, 15, p. 177-202.

BROCHIER J.-L., MANDIER P., ARGANT J. 1991 - Le cône détritique de la Drôme : Une contribution à la connaissance de l'Holocène du Sud-Est de la France. *Quaternaire*, 2, p. 83-99.

CARCAILLET C., THINON M. 1996 - Pedoanthracological contribution to the evolution of upper treeline in the Maurienne Valley (North French Alps): methodology and preliminary data. *Review of Palaeobotany and Palynology*, 91, p. 399-416.

CLARK J. S. 1982 - Point count estimation of charcoal in pollen preparations and thin sections of sediments. *Pollen et Spores*, 24, p. 523-535.

CLARK J.S. 1988 - Charcoal stratigraphic analysis on petrographic thin sections: Recent fire history in northwest Minnesota. *Quaternary Research*, 30, p. 81-91.

CLARK J.S. 1996 - Baseline Biomass Burning Emissions of Eastern North America. In: J.S. Levine eds., Biomass Burning in South America, Southeast Asia and Temperate and Boreal Ecosystems, and the Oil Fires of Kuwait. *Biomass Burning and Global Change*, vol. 2, p. 750-756.

COURTY M.-A. 1982 - *Etude géologique de sites archéologiques Holocènes : définition des processus sédimentaires et post-sédimentaires caractérisation de l'impact anthropique. Essai de méthodologie*. Thèse de 3è cycle, Bordeaux n° 1726, 314 p, 55 fig.

COURTY M.-A. 1990 - *Environnements géologiques dans le nord-ouest de l'Inde. Contraintes géodynamiques au peuplement humain protohistorique*, Doctorat ès Science, Univ. de Bordeaux I, non publié.

COURTY M.-A. 1994 - Le cadre paléogéographique des occupations humaines dans le bassin du haut-Khabur (Syrie du Nord-Est). Premiers résultats. *Paléorient*, 20/1, p. 21-59.

COURTY M.-A., FEDOROFF N. 1982 - Micromorphology of a Holocene Dwelling. *P.A.C.T 7*, II, Strasbourg 1981, p. 257-277.

COURTY M.-A., GOLDBERG P., MACPHAIL R.I. 1989 - *Soils and micromorphology in archaeology*. Cambridge University Press, 344 p.

CWYNAR L. C. 1978 - Recent history of fire and vegetation from laminated sediment of Greenleaf Lake, Algonkin Park, Ontario. *Canadian journal of Botany*, 56, p. 10-21.

D'ANNA A., COURTIN J. 1986 - *Travertins holocènes et sites préhistoriques : exemples dans le Var et les Bouches-du-Rhône. Méditerranée, Aix-en-Provence*, 1/2, p. 31-38.

DEBANO L., EBERLEIN G.E., DUNN P.H. 1979 - Effects of burning on Chaparall soils: I. Soil nitrogen. *Soil Sci. Soc. Am. Journ.* 43, p. 504-509.

DELHON C. 2002 - Phytolith and pedo-anthracological analysis of Holocene pedosedimentary sequence from the middle Rhône valley (France). *BAR, International Series*.

DELHON C., ALEXANDRE A., BERGER J.-F., MEUNIER J.-D., THIEBAULT S., BROCHIER J.-L. submit - Phytoliths assemblages as a promising tool for reconstructing Mediterranean Holocene vegetation, *Quaternary Research*.

DIMBLEBY G. W., SPEIGHT M.C.D. 1969 - *Buried soils. Advancement of Science*, 26, p. 203-205.

EDWARDS K.J., WHITTINGTON G. 2000 - Multiple charcoals profiles in a Scottish Lake: taphonomy, fire ecology, human impact and inference. In: Fire and the palaeoenvironment, A.C. Scott, J. Moore and B. Brayshay eds., *Palaeogeogr., Palaeoclimatol., Palaeoecol.*, 164, n°1-4, p. 67-86.

FALCON-LANG H.J. 2000 - Fire ecology of a carboniferous tropical zone. In: Fire and the palaeoenvironment, A.C. Scott, J. Moore and B. Brayshay eds., *Palaeogeogr., Palaeoclimatol., Palaeoecol.*, 164, n°1-4, p. 339-356.

FEDOROFF N., COURTY M.-A. 1989 - Indicateurs pédologiques d'aridification. Exemples du Sahara, *Bull. Soc. géol. France*, 1, p.

GALOP D. 1998 - *La forêt, l'homme et le troupeau dans les Pyrénées – 6000 ans d'histoire de l'environnement entre Garonne et Méditerranée*. GEODE-Laboratoire d'écologie terrestre, Toulouse, 285 p.

GEBHARDT A. 1990 - *Etude du paléopaysage agricole dans le Nord Ouest de la France. Apport de la micromorphologie*. Thèse de l'Université de rennes 1, 193p.

GIOVANNINI G., LUCCHESI S., GIACHETTI M. 1987 - Effect of heating on some physical and chemical parameters related to soil agreggation and erodibility. *Soil Science*, 146, 4, p. 255-261.

GLASSPOOL I. 2000 - *A major fire event recorded in the mesofossils and petrology of the Late Permian, lower Whybrow coal seam, Sydney Basin*, Australia. p. 357-380.

GREEN R.S.B., CHARTRES C.J., HODGKINSON K. C. 1990 - The Effects of Fire on the Soil in a Degraded Semi-arid Woodland. I. Cryptogram Cover and Physical and Micromorphological Properties, *Aust. J. Soil Res.*, 28, p. 755-777.

GREGUSS P. 1955 - *Identification of living gymnosperms on the basis of xylotomy*. Budapest, 263 p.

GREGUSS P. 1959 - *Holzanatomie der Europaïschen Laubhölzer und Straücher*. Akadémiai Kiado Budapest, 263 p.

GUIBAL F., TEISSIER L. 1998 - Apport de l'analyse dendrochronologique et dendroclimatologique à la connaissance du climat de la vallée du Rhône du Ier s. av. J.C. au IIe s. ap. J.C., *Méditerranée*, t. 90, 4, p. 5-10.

GUILLORÉ P. 1985 - *Méthode de fabrication mécanique et en série de lames minces*. Institut National Agronomique, Paris, Département des sols, 22 p.

HEINZ C., THIEBAULT, S. 1998 - Characterization and Palaeoecological Significance of Archaeological Charcoal Assemblages during Late and Post-Glacial Phases in Southern France. *Quaternary Research*. 50, p. 56-68.

JACQUIOT C. 1955 - *Atlas d'anatomie des bois des conifères*. Centre Technique du Bois, Paris, 135 p, 64 pl. 2 vol.

JACQUIOT C., TRENARD Y., DIROL, D. 1973 - *Atlas d'anatomie des bois des Angiospermes*. Centre Technique du Bois, Paris. 175 p. 2 vol.

JACOBI R.M., TALLIS J.H., MELLARS P.A. 1976 - The Southern Pennine Mesolithic and ecological record. *Journal of Archaeological Science*, 3, p. 307-320.

JONES T.P., CHALONER W.G. 1991 - Les feux du passé, *La Recherche*, 236, vol. 22.

KUHNHOLTZ-LORDAT G. 1938 - *La terre incendiée*. Nîmes, 361 p.

LAGEARD J.G.A., THOMAS P. A., CHAMBERS F.M. 2000 - Using fire scarce and growth release in subfossil Scots pine to reconstruct prehistoric fires. In: Fire and the palaeoenvironment, A.C. Scott, J. Moore and B. Brayshay eds. *Palaeogeogr., Palaeoclimatol., Palaeoecol.*, 164, n°1-4, p. 87-99.

LAIRD L.D., CAMPBELL I.D. 2000 - High resolution palaeofire signals from Christina Lake, Alberta: a comparison of charcoal signals extracted by two different methods. In: Fire and the palaeoenvironment, A.C. Scott, J. Moore and B. Brayshay eds. *Palaeogeogr., Palaeoclimatol., Palaeoecol.*, 164, n°1-4, p. 111-123.

LEVINE S. 1996 - Biomass Burning in South America, Southeast Asia and Temperate and Boreal Ecosystems, and the Oil Fires of Kuwait. In: *Biomass Burning and Global Change*, vol. 2, 873 p.

LITTLE D.J., MITCHELL F.J.G., VON ENGELBRECHTEN S., FARRELL E.P. 1996 - Assessment of impact of disturbance and prehistoric *Pinus sylvestris* on the vegetation dynamics and soil development. In: Uragh Wood, SW Ireland. *Holocene*, 6, p. 90-99.

LOWE J.J. 1993 - Isolating the climatic factors in early-and mid-Holocene palaeobotanical records from Scotland. In: Chambers F. M. eds., *Climate change and human impact on the Landscape*. Chapman and Hall, London, p. 67-82.

MACPHAIL R.I. 1981 - Soil and botanical studies of "Dark Earth". In: M. Jones and G. Dimbleby eds, the environment of man: the Iron Age to the Anglo-Saxon Period, *B.A.R.* 87, Oxford, p. 309-331.

MACPHAIL R.I. 1986 - Palaeosoils in Archaeology ; their role in undestanding Flandrian pedogenesis. In: V.P. Wright eds, *Palaeosoils*, Blackwell Scientific, Oxford, p. 263-290.

MAGNIN F. 1991 - *Mollusques continentaux et Histoire quaternaire des milieux mediterranéens (sud-est de la France, Catalogne),* Thèse de doctorat de 3ème cycle, Université d'Aix-Marseille II, 342 p.

MOORE J. 2000 - Forest fire and human interaction in the early Holocene woodlands of Britain, In: Fire and the palaeoenvironment, A. C. Scott, J. Moore and B. Brayshay eds., *Palaeogeogr., Palaeoclimatol., Palaeoecol.*, 164, n°1-4, p. 43-56.

NICHOLS G.J., CRIPPS J.A., COLLINSON M.E., SCOTT A.C. 2000 - Experiments in waterlogging and sedimentology of charcoal : results and implications, In: Fire and the palaeoenvironment, A.C. Scott, J. Moore and B. Brayshay eds., *Palaeogeogr., Palaeoclimatol., Palaeoecol.*, 164, n°1-4, p. 43-56.

ODE B., ODIOT T. 2001- L'habitat rural de la Moyenne vallée du Rhône aux IVè et Vè siècles. In : *Les campagnes de la Gaule à la fin de l'Antiquité*, colloque AGER IV, Montpellier, 11-14 mars 1998. APDCA eds, Sophia antipolis, p. 225-246.

OZENDA, P. 1964 - *Biogéographie végétale*, Douin, 374 p.

OZENDA P. 1994 - *Végétation du continent européen*, Delachaux et Niestlé, Lausanne, 272 p.

PATTERSON W.A., EDWARDS K.J., MAGUIRE D.J. 1987 - Microscopic charcoal as a fossil indicator of fire. *Quat. Sci. Rev.* 6, p. 3-23.

PONS A., THINON M. 1987 - The role of fire from palaeoecological data. *Ecologia Mediterranea*, 13, p. 1-11.

PROVANSAL M., BERGER J.-F., BRAVARD J.-P., SALVADOR P.G. 2000 - Le régime du Rhône et les mutations des environnements fluviaux du lac de Genève à la mer, dossier hors-série "Le Rhône romain: dynamiques fluviales, dynamiques territoriales" sous la dir. de P. Leveau, *Gallia* 56.

ROUSSEAU L. 1996 - *Reconstitution de l'évolution des agrosystèmes antiques dans la plaine du Tricastin (Vaucluse- Drôme): analyses anthracologiques des sédiments de neuf fossés de la moyenne vallée du Rhône*. Maîtrise de Biologie, des Populations et des Ecosystèmes, USTL Montpellier II, 16p.

SCHWEINGRUBER. F.H. 1978 - *Mikroskopische holzanatomie*. Zürcher A.G. Zug, 226p. 98pl.

SCHWEINGRUBER F.H. 1990 - *Anatomie europaïcher Hölzer*. W.S.L.F.N.P. Haupt. 800 p.

SCOTT D.F., VAN WYCK D.B. - 1990, The effects of wildfire on soil vegetability and hydrological behaviour of an afforested catchment, *Journal of Hydrology*, 121, p. 239-256.

SIGAUT F. 1975 - *L'Agriculture et le feu*, Paris, Mouton, 320 p.

SIMMONS I.G., TOOLEY M.J. 1981 - *The environment in British* Prehistory, London.

SIMMONS I.G., INNES J.B. 1996 - An episode of prehistoric canopy manipulation at North Gill, North Yorkshire, England. *J. Archaeol. Sci.* 23, p. 337-341.

SLAGER S., VAN DER WETERING H.T. J. 1977 - Soil formation in archaeological pits and adjacent loess soils in Southern Germany. *Journal of Archaeological Science*, 4, P. 259-267.

THIEBAULT S. 1988 - L'Homme et le milieu végétal à la fin du Tardiglaciaire et au Postglaciaire, analyses anthracologiques de six gisements des préalpes sud-occidentales, *DAF*, n° 15, 110 p.

THIEBAULT, S. 1999 - Le milieu végétal des derniers chasseurs-cueilleurs en Vercors et Chartreuse - nouvelles données anthracologiques et proposition d'une biozonation. 5e colloque U.I.S.P.P., Grenoble, 18-23 septembre 1995, *L'Europe des derniers chasseurs*, C.T.H.S. p. 589-602.

THINON M. 1992 - *L'analyse pédoanthracologique, aspects méthodologiques et applications*. Thèse de doctorat ès sciences, Université d'Aix-Marseille III, 317 p.

TOLONEN K. 1983 - The post-glacial fire record. In: Wein, R. W and MacLean, D. A. (eds). *The role of Fire in Northern Circumpolar Ecosystems*, John Wiley and Sons, Chichester, p. 21-44.

TRABAUD L. 1982 - *Impact biologique et écologique des feux de végétation sur l'organisation, la structure et l'évolution de la végétation des zones de garrigues du Bas-Languedoc*. Thèse de l'Université de Montpellier II.

TRABAUD L. 1989 - *Les feux de forêts - mécanismes, comportement et environnement*. France-Sélection, 278 p.

TRABAUD L. 1992 - Réponses des végétaux ligneux méditerranéens à l'action du feu. *Pireneos*, 140, p. 89-107.

TRABAUD L. (Ed). 1998 - *Fire Managment and Landscape Ecology*, International Association of Wildland Fire, Fairfield, 334 p.

VANNIERE B., BOSSUET G., GAUTHIER E. 2000 - Susceptibilité magnétique et indices polliniques, marqueurs de l'impact anthropique et de la dynamique fluviale dans la basse vallée du Doubs (Jura, France) entre le Ier et le VIIIe siècle après J.-C. *C.R. Académie des sciences Paris,* Sciences de la terre et des planètes, 331, p. 203-210.

VAN VLIET-LANOE B. 1986 - Le pédocomplexe du dernier Interglaciaire (de 125000 à 75000 BP). Variations de faciès et signification paléoclimatiques, du Sud de la Pologne à l'Ouest de la Bretagne. *Bull. Ass. Fr. Et. Quatern.*, 1-2, p. 139-150.

VAUGHAN A., NICHOLS G.J. 1995 - Controls on the deposition of charcoal: implications for sedimentary accumulations of fusain. *J. Sediment. Res.* A65 (1), p. 129-135.

VERNET J.-L., THIEBAULT S. 1987 - An approach to north-western Mediterranean recent prehistoric vegetation and ecologic implications. *Journal of Biogeography*, 14, p. 117-127.

VERNET J.-L., WENGLER L., SOLARI M.-E., CECCANTINI G., FOURNIER M., LEDRU M.-P., SOUBIES F. 1994 - Feux, climats et végétations au Brésil central durant l'Holocène : les données d'un profil de sol à charbons de bois (Salitre, Minas Gerais). *C.R. Acad. Sci.* Paris, t. 319, série II, p. 1391-1397.

Human activities and fire history since 4500 BC on the northern slope of the Pyrenees: a record from Cuguron (central Pyrenees, France)

Didier Galop[1], Boris Vannière[1] and Michel Fontugne[2]

1)- CNRS-UMR 6565, Laboratoire de Chrono-Ecologie, F-25030 Besançon cedex, France.
2)- CNRS, Laboratoire des Sciences du Climat et de l'Environnement, F-91198 Gif -sur-Yvette cedex, France.
E-mail: didier.galop@univ-fcomte.fr

Abstract

Palynological and charcoal analyses performed at Cuguron in a small peat bog provides a first detailed reconstruction of land use history on the northern slopes of the Pyrenees. This study demonstrates the important role of fire in agrarian practices in this region since the Early Neolithic period and provides preliminary evidence of neolithic forest farming. Several phases of burning were identified. Each phase coincides with the initiation or intensification of farming, and the results suggest that since 4500 BC, fires were mainly due to human activities.

Key-words: *Human activities, Fire history, Slash-and-burn cultivation, Palynology, Charcoal analysis, Pyrenees, Southern France*

Résumé

Une analyse palynologique combinée à l'étude des résidus carbonisées réalisée à Cuguron, dans une petite tourbière, permettent une première reconstitution détaillée de l'histoire de l'environnement, des pratiques agraires sur le piémont du versant nord des Pyrénées. Cette étude montre le rôle important joué par le feu, pour les pratiques agricoles, depuis le Néolithique ancien et apporte les premiers témoignages de l'exploitation de la forêt au Néolithique. Plusieurs phases d'incendie sont identifiées. Chacune coïncide avec le début ou l'intensification de l'exploitation et les résultats suggèrent que depuis 4500 BC les incendies sont principalement d'origine anthropique.

Mots-clés : *Activités humaines, histoire du feu, essartage, palynologie, anthracologie, Pyrénées, Sud de la France.*

Introduction

Today, fire is used for prescribed burnings for highland pasture management (Métailie, 1981), but it played an important role in the history of the Pyrenean agro-systems and in cultural landscape management. Written historical sources provide invaluable descriptions of slash-and-burn cultivation during the medieval and modern periods (Rendu, 1987; Chevalier, 1956; Bonhote, 1992; Galop, 1998); however, discussions concerning the practice of neolithic forest farming and the possibilities of slash-and-burn cultivation still remain on the level of assumptions supported by some ethnographic studies (Sigaut, 1985). Several palynological studies carried out in the upper Aude valley suggest the existence of slash-and-burn cultivation (Reille, 1990; Reille and Andrieu, 1994; Galop, 1998), but the authors do not formulate objective arguments supporting this assumption.

The oldest evidence for human activity-related burning during the Neolithic is recorded at high altitude around 4500-4300 BP on Enveig mountain in the eastern Pyrenees (Rendu *et al.* 1996; Davasse *et al.* 1997; Galop 1998). These fires favoured the deforestation of pine forests for highland zone exploitation in the form of summer farming. However, because of the paucity of information, fire history related to anthropogenic activities in the northern slope of the Pyrenees remains poorly understood. Among the numerous palaeocological research programmes carried out in the Pyrenees over the past twenty years, none has been devoted to the study of the relationship between agrarian practices and fire history.

The study of the peat bog of Cuguron (central Pyrenees) is the first palaeoecological research project that combines palynological and micro-charcoal analysis. For many years (Iversen, 1941), reconstructions of fire history have been based on the measurement of micro-charcoal contained in sedimentary records such as lacustrine sediments (Swain, 1973), soils (Carcaillet, 1998) or peat deposits. While peat deposits provide the most reliable recordings for fire history (Huttunen, 1980; Innes and Simmons, 2000), the choice of the site is significant in the study of anthropogenic firing (Moore, 2000). The study of small sites considered as "local sites" (Jacobson and Bradschaw, 1981) makes it possible to minimize the regional component of the pollen and/or charcoal deposition (Huttunen, 1980). Consequently, the study of a small site like Cuguron should provide a precise reconstruction of small-scale relationships between land use, fire and vegetation history.

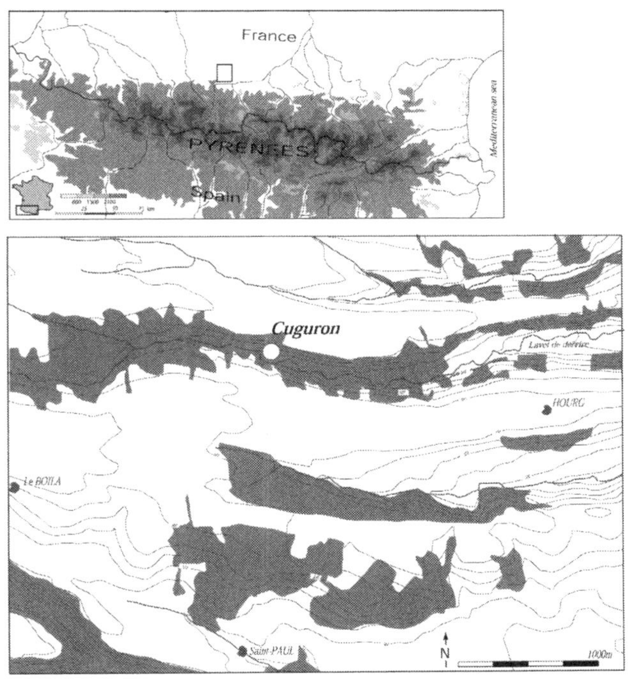

Figure 1: Location of the site of Cuguron.

Site description and archaeological context

The site of Cuguron is a mire basin, approximately 800 m^2 in size and at an altitude of 513 m a.s.l., in the upper part of the Lavet valley, on the north-eastern slopes of the Lannemezan plateau. The vegetation growing on the surface of the mire consists of *Molinia cæruleum* and *Sphagnum*, while *Alnus glutinosa, Populus* and *Salix* have colonised the edges of the mire. Mixed broad-leaved forest has colonised the surrounding slopes and is dominated by *Quercus* and *Corylus*. On the plateau, the landscape is largely deforested and characterised by agricultural areas, meadows and pastures.

The relative lack of archaeological evidence has resulted in a poor understanding of the history of human activity in this area. Some archaeological investigations on the northern Pyrenean piedmont have recorded the presence of human groups during the Early Neolithic period (5600-5300 BC). During the Middle Neolithic, the presence of a necropolis in a mountain valley located near the study area (burials of Arihouat) suggests that the region was more densely populated and settled (Muller, 1995). Late Bronze Age and Iron Age periods were characterised, however, by an expansion of human occupation with sufficient archaeological evidence, such as burial sites in the central part of the northern slopes of the Pyrenees.

Methods

Fieldwork, laboratory and pollen analysis

During the field investigations, core samples were taken with a Russian peat sampler (GIK 0.5 m; 5 cm in diameter). Pollen preparation followed standard methods, using treatment with 10% HCL, 10% KOH, HF, acetolysis and final mounting in glycerine.

Pollen analysis was carried out at 2.5 cm intervals between 1 and 50 cm, 90 and 140 cm, 160-217 cm and at 5 cm intervals in the other section of the core. More than 500 terrestrial pollen grains were counted in each sample. Cyperaceae and spores were systematically excluded from the pollen. *Alnus* was also excluded from the pollen because of its over-representation in the pollen data and because it can mask the dynamics of other taxa (Janssen, 1959; Wiltshire and Edwards, 1993). All pollen types were defined according to Faegri and Iversen (1989), and some identifications required the use of a pollen atlas (Reille, 1992; Moore *et al.* 1991).

The relative pollen diagram is divided into a Local Pollen Assemblage Zone and phases of land use (Lageras, 1996). The phases are based on relative changes in land-use pollen indicators according to Behre (1981) and Latalowa (1992) and on relative changes between trees and open land indicators. Cereal pollen types were defined according to size: all Poaceae pollen with annulus diameter (anl-D) > 9μm and size > 41μm were classified as undifferentiated *Cerealia*-type. Particular attention was given to *Melampyrum, Pteridium* and *Rumex acetosa* because these taxa are regarded as indicators of burning of vegetation associated with slash-and-burn cultivation (Iversen, 1949; Mamakova, 1968; Huttunen, 1980; Tollonen, 1980; Vuorela, 1986; Innes and Simons, 2000).

Charcoal analysis

Microscopic charcoal fragments were counted in the pollen slides. The analysis of the evidence of burning was carried out through microscopic observation in reflected light of wood micro-charcoals present in the pollen slides. Dark remains were more effectively observed by use of the light field (Diot, 1992; Rhodes, 1998; Scott, 2000), and this technique enabled us to identify charcoal fragments by facilitating differentiation between opaque minerals and the vegetal remains. The total surface of charcoal was estimated using a graticule grid with 100 squares 20μm each, and the results are expressed in charcoal influx (mm^2. cm^{-2}. yr^{-1}) (Clark, 1988).

Results and Discussion

Sediment statigraphy

Depth	Description
0-60	Fibrous peat
60-140	Peat
140-180	Peat including wood remains
180-182	Charcoal layer (*Quercus*)
182-217	Peaty loam

Table 1: Sediment stratigraphy.

^{14}C dating and chronology

The Laboratory of Sciences of Climate and Environment (Gif-sur-Yvette, France) provided dates for height samples in order to provide an absolute chronology. Accelerator mass spectrometry ^{14}C and Beta counting were conducted on the materials listed in table 2. The radiocarbon calibration program CALIB version 4.1.2. (Stuiver *et al.* 1998) was used to calibrate the radiocarbon dates.

The percentage pollen diagram was plotted against a linear timescale based on the results of the age/depth calculations in order to compare the chronological sequence of events at the site (fig. 2).

Sample	Depth (cm)	Material	14C-yrsBP (2 sigma)	Calibrated age AD/BC(2 sigma)
GIFA-100280	30	Vegetal remains	530±80	1292(1412)1487 cal AD
GIFA-100281	90	peat	1450±60	439(620,636)679 cal AD
Gif-11010	105-110	peat	2615±60	896(800)557 cal BC
GIFA-100282	125	Vegetal remains	3650±70	2265(2027,1982)1780 cal BC
GIFA-100284	163	wood	4120±70	2885(2835,2625)2471 cal BC
Gif-11011	185-190	peat	4620±95	3638(3368)3032 cal BC
GIFA-100287	207,5	peat	4780±80	3706(3632,3540)3367 cal BC
Gif-11012	213-217	peat	5560±100	4769(4473)4332 cal BC

Table 2: Radiocarbon conventional dates in stratigraphical order in the site of Cuguron (GIFA refer to AMS dating).

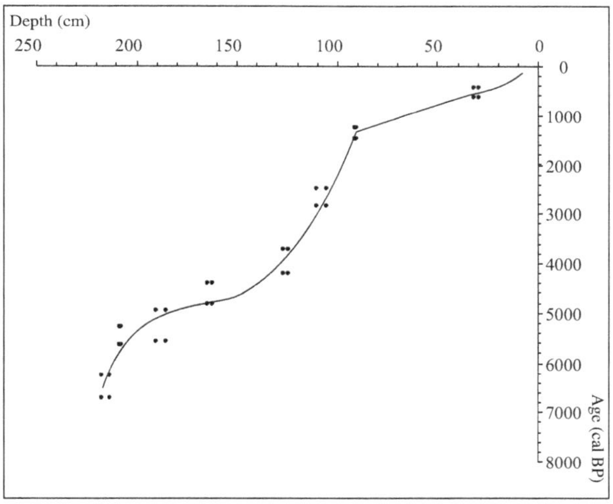

Figure 2: Age/depth ratio of the core from Cuguron based on a polynomial regression (order 3 and 5 with correlation index: 0.96).

Vegetation, fire and land-use history

Interpretation is based on the percentage diagram (fig. 3) and the synthetic diagram plotted on a chronological scale (fig. 4).

Period 1: 4500 BC - *c*. 2800 BC, Early and Late Neolithic periods

Two phases of human activity between the end of the Early Neolithic and first half of the Late Neolithic period were recorded. During the first half (1a) of this phase (4500-3300 cal BC), woodlands were dominated by *Corylus*, *Quercus* and *Betula*, although *Ulmus, Tilia, Fraxinus* and *Alnus* were present. The height frequencies of sprouting trees such as birch and hazel suggest woodland regeneration. They are associated with several pollen indicators of pastoral activities such as *Urtica, Artemisia, Cichorioideae, Rumex* (including both *R. acetosa* and *R. acetosella*) and *Chenopodium* that suggest grazing near the site. In addition, the presence of Poaceae and the occurrences in *Lotus*-type indicate wet grassland near the site.

The first cereals and *Plantago lanceolata* are recorded at 4780±80 BP (3630-3540 cal BC). Direct evidence of cultivation correlates with the stratigraphic record of pollen established for the clearing of *Quercus* and *Corylus*, and an increase in Poaceae. Following this phase of human activity, woodland regeneration is indicated by increases in the percentage pollen values of oak and hazel, and a reduction in Poaceae. This post-farming forest regeneration seems to reveal a combination and flux of activities, including grazing and cereal cultivation. These activities are associated with burning of the vegetation cover and are signalled by high rates of charcoal and by the continual presence of *Melampyrum* and *Pteridium*, which reflect the existence of burnt areas and probably slash-and-burn cultivation. Taking into account the low pollen production and dispersion of *Melampyrum* (Vuorela, 1986), its representation during this phase suggests that it was in close proximity to the vegetation burning. The sedimentation rate indicates that this cultivation phase continued for approximately 300 years. The longevity of this phase indicates periodic settlements near Cuguron that were characterised by short phases of clearance and/or slash-and-burn agricultural activities. Charcoal and pollen analyses indicate that slash-and-burn activities were undoubtedly practiced at this time.

However, the highest charcoal values are recorded when the pollen indicator attests to an absence of agricultural activities. These data indicate a reduction in *Quercus*, *Corylus* and *Betula* following a phase of burning, resulting in the creation of meadowland. This clearing was probably managed for grazing.

The second phase (3300-2800 cal BC) indicates a significant transformation of the local environment (2a). Around 4620 ± 95 BP (*c*. 3370 cal BC), forest regeneration is characterised by an increase in ash, hazel and oak, but especially by a sudden rise in alder (*e.g. Alnus glutinosa*) which colonises the wet meadows close to the site (reduction in Poaceae and *Lotus*). The rapid extension of alder indicates that this taxa, which was present in the surrounding vegetation, benefited from this phase of forest regeneration by spreading out significantly, becoming the primary tree-group colonizer. Similar phenomenon was often observed in Great Britain (Simmons, 1993).

Three episodes of woodland clearing associated with forest farming have been recorded during this phase. At 180 cm (*c*. 3000 cal BC), the pollen data indicate a reduction in the broad-leaved woodland and in sprouting trees, and show a rise in grassland. The presence of cereal pollen and associated weeds of cultivation coincides with evidence of woodland burning (*e.g.* oak). This evidence of burning corresponds to the largest values in microscopic charcoal recorded in the profile and with the presence of oak macro-charcoal in the stratigraphy. In this case, the abundance of micro-charcoal correlates clearly with a local fire (Pitkänen *et al*. 1999), and there is no doubt that this evidence corresponds to a local slash-and-burn cultivation episode. This interpretation is supported by the presence of *Melampyrum*, and a reduction of *Ulmus, Tilia* and *Fraxinus*. According Troels-Smith (1960) and Rasmussen (1990), these trees have been used for leaf foddering. For this reason, the reduction in pollen values at Cuguron may indicate this practice.

This cultivation phase appears to have been a single, short-term event. Forest regeneration by *Betula, Alnus* and *Corylus* is indicated immediately after the site was abandoned. This succession, clearance/cultivation phase/forest regeneration, can be compared to the "landnam" model described by Iversen (Iversen, 1949). Two more phases of clearance associated with agro-pastoral activities are recorded at 170 cm and 165 cm, before 4120 ± 70 BP (2835-2625 cal BC). Both of these phases appear to be less intense than the first one, possibly reflecting increased distance of human activity from the mire basin. Pollen and charcoal taphonomic models support this interpretation and suggest that within forested areas, the dense forest cover inhibited the transportation of microfossils to the mire basin. In addition, lower micro-charcoal concentrations are likely to reflect burning of vegetation at a significant distance from the sampling site.

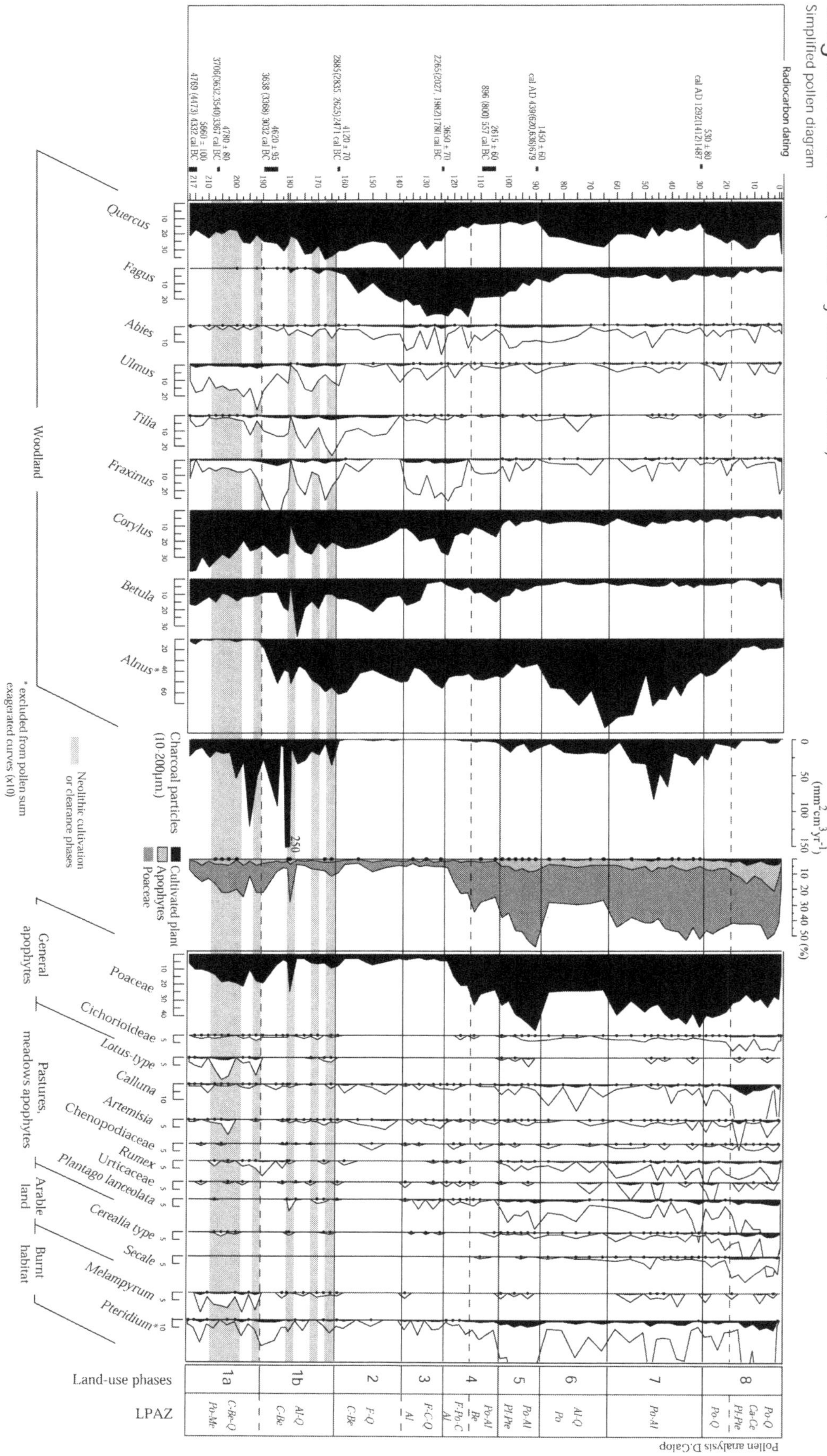

Figure 3: Relative simplified pollen diagram from Cuguron.

Between 4500 BC and 2800 BC, the landscape near Cuguron was subject to phases of intensive forest farming, including agriculture and grazing based on shifting cultivation using slash-and-burn. In this forest agro-system, cultivation phases of short duration resulted in disturbances in the local woodland cover. Nevertheless, the stratigraphic records of pollen indicate that the composition and structure of the forest cover remained largely unchanged. Rapid woodland regeneration following cultivation, however, would have favoured the development of coppice woodland.

The presence of charcoal in the sedimentary sequence at Cuguron during post-cultivation phases indicates that woodlands were cleared for temporary farming over a wide area. Several pollen indicators of woodland grazing (*e.g. Rumex, Urtica*, Chenopodiaceae), suggesting that agricultural activities shifted in emphasis from cereal cultivation to a pasture system, support this interpretation.

Period 2: *c.* 2800 BC - *c.* 2300 BC
The absence of charcoal particles seems to indicate a reduction in vegetation burning during this period. This is supported by the absence of direct evidence for cereal cultivation and the lack of pollen indicators of Late Neolithic human activity. The decline in human activities coincides with the development of *Fagus* at the site after 4120 ± 70 BP. This result confirms other pollen data for the Pyrenees, which indicate an expansion of *Fagus* before 4000 BP (Jalut, 1984; Galop and Jalut, 1994; Jalut *et al.* 1996).

Period 3: *c.* 2300 BC - *c.* 1750 BC
During the transition between the Late Neolithic and the Early Bronze Age (*c.* 2300 cal BC), the pollen data indicate a decrease in *Quercus, Ulmus, Betula* and, to a lesser extent, *Alnus*. At the same time, ash, hazel and later alder and beech expand in the area, indicating the colonisation of secondary woodland and the expansion of beech within the mixed oak forest. These episodes of clearing are broadly synchronous with the renewal of farming practices suggested by the occurrences of *Cerealia, Plantago lanceolata* and *Urtica*. While, the presence of ribwort plantain announces the existence of trampled and opened areas, the low frequencies in Poaceae indicate that these open areas were somewhat limited. The absence of charcoal in the sedimentary sequence indicates that the cultivation was probably some distance from the sampling site.

Period 4: *c.* 1750 - *c.* 600 BC
The pollen data recorded during the Bronze Age and the beginning of the Iron Age period indicate woodland clearing and substantial change in forest composition. Between 3650 ± 70 BP (2027, 1982 cal BC) and 2615 ± 60 BP (800 cal BC), the broad-leaved forest taxa decrease. The pollen values of *Betula, Corylus* and Poaceae indicate that secondary woodland regeneration occurred as a consequence of temporary clearing. The increase in pollen from herbs and indicators of grazing (*Plantago lanceolata, Rumex, Urtica*, Chenopodiaceae), combined with the absence of cereal pollen, suggests that woodlands were temporarily cleared as a result of a growing need for pasture. The absence of charcoal during most of this period does not indicate that the clearance episodes occurred at a great distance from the site because of the open structure of the vegetation cover. Indeed, the presence of charcoal and cereal pollen (*Cerealia, Secale*) towards the end of the period suggests extensive human activity.

Between the Late Neolithic and the beginning of Iron Age period, the reduction in charcoal particles during episodes characterised by human activity is enigmatic. These episodes indicate great differences in the role of fire in land use and imply the use of clearing methods other than fire. Similar observations were made for the same period in western Germany (Clark *et al.* 1989; Rösh, 1996) and in Scotland (Edwards and Whittington 2000). In both cases, the Late Neolithic and Bronze Age periods were characterised by a reduction in burning, changes in agronomic practices and abandoning of slash-and-burn. It could be speculated that the introduction of metalworking led to a reduction in the need for vegetation burning (Clark *et al.* 1989), although other methods such as girdling or ring barking may have led to the death of trees and/or favoured coppice woodland (Göransson, 1986).

Period 5: *c.* 600 BC - AD 630
From the second half of Iron Age to the Early Middle Ages periods, pollen data indicate extensive deforestation associated with a rapid rise in indicators of cultivation and grazing. The increase in charcoal, as well as the presence of *Melampyrum* and the increase in *Pteridium*, indicate an expansion of burnings and slash-and-burn during this period. This intensification of human pressure, which occurs after 800 cal BC, was responsible for a significant transformation of the forest cover. The reduction of *Fagus* and *Quercus*, which began during the previous period, continued and was accompanied by a reduction in *Betula, Corylus* and *Alnus* woodland. The absence of post-farming woodland colonisation indicates that the agricultural system had been substantially modified by this time, and probably consisted of permanent fields, increased productivity and shortened fallow periods (in order to accelerate cultivation turnover). The pressure for grazing land is reflected in the increase of ribwort plantain. *Rumex* and Cichorioideae can also hinder the possibility of forest recolonisation. Our results indicate that during this period there was a significant change from a shifting agro-system to a permanent agro-system. This change was responsible for a significant transformation in the forest cover as a response to regional demographic pressure around Mediterranean areas at this time (Brochier *et al.* 1998; Chabal, 1998). Here, the model that relates agricultural systems and demographic pressure (Boserup, 1970) seems to be appropriate.

Period 6: AD 630 - *c.* AD 1000
During the Early Middle Ages, from 630 AD to the beginning of the 11th century, the pollen record indicates that oak and alder dominated the woodland cover, with other arboreal taxa forming only a minor component. It is likely that alder invaded the meadows close to the site, resulting in a reduction in grassland. The regular occurrences of *Cerealia, Secale*, as well as the values of *Plantago lanceolata*, testify to the existence of large agro-pastoral activities. This rise in agrarian activities during the 7th century is recorded across the northern slope of the Pyrenees (Galop, 1998; 2000). The rates of charcoal particles remain stable and suggest the regional existence of fires. The increase in *Quercus* does

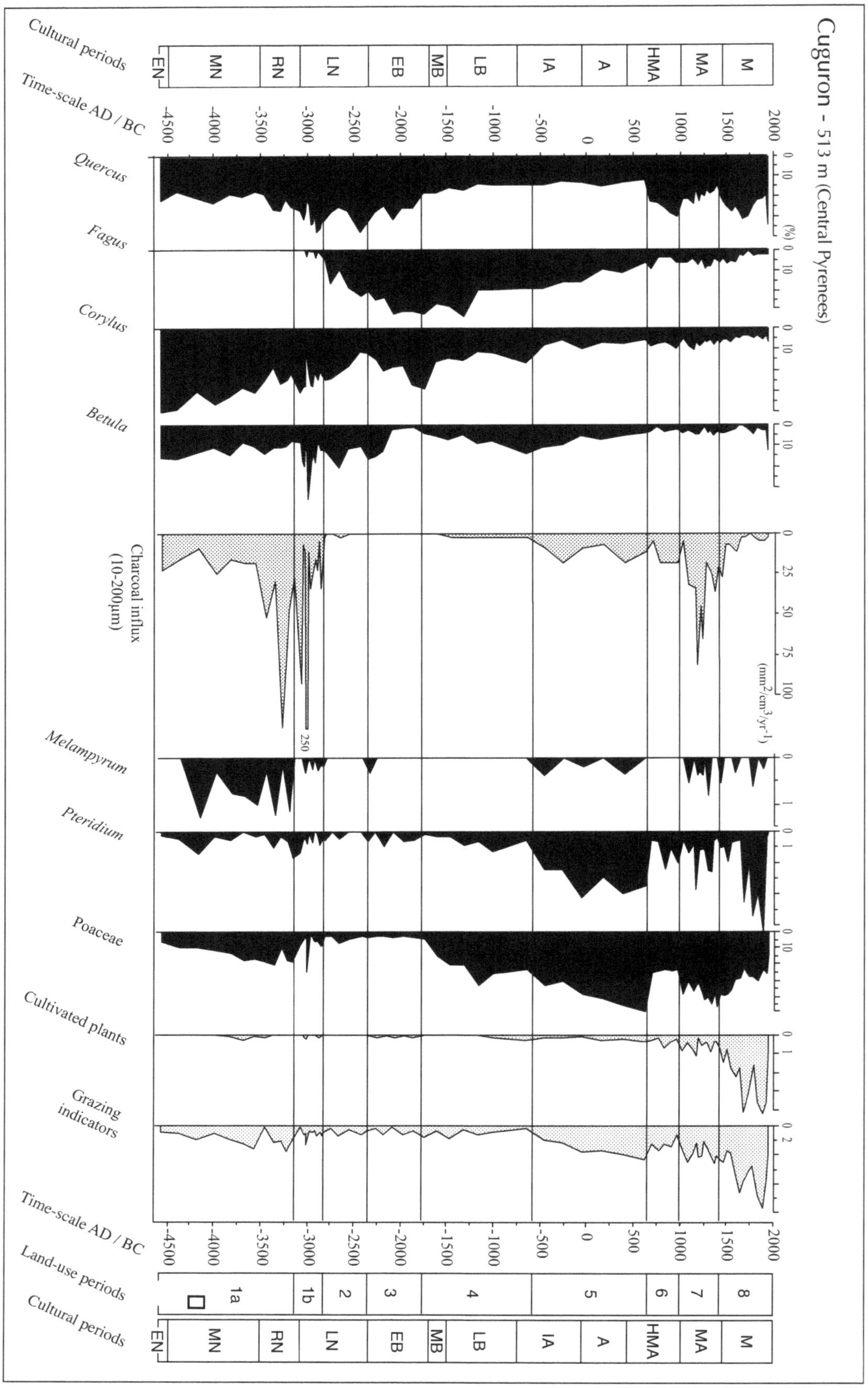

Figure 4: Synthetic pollen diagram plotted against chronological scale.

not necessarily represent an extension of oak forest, but may be interpreted as the result of a better flowering, favoured by forest management that included pollarding. These practices are recorded in Switzerland (Richoz and Gaillard, 1989) and in the Pyrenees by many written historical sources. This method of woodland management is not intended to produce wood (Rackham, 1988), but rather to increase productivity of acorns for pig feed. The existence of wooded grazed areas is supported by the increase in *Calluna* pollen, suggesting the formation of woodland heath during this period.

Periods 7; 8: *c.* AD 1000 - present

Pollen data recorded for the Middle Ages indicate a strong increase in human influence in the vicinity of the Cuguron peat bog. The high frequencies of indicators of grazing and cultivation correspond to a phase of woodland reduction, especially oak and alder. This expansion of farming is recorded throughout the region during the period between 900 AD and 1000 AD (Galop, 1998; 2000). The decline in *Alnus* indicates that forest bordering valleys and wet meadows were cleared to encourage the expansion of meadowland. The renewal of slash-and-burn cultivation is indicated by a peak in charcoal influx during the 12-13th century and the expansion of pollen indicators associated with burning notably *Pteridium* and *Melampyrum* which suggest new slash-and-burn cultivation not far from the site. This intensification of burning coincides with historical records of known demographic pressure (Bonnassie, 1989; Galop, 1998).

From 530±80 BP (1412 AD) to present day, this period indicates extensive human pressure on the landscape. The decrease in alder suggest that wetlands were largely deforested during this period as the need for new grazing land (meadows and wet grassland) became greater The expansion of heath land and woodland pastures is indicates by the consistently high pollen values of oak. The gradual reduction in charcoal concentration throughout the period indicates a reduction in burning and the need to clear woodland by this method. The presence of charcoal may therefore suggest the use of fire for pasture management, such as removal of heath.

Conclusion

The results of palynological and charcoal analyses performed at Cuguron provides the first detailed reconstruction of land use history on the northern slope of the Pyrenees. The record demonstrates the role played by fire in the agrarian practices in this area of the Pyrenees since the Early Neolithic and provides the first evidence for neolithic forest farming using shifting and slash-and-burn cultivation.

According to the palynological and charcoal data, each phase of burning coincided with the initiation or intensification of farming and, for this reason, it is possible to conclude that burning of the vegetation cover beginning in 4500 BC was at a local scale mainly due to human activities. The results are similar to Swiss records (Tinner *et al.* 1999) for the use of fire by human activity since 2300 BC. Three stages in firing episodes have been recognized. The first stage is recorded between 4500 BC and 2700 BC and corresponds to burning by neolithic forest farming as part of a system of localised slash-and-burn cultivation. Small scale clearing of broad-leaved trees was always followed by rapid regeneration of post-occupation secondary woodland (*Betula, Corylus, Alnus, Fraxinus*) and the re-establishment of the mixed oak forest. This succession indicates a dynamic and extensive shifting agro-system.

Despite the palynological evidence of forest removal and large anthropogenic activities, a decrease in fires is clearly recorded during the Late Neolithic and the Bronze Age (2700 BC - 630 BC). This can be interpreted as a change in forest management and in the methods of clearing woodland, perhaps corresponding to the use of metal tools.

The second stage began in 600 BC, in the second half of the Iron Age. During this period, an increase in burning is linked with the spread of agro-pastoral activities. Pollen records indicate a reduction in the forest cover; the absence of woodland regeneration indicates the establishment of permanent fields, the reduction of fallow periods and extensive modification of the landscape.

The last stage is recorded during historical times, between the 11th and 16th centuries. The increase of an influx of charcoal and the reduction in oak woodland reflect deforestation during the medieval agrarian expansion. After this episode, the decrease in charcoal can be interpreted as a reduction of fires during the last few centuries. At this time, the scarcity of forested areas and an agrarian system based on permanent fields and meadows led to a drastic reduction in the use of fire. The low charcoal concentrations for this period are therefore indicative of regional burning of the vegetation cover, the use of domestic fires and burning of pasture-land to encourage grazing.

ACKNOWLEDGMENTS
The authors are indebted to Jean-Jacques Deboubat and Jean Michel Pardé for their participation in fieldwork; to Martine Paterne for dating; and to Nick Branch for linguistic corrections and useful comments. The program "Exploitation des ressources naturelles dans les Pyrénées durant l'Antiquité" (directed by Robert Sablayrolles, UTAH-CNRS) Service Régional de l'Archéologie de Midi-Pyrénées (Ministère de la Culture) provided financial support.

References

BATARDY C., BUCHSENSCHUTZ O., DUMASY F. (eds), 2001 - Le Berry Antique - Atlas 2000. Supplément n° 21 à la *Revue Archéologique du Centre de la France*, Tours : 192 p.

BEHRE K.-E. 1981 - The anthropogenic indicators in pollen diagrams, *Pollen et Spores*, vol. XXIII, 2, p. 225-245.

BONNASSIE P. 1989 - La croissance agricole du haut Moyen Age dans la Gaule du midi et le nord-est de la péninsule ibérique. *La croissance agricole du haut Moyen Age*, Flaran 10, p. 13-35.

BONHÔTE J. 1992 - *Forges à la catalane et évolution forestière dans les Pyrénées de la haute-Ariège. Pour une histoire de l'environnement.* Thèse de Doctorat, Université de Toulouse-le-Mirail, CIMA URA 366 CNRS. 425 p.

BOSERUP E. 1970 - *Evolution agraire et pression démographique*, Flammarion, Paris.

BROCHIER J.-E., CLAUSTRE F., HEINZ C. 1998 - Environmental impact of Neolithic and Bronze age farming in the eastern Pyrenees forelands, based on multidisciplinary investigations at La Caune de Bélesta (Bélesta Cave), near Perpignan, France. *Veget. Hist. Archaeobot.*, 7, p. 1-9.

CARCAILLET C. 1998 - A spatially precise study of Holocene fire history, climate and human impact within the Maurienne valley, North French Alps. *Journal of Ecology*, 86, p. 384-396.

CHABAL L. 1997 - Forêts et sociétés en Languedoc (Néolithique final, Antiquité tardive). *Documents d'Archéologie Française* n°63, 189 p.

CHEVALIER M. 1956 - *La vie humaine dans les Pyrénées ariégeoises*. Paris, Génin, 1061 p.

CLARK J.S. 1988 - Stratigraphic charcoal analysis on petrographic thin sections: application to fire history in northwestern Minnesota. *Quaternary Research*, 30, pp. 81-91.

CLARK J.S., MERKT, J., MÜLLER H. 1989 - Post-glacial fire, vegetation, and human history on the northern alpine forelands, south-western Germany, *Journal of Ecology*, 77, p. 897-925.

CLARK J.S., ROYALL P.D. 1995 - Particle-size evidence for source areas of charcoal accumulation in late Holocene sediments of Eastern North American lakes. *Quaternary Research*, 43, p. 80-89.

CLARK R.L. 1982 - Point count estimation of charcoal in pollen preparations and thin sections of sediment. *Pollen et Spores*, 24, p. 523-535.

DAVASSE B., GALOP D., RENDU C. 1997 - Paysages du Néolithique à nos jours dans les Pyrénées de l'Est d'après l'écologie historique et l'archéologie pastorale. *La dynamique des paysages protohistoriques, antiques, médiévaux et modernes*. Editions APDCA, Sophia Antipolis. p. 577-599.

DIOT M.-F. 1992 - Les microcharbons, éléments du palynofaciès. *Bull. Soc. bot. Fr.*, 139, Actual. Bot., 2-4, p. 265-272.

EDWARDS K.J., WHITTINGTON G. 2000 - Multiple charcoal profiles in a Scottish lake : taphonomy, fire ecology, human impact and inference, *Palaeogeography, Palaeoclimatology, Palaeoecology*, 164, p. 67-86.

EDWARDS K.J. 1993 - Models of mid-Holocene forest farming for North-West Europe, *Climate change and human impact on the landscape*, F.M. Chambers editor, Chapman & Hall, p. 133-145.

FAEGRI K., IVERSEN J. 1989 - *Textbook of pollen analysis*. 4th edition, John Wiley & Sons.

GALOP D., JALUT G. 1994 - Differential human impact and vegetation history in two adjacent Pyrenean valleys in the Ariège basin, southern France, from 3000 BP to the present. *Veget. Hist. and Archaeobotany*, 3, p. 225-244.

GALOP D. 1998 - *La forêt, l'homme et le troupeau dans les Pyrénées–6000 ans d'histoire de l'environnement entre Garonne et Méditerranée*, GEODE-LET-FRAMESPA, Toulouse, 285 p.

GALOP D. 2000 - Les apports de la palynologie à l'histoire rurale. La longue durée des activités agro-pastorales pyrénéennes. *Etudes rurales*, 153-154, p. 127-138.

GÖRANSSON H. 1986 - Man and the forest of nemoral broadleafed trees during the stone age, in Nordic late Quaternary biology and ecology (L.-K. Königsson, Ed.), *Striae*, 24, p. 143-152.

HUTTUNEN P. 1980 - Early land use, especially the slash-and-burn cultivation in the commune of Lammi, southern Finland, interpreting mainly using pollen and charcoal analyses. *Acta Bot. Fennici,* 113, p. 1-45.

INNES J. B., SIMMONS I.G. 2000 - Mid-Holocene charcoal stratigraphy, fire history and palaeoecology at North Gill, North York Moors, UK. *Palaeogeography, Palaeoclimatology, Palaeoecology*, 164, p. 177-191.

IVERSEN J. 1949 - *The influence of prehistoric man on vegetation,* Danmarks Undersogelse IV, 3, 6, p. 1-23.

JACOBSON G. L., BRADSCHAW, R.H.W. 1981 - The selection of sites for palaeovegetational studies. *Quaternary Research,* 16, p. 80-96.

JALUT G. 1984 - *L'action de l'homme sur la forêt montagnarde dans les Pyrénées ariègeoises depuis 4000 BP d'après l'analyse pollinique*. Actes du 106eme congrès des sociétés savantes, CTHS, Paris, p. 163-172.

JALUT G., GALOP D. 1994 - Differential human impact and vegetation history in two adjacent Pyrenean valleys in the Ariège basin, southern France, from 3000 BP to the present, *Vegetation History and Archeobotany*, 3, p. 225-244.

JALUT G., AUBERT S., GALOP D., FONTUGNE M., BELET J.-M. 1996 - France, type region F-zg and F-r, the northern slope of the Pyrenees, in Palaeoecogical events during the last 15000 years – *Regional synthesis of palaeoecological studies of lakes and mires in Europe* (B.E. Berglund, H.J.B. Birks, M. Ralska-JASIEWICZOWA, H.E. Wright, Edts.), J. Wiley & Sons, p. 612-632.

JANSSEN C.R. 1959 - *Alnus* as a disturbing factor in pollen diagrams. *Acta Bot. Neerland*, 8, p. 55-58.

LAGERAS P. 1996 - Farming and forest dynamics in a agriculturally marginal area of southernn Sweden, 5000 BC to present: a palynological study of lake Avegöl in the smaland Uplands. *The Holocene*, 6, 3, p. 301-314.

LATALOWA M. 1992 - Man and vegetation in the pollen diagrams from Wolin Island (NW Poland), *Acta Palaeobotanica*, 32, 1, p. 123-249.

MACKLIN M.G., BONSALL C., DAVIES F.M., ROBINSON M.R. 2000 - Human-environment interactions during the Holocene: new data and interpretations from the Oban area, Argyll, Scotland. *The Holocene*, 10 (1), p. 109-121.

MAMAKOVA K. 1968 - Lille bukken and Lerøy - two pollen diagrams from Western Norway. *Arbok Universitet i Bergen Naturvitenskapelig*,4, p. 5-42.

MÉTAILIÉ J.-P. 1981 - *Le feu pastoral dans les Pyrénées centrales* (Barousse, Oueil, Larboust). CNRS, Paris, 293 p.

MOORE J. 2000 - Forest fire and human interaction in the early Holocene woodlands of Britain. *Palaeogeography. Palaeoclimatolology. Palaeoecology*, 164 (1-4), p. 125-137.

MOORE P.D., WEBB J.A., COLINSON M.E. 1991 - *Pollen analysis*, Blackwell scientific publications, Oxford, 216 p.

MULLER A. 1985 - La nécropole en "cercles de pierre" d'Arihouat à Garin (Haute-Garonne) *Archéologie* n°1, Ed. Vésuna, Perigueux.

PITKÄNEN A., HUTTUNEN P. 1999 - A 1300-year forest-fire history at a site in eastern Finland based on charcoal and pollen records in laminated lakes sediment. *The Holocene*, 9 (3), p. 311-320.

PITKÄNEN A., LETHONEN H., HUTTUNEN P. 1999 - Comparison of sedimentary microscopic charcoal particle records in a small lake with dendrochronological data: evidence for the local origin of microscopic charcoal produced by forest fire of low intensity in eastern Finland. *The Holocene*, 9, p. 559-567.

RACKHAM O. 1988 - Trees and woodland in a crowded lanscape - the cultural landscape of the British isles, *in The cultural landscape. Past, present and future* (H.H. Birks, H.J.B. Birks, P.E. Kaland, D. Moe), Cambridge University Press, p. 53-77.

RASMUSSEN P. 1989 - Leaf-foddering of live stock in the Neolithic: Archaeobotanical evidence from Weier, Switzeland, *Journal of Danish Archaeology*, 8, p. 51-71.

REILLE M., ANDRIEU V. 1994 - Vegetation history and human action in Ariège (Pyrénées, France), *Dissertationes Botanicae*, 234, p. 413-422.

REILLE M. 1990 - Recherches pollenanalytiques dans l'extrémité orientale des Pyrénées : données nouvelles de la fin du glaciaire à l'actuel. *Ecologia Mediterranea*, XVI, p. 317-357.

REILLE M. 1992 - *Pollens et spores d'Europe et d'Afrique du nord*. Laboratoire de Botanique historique et palynologique, URA 1152 CNRS, Marseille.

RENDU C. 1987 - Quelques jalons pour une histoire des forêts en Cerdagne : le massif d'Osseja entre 1030 et 1430. Etudes Roussillonaises offertes à P. Ponsich, Perpignan, *Le Publicateur*, p. 245-251.

RENDU C., CAMPMAJO P., DAVASSE B., GALOP D., CRABOL D. 1996 - Premières traces d'occupation pastorale sur la montagne d'Enveig. *Travaux de Préhistoire Catalane*, 8, p. 35-43.

RHODES A.N. 1998 - A method for the preparation and quantification of microscopic charcoal from terrestrial and lacustrine sediment cores. *The Holocene*, 8, p. 113-117.

RICHOZ I., GAILLARD M.-J. 1989 - Histoire de la végétation de la région neuchâteloise de l'époque néolithique à nos jours. Analyse pollinique d'une colonne sédimentaire prélevée dans le lac de Neuchâtel (Suisse). *Bull. Soc. Vaud. Sc. Nat.*, 79, (4), p. 355-377.

RÖSCH M. 1996 - New approaches to prehistoric land-use reconstruction in south-western Germany, *Vegetation History and Archeobotany*, 5, p. 65-79.

SCOTT A.C. 2000 - The Pre-Quaternary history of fire. *Palaeogeography, Palaeoclimatolology, Palaeoecology*, 164 (1-4), p. 281-330.

SIGAUT F. 1975 - *L'agriculture et le feu. Rôle et place du feu dans les techniques de préparation du champ de l'ancienne agriculture européenne*. EHESS, Mouton.

SIMMONS I.G. 1993 - Vegetation change during the Mesolithic in the British Isles: some amplifications. *Climate change and human impact on the landscape*. F.M. Chambers editor, Chapman & Hall, p. 109-118.

STUIVER M., REIMER P.J., BARD E., BECK J.W., BURR G.S., HUGHEN K.A., KROMER B., MC CORMAC F.G., PLICHT J., SPURK M. 1998 - INTCAL98 Radiocarbon age calibration 24,000 - 0 cal BP. *Radiocarbon*, 40 (3), p. 1041-1083.

SWAIN A.M. 1973 - A history of fire and vegetation in Northeastern Minnesota as recorded in lake sediments. *Quaternary Research*, 3, p. 383-396.

TINNER W., HUBSCHMID P., WEHRLI M., AMMANN B., CONEDERA M. 1999 - Long-term forest fire ecology and dynamiques in southern Switzerland. *J. Ecol*, 87, p. 273-299.

TOLONEN K. 1986 - *Charred particles analysis, in Handbook of Holocene Palaeoecology and Palaeohydrology* (B. E. Berglund, Ed.), John Wiley & Sons Ltd. p. 485-496.

TOLONEN M. 1985 - Palaeoecological record of local fire history from a peat depuis in SW *Finland, Ann. Bot. Fennici*, 22, p. 15-29.

TROEL SMITH J. 1960 – Ivy, mistletoe and elm. Climate indicator / fodder plants. *Denmark Geologiska Undersogelse*, IV, 4, p. 1-32.

VANNIÈRE B. 2001 - *Feu, agro-pastoralisme et dynamiques environnementales en France durant l'Holocène. Analyse du signal incendie, approches sédimentologiques et étude de cas en Berry, Pyrénées et Franche-Comté*. Thèse de l'Institut National Agronomique de Paris-Grignon, 329 p.

VUORELA I. 1986 - *Palynological and historical evidence of slash-and-burn cultivation in south Finland. Anthropogenic indicators in pollen diagrams* (K.-E. Behre, Ed.), A.A. Balkema, p. 53-64.

WILTSHIRE P.E.J., EDWARDS K.J. 1993 - Mesolithic, early neolithic, and later prehistoric impacts on vegetation at a riverine site in Derbyshire, *England, Climate change and human impact on the landscape*, F.M. Chambers editor, Chapman & Hall, p. 157-168.

Prehistoric Holocene fires at the Causse Méjean (Lozère, France), climatic or human impact?

Frédéric Bazile, Paule Ogereau, Jean-Louis Vernet and Arnaud Martin
Laboratoire de Paléoenvironnements, Anthracologie et Action de l'Homme, UMR 5059, Université de Montpellier II, Institut de Botanique, 163 rue A. Broussonet, 34090 Montpellier cedex (France)
E-mail: f_bazile@crit.univ-montp2.fr; ogereau@crit.univ-montp2.fr; vernet@crit.univ-montp2.fr; martin@crit.univ-montp2.fr

Abstract
Charcoal, the remnants of ancient fires, have been discovered in several sinks in the Causse Méjean, a submediterranean mountain located in the Causses region of the southern Massif Central, France. The charcoal was sampled in the eastern asylvatic or deforested area. The first ^{14}C AMS dates carried out on *Pinus* cf. *sylvestris* charcoal, cover the period of 4805-2990 BP. The results show the development of a pine forest during the mid to late Holocene transition and perhaps the existence of particular conditions – climatic and/or anthropogenic - that would have favoured the initiation of forest fires. The presence of several fire events, affecting the forests of the north-western part of the Causse Méjean during the second half of the Holocene, is also proved. The origins of the present grassland vegetation seem to date back to the Bronze Age.

Key-words: *Mediterranean, Holocene, Mountains, Forest, Grassland, Charcoal, Bronze Age.*

Résumé
Des charbons de bois, témoins d'anciens feux, ont été découverts dans des dolines de la partie asylvatique ou déforestée des Grands Causses subméditerranéens. Les premières datations AMS ^{14}C ont été effectuées sur des charbons de *Pinus* cf. *sylvestris* ; elles couvrent la période 4805-2990 BP. Ceci est la preuve du développement de forêts de pins à la transition Holocène moyen-Holocène récent et, sans doute, aussi de conditions spéciales au déclenchement des feux de forêt. Les résultats indiquent aussi plusieurs phases de feux affectant les forêts alors étendues au nord-ouest des Causses au cours de la seconde partie de l'Holocène. On montre aussi que les pelouses actuelles remontent à l'âge du Bronze.

Mots-clés : *Méditerranéen, Holocène, montagnes, forêts, pelouses, charbons, Age du Bronze.*

Introduction

Research based on the analysis of charcoal from archaeological contexts has contributed to an increased knowledge and understanding of ancient environments and anthropogenic processes, in the Mediterranean area in general and in the Causses region in particular (Krauss-Marguet, 1981; Vernet 1968, 1972, 1981, 1985, 1995, 1997). However, this research has been limited to archaeological sites. In the Causses area, the most interesting of them, La Poujade, was found in a deep adjacent valley (Krauss-Marguet, 1981).

Recently, the discovery of charcoal, *i.e.* the remnants of ancient fires, from karstic deposits (sinks) in the Causse Méjean has enriched and completed the data obtained from archaeological sites. This may allow us to extend our research and to estimate forest density (Vernet *et al.* 1994), composition and structure as well as to establish a fire chronology by means of ^{14}C AMS dating.

The Causses is characterised by a submediterranean environment (Vernet, 1985). Moreover, it is situated in a transition zone where Euro-Siberian, subalpine, steppic and oromediterranean floral elements coexist. This configuration is the result of different movements or plant migrations since the last glacial period (Van Campo and Aymonin, 1962). The floristic diversity is accompanied by a diversity of landscapes (Vanden Berghen, 1963). Among forests and woodlands, the pine woodlands of *Pinus sylvestris* constitute a specific ecological group. The woodlands are dense in the western part of the Causse Méjean, in the Causse Noir, in the north-east of the Larzac and in a major part of the Causse de Sauveterre. On the contrary, very largely denuded sectors are represented in the eastern part of the Causse Noir, west of the Larzac and especially in the eastern area of the Causse Méjean where the limestone table reaches its maximal altitude at 1300 m.

Open grasslands are characteristic of the landscapes of the Causses. They are close-cropped with the occasional presence of low woody plants. The growth height of the grasslands does usually not exceed 10 cm, excluding the floral elements, and the root is limited to 5-15 cm depht. This situation could be due to various factors:
(a) the cold winters and the dry summers (steppic conditions) that characterise the area;
(b) the presence of superficial arenaceous soils without water reserves with an excess of Ca^{++}, inhibiting the absorption of humidity;
(c) previous exploitation and overgrazing.

Today, these herbaceous formations, consisting mainly of *Festuca duriuscula* and *Stipa pennata* growing on shallows red clayish soils or on dolomite, are considered as the final state of the forest degradation caused by human exploitation as well as by the grazing of sheep and goats. The precise terms of this situation, if these are largely dependent of human action, are not however clearly established. In fact, the alternation between open herbaceous and more dense arboreal formations in the Causses region might be quite ancient, *i.e.* go back to prehistory (Lorblanchet, 1963). But on this basis, is it possible to record a unique and anthropogenic origin?

Remains of ancient fires

As charcoal fragments can provide us with palaeoecological information, this type of macrobotanical remains were looked for and found in the asylvatic part of the Causse Méjean. The main geological features sampled consisted of karstic depressions, sinks, "uvalas" or "poljes" and their complex fillings, as these are likely to contain concentrations of

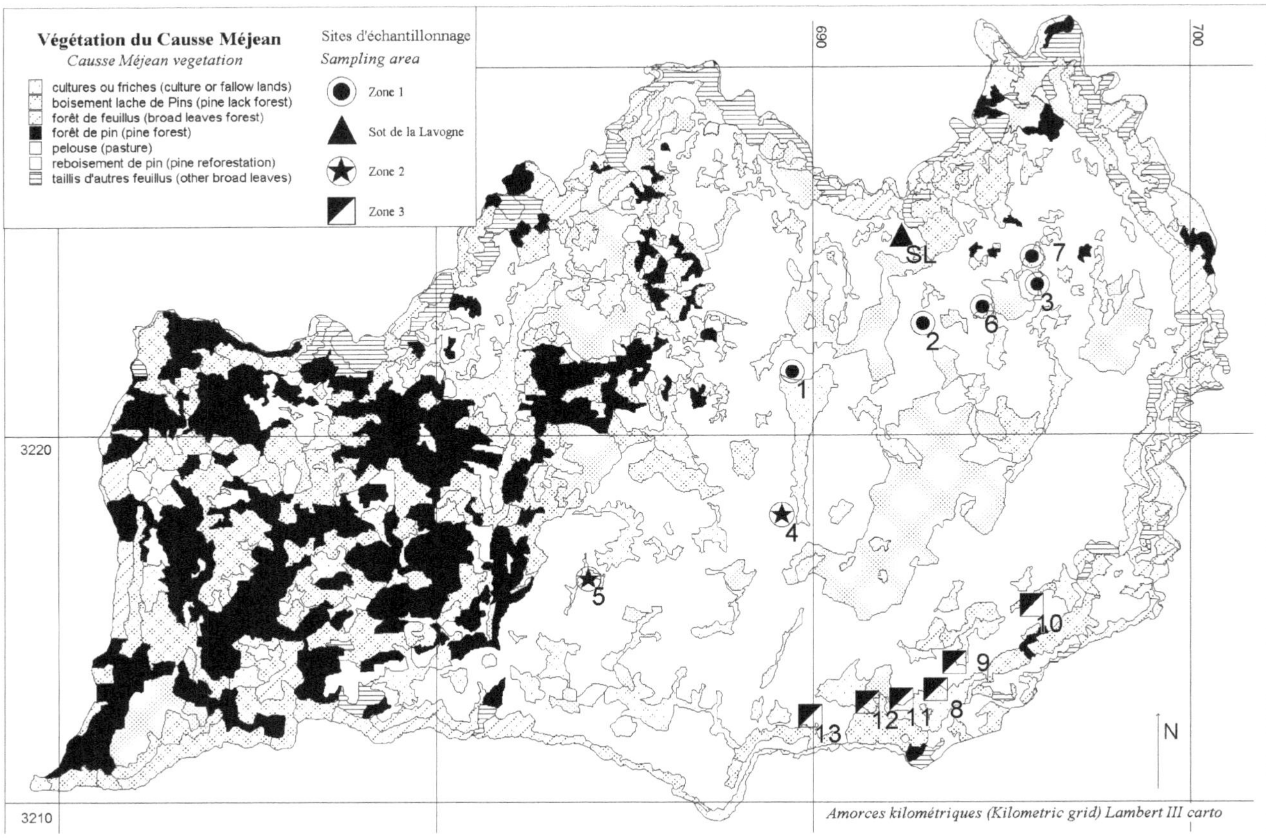

Figure 1: The opposition between forested and unforested areas of the Causses and location of the samples sectors

material. Ancient bed formations, such as deposits of "terra rossa", have been distinguished from superficial deposits limited to recent soils, probably Quaternary to Holocene: humus bearing limestones, rendzinas, humo-limestones soils, humo-calcic soils and dolomitic rendzinas.

Three zones including 13 sites and 18 levels, have been studied:

- a sector located to the north-east of the departemental road 16 (D. 16) departing from Chanet polje,
- a central sector along the D. 63 from Nivoliers to Drigas,
- a south-eastern sector along the foothills of the Causse, from Frépestel to Nîmes le Vieux. Moreover, a sounding was done in a sink hole, the "Sot de la Lavogne" (Quilès, 2000; Quilès *et al.* submitted) (fig. 1).

At least thirty litres of soils have been sampled in order to obtain representative data. Each sample was sieved through a 2 mm mesh, the refusal was dried and then the charcoal fragments were extracted with the help of a pair of "Brussels" tweezers. The identifications are based on the comparaison with a reference collection of oriented sections belonging to the laboratory of the UMR 5059 (Montpellier). The anatomy of the charcoal fragments was observed with the help of a reflecting light microscope.

Density of charcoal

402 pieces of charcoal were extracted from 500 litres of soil, sampled from 18 levels. All the stations from which samples were taken are localised in grasslands or herbaceous formations, dominated by *Stipa pennata,* or in "matorrals" with *Buxus sempervirens*. There is no correlation between the mass of charcoal and the current composition of the vegetation. The present vegetation is always herbaceous.

The frequencies of charcoal proved to be variable from one zone to another:
1.17 charcoal/litre in zone 1
0.725 ch/litre in zone 3
0.116 ch/litre in zone 2.

In the present state of our research these differences can be imputed to local pedological, edaphic and geomorphologic conditions or to changes in the forest shelter.

Forest composition

The proportion of indeterminable, small or vitrified charcoal pieces (35%) is acceptable. Gymnosperms and Angiosperms appear in comparable proportions (48/52%). Twenty-seven taxa have been identified, mostly to the genus level. Among the Gymnosperms, pine (fig. 2) dominate over juniper (85/5%). The diversity of taxa is greater among the Angiosperms, including (fig. 3) *Fagus sylvatica (29%)*, Pomoïdea (25%), Salicaceae (7%), *Quercus* cf. *pubescens* (deciduous oaks, 4%), *Buxus sempervirens* (4%), *Betula* cf. *verrucosa* (4%), Oleaceae (2%), *Ulmus* (2%), *Acer* (1%), Cistaceae (1%), *Ilex aquifolium* (1%) and *Prunus* cf. *avium* (1%).

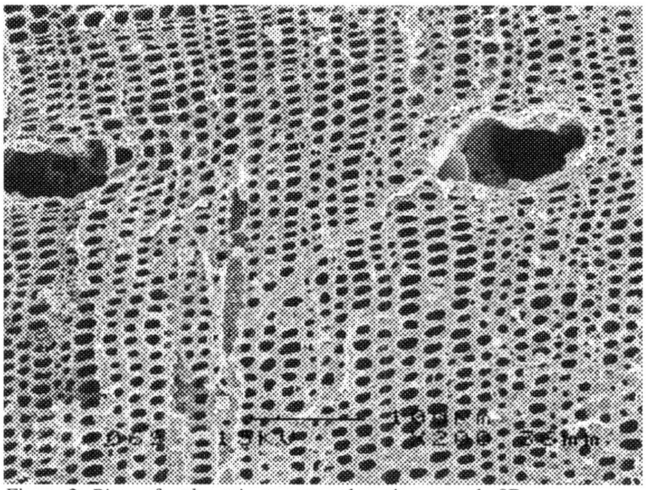

Figure 2: *Pinus* cf. *sylvestris*, transversal section, sample 8B.

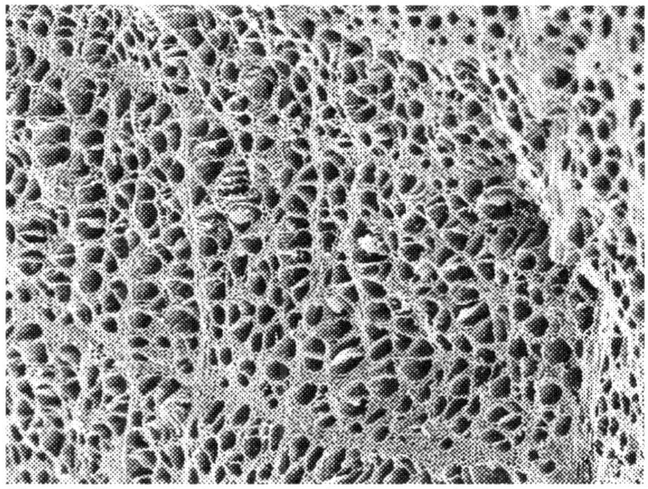

Figure 3: *Fagus sylvatica*, transversal section, sample 7.

Three zones are clearly distinguished (fig. 1) on the basis of the frequency of the various taxa (number of fragments):
- Zone 1: *Pinus* type *sylvestris* is dominant (74%) followed by *Juniperus* (4%), *Quercus* (deciduous, 3%), Rosaceae/Pomoïdea (3%), *Buxus sempervirens* (1%), *Betula* (1%), Oleaceae (1%) and *Ulmus* (1%).
- Zone 2: Charcoal fragments are very scarce (2 fragments of *Pinus* and 3 of *Juniperus*).
- Zone 3: The taxonomic diversity increases with the presence of *Fagus sylvatica* (34%), Rosaceae/Pomoïdea (26%), Salicaceae (7%), *Buxus sempervirens* (4%), *Betula* (3%), *Ulmus* (2%), *Pinus* cf. *sylvestris* (2%), *Quercus* (deciduous, 2%), Oleaceae (2%), *Ilex aquifolium* (1%), *Acer* (1%) and *Prunus* cf. *avium* (1%).

The taxonomic list obtained in zone 1 may correspond to a pine forest growing on fertile soils (Vanden Berghen, 1963).

Ecological significance and chronology

Our results demonstrate the existence of a heterogeneous forest cover in the eastern part of the Causse Méjean, an area usually known as asylvatic or deforested. The existence of a pine forest in zone 1 seems particularly clear. Vanden Berghen (1963) distinguishes three categories of pine communities:
(a) Pioneer pine forests with *Juniperus phoenicea*: this formation is further characterised by a low biodiversity and the growth of trees less than 4 m tall, on juvenile soils and dolomites;
(b) Open pine woodlands including the growth of steppic grasses;
(c) Mossy pines, up to 12 m tall, growing on deep soils. This formation shows moreover diversified woody flora (10 species in the *Cephalanthero-pinetum*).

According to the taxa identified by the charcoal analysis, zone 1 would have been covered by a forest of mossy pines growing on deep soils. Deciduous oak is scarce. Our results could reflect the development of the forest cover on the Causse in relation to the Holocene optimum.

A preliminary series of ^{14}C AMS dates has been obtained (tabl. 1). The taxonomic affiliation of the material used for radiocarbon dating is very important as, even in case of bioturbation, the chronology of fire events will remain valide.

The dates obtained from charcoal fragments identified as *Pinus* cf. *sylvestris* cover a period from 4805 BP to 2990 BP. These results suggest the development of a pine forest during the transition from the mid to the late Holocene. The earliest date comes from Chanet 1 and is based on charcoal pieces belonging to three different taxa. Therefore, the possibility of a mixture of charcoal cannot be discarded. Homogeneous dates have been obtained from the samples from Chanet 3, despite the diversity of material submitted to ^{14}C analysis. It seems however that the age of the charcoal is independent of the fragmentation state. Other charcoal fragments provide evidence of the presence of a beech forest in an area close to the south-eastern limit during historical times.

The "Sot de la Lavogne"

In order to obtain further anthracological information, sediment samples were taken from a sink hole situated next to a burial site at Sot de la Lavogne (Quilès, 2000). Archaeological work was previously carried out at the site under the supervision of G. Fages (Fages and Courtaud, 1996, 2000).

Zone	Sampling sites	Number	References and 14C dates BP	Identified charcoal	Charcoal number	Charcoal weight
1	Chanet 1	1	GifA 99689 :5620BP ±70	*Juniperus* *Pinus* Undetermined	1 8 2	30mg
	Chanet 2	2	GifA 99500 : 4180BP ±60	*Pinus*	48	238mg
	Chanet 3	3	GifA 99501 : 2990BP ±60	*Pinus*	7	25 mg
	Chanet 3	4	GifA 99502 : 3100BP ±60	*Pinus*	1	44 mg
	Chanet 3	5	GifA 99503 : 3020BP ±60	*Pinus*	1	15 mg
	Chanet 3	6	GifA 99504 : 3040BP ±70	*Pinus*	28	100 mg
	Mativet	7	GifA 99505 : 4805BP ±50	*Pinus*	27	106 mg
3	Villeuneuve	8	GifA 99506 :170BP ±60	*Fagus*	12	66 mg
	Aures	9	GifA 99507 : 30BP ±50	*Fagus*	6	27 mg
	Frépestel	10	GifA 99508 : 10BP ±60	*Fagus*	9	62 mg

Table 1: ^{14}C dating.

Ten litres of sediment were collected from each 10 cm layer. A total of 1204 charcoal fragments were recovered. While they were particularly abundant at the bottom of the sequence they were scarce in the upper layers. Several phases were identified:

(a) Phase Sot 1: This phase is characterised by a layer of dark sediment. *Pinus* cf. *sylvestris* dominates the charcoal record (*c*. 85%). *Fagus* and *Quercus* (deciduous) are also noted.

(b) Phase Sot 2: As the concentration of charcoal was low, we can only consider general trend. *Pinus* might have been less important than in the previous phase, here substituted by *Juniperus* and Rosaceae species. Deciduous *Quercus* seems to disappear by the end of the phase Sot 2. The results might reflect a phase of forest devastation followed by the introduction of a scrub vegetation including *Juniperus* and Rosaceae. These changes may have taken place after the Chalcolithic/Bronze Age.

(c) Phase Sot 3: This phase is very recent. However, the trend observed in phase 2 is confirmed: *Pinus* is either very rare or disappears completely, while *Juniperus* and Rosaceae are still present. The spread of *Buxus* is noted. At the archaeological site of La Poujade the results of the charcoal analysis suggest an even earlier spread of *Buxus*, around 4000 BP (Krauss-Marguet, 1981).

The forest and the prehistoric settlement

The spread of human settlement in the Causse Méjean began during the Neolithic period (Hugues, 1950; Prunières, 1872; Temple, 1936) and attained its peak during the Late Neolithic/Chalcolithic periods (Fages, 1972, 1997; Lorblanchet, 1965). According to Lorblanchet, the presence of the pine forest dates back at least to the Roman period and probably even earlier, *i.e.* to the Neolithic. This assumption is based on the distribution of archaeological features, such as megalithic monuments.

The results obtained at "Sot de la Lavogne" and from other sinks that were sampled indicate the existence of several fire events, affecting the pine forests of the northern part of the Causse around 4800-3000 BP.

The main phase of forest fires is correlated with the second part of the Holocene. Preliminary ^{13}C measurements on archaeological charcoal (Vernet *et al.* 1996) from lower parts of the garrigue show the presence of a dry period at this time (*c*. 4000-2000 BP).

The present results are important for our understanding of how humans used their environment in the past. We suggest that during the second part of the Holocene, a dry period, with frequent fires was probably the major cause of deforestation. Man would thus invest an open herbaceous landscape, where sheep and goats could graze. Later, during the Bronze Age and historic times, the deforestation processes were further accelerated by grazing and agriculture.

Present pine forest dynamics

The present progression of pine forest, a consequence of less intensive land use, is studied from aerial photographies. The data have been computerised with the help of the "Mapinfo program". The results show that the present-day forest extension is not related to former dense forests in the western parts of the Causse (Cohen *et al.* 1992). Extensive forests have developed as a result of seed dispersal from isolated trees. A thorough investigation of the present-day distribution of pine is needed in order to assess the possibilities of a "landscape recovery" in the Causse Méjean, now when the anthropogenic pressure is much reduced.

AKNOWLEDGEMENT
The present research is supported by the PEVS program (Espaces boisés-Espaces ouverts dans les Grands Causses) of the CNRS and the Languedoc-Roussillon region. The AMS datings have been realised with the Tandétron of the Gif-sur-Yvette Centre des faibles radioactivités. Gilbert Fages made discover us the sink hole of the "Sot de la lavogne". Many thanks also to all the students who participate in the program : David Quilès, Virginie Rohr and Sylvain Lhuillier.

References

COHEN M., HOTYAT M., ASSINARE J.-L. 1992 - Embroussaillement et boisement sur le Causse Méjean entre 1965 et 1992, Université Paris VII, *Trav. Lab. Géographie physique* p. 1-34.

FAGES G. 1972 - Fonds de cabanes de l'âge du Fer sur le Causse Méjean, *Revue du Gévaudan*, Nelle série, 18-195-16.

FAGES G. 1977 - L'habitat en Lozère au Bronze final et aux âges du Fer, XXI *Coll. AFEAF,* Résumé 1p.

FAGES G., COURTAUD P. 1996 - *Le Sot de la Lavogne*. Rapport de fouille. DRAC Montpellier.

FAGES G., COURTAUD P. 2000 - L'abri sépulcral du Sot de la Lavogne (Montbrun, Lozère). *IVème rencontres méridionales de Préhistoire récentes*, Nîmes 28-29 octobre 2000, préactes, p. 62 - 63, 1 fig.

HUGUES C. 1950 - Préhistoire du Causse Méjean oriental, *Congrès préhistorique de France*, XIIIème session p. 358-385.

KRAUSS MARGUET I. 1981 - Analyse anthracologique du gisement postglaciaire de La Poujade (Millau, Aveyron), *Paléobiologie continentale*, XII, 1, p. 93-110.

LORBLANCHET M. 1965 - Contribution à l'étude du peuplement des Grands Causses, *Bull. Soc. préhist. Fr.*, 3, p. 667-712.

MARRES P. 1935 - *Les Grands Causses*. Etude de géographie physique et humaine, Tours.

PRUNIERES P. 1872 - Distribution des dolmens dans la Lozère, *Matériaux*, VIII, p. 475-486.

QUILES D. 2000 - *Paléofeux dans la région asylvatique du Causse Méjean*, DEA Paris I, 42 p.

QUILES D., ROHR V., JOLY K., LHUILLIER S., OGEREAU P., MARTIN A., BAZILE F., VERNET J.-L. - Les feux préhistoriques holocènes en montagne sub-méditerranéenne : Premiers résultats sur le Causse Méjean (Lozère, France). Soumis *C.R Acad. Sc. Paris*.

TEMPLE P. 1936 - La Préhistoire du département de l'Aveyron. *Cahiers d'Histoire et d'Archéologie*, Rodez.

VAN CAMPO M., AYMONIN G. 1962 - Le problème de l'histoire de la flore et de la végétation dans les Cévennes méridionales vu sous l'angle de l'analyse pollinique, *Flora*, 152, p. 679-688.

VANDEN BERGHEN C. 1963. - Etude sur la végétation des Grands Causses du Massif Central de France, *Soc. roy. Bot. Belgique*.

VERNET J.-L. 1968 - La couverture forestière des Causses du Massif Central de France au Postglaciaire, d'après l'étude des charbons de bois préhistoriques, *Bull. Assoc. fr. Et. Quaternaire*, 16, p. 181-198.

VERNET J.-L. 1972 - Nouvelle contribution à l'histoire de la végétation holocène des Grands Causses d'après les charbons de bois, *Bull. Soc. bot. Fr.*, 119, p. 169-184.

VERNET J.-L. 1981 - Le Milieu naturel au Quaternaire dans les Causses et vallées périphériques. *Paléobiologie continentale*, XII, 1, 277 p.

VERNET J.-L. 1985 - Ecologie des Causses au Quaternaire, *Bull. Soc. lang. Géographie*, 19 (3/4), p. 265-286.

VERNET J.-L. 1995 - Végétations et paléoécologie des Grands Causses, in : *Grands Causses, Féd. Vie et Sauvegarde du pays des Grands Causses* (éd.), p. 91-105.

VERNET J.-L. 1997 - *L'Homme et la Forêt méditerranéenne de la Préhistoire à nos jours*. Errance, Paris, 247 p.

VERNET J.-L., WENGLER L., SOLARI M.-E., CECCANTINI G., FOURNIER M., LEDRU M.-P., SOUBIES F. 1994 - Feux, climats et végétations au Brésil central durant l'Holocène : les données d'un profil de sol à charbons de bois (Salitre, Minas Gerais), *C.R. Acad. Sci. Paris*, II 319 p. 1391-1397.

VERNET J.-L., PACHIAUDI C., BAZILE F., DURAND A., FABRE L., HEINZ C., SOLARI M.-E., THIEBAULT S. 1996 - Le δ C13 des charbons de bois préhistoriques et historiques méditerranéens, de 35000 BP à l'actuel. Premiers résultats. *C.R. Acad. Sci. Paris*, 323 (II a), p. 319-324.

Quantitative anatomical criteria for discriminating wild grapevine (*Vitis vinifera* ssp. *sylvestris*) from cultivated vines (*Vitis vinifera* ssp. *vinifera*)

Jean-Frédéric Terral

CNRS-UMR 5059, Paléoenvironnements, Anthracologie et Action de l'Homme
Institut de Botanique, 163, rue A. Broussonet - F-34090, Montpellier, France.
E-mail: terral@crit.univ-montp2.fr

Abstract

The origin of grapevine cultivation (*Vitis vinifera*) in the western Mediterranean remains enigmatic in spite of many archaeological, historical and genetic investigations. Without questioning the diffusionist model, it is clear that, in the current state of research, nothing contradicts the hypothesis of an indigenous emergence of vine cultivation and domestication in the western Mediterranean basin. In order to approach this problem, a preliminary study based on the search for discriminatory criteria between wild and cultivated vines was carried out in collaboration with the INRA-ENSAM grapevine collection ("Le Domaine de Vassal", France). Quantitative eco-anatomical data obtained show that it is possible to distinguish (1) wild forms from cultivated varieties, (2) wild individuals growing in natural conditions from wild cultivated individuals and, (3) immature wood from vine shoots and mature wood from old stems. In the near future, we aim to identify archaeological vine charcoal using these criteria and to contribute to a better knowledge and understanding of the history of grapevine cultivation.

Key-words: *western Mediterranean, vine cultivation, indigenous emergence, domestication, charcoal.*

Résumé

Les origines de la culture de la vigne (*Vitis vinifera*) en Méditerranée occidentale reste énigmatique, malgré les nombreuses recherches archéologiques, historiques et de génétique moléculaire. Sans remettre en cause le modèle diffusionniste, aucun argument ne s'oppose en l'état actuel des recherches à l'hypothèse d'une émergence autochtone de la culture et de la domestication de la vigne. Dans le cadre de cette problématique, une étude préliminaire fondée sur la recherche de critères anatomiques de discrimination entre la vigne sauvage de la vigne cultivée a été réalisée en collaboration avec le Conservatoire mondial de cépages de la station INRA-ENSAM ("Le Domaine de Vassal", France). Les résultats obtenus montrent qu'il est possible de distinguer (1) les formes sauvages des cépages, (2) les individus sauvages se développant en conditions naturelles des individus sauvages mis en culture et, (3) de différencier également le bois immature de sarments et le bois mature de la tige principale. A l'avenir, nous pourrons être capables d'identifier des fragments de charbon de bois dans la perspective de mieux contribuer à la connaissance et à la compréhension de l'histoire de la culture de la vigne.

Mots-clés : *Méditerranée occidentale, culture de la vigne, émergence autochtone, domestication, charbon de bois.*

Introduction

Distributed over Europe and around the Mediterranean basin in mesophilous vegetations, *Vitis vinifera* constitutes currently a complex of wild forms (*Vitis vinifera* ssp. *sylvestris*), cultivated varieties (cultivars or clones) classed as *Vitis vinifera* ssp. *vinifera* and feral forms. Consequently to human influence on primary formations and to the *Phylloxera* crisis of the XIXth century, populations of wild grape are confined to small forest or riparian zones (Mathon, 1981). In the Languedoc region (France), several individuals still remain near the Pic St Loup mountain, growing in a mixed-oak forest dominated by *Quercus ilex* (holm oak) and *Quercus pubescens* (pubescent oak) (fig. 1). In the Languedoc region, borders of rivers are often colonised by feral forms originating from pips disseminated by birds. These forms derive directly from cultivated individuals or hybridisation between cultivated clones, other feral forms and wild plants. Based on phenological criteria, distinction between wild and feral individuals remains difficult (see for example Viala and Vermorel, 1910 and Stummer, 1911 for actual *Vitis* and Mangafa and Kotsakis, 1996; Jacquat and Martinoli, 1999 for archaeological material). Nevertheless, native wild vines are dioecious and feral individuals have hermaphroditic flowers (see Huglin, 1986 for a review).

Together with the olive tree (*Olea europaea*) and fig (*Ficus carica*), the grapevine is historically and economically one of the most important fruit trees of the Mediterranean basin (Zohary and Spiegel-Roy, 1975; Van Zeist, 1992). The origin of vine cultivation is documented both by historical and archaeological data. Vine domestication is reported to have first started in Palestine, Syria and the Aegean area around 3200 BC. Thereafter, during the third millennium BC, viticulture has been developed, as discoveries of numerous remains of *Vitis* charred wood and pips and remnants of grape-decorated pottery and vessels for wine storage seem to indicate (Renfrew, 1973; Zohary and Hopf, 1993). From the eastern Mediterranean, viticulture practises and products were largely imported to the western part of the Mediterranean basin and to throughout Europe by Phoenicians, Greeks and Romans. Finally, beginnings of vine cultivation in France seem to date back to the last millennium BC (Van Zeist, 1980).

The history of viticulture is comparable to that of the olive. In fact, it was found that olive cultivation has occurred independently in several geographic areas outside the Near East (Besnard and Bervillé, 2000) and particularly in the north-western Mediterranean regions (Terral, 1999; Terral and Arnold-Simard, 1996). In regard to the wide distribution of wild vine grape around the Mediterranean basin and in Europe, the hypothesis of early vine cultivation in the north-western Mediterranean, long before the introduction of viticulture by classical cultures, may be put forward. Nevertheless, in the current state of research, it is impossible to confirm this hypothesis, on archaeological or ethnobotanical grounds. More generally, archaeobotanical discoveries alone, related to chronocultural periods will not allow us do deduce the appearance of cultivation and domestication. In this context, we have developed a new approach which may contribute in the future to the knowledge and understanding of the origins of vine cultivation, not only by means of archaeological and ecological research but also through botanical and biological investigations. In this way,

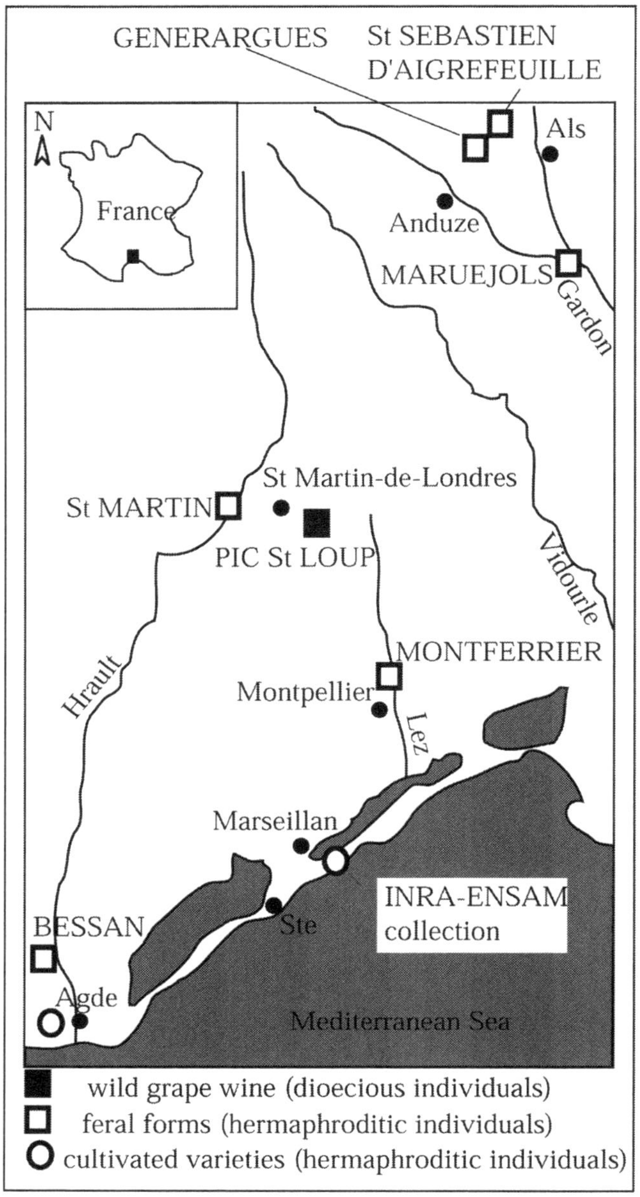

Figure 1: Geographic location of grapevine populations.

a comparative quantitative eco-anatomical study of modern grapevines was carried out in order to show the influence of cultivation on wood anatomy.

Material

The woody material sampled from living plants came from 12 localities (fig. 1): 2 located near the Pic St Loup mountain (France), 1 from Romania and 1 from the INRA-ENSAM collection for wild vine (tabl. 1); 6 concerned feral forms and 3 cultivated clones (tabl. 2). Samples from cultivated varieties were essentially collected at the INRA grape collection ("Le Domaine de Vassal", France).

Wood was collected from two or three different spots on each studied plant, in order to test the intra-individual variability in wood (fig. 2). This sampling protocol has provided 103 individual samples, 18 for the wild (7 plants), 38 for feral forms (13) and 47 for cultivated varieties (18). Finally,

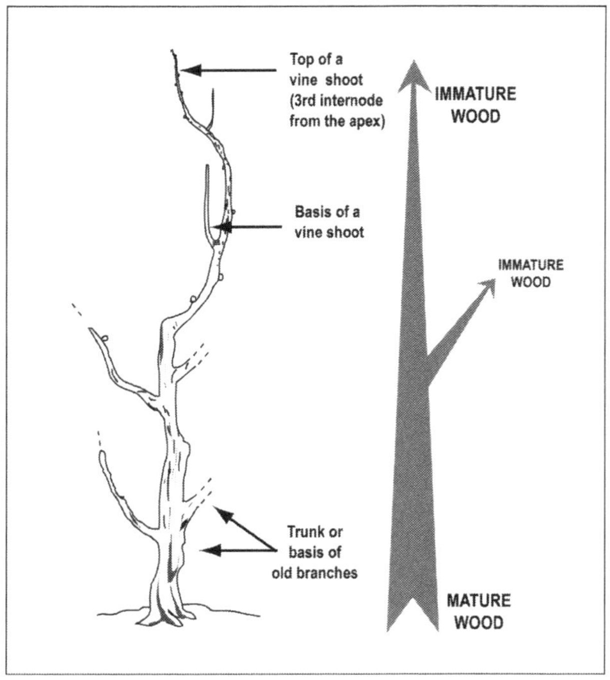

Figure 2: Wood sampling at different heights of the plant.

these samples were dried for several weeks and charred in an electric oven without oxygen at about 450°C.

Method

Charred samples were broken as to expose the transversal section and the anatomy of *Vitis vinifera* was observed using a reflected-light microscope connected to an image analysis system (OPTIMAS 5.2). Concerning qualitative characters, wild, feral and cultivated forms of grapevine have a similar a ring-porous wood (differentiation early - late-wood) (Schweingruber, 1978). Vessels are large and solitary in early-wood, while they are small and grouped in

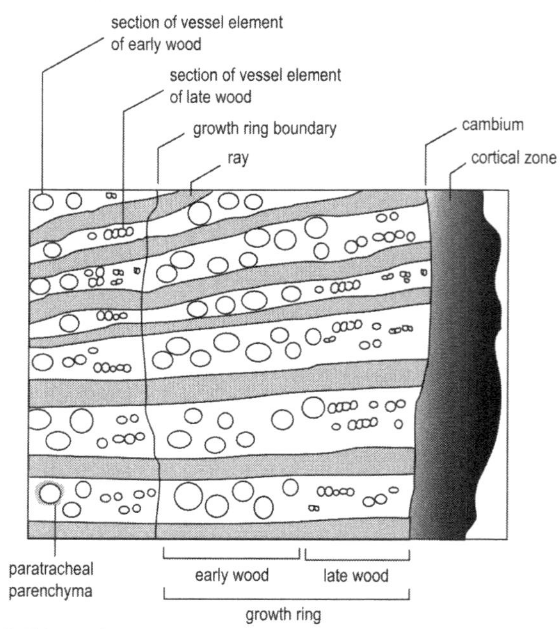

Figure 3: Schematic representation of the wood anatomy of vine and characters measured in this study.

Localities/nb. individual	Type of woody sample	Vessel area of early wood μm^2	Vessel area of late wood μm^2	Vessel densiy in early wood N / mm^2	Vessel densiy in late wood N / mm^2	Growth ring width µm	Width of early wood µm	Width of late wood µm	Width of ray µm	Frequence in ray
Pic St Loup 1	old branch	27084,41	1241,38	14,25	107,83	521,79	366,83	219,48	99,53	0,88
	basis of shoot	11899,28	1395,84	25,19	207,11	491,14	371,16	184,38	93,79	0,75
	top of shoot	6485,41	668,40	46,40	222,01	461,68	298,22	129,64	54,84	0,88
Pic St Loup 2	old branch	25642,36	1295,96	13,37	146,77	538,94	366,96	209,76	95,95	0,90
	basis of shoot	12662,72	875,90	25,77	219,72	391,49	247,69	110,63	80,26	0,86
	top of shoot	6315,10	454,81	53,58	212,82	350,04	159,77	190,27	57,03	0,86
Pic St Loup 3	old branch	10315,75	1048,78	26,03	182,43	456,76	258,02	178,18	77,37	0,86
	basis of shoot	8225,53	1359,40	33,40	125,22	845,90	686,85	264,55	43,62	0,88
	top of shoot	4151,80	1095,68	105,05	196,39	371,84	266,73	64,71	48,26	0,81
Pic St Loup 4	old branch	23012,52	1212,85	15,63	177,26	555,36	390,43	233,60	98,27	0,89
	basis of shoot	11025,62	989,63	30,58	209,63	752,01	568,30	282,31	79,35	0,78
	top of shoot	5468,39	702,96	70,29	212,85	404,95	261,58	113,71	50,25	0,80
Pic St Loup 5	old branch	20001,23	1125,91	18,05	122,02	600,12	408,62	233,57	85,26	0,91
	basis of shoot	10589,10	1098,96	25,63	189,42	599,26	379,14	169,34	58,02	0,85
	top of shoot	6221,25	621,03	87,59	196,36	402,62	183,77	218,85	52,11	0,82
Générargues 1	old branch	12273,34	656,39	17,84	182,87	916,77	864,68	207,06	90,28	0,93
	basis of shoot	9215,37	491,29	31,42	377,01	863,17	845,08	90,29	82,19	0,82
	top of shoot	3301,46	267,64	99,55	393,90	424,05	392,37	65,96	69,05	0,70
Générargues 2	old branch	14378,74	422,75	25,01	234,31	1099,85	960,80	138,20	85,54	0,95
	basis of shoot	5751,44	540,17	29,71	175,53	1071,63	885,89	187,14	89,70	0,89
	top of shoot	5059,19	568,69	67,23	250,02	547,95	440,85	68,78	89,96	0,69
Générargues 2	old branch	12032,26	396,26	31,59	150,27	745,46	523,71	164,83	95,78	0,89
	basis of shoot	6852,03	359,65	33,09	198,27	1153,77	998,38	188,66	84,03	0,85
	top of shoot	5001,90	380,01	60,13	212,59	933,68	780,05	91,35	77,09	0,71
Maruéjols 1	old branch	25117,45	1286,08	15,57	83,20	685,45	505,52	153,51	83,54	0,93
	basis of shoot	16035,75	817,29	28,68	262,04	531,09	449,49	175,22	82,60	0,90
	top of shoot	5549,36	363,70	63,58	165,95	465,03	425,47	62,36	53,64	0,40
Maruéjols 2	old branch	12292,70	618,94	22,30	242,21	641,22	467,25	152,37	86,50	0,90
	basis of shoot	4840,04	239,06	97,57	291,11	429,61	348,91	63,06	84,65	0,62
	top of shoot	4607,76	276,90	88,83	297,07	406,36	328,70	78,00	64,12	0,58
St Sébastien	old branch	10680,78	648,00	21,57	174,63	702,02	423,59	182,74	88,72	0,96
d'Aigrefeuille 1	basis of shoot	5425,00	499,43	51,26	200,23	539,38	429,22	87,09	50,64	0,62
	top of shoot	1999,90	432,30	130,49	402,90	404,90	372,81	36,17	53,50	0,56
St Sébastien	old branch	14695,82	864,24	25,82	143,40	940,77	830,74	275,24	83,74	0,70
d'Aigrefeuille 2	basis of shoot	3264,46	277,93	45,53	278,41	764,61	692,75	76,59	66,09	0,60
	top of shoot	2845,45	250,01	54,12	311,58	398,29	328,36	70,40	59,63	0,57
Monferrier 1	old branch	22140,98	1099,72	16,21	176,90	364,70	262,55	140,61	109,41	0,90
	basis of shoot	8730,52	766,53	56,34	184,86	609,02	522,41	92,15	74,22	0,82
	top of shoot	8446,77	748,45	38,74	220,31	1062,05	974,37	157,67	58,48	0,72
Monferrier 2	old branch	17891,79	1361,11	18,38	139,60	616,08	350,01	203,40	93,74	0,89
	basis of shoot	9609,05	713,15	52,42	322,41	1025,53	772,08	288,73	70,58	0,85
St Martin 1	old branch	19250,36	950,54	28,56	150,29	714,26	673,67	101,32	100,23	0,95
	basis of shoot	10256,28	699,25	44,56	200,28	658,01	644,22	68,83	80,12	0,66
	top of shoot	5200,20	396,01	58,02	220,63	499,02	461,74	77,62	50,25	0,54
St Martin 2	old branch	18527,23	1100,58	18,02	174,05	812,29	709,60	102,07	90,28	0,77
	basis of shoot	9656,39	801,26	39,99	199,28	777,45	642,70	105,77	56,67	0,65
	top of shoot	2566,67	412,25	50,11	222,39	401,39	322,94	50,38	38,36	0,54
Bessan1	old branch	17823,15	888,01	20,21	151,85	910,28	639,50	201,27	89,99	0,95
	basis of shoot	8999,65	765,39	46,96	180,29	840,36	727,18	107,41	70,15	0,74
	top of shoot	2800,36	552,32	58,25	210,96	502,28	419,63	49,14	50,12	0,49
Bessan 2	old branch	20054,82	985,28	23,58	128,36	845,69	623,70	189,40	100,28	0,76
	basis of shoot	9025,38	555,17	84,01	180,88	701,37	593,61	131,40	80,29	0,67
	top of shoot	3890,33	485,02	97,85	201,96	396,32	298,26	53,15	44,01	0,60
Roumania*	old branch	16429,20	386,51	18,31	204,73	629,74	451,78	172,94	85,94	0,86
INRA-collection	basis of shoot	12302,16	247,62	30,22	455,67	1352,94	1213,50	143,30	60,68	0,50
	top of shoot	4877,29	291,19	102,04	514,87	626,70	529,51	74,39	57,85	0,56

Table 1: Quantitative eco-anatomical results from the analysis of samples from wild and feral vines.

radial lines in the late-wood. Vine wood has paratracheal parenchyma and, multiseriate and heterogeneous rays.

According to the anatomy of grapevine wood (fig. 3), 9 characters were measured in the transverse section, following a radial line and crossing several growth rings. For each sample, several measurements were necessary in order to obtain a good estimation of each criterion as is shown in figure 4: growth ring width (GRW, µm, 20 measurements); early-wood width (EWW, µm, 20 measurements); late-wood width (LWW, µm, 25 measurements); vessel area of early-wood (VAE, µm, 30 measurements); vessel area of late-wood (VAL, µm, 25 measurements); vessel density in early-wood (VDE, number of vessels per square mm, 15 measurements); vessel density in late-wood (VDL, N/mm, 10-15 measurements); ray width (RW, µm, 15 measurements); frequency of rays (FR, number of rays per radial line of vessel, 10 measurements).

Data from the quantitative eco-anatomical analysis were treated by principal component analysis (PCA), carried out on 103 samples and including 9 quantitative variables (9 anatomical criteria).

Geographical location	Variety	Type of wood sample	Vessel area of early wood μm^2	Vessel area of late wood μm^2	Vessel densiy in early wood N / mm^2	Vessel densiy in late wood N / mm^2	Growth ring width μm	Width of early wood μm	Width of late wood μm	Width of ray μm	Frequence in ray
INRA collection	Aetonikis	basis of shoot	1939,75	509,88	81,05	165,45	1864,02	1751,93	271,17	82,29	0,55
INRA collection	Aetonikis	top of shoot	1015,78	339,90	183,70	365,43	598,53	100,97	416,24	68,36	0,47
Agde	Aramon	old branch	10025,33	620,28	25,03	178,10	758,21	550,06	200,90	74,23	0,50
INRA collection	Aramon	old branch	12032,56	599,25	26,68	188,03	548,26	425,82	120,87	71,09	0,48
INRA collection	Aramon	basis of shoot	7465,57	567,49	48,91	190,88	1909,30	1699,34	337,43	131,23	0,53
INRA collection	Aramon	top of shoot	1345,32	347,25	184,63	281,36	515,24	380,92	103,76	67,35	0,47
INRA collection	Asmi noir	basis of shoot	10656,35	731,23	22,14	204,57	1731,54	1409,81	143,44	81,39	0,71
INRA collection	Asmi noir	top of shoot	1427,29	185,39	167,68	451,28	404,08	284,09	129,11	63,24	0,62
Agde	Carignan	old branch	7906,86	395,34	23,59	221,96	1500,93	896,33	623,25	95,67	0,65
INRA collection	Carignan	old branch	8534,62	451,48	21,58	229,29	2041,91	1081,63	988,05	101,27	0,75
INRA collection	Carignan	basis of shoot	5305,42	368,50	46,97	247,51	1655,04	1288,79	375,17	114,18	0,64
INRA collection	Carignan	top of shoot	1318,51	186,06	203,61	434,45	496,61	352,71	158,52	61,04	0,38
INRA collection	Cinsaut	old branch	11850,67	789,06	30,00	200,73	901,56	685,39	282,04	89,59	0,49
INRA collection	Cinsaut	basis of shoot	9313,68	849,43	26,88	183,65	2526,08	2279,65	298,12	129,34	0,78
INRA collection	Cinsaut	top of shoot	3527,57	444,25	99,25	364,43	416,76	357,68	71,74	69,06	0,64
INRA collection	Corinthe noir	basis of shoot	18795,75	1239,46	22,22	95,21	1064,06	909,95	116,47	73,56	0,54
INRA collection	Corinthe noir	top of shoot	4986,91	276,76	62,85	300,00	501,27	385,27	106,06	69,97	0,41
Agde	Grenache	old branch	10253,33	510,85	32,23	252,05	699,02	510,56	105,96	78,99	0,45
INRA collection	Grenache	old branch	11590,40	500,29	30,24	199,80	783,64	655,56	105,68	82,30	0,48
INRA collection	Grenache	basis of shoot	8543,52	401,43	38,28	214,54	2086,15	1734,52	358,05	99,25	0,50
INRA collection	Grenache	top of shoot	3961,60	305,17	84,92	363,99	881,68	601,87	268,35	67,76	0,60
INRA collection	Gros vert	basis of shoot	3480,59	404,48	79,15	186,37	908,05	729,37	107,84	94,05	0,62
INRA collection	Gros vert	top of shoot	2328,89	402,84	126,61	270,68	669,97	482,35	177,96	68,05	0,85
INRA collection	Macabeu	basis of shoot	8675,05	632,71	33,61	169,90	2274,86	2107,49	159,48	83,82	0,60
INRA collection	Macabeu	top of shoot	2673,81	267,67	82,28	211,24	770,14	574,37	66,76	178,98	0,42
INRA collection	Marsanne	basis of shoot	5447,09	467,31	36,98	206,91	2018,35	1816,14	231,40	88,97	0,70
INRA collection	Marsanne	top of shoot	4626,98	237,71	57,35	327,67	1091,72	936,52	159,61	71,18	0,60
Agde	Mourvèdre	old branch	6012,89	550,19	33,96	150,85	801,96	611,26	185,20	68,20	0,61
INRA collection	Mourvèdre	basis of shoot	3254,28	547,97	45,41	152,83	1558,35	1344,38	269,01	71,04	0,56
INRA collection	Mourvèdre	top of shoot	4684,16	698,04	58,27	205,47	1287,07	1045,38	154,37	70,12	0,55
INRA collection	Muscat d'Alexandrie	basis of shoot	5162,65	408,78	51,07	211,27	1462,80	1324,78	138,84	74,39	0,50
INRA collection	Muscat d'Alexandrie	top of shoot	4925,69	380,92	58,29	297,12	780,95	662,48	76,96	52,20	0,64
INRA collection	Nebbiolo	basis of shoot	9350,80	460,55	42,68	305,81	1561,72	1381,80	134,00	73,79	0,60
INRA collection	Nebbiolo	top of shoot	1610,41	73,79	228,97	569,29	434,28	347,31	69,61	65,04	0,42
INRA collection	Sangionese	basis of shoot	6754,68	423,96	70,48	475,00	1062,46	917,26	168,24	76,86	0,67
INRA collection	Sangionese	top of shoot	4336,29	392,76	102,78	460,10	734,87	619,18	117,43	71,54	0,63
INRA collection	Sultanine	old branch	8382,53	798,04	29,48	120,90	803,89	454,18	477,08	87,78	0,69
INRA collection	Sultanine	basis of shoot	6484,76	892,98	40,40	150,72	2679,11	2026,69	591,05	109,92	0,60
INRA collection	Sultanine	top of shoot	3595,54	531,44	107,72	239,61	1091,58	834,03	230,79	80,95	0,62
INRA collection	Syrah	old branch	6101,87	764,77	53,15	115,18	815,78	543,65	290,16	118,60	0,50
INRA collection	Syrah	basis of shoot	3558,32	675,11	50,20	166,86	1949,30	1512,06	359,02	67,88	0,50
INRA collection	Syrah	top of shoot	3824,92	398,14	58,03	244,88	1319,83	1129,48	193,27	50,45	0,63
INRA collection	Vermentino	basis of shoot	8454,14	730,05	25,15	122,13	2948,16	2396,04	495,54	95,07	0,78
INRA collection	Vermentino	top of shoot	1857,05	168,02	168,62	349,30	488,96	347,89	71,72	59,53	0,57
Montjézieu*	Negret	old branch	7913,65	594,93	37,86	324,62	707,04	495,47	172,66	87,58	0,70
Montjézieu*	Negret	top of shoot	4759,60	436,37	76,77	309,78	644,38	557,62	142,92	70,55	0,57
Montjézieu*	Negret	basis of shoot	3344,12	207,75	87,22	314,10	598,01	542,25	59,33	61,52	0,57

Table 2: Quantitative eco-anatomical results from analysis of samples from cultivated vines.

Results and discussion

Statistical analysis

The multivariate statistical analysis was carried out on 97 samples (samples located in the geographical area of our study, fig. 1). Six samples (from a Romanian wild plant, a cultivated wild vine and a clone abandoned for about 30 years) were used as additional statistical individuals. They were compared with individuals according to their distribution in the PCA.

On the principal axes 1 and 2, the significant variables contributing to the representation of individuals include 3 groups of 2 variables that are strongly correlated: growth ring width (GRW) - width of early-wood (EWW); vessel area in early-wood (VAE) - vessel area in late-wood (VAL); vessel density in early-wood (VDE) - vessel density in late-wood (VDL) (fig. 5).

Two groups of individuals are clearly separated by GRW and EWW on axis 2 (27.9 % of inertia): the cultivated vine group including samples with large growth rings and the group composed by wild and feral vines (fig. 5). Among the second group (wild type), it is interesting to note the discrimination according to the axis 1 (41 % of inertia) between mature woody samples from old branches and immature wood from shoots. The criteria "vessel density" and "vessel area" contribute to this discrimination. Samples from old branches have large vessels and low vessel density whereas samples from shoots are characterised by narrow vessels and a higher vessel density (fig. 5).

Moreover, taking into account this last result, we have carried out on the cultivated individuals (N=47) a new principal component analysis in order to test the variability of anatomical characters in relation to wood maturity (Fig. 6). Results of this statistical analysis showed that the character "vessel area in early-wood" separates immature wood samples (narrow vessels) from mature woody samples (large vessels).

Finally, anatomical measurements from additional individuals were confronted to the model provided by the first

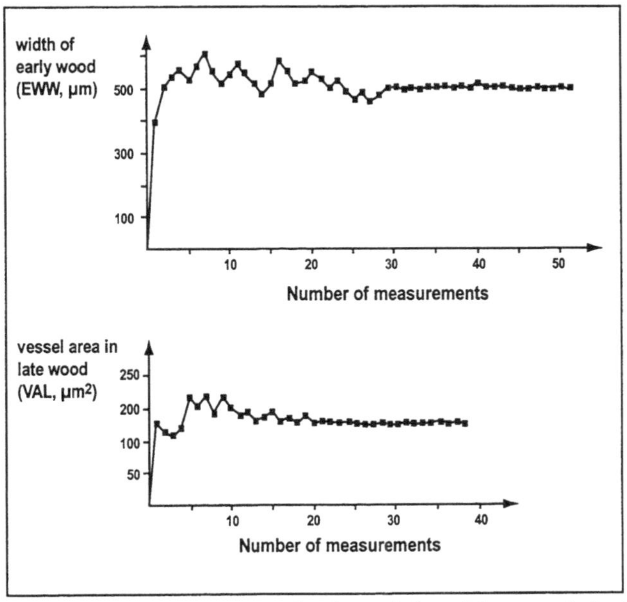

Figure 4: Repeated mean curves showing variation of the anatomical criteria "width of early wood" and 'vessel area in late wood' in relation to the number of measurements.

PCA (fig. 7). The two samples from a wild plant cultivated at the INRA-ENSAM collection was classified as belonging to the cultivated vine group. The single sample of a Romanian wild vine and the three specimen from a variety uncultivated since 30 years were affiliated to the wild-type group.

Interpretation

Throughout the life of a tree, its wood records the influence of various factors, natural or anthropogenic, intrinsic or extrinsic, accidental or otherwise. Human influence on plants such as cultivation practises, is a parameter which also modulates anatomical characteristics of the tree (see for example Terral and Arnold-Simard, 1996).

This present study has shown that cultivation practises affect the grapevine wood. "Growth width ring" and "width of early-wood" allow a discrimination between wild and cultivated *Vitis*. Favourable conditions and practices of cultivation such as fertilisation, adequate spacing, thinning, pruning and irrigation permit a greater growth and development of a tree compared to the wild individuals growing in "natural" conditions. Moreover, feral forms and wild vines cannot be separated on the basis of their anatomy, and it is interesting to note that native wild individuals present cultivated anatomical characteristics when they grow in similar conditions to those prevailing under cultivation. Cultivars that have been abandoned and thus grow under conditions without human influence acquire the anatomy of wild individuals. This last results seems to confirm that variation in size of "growth ring width" and "width of early-wood" are modulated by environmental factors and do not record genetic parameters (selection practises).

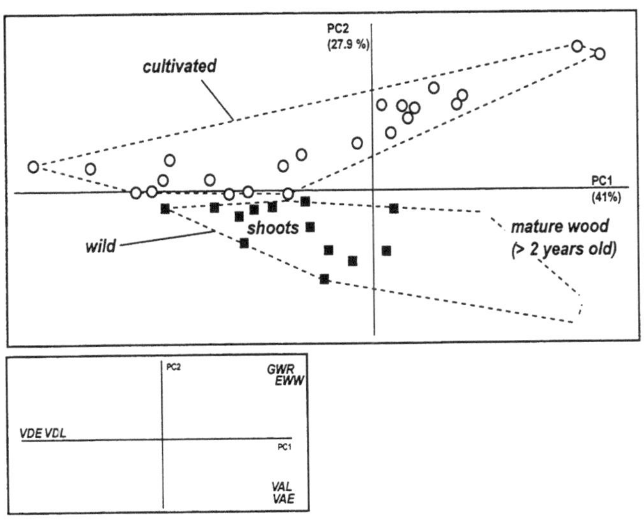

Figure 5: Representation of Principal Component Analysis (PCA) of vine wood samples in principal plane 1-2.

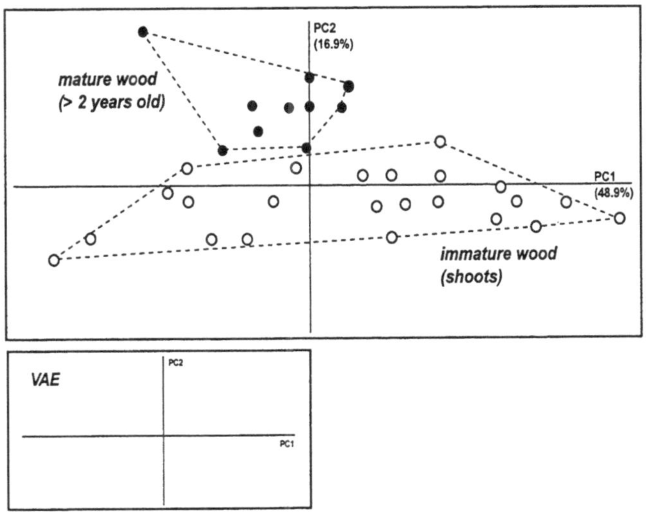

Figure 6: Representation of PCA of wood samples from cultivated vines in principal plane 1-2.

The other criteria such as "vessel density" and "vessel area" are on the contrary characteristic of the stage of maturity at which wood was sampled. In our case, vessels are narrower in immature wood of vine shoots than in mature wood of old stems and roots (fig. 6). This seems to represent an increase in volume of the water conducted in vessels, from root to twigs (Carlquist, 1988). Vine wood shows a decrease in "vessel density" and "vessel area" for wild individuals and a decrease in "vessel area in early wood" for cultivated specimens. Variations of these two criteria have allowed us

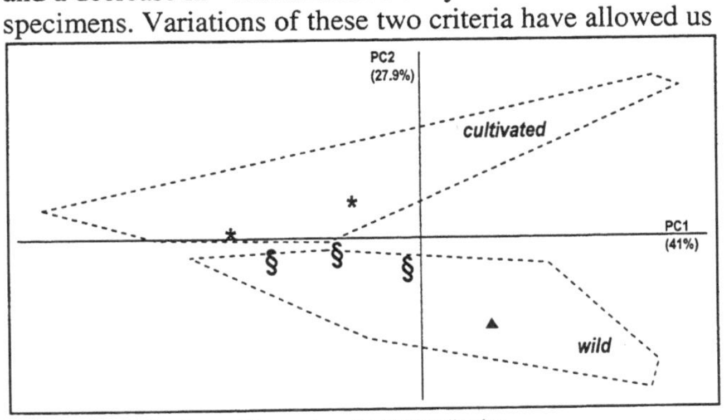

* *cultivated dioecious wild grape from the INRA collection*
▲ *wild grape from Romania (ripicolous forest)*
§ *varieties uncultivated since 30 years*

Figure 7: Comparison of additional wood samples to the statistical model obtained from PCA carried out on vine wood samples in principal plane 1-2.

to distinguish, between shoots and old stems both in wild and cultivated vine wood.

Conclusion and perspectives

The quantitative eco-anatomical analyses carried out allowed us to (1) discriminate wild forms (native and feral forms) from cultivated varieties (cultivars), (2) differentiate specimens according to their state of wood maturity (shoots and old stems). These results have shown that anatomical differences are not genetically fixed but determined by cultivation practises. In this study, each sample of the reference collection comes from a restricted region; it was clearly shown that wood anatomical differences observed were not a result of climatic conditions. In our case, variable wood anatomy characteristics seem to be the result of anthropogenic factors (cultivation).

In the near future, the application of this method to a larger sample from other geographical areas may allow us to characterise more accurately quantitative anatomical differences between wild and cultivated individuals, all while considering the different positions of the wood samples in the plant and ecological parameters. From an archaeological, ethnobotanical and historical perspective, we will be able to identify with precision archaeological charcoal. Then, through biological models established on the basis of modern populations, we will be able to contribute to the knowledge and the understanding of the history of grapevine cultivation.

By combining biological, cultural, archaeological and palaeoenvironmental studies, we will investigate more fully the agricultural societies, their functioning (in a socio-cultural sense, relations with others societies, exchanges and trade) and their mechanisms of production (agricultural systems, farming practises, treatment of products and storage).

ACKNOWLEDGEMENT
I am grateful to Thierry Lacombe (INRA-ENSAM, Domaine de Vassal, Marseillan Plage, France) for his help and assistance in the elaboration of the extant reference collection of cultivated grapevine wood and Lucie Chabal for her collaboration in the search for wild vines. I wish to thank K. Neumann and G. Willcox for their constructive comments.

References

BESNARD G., BERVILLE A. 2000 - Multiple origins for Mediterranean olive (*Olea europaea* L. ssp. *europaea*) based upon mitochondrial DNA polymorphisms. *Comptes Rendus de l'Académie des Sciences*, Paris, Sciences de la vie, 323, p. 173-181.

CARLQUIST S. 1988 - *Comparative wood anatomy. Systematic, Ecological and evolutionary aspects of Dicotyledon wood.* Berlin - Heidelberg, Springer Verlag, 436 p.

HUGLIN P. 1986 - *Biologie et écologie de la vigne.* Lausanne, Payot, 372 p.

JACQUAT C., MARTINOLI D. 1999 - *Vitis vinifera* L.: wild or cultivated? Study of the grape pips found at Petra, Jordan ; 150 BC - AD 40. *Vegetation History and Archaeobotany*, 8, p. 25-30.

MANGAFA M., KOTSAKIS K. 1996 - A new method for the identification of wild and cultivated charred grape seeds. *Journal of Archaeological Science*, 23, p. 409-418.

MATHON C.C. 1981 - *L'origine des plantes cultivées.* Phytogéographie appliquée. Paris, Masson, p. 109-114.

RENFREW J.M. 1973 - *Palaeoethnobotany. The prehistoric food plants of the Near East and Europe.* London, Methuen & Co Ltd, 248 p.

SCHWEINGRUBER F.H. 1978 - *Mikroskopische Holzanatomie.* Zürcher, Eidgenössische Anstalt für das fortliche Versuchswesen, 226 p.

STUMMER A. 1911 - *Zur Urgeschichte der Rebe und des Weinbaues.* Mitteilungen der Anthropologischen Geysellschaft in Wien, 41, p. 283-296.

TERRAL J.-F. 1999 - La morphométrie à la recherche des origines de la culture et de la domestication des arbres fruitiers : Le modèle de l'olivier en Méditerranée nord-occidentale. *Revue d'Archéométrie*, 23, p. 101-126.

TERRAL J.-F., ARNOLD-SIMARD G. 1996 - Beginnings of olive cultivation in relation to Holocene bioclimatic changes. *Quaternary Research*, 46, p. 176-185.

VAN ZEIST W. 1980 - Aperçu sur la diffusion des végétaux cultivés dans la région méditerranéenne. In Colloque de la fondation L. Emberger. "La mise en place, l'évaluation et la caractérisation de la flore et de la végétation circum-méditerranéenne". *Naturalia Monspeliensa*, n° hors série, p. 129-145.

VAN ZEIST W. 1992 - *The origin and development of plant cultivation in the Near East.* Japan Review, 3, p. 149-165.

VIALA P., VERMOREL V. 1910 - *Ampélographie.* Tome I. Paris, Masson.

ZOHARY D., SPIEGEL-ROY P. 1975 - Beginnings of fruit growing in the Old World. *Science*, p. 319-327.

ZOHARY D., HOPF M. 1993 - *Domestication of plants in the Old World.* Oxford, Clarendon Press, 2nd edition. 279 p.

Anthracology and radiochronology of the Upper Pleistocene in the loessic areas of Eurasia

Freddy Damblon and Paul Haesaerts
Institut Royal des Sciences Naturelles de Belgique
Rue Vautier, 29, B-1000 Bruxelles, Belgique.
E-mail: damblon@naturalsciences.be

Abstract
The climatic oscillations of the Upper Pleistocene can be recognised in the stratigraphy of the Eurasian loess in the form of various palaeosoils. The chronology of their succession and the reconstruction of past glacial environments can be supported by anthracological analysis. The following paper exposes some methodological problems linked to the choice of informative deposits, in archaeological as well as in natural contexts, to the extraction, the preparation, the identification and the selection of the charcoal material intended for radiocarbon dating. Finally, an overview of the results obtained from western Europe to central Siberia is presented; they concern on the one hand the stratigraphic and geographical distribution of taxa and on the other hand their chronostratigraphical distribution.

Key-words: *Upper Pleistocene, Eurasian loess, Anthracological analysis.*

Résumé
Les oscillations climatiques du Pléistocène supérieur sont reconnaissables dans la stratigraphie des formations loessiques de l'Eurasie sous forme de divers paléosols. La chronologie de leur succession et la reconstitution des environnements trouve un support essentiel dans l'analyse anthracologique. L'article expose quelques problèmes méthodologiques liés au choix des sites, tant archéologiques que naturels, ainsi qu'à l'extraction, la préparation, l'identification et la sélection des charbons de bois destinés aux datations radiocarbones. Il donne ensuite une vue générale des résultats obtenus depuis l'Europe occidentale jusqu'en Sibérie centrale et qui concernent d'une part la distribution stratigraphique et géographique des taxons et d'autre part leur distribution chronostratigraphique.

Mots-clés : *Pléistocène supérieur, Eurasie, Loess, Anthracologie.*

Introduction

In the course of the Upper Pleistocene (Isotopic Stages 5 to 2), important deposits of loess accumulated on the Eurasian continent, more especially between western Europe and central Siberia. In these deposits, organic material suitable for dating may abound but is unequally distributed in time and space. The aim of the research undertaken in this area since 1993 is the identification of the successive climatic oscillations during the Upper Pleistocene, their dating and the reconstruction of the palaeoenvironments in the continental loessic field.

Methodology

The basic principle consists in selecting long sequences that include enough climatic markers and high quality organic material for dating. These conditions were often found within some palaeolithic sites rich in charcoal and bones and in some particular geomorphological conditions, within natural deposits.

In the loess-like deposits, the analysis of fossil pollen appeared rather deficient, notably because of the difficulty to control the origin of the various supplies of pollen, in part of local origin, but often reworked, corroded or bioturbated. However, locally, as in Kurtak (south of Krasnoyarsk, central Siberia: fig. 1b, n° 14), some particular conditions exceptionally allowed the conservation of local palynological records in excellent agreement with xylological and anthracological material (in preparation).

Thus, the study of the organic material (charcoal and bone) was undertaken with the main objective of ^{14}C dating and as a complement to the reconstruction of past environments. Unfortunately the chronology of loess deposits is hampered by a recurrent lack of reliable radiocarbon dates. A minute examination of the results in the literature and correspondence concerning the sites in this project, led to the conclusion that a lot of dates remained dubious for various reasons such as the position of the sample in the stratigraphy; a contamination of the dated material by various agents like rootlets or humics; an intrusion of recent charred bone or charcoal in the fossil assemblage; an injudicious selection of material, for example porous bone, for dating; mixed samples with charcoal and charred bone debris; an insufficient quantity of carbon material destined to radiometric measurements; some errors in calculation or transcription of the results.

In order to avoid such drawbacks and to aim at a maximum of reliability and precision of radiocarbon dates, the following methodological rules were adopted: precise knowledge of the microstratigraphy and geometry of the deposits in each station or site; minute sampling of the carbon material, preferably on sections or bankets, from a well identified stratigraphic unit (sampling from excavation surfaces was avoided to lower the risk of mixing successive layers); selection of the best material for dating, essentially charcoal or eventually dense bone; implementation of a procedure for the physical and chemical treatment of the material leading to elimination of the pieces contaminated by rootlets, to the preparation of the purest and the least contamined sample for dating; taxonomic identification of the material; selection of the best dating system (conventional or AMS) adapted to the available weight of carbon, to the type of material and its state of conservation; radiometric measurements following standard procedures in the same laboratory (Groningen) using efficient equipment; and finally critical analysis of the results with regard to the sequential stratigraphy, the palaeoenvironment and other chronological data. In this perspective, a strict procedure was systematically adopted to gain the reliability and precision goal (Damblon and Haesaerts, 1997).

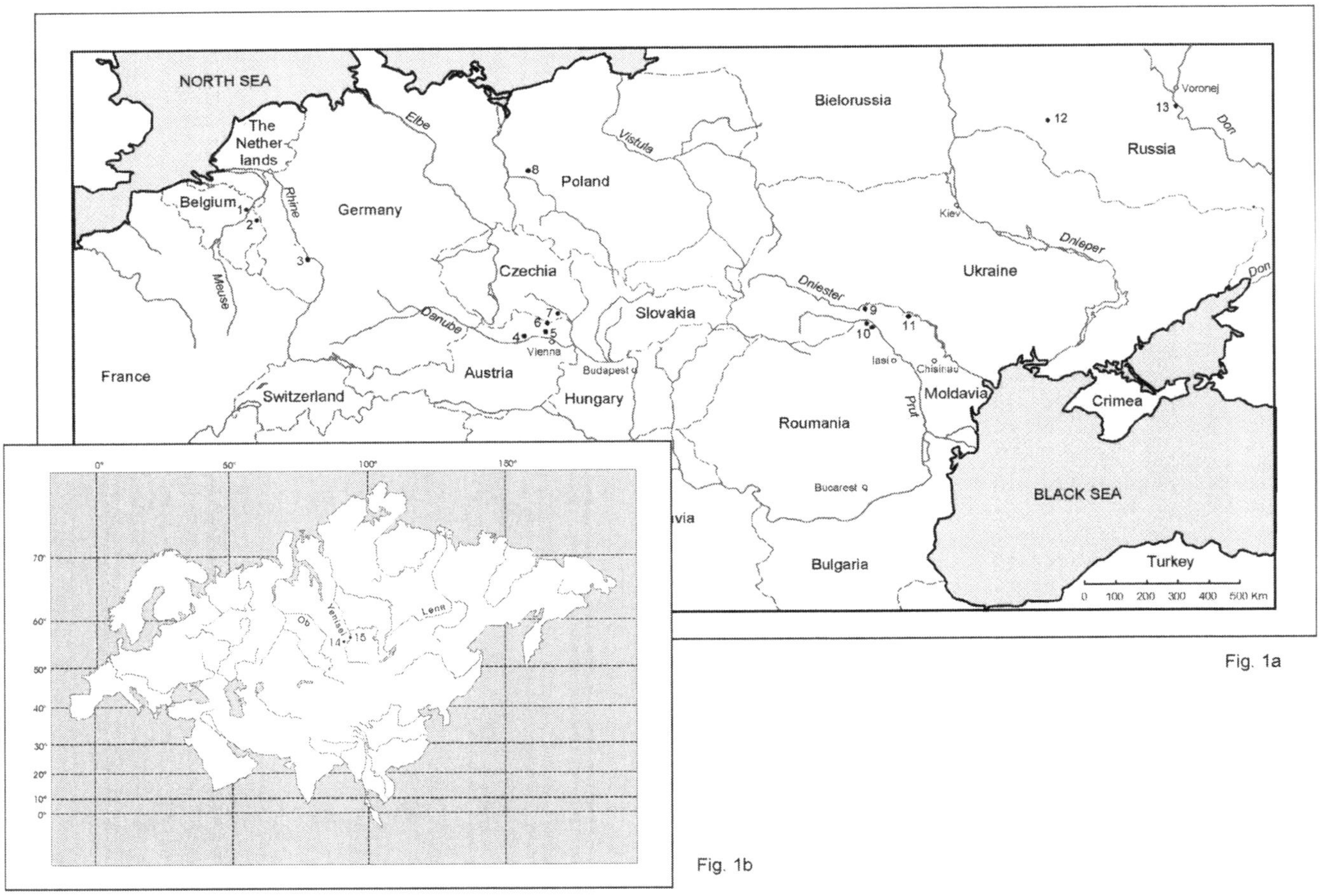

Fig. 1a

Fig. 1b

Figures 1a-b: Distribution of the sites having provided anthracological data in the loessic belt of Eurasia.
Fig. 1a: n° 1: Remicourt, Rocourt and Hezerwater, n° 2: Trô-Maret, n° 3: Wallertheim, n° 4: Willendorf and Schwallenbach, n° 5: Alberndorf, n° 6: Hollabrunn, n° 7: Dolni Vestonice and Pavlov, n° 8: Siedlnica, n° 9: Molodova, n° 10: Mitoc Malu Galben and Crasnaleuca, n° 11: Cosautsi, n° 12: Zeleznogorsk, n° 13: Kostienki.
Fig.1b. n° 14: Kurtak , n° 15: Listvenka and Afontova Gora.

In this present paper, it is not possible to give all details on the charcoal analyses but, in the perspective of gaining safe radiometric data, it can be pointed out that effort was focused on collection of charcoal from sections, in a precise stratigraphic position, rather than on extensive sampling from wide surfaces (Damblon *et al.* 1996). This is the reason why table 2 gives only the presence of the various taxa by period.

The sites

According to the criteria exposed above, and among the 54 sites investigated, about 20 Upper Pleistocene sites have been selected for charcoal analysis and ^{14}C dating in the vast loessic domain of Eurasia (fig. 1a and b) comprising Belgium (Meuse basin), Germany (middle Rhine basin), Austria and Czech Republic (middle Danube basin), the East-Carpathian area (Prut and Dniester basins), the great Russian plain (Don and Dnieper basins), and finally central Siberia (Yenisei basin).

The choice of the anthracological material and its exploitation for dating

The majority of the published radiocarbon dates from the loessic sites of the Upper Pleistocene has been obtained on bone and charcoal (see notably: Soffer 1985, 1986; Street *et al.* 1994; Svezhentsev, 1993; Damblon *et al.* 1996; Damblon and Haesaerts, 1997). Besides the problems of stratigraphy, the precision and reliability of the results presented and discussed by the different authors vary considerably due to various factors depending on the quality of the material preserved and on geochemical interactions with the sediment (Mook and Waterbolk, 1985; Bowman, 1990; Lanting and van der Plicht, 1990; Damblon and Haesaerts, 1997).

In a general manner, the question of compatibility of the dates on bone and charcoal is questioned, considering that dating humates from pedological horizons appears too hazardous. In order to illustrate the complexity of the problem, two extreme opposite examples will be exposed, the first from Kostienki, the second from Cosautsi.

Kostienki is a large Upper Palaeolithic site, situated on the Don river and comprising not less than 21 local stations (Praslov and Rogachev, 1982). At Kostienki 1, the Cultural layer I (Gravettian) provided numerous mammoth teeth and bones from which 43 radiocarbon dates running from about 24,000 to 18,000 BP (fig. 2) were obtained (Praslov and Rogachev, 1982; Svezhentsev, 1993; Sinitsyn, 1996; Sinitsyn *et al.* 1997). This wide range of dates complicates

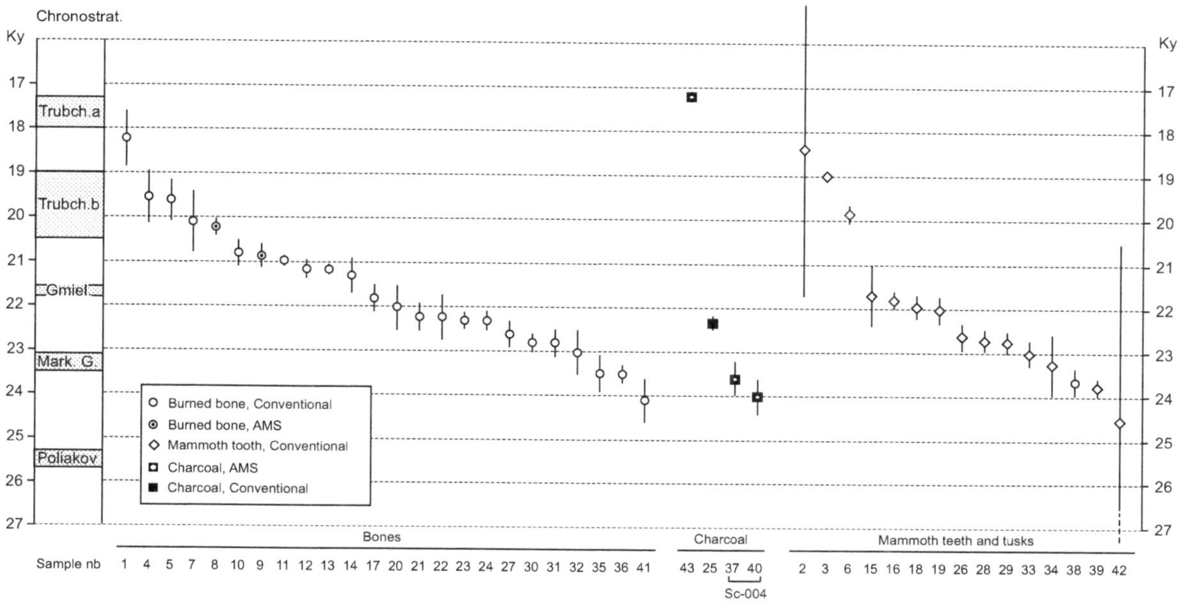

Figure 2: Kostienki 1, Cultural layer I (central Russia). Distribution of the ^{14}C dates on bones, teeth and charcoal. The AMS-dates nb 37 and 40 on charcoal are close to the oldest dates on bone material. The date nb 25 comes from an unsorted charcoal sample. On the contrary, the date nb 43 was obtained on a selection of *Picea* charcoal but from no clear stratigraphic position.

the chronological interpretation of the Cultural layer. Therefore, two samples of charcoal well positioned at the base of Cultural layer I by the excavator (N. Praslov) were identified by us and a selection of *Picea* pieces were submitted to AMS-dating at the Groningen laboratory (fig. 2, n° 37 and 40). The results from selected *Picea* charcoal pieces confirm the oldest ages obtained on teeth and bones. How can we explain the wide spread of the dates on mammoth material? Amongst various factors that could have influenced the radiocarbon age of bone and tooth material, the position of Layer I just below the present chernozem appears as a major source of contamination of the collagen by humates difficult to eliminate totally in the laboratory. On the contrary, the chemical treatment of the charcoal with sodium hydroxide permits the total elimination of the contaminants from the inert carbon. Consequently, the two dates around 24,000 BP on charcoal confirm the oldest ^{14}C ages for the Cultural layer I of Station 1. Nevertheless, two other dates on charcoal are unreliable due to unsorted material (n°25) or to unclear stratigraphic origin (n°43).

The opposite example comes from the Upper Palaeolithic site of Cosautsi, situated on the Dniester in the Republic of Moldova (Borziak 1991, 1993; Haesaerts *et al.* 1998). In the present perspective, the main interest of the site lies in a 14 m sequence of loess and sandy loess that includes 21 Gravettian successive occupation layers, essentially between 12 m and 8 m in depth. Such exceptional conditions have favoured a good conservation of charcoal and bone material, preserved from any pollution and disturbance. Therefore, our team collected in parallel new samples of charcoal and of bones in large borings following the successive Cultural layers 9 to 2c which were dated to between 19,410 and 17,620 BP (fig. 3). The dates obtained at Groningen on selected *Picea* charcoal pieces (Damblon *et al.* 1996; Damblon and Haesaerts, 1997) and at Oxford on reindeer dense bones (Otte *et al.* 1996) show coherent and comparable ages with the same general line of distribution (fig. 3), even if the line of dates on bones appears more crushed that the one on charcoal. A slight contamination of collagen by humates could probably explain this phenomenon. A downward shift of certain dates on charcoal because of an "old wood effect" seems hardly perceptible, considering the age of the samples and the errors of measurement in the laboratory. Anyway, these results show that excellent ^{14}C dates can be obtained on bones when they are sufficiently dense and when the collagen was kept safe from contamination. Still, cross dating with charcoal is needed to control the reliability and accuracy of the dates on bone.

From these considerations it results that charcoal appeared as the most reliable material for radiocarbon dating because of its elevated content in uncontaminated or, in any case, purifiable inert carbon. Moreover, charcoal may be rather easily identified and consequently taxonomically selected for dating (in the present case with a preference for conifer wood), a process which further improves the reliability and accuracy of the ^{14}C dates.

Identification of charcoal in samples

Today, the charcoal samples collected in loessic archaeological context and submitted to dating are rarely taxonomically identified. And yet, this process is essential for loess deposits in order to detect possible intrusive contamination. A good example of such disturbance comes from Cultural layers I and III of Kostienki 12 in which we discovered charcoal remains of broad-leaved deciduous trees (*Quercus, Fraxinus*, Malaceae) with charcoal of conifers (*Picea, Pinus*). Such a mixture could have important palaeoecological and climatic implications. Therefore, deciduous and conifer charcoal were submitted to separate dating (tabl. 1). The results clearly showed that charcoal from deciduous tree species must be related to the Holocene whilst conifer charcoal provided ages in clear accordance with the site's pleniglacial sequence. These

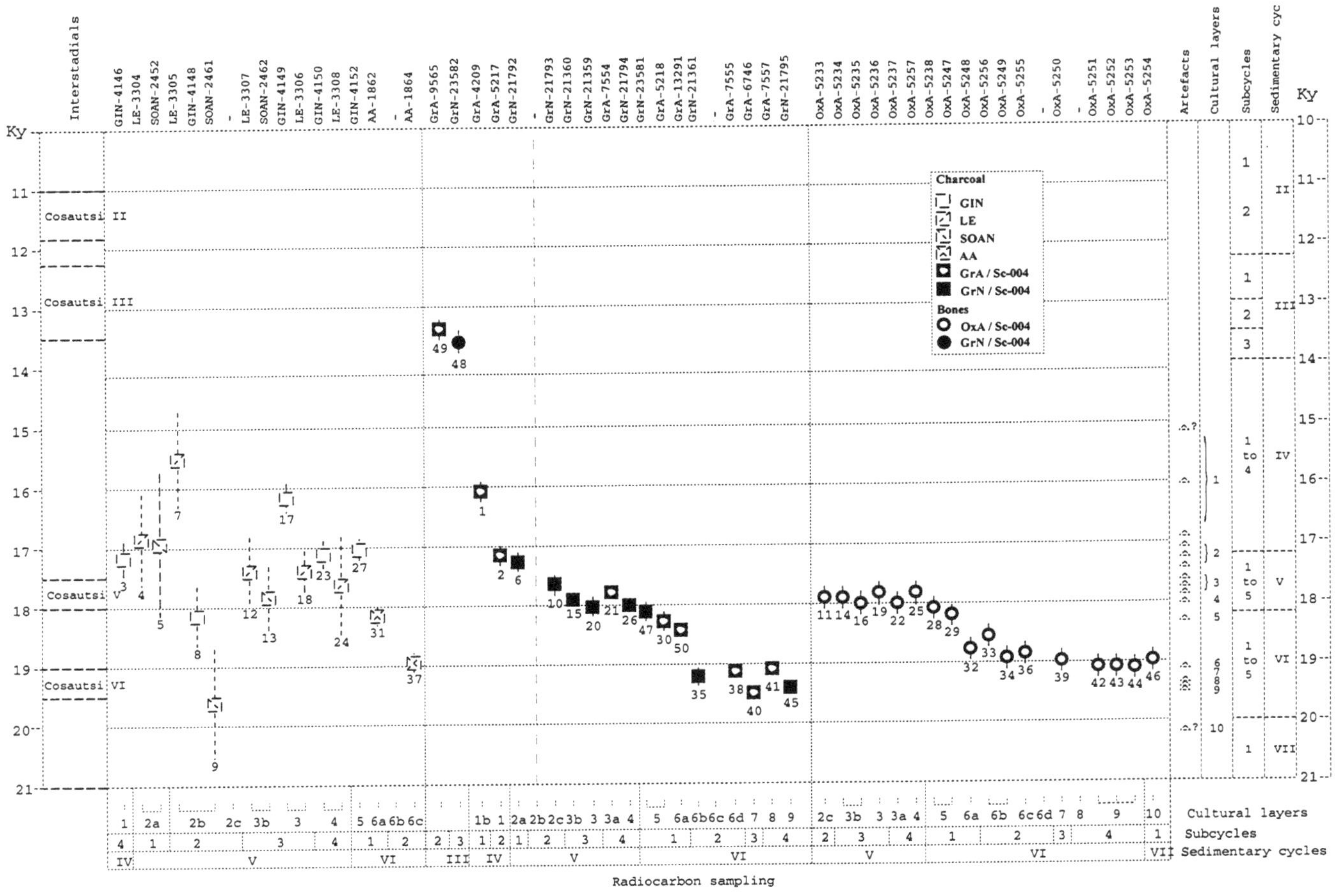

Figure 3: Cosautsi (Moldavia). Chronostratigraphy and chronology of the site. Distribution of the dates by sedimentary unit and by cultural layer. First group: published Russian and American dates; second group: Groningen dates on charcoal; third group: Oxford dates on bones.

isolated associations of conifers and broad-leaved deciduous trees therefore result, at least for Kostienki, from reworking or intrusive phenomena. As a consequence, a general preference was given to conifer charcoal for dating, excepted when charcoal from small-leaved trees appeared as the main component of the sample, as in the case of *Salix* in Afontova Gora (Siberia). Of course, some intrusive movements of conifer charcoal fragments cannot be excluded but this is rapidly suspected by puzzling ^{14}C results occurring in the stratigraphical sequence. Post-depositional redistribution of conifer charcoal may also occur on deep slopes leading to some inversions of ^{14}C dates as stated locally in some sites. Only series of ^{14}C dates in sequences allow a clear identification of such disturbances.

Concerning the question of charcoal identification, the major problem was to make a clear distinction between *Picea* and *Larix* (Schweingruber, 1990; Anagmost *et al.* 1994; Talon, 1997). Charcoal was ascribed to *Larix* when all anatomical features occurred in the pieces (contrasted transition from early wood to late wood, double line of areolated pits along the tracheids, areolated pits with flat walls in cross fields). Yet, we observed that the features of the *Picea* type and those of the *Larix* type could be found within a single charcoal piece, which confirms the high intraspecific anatomical variability within those conifers. It is therefore likely that *Larix* was somewhat underestimated, but this is certainly of no importance for ^{14}C dating and probably of low importance concerning the environment in a pleniglacial context.

Cult.layer	material	date nb	14 C age
Layer I	*Quercus* charcoal	GrA- 6865	1,340 ± 50
Layer I	*Quercus* charcoal	GrA- 9251	1,400 ± 40
Layer I	*Picea* charcoal	GrA- 5552	28,500 ± 140
Layer III	Malaceae charcoal	GrA- 22752	430 ± 40
Layer III	*Picea* charcoal	GrA- 5551	36,280 +360/-350

Table 1: Kostienki 12 (central Russia), Cultural layers I and III. 14C dates on selected charcoal of conifers and broad-leaved deciduous taxa.

Anthracological results

Anthracological data from Holocene sites, mainly from Neolithic and Iron age, with their typical assemblages of broad-leaved deciduous tree, are not presented in this paper. They were collected independently and used as a reference for comparison with older assemblages. The majority of the Upper Pleistocene charcoal samples was collected within pleniglacial deposits, a few were retrieved from Early and Late Glacial deposits.

The results presented in table 2 clearly show the distinct predominance of conifers, especially *Picea*, *Pinus* (t. *sylvestris* and t. *cembra*) and *Larix*, in the pleniglacial deposits throughout the whole loessic area from the Danube basin to central Siberia whereas no charcoal was observed in the western part of Europe. Moreover, concerning the pleniglacial deposits, no major differences in the woody components of the assemblages from interstadial soil horizons and from stadial loess deposits were noticed. Of course, this reduces the direct climatic significance of the anthracological records in pleniglacial loess in contrast to Late-Glacial or pleniglacial records in southern Europe (among others: Vernet, 1980; Vernet and Thiébault, 1987; Thiébault, 1988; Heinz, 1991; Uzquiano, 1992; Heinz and Barbaza, 1998; Ntinou and Kotjabopoulou, 2002). Besides, this fact weakens the hypothesis that man possibly selected his wood according to the species. It is therefore clear that these assemblages are qualitatively representative of the landscape's arboreal component in loessic environments, even if relative changes did happen in the density of the arboreal cover. The high frequency of twisted wood structures seems correspond to stunted trees.

Concerning deciduous trees, various small-leaved taxa, such as *Betula*, *Populus* and *Salix*, were more locally represented, usually in low amounts. In that respect, it is worth to point out that all studied sites are situated along major water courses and tributaries which can be considered as favourable for tree growth, in the form of galery forests in a periglacial environment.

Regions		Meuse Basin				Rh	Middle Danube Basin					East Carpath.			Central Russian Plain						Central Siberia				
Countries		Be	Be	Be	Be	Ge	Au	Au	Au	Au	Cz	Uk	Ro	Mo	Ru	Ru	Ru	Ru	Ru	Ru	Ru	Ru	Ru	Ru	Ru
Site nb:		1	1	1	2	3	4	4	5	6	7	9	10	11	12	13	13	13	13	13	14	14	14	15	15
Sites	Material	Hezerwater	Remicourt	Rocourt (horiz. bl.)	Trô-Maret	Wallertheim	Willendorf II	Schwallenbach	Alberndorf	Hollabrunn	Pavlov I / 57	Molodova V	Mitoc Malul Galben	Cosautsi	Zeleznogorsk	Kostienki 1	Kostienki 12	Kostienki 14	Kostienki 21	Borschchevo 2	Kurtak P 31	Kurtak Chani	Kurtak Kammeni Log	Listvenka	Afontova Gora
Charcoal																									
Picea sp.	C	-	E	-	-	E	MU	M	M	E	M	M	MU	U	-	M	M	M	M	M	M	M	M	U	T
Picea obovata	N	-	-	-	-	-	-	-	-	-	-	-	-	-	-	-	-	-	-	-	M	M	M	-	-
Picea / Larix	C	-	-	-	-	-	-	-	M	-	-	M	-	U	-	-	-	-	-	-	-	-	M	-	-
Larix / Picea	C	-	-	-	-	-	M	-	-	-	M	M	-	-	-	-	-	-	M	-	-	-	M	-	-
Larix sp.	C	-	-	-	-	-	-	-	-	-	-	-	-	-	-	-	-	-	-	-	-	-	M	-	T
Juniperus sp.	C	-	E	-	E	-	-	-	-	-	-	-	-	-	E	-	-	-	-	-	-	-	-	-	-
Pinus t. cembra	C	-	-	-	-	-	MU	M	M	-	M	M	U	U	-	M	-	M	-	-	M	M	M	-	-
Pinus t. sylvestris	C	E	-	-	E	E	MU	-	M	-	-	M	-	-	-	-	-	-	-	-	-	-	-	-	-
Pinus sp.	C	-	-	-	-	-	MU	-	M	-	-	-	-	-	-	-	-	-	-	-	-	-	-	-	-
Betula sp.	C	E	E	E	E	E	M	-	M	-	-	-	U	-	-	-	-	-	-	-	-	M	-	-	-
Betula cf. nana	C	-	-	-	-	-	-	-	-	-	-	-	-	-	-	-	-	-	-	-	-	-	-	-	T
Salix sp.	C	-	E	-	-	-	M	-	-	-	-	-	-	U	-	-	-	M	-	-	-	M	-	-	T
Populus sp.	C	-	-	-	-	-	-	-	-	-	M	-	-	-	-	-	M	M	-	-	-	M	-	U	-
Populus / Salix	C	-	-	-	-	E	-	-	-	-	-	-	-	-	-	-	-	-	-	-	-	-	-	-	-
Alnus sp.	C	-	-	-	-	-	-	-	-	-	-	-	MU	-	-	-	-	-	-	-	-	-	-	-	-
cf. Rhamnus	C	-	-	-	-	-	-	-	-	-	-	-	-	-	-	-	-	-	-	-	-	-	-	-	T
Lonicera sp.	C	-	-	-	-	-	M	-	-	-	-	-	-	-	-	-	-	-	-	-	-	-	-	-	-
Quercus t. petr./rob.	C	-	-	-	-	E	-	-	-	-	-	-	-	-	-	-	-	-	-	-	-	-	-	-	-
cf. Cytisus	C	-	-	-	-	-	-	-	-	-	-	-	-	-	-	-	-	-	-	-	-	-	-	-	T
Prunus sp.	C	-	-	-	-	E	-	-	-	-	-	-	-	-	-	-	-	-	-	-	-	-	-	-	-
Fruits, seeds, leaves																									
Prunus avium	L	-	-	-	-	E	-	-	-	-	-	-	-	-	-	-	-	-	-	-	-	-	-	-	-
Sambucus cf ebulus	F	-	-	-	-	-	-	-	M	-	-	-	-	-	-	-	-	-	-	-	-	-	-	-	-
Chenopodiaceae	C	-	-	-	-	-	M	-	-	-	-	-	-	-	-	-	-	-	-	-	-	-	-	-	-
Apiaceae	F	-	-	-	-	-	M	-	-	-	-	-	-	-	-	-	-	-	-	-	-	-	-	-	-
Galium	F	-	-	-	-	-	M	-	-	-	-	-	M	-	-	-	-	-	-	-	-	-	-	-	-
Medicago	S	-	-	-	-	-	M	-	-	-	-	-	M	-	-	-	-	-	-	-	-	-	-	-	-
Vicia	S	-	-	-	-	-	M	-	-	-	-	-	-	-	-	-	-	-	-	-	-	-	-	-	-
Lychnis flos cuculi	S	-	-	-	-	-	M	-	-	-	-	-	-	-	-	-	-	-	-	-	-	-	-	-	-
Caryophyllaceae	S	-	-	-	-	-	M	-	-	-	-	-	M	-	-	-	-	-	-	-	-	-	-	-	-
Rumex	F	-	-	-	-	-	M	-	-	-	-	-	-	-	-	-	-	-	-	-	-	-	-	-	-
Polygonum aviculare	F	-	-	-	-	-	-	-	M	-	-	-	M	-	-	-	-	-	-	-	-	-	-	-	-
Picris	F	-	-	-	-	-	-	-	M	-	-	-	-	-	-	-	-	-	-	-	-	-	-	-	-
Poaceae	F	-	-	-	-	E	-	-	-	-	-	-	M	-	-	M	-	-	-	-	-	M	M	-	T
cf. Cyperaceae	F	-	-	-	-	-	-	-	-	-	-	-	M	-	-	-	-	-	-	-	-	M	-	-	-
Carex	F	-	-	-	-	-	-	-	-	-	-	-	-	-	-	-	-	-	-	-	-	-	-	U	-
Potamogeton	F	-	-	-	-	-	M	-	-	-	-	-	-	-	-	-	-	-	-	-	-	-	-	-	-
X roots	R	-	-	-	-	-	-	-	-	-	-	-	M	-	-	-	-	-	-	-	-	-	-	-	-

Table 2: The loessic plain. Anthracological data. Distribution of the various taxa by region, by country and by period. Legend: Countries. Be: Belgium; Ge: Germany; Au: Austria; Ro: Roumania; Mo: Moldova; Uk: the Ukraine; Ru: Russia; Regions. Rh. : Rhine basin; Material: C= charcoal; N= charred needle; F= charred fruit; S= charred seed; L= leave; R= root; Periods: H: Holocene; T: Late Glacial; U: Upper Pleniglacial; M: Middle Pleniglacial; L: Lower Pleniglacial; E: Early Glacial. Intrusive material has been excluded from the table.

A main conclusion of our observations is the relative poorness in arboreal taxa and the very uniform picture of the landscape throughout the loessic belt between the Danube basin and central Siberia during the Upper and even the Middle Pleniglacial, with *Picea, Larix* and *Pinus* (mainly *P. cembra*) as major tree taxa, the western loess area of Europe appearing rather free of trees.

By contrast, charcoal fragments extracted from Early Glacial deposits in western Europe (especially Wallertheim) constitute quite diversified assemblages that associate remains of broad-leaved trees (*Quercus, Prunus*) with remains of small-leaved trees (*Betula*) and conifers (*Picea, Pinus*). Such a mixture appears in accordance with mid-temperate to boreal environments.

The anthracological results also show meaningful geographical differences in the quantities of charcoal preserved in the loess deposits. Thus, in western Europe, the loess-like deposits of the Pleniglacial contain no charcoal (tabl. 2). On the contrary, further to the East, the deposits of that period are rich in charcoal, although their abundance decreases strongly at the beginning of the Upper Pleniglacial. This difference is assigned to the climatic gradient between atlantic Europe (zone of the tundra-steppes) and continental Europe (zone of the steppe-forests with conifers). Other variations in the quantities of charcoal have been observed again inside the successive periods of 45-26.000 and of 26-14.000 BP as well according to the regions as to the types of sites. Some basins, notably of the Prut and the Dniester, seem to have been especially favourable for the persistence of conifers during almost the whole Upper Pleniglacial. The proximity of the Carpathians undoubtedy favoured the persistence of refuges but for conifers and small-leaved trees only, probably not for broad-leaved taxa in that part of the loess domain. This seems in contrast to the western foothills of the Carpathian basin where *Carpinus betulus* charcoal has been dated around 27,000 BP (Willis *et al.* 2000; Rudner and Sümegi, 2001). In fact, the persistence of broad-leaved deciduous trees in shelter stations during the last Pleniglacial in the loessic plain could not be demonstrated by our anthracological data.

All together, charcoal of conifers collected in precise stratigraphic positions and selected preferentially, provided more than 225 reliable radiocarbon dates which allowed us to control the chronology of some 17 positive climatic events, mainly recorded as humic horizons, in the loess deposits of Eurasia between 45,000 and 10,000 BP, but only a few data refer to periods later than 25,000 BP.

ACKNOWLEDGEMENTS
This work is a contribution to the Belgian programme Service Centres and Research Networks (Sc-004 and Sc-09) projects initiated by the Federal Office for Scientific, Technical and Cultural Affairs (OSTC). It was also supported by the INTAS projects n° 93-1963 and 96-0072. The authors are grateful to Drs N. Praslov and A. Sinitsyn (St-Petersburg), Prof. I. Borziak (Chisinau) for their collaboration and discussion in the field, to Dr W. Schoch (Birmensdorf), Dr M. Thinon (Marseille), Dr C. Heinz (Montpellier) for their advices on charcoal analysis, to E. Dermience, H. De Potter and C. Herzé (I.R.S.N.B.) for their help in the technical achievement of the projects.

References

ANAGMOST S.E., MEYER R.W., DE ZEEUW C. 1994 - Confirmation and significance of Bartholin's method for the identification of the wood of *Picea* and *Larix*. *IAWA Journal*, t. 15, fasc. 2, p. 171-184.

BORZIAK I. 1991 - Quelques données préalables sur l'habitat tardipaléolithique pluristratifié de Cosseoutsy sur le Dniestr moyen. in : V. CHIRICA (éd.), *Le Paléolithique et le Néolithique de la Roumanie en contexte européen*. Bibliotheca Archaeologica Iassiensis, Iasi, 4, p. 56-71.

BORZIAK I. 1993 - Les chasseurs de renne de Kosoioutsy, site paléolithique tardif à plusieurs niveaux sur le Dniestr moyen (rapport préliminaire). *L'Anthropologie*, (Paris), 97(2-3), p. 331-336.

BOWMAN S. 1990 - *Radiocarbon Dating*. British Museum Publications, London, 64 p.

DAMBLON F. 2002 - *Charcoal and other plant macro-remains from Wallertheim (Mainz, Germany) and their contribution to Middle Palaeolithic environmental reconstruction*. In press.

DAMBLON F., HAESAERTS P. 1997 - Radiocarbon chronology of representative Upper palaeolithic sites in the Central European Plain: a contribution to the Sc-004 project. *Préhistoire européenne*, t. 11, p. 255-276.

DAMBLON F., HAESAERTS P., VAN DER PLICHT J. 1996 - New datings and considerations on the chronology of Upper Palaeolithic sites in the Great Eurasiatic Plain. *Préhistoire Européenne*, t. 9, p. 177-231.

HAESAERTS P., BORZIAK I., VAN DER PLICHT J., DAMBLON F. 1998 - Climatic events and Upper Palaeolithic chronology in the Dniestr basin: new ^{14}C results from Cosautsi. In W.G. MOOK and J. VAN DER PLICHT (eds.) Proc. 16th Int. ^{14}C Conf., Groningen, *Radiocarbon*, t. 40, n° 2, p. 649-657.

HEINZ Ch. 1991 - Upper Pleistocene and Holocene vegetation in the south of France and Andorra. Adaptations and first ruptures: New charcoal analysis data. *Review of Palaeobotany and Palynology*, 69, 299-324.

HEINZ Ch., BARBAZA M. 1998 - Environmental changes during the Late Glacial and Post-Glacial in the Pyrenees (France): new charcoal analysis and archaeological data. *Review of Palaeobotany and Palynology*, 104, p. 1-17.

LANTING J.N., VAN DER PLICHT J. 1994 - ^{14}C-AMS: Pros ans Cons for Archaeology. *Palaeohistoria*, t. 35/36, p. 1-12.

MOOK W.G., WATERBOLK H.T. 1985 - *Radiocarbon Dating. Handbooks for Archaeologists*, t. 3, 65 p.

NTINOU M., KOTJABOPOULOU E. 2002 - Charcoal analysis at the Boila Rockshelter: woodland expansion during the Late Glacial in Epirus, Northwest Greece. *Acts 2d Intern. Meeting of Anthracology*, Paris, 13th-16th Sept. 2000, p.79-86.

OTTE M., LOPEZ BAYON I., NOIRET P., CHIRICA V., BORZIAC I. 1996 - Recherches sur le Paléolithique supérieur de la Moldavie. *Anthropologie et Préhistoire*, t. 107, p. 45-80.

PRASLOV N.D., ROGACHEV A.N. 1982 - *Palaeolithic of the Kostenki-Borschchevo area on the River Don 1879 - 1979*. Academy of Sciences of the USSR, Leningrad, 285 p.

RUDNER Z.E., SÜMEGI P. 2001 - Recurring Taiga forest-steppe habitats in the Carpathian Basin in the Upper Weichselian. *Quaternary International*, 76/77, p. 177-189.

SCHWEINGRUBER F.H. 1990 - *Anatomie europäischer Hölzer*. WSL/FNP. Haupt, Stuttgart, 800 p.

SINITSYN A.A., PRASLOV N.D., SVEZHENTSEV Yu.S., SULERSHITSKIY L.D. 1997 - Radiocarbon Chronology of the Upper Palaeolithic of Eastern Europe. In Sinitsyn, A.A. and N.D. Praslov (eds) - *Radiocarbon Chronology of the Palaeolithic of Eastern Europe and Northern Asia. Problems and Perspectives.* Russian Academy of Sciences. Institute of the History of Material Culture. Saint-Petersburg, p. 21-66.

SINITSYN A.A. 1996 - Kostenki 14 (Markina Gora): data, problems and perspectives. *Préhistoire Européenne*, t. 9, p. 273-313.

SOFFER O. 1985 - *The Upper Palaeolithic of the Central Russian Plain*. Studies in Archaeology, Academic Press, 539 p.

SOFFER O. 1986 - *Radiocarbon accelerator dates for Upper Palaeolithic sites in European U.S.S.R.* Oxford University Commitee for Archaeology. Monograph. t. 11, p. 109-115.

STIEBER J. 1967 - A Magyarorszagi felsopleisztocén vegetaciotörtenete az anthrakotomiai eredmenyek (1957-IG) tükrében. *Földani Közlony*, 97, p. 308-317.

STIEBER J. 1968 - Anthrakotomische Untersuchungen an Holzkohlen der Lösswand bei Solymar: *Acta Botanica Academiae Scientiarum Hungaricae*, 14, p. 165-173.

STREET M., BAALES M., WENINGER B. 1994 - Absolute Chronologie des späten Paläolithikums und des Frühmesolithikums im nördlichen Rheinland. *Archäologisches Korrespondenzblatt*, t. 24, p. 1-28.

SVEZHENTSEV Y.S. 1993 - Radiocarbon chronology for the Upper Palaeolithic sites on the East European Plain. In: O. Soffer and N.D. Praslov (eds), *From Kostenki to Clovis: Upper Palaeolithic - Paleo-Indian Adaptations*. Plenum Press, New-York, p. 23-30.

TALON B. 1997 - Etude anatomique et comparative de charbons de bois de *Larix decidua* Mill. et de *Picea abies* (L.) Karst. *C. R. Acad. Sci. Paris. Sciences de la vie*, t. 320, p. 581-588.

THIEBAULT S. 1988 - L'homme et le milieu végétal à la fin du Tardiglaciaire et au Postglaciaire: analyses anthracologiques de six gisements des Préalpes sud-occidentales. *Documents d'Archéologie Française,* 15, 110 p.

UZQUIANO P. 1992 - L'homme et le bois au paléolithique en région cantabrique, Espagne. Exemples d'Altamira et d'El Buxu. *Bulletin de la Société botanique de France*, 139, 361-372.

VERNET J.-L., THIEBAULT S. 1987 - An approach to northwestern Mediterranean recent prehistoric vegetation and ecologic implications. *Journal of Biogeography*, 14, p. 117-127.

VERNET J.-L. 1980 - La végétation du Bassin de l'Aude, entre Pyrénées et Massif central, au Tardiglaciaire et au Postglaciaire d'après l'analyse anthracologique. *Review of Palaeobotany and Palynology*, 30, p. 33-55.

WILLIS K.L., RUDNER E., SÜMEGI P. 2000 - The Full-Glacial Forests of Central and Southeastern Europe. *Quaternary Research*, t. 53, p. 203-213.

Analysis of botanical remains from Upper Palaeolithic sites in northern Italy

Alfio Maspero

Laboratorio di Archeobiologia,
Musei Civici di Como, Piazza Medaglie d'Oro 1 - 22100, Italy.
E-mail: almasp@tin.it

Abstract
This short paper intends to present the anthracological studies that have been carried out at Upper Palaeolithic sites in northern Italy. The review of the data shows that species adapted to the alpine climatic conditions and which are now absent from the Prealps, were present along the southern alpine fringes in the past. Some problems concerning the stratigraphy and the identification methodology will also be discussed here.

Key-words: *Charcoal analysis, northern Italy, Upper Palaeolithic.*

Résumé
Ce court article tente d'établir la liste des études anthracologiques réalisées sur des gisements datés du Paléolithique supérieur et situés en Italie du nord. L'évaluation des données montre que des espèces adaptées aux conditions climatiques alpines, absentes aujourd'hui des Préalpes, étaient présentes tout au long de la frange sud des Alpes. Quelques problèmes stratigraphiques et méthodologiques seront également discutés.

Mots-clés : *Charbons de bois, Italie du nord, Paléolithique supérieur.*

Introduction

Most of the material from sites mentioned in this paper has been examined in the Laboratory of Archaeobiology of the Museum of Como. The sites are located on the south-eastern fringes of the alpine region, from 100 m to more than 1000 m a.s.l. Chronologically, they span a period of more than 4000 years, from the beginning of the late Epigravettian to the early Mesolithic period (Woillard, 1980; De Beaulieu and Reille, 1984; Cremaschi *et al.* 1990). Without pretending to be exhaustive, this paper enumerates the sites of northern Italy which can be dated back to the Upper Palaeolithic and from which macro botanical remains have been examined. A review of the data obtained by the analysis of macrobotanical remains could be useful both for checking the methods of analysis and for comparing the actual data (Nisbet, 1983; Paganelli, 1984; Cattani and Renault-Miskovsky, 1989). Palaeolithic sites usually have a complicated stratigraphy which requires a detailed description. This is however not the aim of this short paper; the descriptions of the archaeological sequences can be found in the bibliography (tabl. 3).

Material and methods

The dating of previously identified charcoal pieces could in some cases avoid chronological problems that frequently affect the anthracological analysis at palaeolithic sites. This method has very often yielded a wide chronological range for a single archaeological layer (tabl. 1); in the Epigravettian levels at Bus de la Lum, two charcoal fragments from the same archaeological level were dated to 10.430 ± 50 BP (*Larix* sp.) and 9120 ± 150 BP (*Tilia* sp.) respectively (fig. 1 and 2). Another example comes from layer T6 of the Palughetto bog (tabl. 2), where the wide range of radiocarbon dates (from 9495 to 6414 BP) are in contrast to the archaeological material which can be dated to the late Epigravettian and early Mesolithic period. On the other hand, a perfectly reliable chronological scanning has been obtained by the radiocarbon dating of organic remains (branches, cones and peat) preserved in a peat bog situated below the archaeological site (Peresani, 2001).

To the table enumerating the sites and their bibliographic references, we have added sites in karstic caves and the open air deposits of Val Lastari, Velo Val d'Astico and Cividate Camuno. The sites of Como S. Abbondio and Fornaci di Revine which are not archaeological sites have also been mentioned (tabl. 3).

Concerning the identification work, only the most important problem, *i.e.* the identification of *Larix decidua* Mill. and *Picea abies* (L.) Karst, will be discussed here. The generic name *Picea/Larix* has been used in this article. On all sites, except for the Biarzo shelter, some of the charcoal fragments identified as *Picea/Larix* belong in fact to the genus *Larix* and can be identified as *Larix decidua*. At Biarzo some anatomical characteristics of the few and small charcoal fragments attributed to *Picea/Larix*, would allow the identification of *Picea abies* (Bartholin, 1979; Schweingruber, 1990; Anagnost *et al.* 1994; Talon, 1997). The key characteristics used for the distinction between the two species are the bordered pits of the ray tracheids and the biseriate pits in the longitudinal tracheids (fig. 6, 7, 8, 9). An attempt has been made to estimate the reliability of the identification of charcoal, depending if the wood was retrieved from the trunk or from big or small branches. The characteristics that allow an easy identification of *Larix decidua* (biseriate pits in the longitudinal tracheids) are often visible in charcoal from trunks or big branches while charcoal pieces showing little earlywood do not allow the distinction. Some authors maintain that the presence/absence of bordered pits in the ray tracheids always allows a distinction between the two species (Talon, 1997), others consider that this is impossible to do on charred material (Bartholin, 1979). In any case, this key characteristic may be very difficult to use in the study of charcoal from Palaeolithic sites.

Considerations

Given the shortage of data, it is impossible to come to any conclusion and this review cannot be anything but a short collection of very scattered data both in a spatial and chro-

Settlement	age 14C BP	Lab. code	Material	Species
Riparo Biarzo (PN) 150 m asl	11.100 ± 125	R 1850	charcoal	
Bus de la Lum	10430 ± 50	UtC 8912	charcoal	*Picea /Larix*
	9120 ± 150	UtC-10759	charcoal	*Tilia sp.*
Fornaci di revine	14370 ± 115		wood	*Larixsp.*
	14765 ± 135		wood	*Larixsp.*
Ripari Villabruna (BL) 500 m	12.040 ± 150	R 2023	charcoal	*Pinus sylv/mont*
	12.040 ± 125	R 2022	charcoal	*Pinus sylv/mont*
Val Lastari (VI) 1060 m	10280 ± 110	UtC 2685 SMA	charcoal	*Picea /Larix*
	11.010 ± 90	UtC 2041 SMA	charcoal	*Picea /Larix*
	11390 ± 110	UtC 1773 SMA	charcoal	*Picea /Larix*
	11800 ± 150	UtC 2687 SMA	charcoal	*Picea /Larix*
	13450 ± 130	UtC ???? SMA	charcoal	*Picea /Larix*
Riparo Dalmeri (TN) 1240 m	10800 ± 110	Rome-425 RD 1	charcoal	undet.
	11000 ± 115	Rome-426 RD 1	charcoal	undet.
	11250 ± 100	Rome-657	charcoal	undet.
	11260 ± 100	KI 2 - 3634	charcoal	undet.
	11550	UtC-Nr.5040 RD-5	charcoal	undet.
Riparo la Cogola	9450 ± 60	UtC 9284	charcoal	undet.
	9820 ± 60	UtC 9285	charcoal	undet.
	10380 ± 70	UtC 9286	charcoal	undet.
	10640 ± 60	UtC 9287	charcoal	undet.
Velo (VI) 330 m asl	13029 ± 108	Gx 19893 ASM	charcoal	*Picea /Larix*
Riparo Tagliente (VR) 250 m	12000 ± 400	ST - R - 604	charcoal	undet.
	12040 ± 170	ST - R - 371	charcoal	undet.
	12650 ± 160	SMA Oxa 3530	charcoal	undet.
	13070 ± 170	SMA Oxa 3531	charcoal	undet.
	13270 ± 170	SMA Oxa 3532	charcoal	undet.
	13330 ± 160	ST - R - 605	charcoal	undet.
	13430 ± 180	ST - R - ???	charcoal	undet.
Riparo Soman (VR) 115 m	10370 ± 110	Gd 6163	charcoal	undet.
	10470 ± 150	Gd 6159	charcoal	undet.
	10510 ± 180	Gd 4511	charcoal	undet.
	11880 ± 170	Gd 6158	charcoal	undet.
Cividate Camuno (BS) 271 m	13805 ± 440		charcoal	undet.
Romagnano	9830 ± 90		charcoal	undet.
	9490 ± 80		charcoal	undet.
	9420 ± 60		charcoal	undet.
	9580 ± 250		charcoal	undet.
Bus dei Lader				
Como S. Abbondio	11730 ± 180	wood	wood	*Larix decidua*

Table 1: Radiocarbon dates of the mentioned settlements given in conventional ^{14}C years BP (before AD, 1950 uncalibrated). Most of the dates were obtained on charcoal pieces of which some were identified. Charcoal from hearths is a common source for radiocarbon dating but often such material is destroyed before any identification has been done, a procedure that deprives us of useful palaeobotanical information on well-dated material. At Fornaci di Revine only *Larix* sp. has been noted and at Romagnano *Pinus* sp. type *sylvestris* and *Picea* sp./*Larix* sp. have been identified.

nological sense. However, within this assemblage of palaeobotanical and chronological data, which require further checking and integration, some facts can be considered. Species such as *Larix decidua*, *Betula* sp. and *Pinus sylvestris/mugo* (fig. 3) have been found on sites located at altitudes of less than 300 m a.s.l.; in the Soman, Tagliente and Biarzo shelters and at the open-air sites of Cividate Camuno and Velo Val d'Astico (tabl. 1, 4). This is corroborated by the results obtained by the analysis at Fornaci di Revine and Romagnano (Follieri, 1975 *in* Cattani, 1992; Aa Vv, 1992). The data suggest the presence of a belt of coniferous forests along the southern fringe of the central Alps. The dating of the Soman shelter further suggests that these forests remained at low altitudes for a long period of time. However, as archaeological data that could confirm this is still lacking, the evidence should be taken cautiously.

There are many indications that species closely connected with alpine climatic conditions and now absent from the Prealps, such as *Pinus cembra* (fig. 4), were present along the southern Alpine fringes. *Pinus cembra* has been noted at many sites such as Ganna lake, Annone Lake and Cerete by pollen analysis, which cannot however prove beyond doubt the local presence of the cembran pine (Schneider, 1978; Wick, 1996; Ravazzi *et al.* 1996). A few, small charcoal fragments of *Pinus cembra* have been retrieved from the Tagliente and La Cogola shelters, which are located far from the inner parts of the Alps where the species grows

Palughetto (TV) 1040 m asl				
Archaeological levels				
6414 ± 203	R-3130	charcoal	undet.	T6
7633 ± 85	R-3129	charcoal	undet.	T6
8585 ± 75	R-3122	charcoal	undet.	T6
8800 ± 63	R-3123	charcoal	undet.	T6
9495 ± 150	GX-21231	charcoal	undet.	T6
Macro remains of the forest				
10640 ± 95	R-3107	branch	undet.	T6base
11679 ± 91	R-3108	branch	undet.	T6base
10162 ± 89	R-2587	peat		T9top
10795 ± 165	H-4755	peat		T9
11603 ± 84	R-2591	3 cones	*Picea abies*	T9
12205 ± 245	H-4757	peat		T10
12000 ± 340	GX-21230	cone	*Picea abies*	T10
12260 ± 160	GX-20588-AMS	peat		T10base
12720 ± 160	H-2148	cone	*Pinus mugo*	T11

Table 2: Radiocarbon dates obtained on archaeological charcoal and on other botanical remains at the site of Palughetto.

Biarzo shelter	Castelletti *et al.* 1996
Fornaci di Revine	Casadoro and Terenziani, 1976; Casadoro *et al,* 1975
Palughetto	Avigliana *et al* . 1999
Bus de la Lum	Peresani *et al.* 1998
Villabruna shelter	Aa. Vv., 1992
Dalmeri shelter	Dalmeri and Lanzinge,. 1992; Castelletti and Maspero, 1992
La Cogola shelter	
Val Lastari	Maspero, 1996; Aa. Vv., 1992;
Velo Val d'Astico	
Tagliente shelter	Maspero, 1996
Soman shelter	Broglio and Lanziger, 1985
Bus dei Lader	Biagi, 1976; Castelletti and Maspero, 1992
Cividate camuno	Poggiani Keller, 1990
Como S. Abbondio	Castelletti and Orombelli, 1986

Table 3: Bibliographical references to the sites mentioned in the text.

today. *Larix decidua* (fig. 5), a frequent species in the palaeolithic hearths of northern Italy, is now located at high altitudes. It is a typical alpine species which is present both in the Alps and as very isolated populations in Poland and in the Carpathian Mountains. The difficulties in identifying larch pollen has probably obscured our understanding of its wide distribution in the past.

There must have been an early occupation of the Alps, given that sites located at over 1000 m a.s.l. (Val Lastari, Dalmeri shelter) and at the bottom of some alpine valleys (Cividate Camuno in Camonica valley and Villabruna shelter in Cismon valley) show an early presence of coniferous forests (tabl. 1, 4).

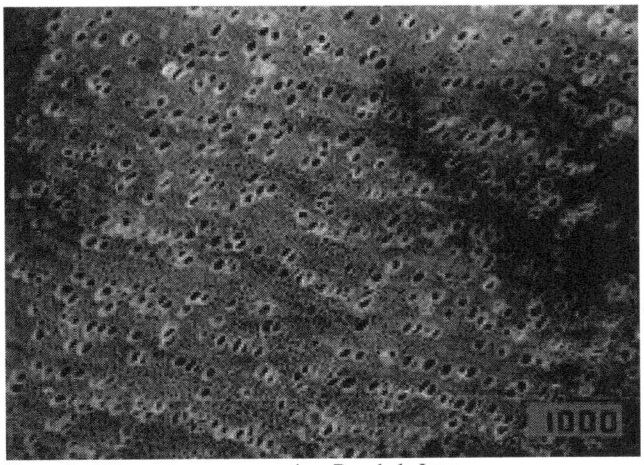

Figure 1: *Tilia* sp., transverse section, Bus de la Lum.

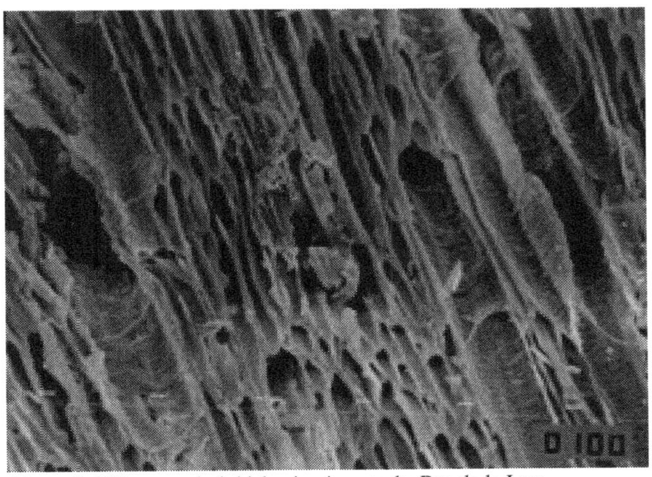

Figure 2: *Tilia* sp., spiral thickening in vessels, Bus de la Lum.

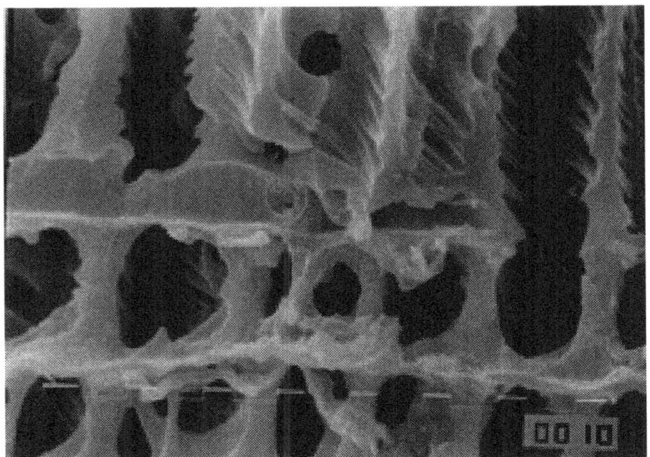

Figure 3: *Pinus sylvestris/mugo*, dentate walls of a ray tracheid, La Cogola Shelter.

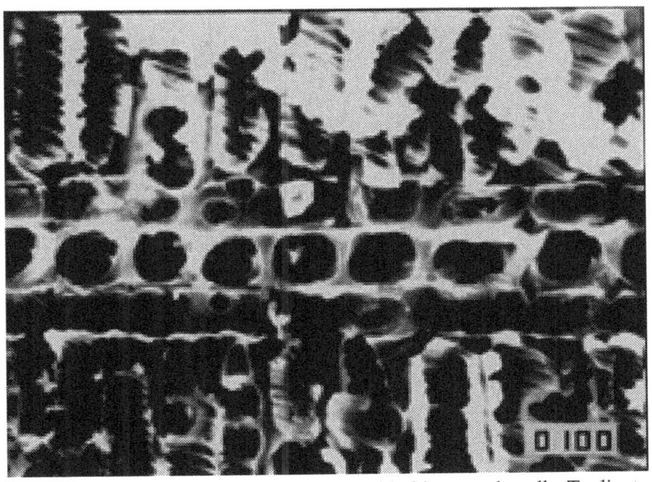

Figure 4: *Pinus* cf. *cembra*, ray tracheids with thin smooth walls, Tagliente shelter.

	Riparo Biarzo 5a	5b	5c	Bus de la Lum US2	Palughetto T6	Riparo Dalmeri 26b1	26c	26d	Val Lastari 3b	3c	3d	3e	3f	SIII	SV	SIV	Velo
Betula sp.	2	1	4		28							2	1				
Coniferae	1		2	6	7		6	4		1	1	2	2				8
Fagus sp.					1												
Hippophae Rhamnoïdes					2												
Juniperus																	
Larix/Picea	2	3	4	16	99	43	48	19	25	20	35	19	15	10	1		9
Latifolieae				1	3				2		1						
Latifolieae 1				2													
Pinus cembra						1											
Pinus sylv/mugo	40	57	98			9	6	7	4	9	23	94	21	19	28	10	5
Pinus sp.																	
Pomoïdea					1												1
Populus				1		3											
Prunus sp.																	
Rhamnus											1						
Salix					2	1			9	4	2		1	1			
Salix/Populus																	
Tilia sp.				13													
Total	45	61	108	39	143	57	60	30	40	34	63	117	40	30	29	10	24
		214					147					363					

	Riparola la Co.	Riparo Tagliente 10b	10c	10d	10e	10f	12	13	14	Riparo Soman US131	US132	US133	Ripari villabruna US16	US17	US18	Cividate US
Betula sp.				2												
Coniferae	3	1	2	1		3	4	6	5							
Fagus sp.																
Hippophae Rhamnoïdes																
Juniperus		1	1													
Larix/Picea	6	9	13	45	3	36	48	75	79	6	23	6				
Latifolieae																
Latifolieae 1																
Pinus cembra							1		3							
Pinus sylv/mugo	21	18	41	48	26	22		7	2	27	10	27	34	31	35	48
Pinus sp.			1	1		1										
Pomoïdea																
Populus																
Prunus sp.						2										
Rhamnus			2													
Salix																
Salix/Populus																12
Tilia sp.																
Total	30	29	60	96	29	64	53	88	89	33	33	33		100		60
					507						99					

Table 4: Results of the identification of charcoal at various sites.

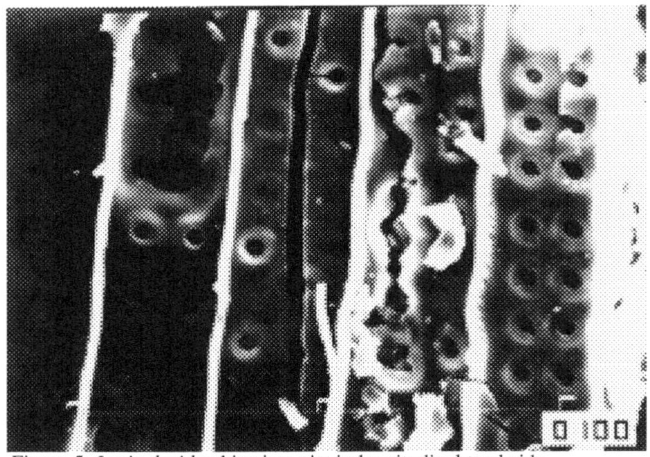

Figure 5: *Larix decidua* biseriate pits in longitudinal tracheids.

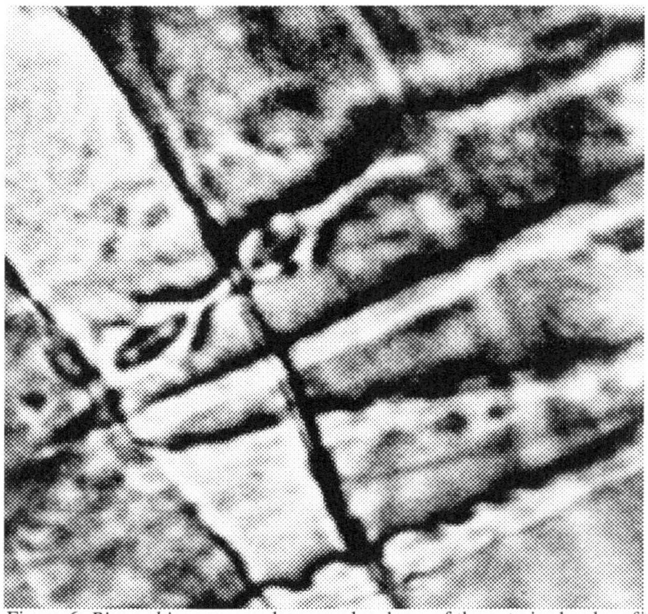

Figure 6: *Picea abies*, more or less angular shape of the exterior border of the ray tracheids pits in transverse section, wood from the peat bog of Palughetto.

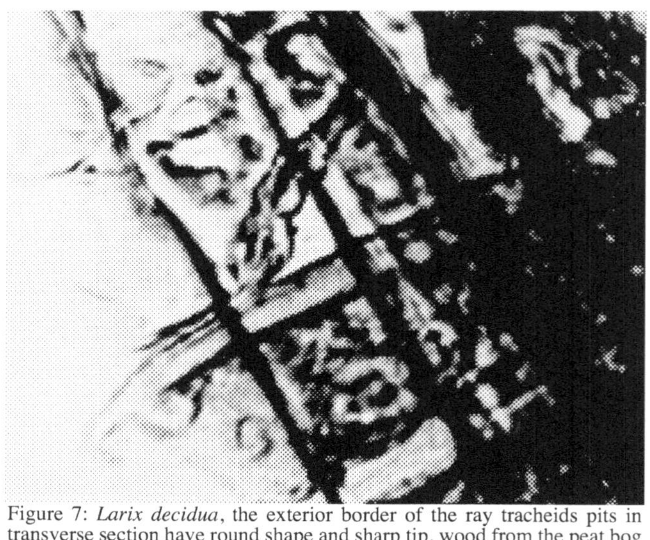

Figure 7: *Larix decidua*, the exterior border of the ray tracheids pits in transverse section have round shape and sharp tip, wood from the peat bog of Palughetto.

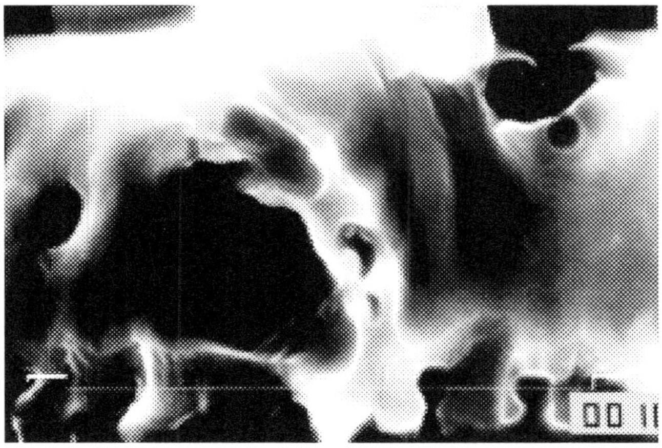

Figure 8: Shape of the exterior border in ray tracheids pits which might be classified as type P-2, according to B. Talon (Talon, 1997), Biarzo shelter.

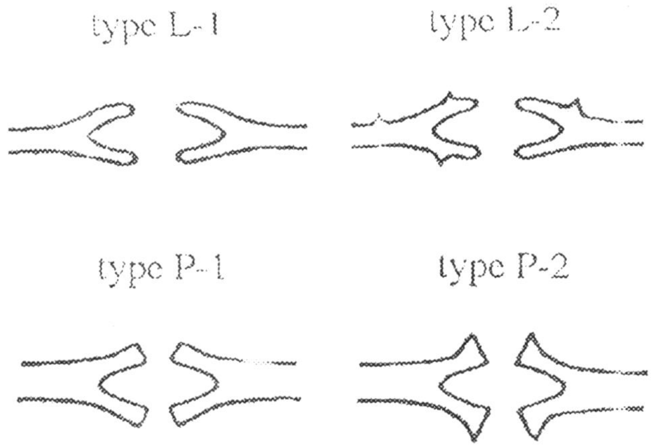

Figure 9: The four types of bordered pits described by B. Talon who considers the "presence of type P-2 pits in the ray tracheids of latewood as a criteria to identify *P. abies*".

References

ANAGNOST S.E., MEYER R.W., ZEEUW C. de 1994 - Confirmation and significance of Bartholin's method for the identification of the wood of *Picea* and *Larix*. *IAWA Journal*, 15, p. 171-184.

BARTHOLIN T. 1979 - The *Picea-Larix* problem. *IAWA Bull.* 1, p. 8-10.

BIAGI P., CREMASCHI M., NISBET R. 1985 - Palaeoecological implications for the Later Prehistory of Northern Italy. In MALONE C., STODDART S (eds.), *Papers in Italian Archaeology* IV, Part II, Prehistory, Oxford, p. 272-281.

BROGLIO A., LANZIGER M. 1985 - I risultati preliminari degli scavi al Riparo Soman presso Ceraino In Valdadige. *Atti dei primo convegno archeologico sulla Valdadige meridionale*, p. 10-28.

CASADORO G., TERENZIANI F. 1976 - Analisi istologica dei tronchi. In CASADORO G. *et al.* Un deposito tardowürmiano con tronchi subfossili alle fornaci di Revine (Treviso). *Boll. Comit. Glac. It.*, 24, p. 48-50.

CASADORO G. *et al.* 1975. - Notizie preliminari su una foresta fossile riaffiorata nella zona di Revine (Viflorle Veneto - Treviso). In: *I'. botanico Italiano*, 7, p. 23-24.

CASTELLETTI L., MASPERO A. 1992a - (Abri Soman) Analyse anthracologique, (Grotte Abri de Fumane) Analyse anthracologique, (Abri Tagliente) Analyse anthracologique, (Les Abris Villabruna) Analyse anthracologique, (Val Lastari) Analyses anthracologiques, (Abri Dalmeri) Analyses anthracologiques. In: AA.VV. Colloque International *Adaptations au milieu montagnard au Paléolithique superieur et au Mésolithique*, Trento 5-11 ottobre 1992, Livret-Guide des excursions, p. 27, 36-37, 51, 76, 82-83, 91. Trento.

CASTELLETTI L., MASPERO A. 1992b - Antracologia degli insediamenti paleolitici nella penisola italiana. Nuovi dati per la ricostruzione del Paleoambiente durante il Würm. Actes des colloque: Les Charbons de Bois, les anciens écosystèmes et le rôle de l'homme. *Bull. Soc. bot. Fr.* 139, 2/3/4, p. 297-309.

CASTELLETTI L., OROMBELLI G. 1986 - Una nuova data ^{14}C per la storia della deglaciazione del Bacino del Lago di Como. Geografia Fisica e Dinamica, *Quaternaria*, 9, p. 56-58.

CATTANI L., RENAULT-MISKOVSKY J. 1989 - La réponse des végétations aux variations climatiques quaternaires autour des sites archéologiques du sud de la France et du nord est de l'Italie. *Il Quatemario*, 2, p. 147-170.

CREMASCHI M. *et al.* 1990 - Loess in northern and central Italy: a loess basin between the Alps and the Mediterranean Region. *Quaderni di geodinamica alpina e quaternaria*, 1.

DE BEAULIEU J.L., REILLE M. 1984 - A long Upper Pleistocene pollen record from les Echets near Lyon (France). *Boreas*, 13, p. 111-132.

NISBET R. 1983 - Ventimila anni di storia forestale. Aspetti evolutivi del paleopaesaggio. 165-168. In: *Preistoria nella Liguria Orientale*, Chiavari. p. 7-12.

PAGANELLI A. 1984 - Storia climatico forestale del Pliocene e del Quaternario. In: *il Veneto nell'antichità -preistoria e protostoria*, ASPES A. (a cura di) Banca Popolare di Verona, Verona, p. 69-94.

PERESANI *et al.* 1999 - Bus de la Lum and the Epigravettian occupation of the Venetian pre-Alps during the Younger Dryas. *Rivista di Scienze Preistoriche L.* 1999 - 2000 p. 103-132.

POGGIANI KELLER R. 1990 - Cividate Camuno (BS) via Palazzo, resti di insediamento Paleo-Mesolitico e Neolitico. *Notiziario della Soprintendenza Archeologica della Lombardia*, 1988-89, p. 27-28.

SCHNEIDER R., TOBOLSKI K. 1983 - Palynologische und stratigraphische Untersuchungen im Lago di Ganna (Varese, Italien). *Botanica Helvetica*, 93 p. 115-122.

SCHNEIDER R., TOBOLSKI K. 1985 - Late-glacial and Holocene environments of a lake in the southern alps. *Diss. Bot.* 87, p. 229-271.

SCHWEINGRUBER F. H. 1990 - *Anatomie Europäischer Hölzer. Anatomy of European woods*. Eidgenöss. Forschungsanst. f. Wald, Schnee u. Lanschaft, Birmensdorf. Verlag Paul Haupt, Berne.

TALON B. 1997 - Etude anatomique et comparative de charbons de bois de *Larix decidua* Mill. et de *Picea abies* (L.) *Karst C. R. Académie des sciences Paris*, Sciences de la vie 320, p. 581-588.

WOILLARD G. 1980 -The pollen record of Grand Pile (NE France) and the climatic chronology through the last interglacial-glacial cycle. In: Problèmes de stratigraphie quaternaire en France et dans les pays limitrophes, Chaline J. (Ed.). *Suppl. Bull. A.F.E.Q.*, 95, 103 p.

Charcoal analysis at the Boila rockshelter: woodland expansion during the Late Glacial in Epirus, north-west Greece

Maria Ntinou[1] and Eleni Kotjabopoulou[2]

1) Wiener Laboratory, American School of Classical Studies at Athens, 54, Souidias Str., Athens 106 76, Greece.
2) Ephoreia of Palaeoanthropology-Speleology, Ardittou 34B, Athens 116 36, Greece.
E-mail: Eleni.Kotjabopoulou@eps.culture.gr
E-mail: Maria.Ntinou@uv.es

Abstract

This paper presents the analysis of dispersed charcoal recovered from systematic excavations at the Late Glacial rock shelter site of Boila in the mountainous region of Epirus, north-west Greece. The qualitative and quantitative composition of the assemblages is characterized by a notable diversity of plant formations. The expansion of pioneer woodland formations is attested as early as 14,500 BP. The charcoal profile is further discussed in light of the palynological diagrams available in the same geographical region. An important point that the present analysis brings forward is the regional character of the vegetation evolution in the southern Balkan peninsula, which does not, however, comply with the neat subdivision of northern European into biozones of the Late Glacial. Within a framework of macroclimatic fluctuations, other factors, such as competition among species and landscape diversity, emerge as decisive in shaping the vegetation evolution in this part of Europe.

Key-words: *Charcoal analysis, Late Glacial, Epirus, North-west Greece.*

Résumé

L'analyse des charbons de bois dispersés, récupérés lors des fouilles systématiques de l'abri de Boila, situé au nord-ouest de la Grèce est présentée ici. Le site se trouve dans la région montagneuse de l'Epire et la stratigraphie s'étend chronologiquement de *c.* 14,500 à *c.* 10,000 BP. La composition qualitative et quantitative des niveaux examinés est caractérisée surtout par la diversité écologique des communautés végétales représentées. L'expansion des espèces forestières pionnières apparaît déjà à partir de *c.* 14,500 BP. Les données anthracologiques sont discutées à la lumière des données palynologiques, abondantes dans la région en question. La présente étude permet de souligner l'importance du caractère régional de l'évolution des végétations dans la Péninsule Balkanique du Sud. Cette évolution ne correspond pas nettement à la biozonation du Tardiglaciaire proposée pour l'Europe du Nord. Dans un cadre macro-climatique dynamique, on constate que l'évolution de la végétation repose aussi sur des paramètres comme la compétition entre les espèces et la variabilité géographique (terrain, altitude, géologie etc.).

Mots-clés : *Anthracologie, Tardiglaciaire, Epire, Grèce du nord-ouest.*

Introduction

A long-standing tradition of integrating palaeoenvironmental studies with the history of foraging exploitation patterns during the Pleistocene characterizes the Palaeolithic research design in Epirus, north-west Greece (*e.g.* articles in Bailey, 1997; Higgs and Webley, 1971). In particular, a prolific data archive, based on pollen sites of both large and small basins at various altitudes and topographic settings, documents the evolution of regional vegetation extensively (Bottema, 1974; Turner and Sanchez-Goñi, 1997; Tzedakis, 1993, 1994; Willis, 1992a, 1992b, 1997). The area presents special interest in the reconstruction of the vegetation history of Greece and the Balkans, since it has been associated with the existence of long-term tree refugia (Bennett *et al.* 1991; Bottema, 1974; Tzedakis, 1993; Willis, 1994).

This paper presents the results of charcoal analysis from the Late Glacial rock shelter site of Boila in the Voidomatis River basin in the Pindus mountain range, Epirus. To date, this is the first such analytical inquiry undertaken at a Palaeolithic site in this part of the southern Balkan peninsula (see Kotjabopoulou *et al.* 1997). Our aim is to discuss the palaeoecological information embedded in the dispersed charcoal material from Boila. The hunter-gatherer occupation at this site has been characterized by ephemeral and/or sporadic activity. In this respect, the available charcoal assemblages constitute an especially valuable palaeoenvironmental depository for reconstructing the vegetation structure(s) on a local scale. Moreover, the stratigraphic sequence at Boila is well contained with a series of conventional and AMS radiocarbon dates, which contribute decisively to the chronological refinement of the vegetation scheme at the regional level.

Boila rockshelter in the Voidomatis river catchment

Boila is located in north-east Epirus at an elevation of *c.* 410 m on the southern bank of the Voidomatis river at the mouth of the lower Vikos gorge (fig. 1).

Tectonic activity, geological variability, physiographic diversity and marked altitudinal variation (400 m - 2500 m) characterize the complex landscape configuration of the

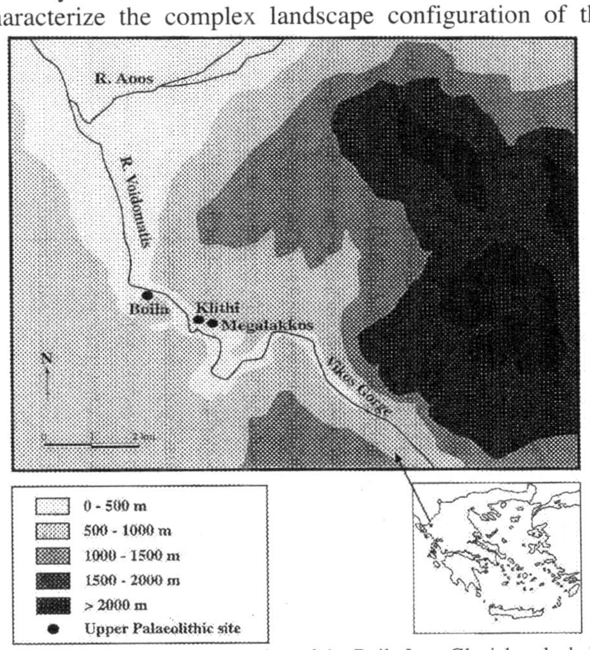

Figure 1: Map showing the location of the Boila Late Glacial rock shelter within the Lower Vikos Gorge. Other Late Palaeolithic sites are also indicated (after Bailey, 1997).

drainage system in question (Lewin *et al.* 1991; Bailey *et al.* 1997; Macklin *et al.* 1997). In such a setting, climatic particularities are to be expected. The climate is Mediterranean with continental influences. Mean annual temperatures range from 2 °C in January to 20.6 °C in July. The dry summer months are frequently interrupted by thunderstorms often of orographic origin. The climate can be described as humid/hyper humid, with mean annual precipitation oscillating between 800 mm and 2000 mm.

In the light of the factors cited above, plant formations are also quite varied. Vegetation has been recovering quickly in recent decades, since the declaration of the area as a National Park. Vegetation within the catchment is especially rich, combining riverside elements, Mediterranean and European species, as well as Balkan endemics (see Polunin, 1980; Bailey *et al.* 1997). The detailed composition of the present day vegetation within the lower Vikos gorge is described by Ntinou (2000a).

The site and the provenance of charcoal assemblages

The Boila rock shelter is formed by Eocene limestone and the overhang has been subject to erosion. The surface of the deposits extends over 80 m² and lies 11 m above the present river channel. Significantly, the sequence containing cultural material overlies alluvial gravels of a Late Würm date (Kotjabopoulou *et al.* 1997; Lewin *et al.* 1991). The site was made available for use only after the Voidomatis had incised the glacio-fluvial terraces and left this location above the highest Late Glacial flood level (Woodward *et al.* 2001). The late Upper Palaeolithic industrial and faunal record at Boila is discussed in its wider context by Kotjabopoulou *et al.* (1999) and need not be repeated here. Suffice it to mention that aspects of the cultural assemblages suggest that Palaeolithic human groups made use of resources available both in the gorge and the more open terrain to the north and north-west. As things stand at present, the Boila record provides a glimpse into foraging activities carried out by small groups or groups of restricted composition (*ibid.*) at a time when other sites in the Voidomatis catchment were used infrequently (Bailey, 1997).

The basic stratigraphic and chronological scheme at Boila to which the charcoal assemblages correspond is shown in table 1. For all stratigraphic units, the analysis is based on dispersed charcoal. Almost all of the charred wood fragments found in the excavated deposits were recovered by screening water through mesh with openings measuring 2 mm. Material originating from circumscribed hearth features is omitted -such contexts are more pertinent to an ethnobotanical inquiry (Badal, 1992; Chabal, 1988). An exception to this approach is the consideration of Unit IV, which is a sedimentary unit comprised partly of strong anthropogenic imprint (Kotjabopoulou *et al.* 1999). As the depositional history of this unit is still under study, any interpretation is subject to revision.

Overall, the non-hearth contexts at Boila were not especially rich in charcoal. Samples recovered during the excavation of each unit have been put together to form charcoal assemblages. In this sense, each analysed dispersed charcoal assemblage corresponds to one stratigraphic unit. The only exception is Unit II, which contains a high charcoal density. For this reason, the charcoal assemblage from this stratigraphic unit has been divided into two anthracological spectra, lower and upper (tabl. 1). In total, the charcoal material is divided into six spectra.

Stratigraphic unit	Lithology	Date (BP) 14C concentional, ESR, TL	Date (BP) 14C AMS	Charcoal spectrum
IV	silt and deposit			6
IIIb	Mainly matrix supported screes	11 173±453 (DEM-371)	10 190±90 (OxA-5243)	5
IIIa	Mainly matrix supported screes	12 233±221(DEM-501) 12 901±157 (DEM-502)	12 480±120 (OxA-5241)	4
II	Clayey silt layer devoid of coarse clasts	13 709±238 (DEM-622))	13 810±130 (OxA-5246)	3
Ib	Flood Unit (Alluvial sand and silt layers)		13 960±130 (Beta-109162) 14 310±100 (Beta-109187)	2
Ia	Upper part of Aristi Unit (Coarse alluvial gravels)	24 300±600 (ESR)* 28 200±7100 (TL)*		1

Table 1: Stratigraphic and chronological scheme of the Boila Late Glacial rock shelter (after Kotjabopoulou *et al.* 1999 with additions). Correlations are made with the main lithological characteristics of the distinct stratigraphic units (Kotjabopoulou *et al.* 1999; Macklin *et al.* 1997; Woodward *et al.* 2001) and the charcoal analysis spectra of the charcoal diagram.
* The ESR and TL dates from the upper part of the Aristi Unit fluvial terraces are after Macklin et al. 1997: 352-353.

Analysis and discussion of charcoal assemblages

A total of 13 angiosperm and 2 conifer families are represented in the charcoal fragments analysed (tabl. 2). A good number of these remained unidentified (23%). From the identified component, 31 taxa were specified while one is indeterminate. The results of the analysis are presented in the charcoal diagram, which is divided into six spectra (fig. 2).

In general terms, the charcoal analysis results of the Late Glacial rock shelter of Boila present a rich list of plants. All of the diagram spectra are of similar composition. The qualitative homogeneity is demonstrated by 14 taxa common to all spectra (tabl. 2).

In quantitative terms, a clear distinction is observed between the five older spectra and the younger one. This is evident in the evolution of two taxa, namely *Betula* sp. and *Cornus* sp. The former taxon dominates the sequence from spectrum 1 to 5, while the latter is the most frequent in spectrum 6.

Acer sp. (maple)		*Phillyrea-Rhamnus*	*
Alnus sp. (alder)	*	*Pinus nigra-P. sylvestris* (black pine-Scots pine)	*
Alnus-Corylus		*Pinus* sp.	
Betula sp. (birch)	*	*Pistacia terebinthus* (pistachio)	
Betulaceae		*Pistacia* sp.	*
Carpinus orientalis (oriental hornbeam)	*	*Platanus orientalis* (plane tree)	
Carpinus-Ostrya (hornbeam-hop hornbeam)	*	*Prunus* cf., *P. amygdalus* (almond tree)	
Conifer		*Prunus* sp.	*
Cornus sp. (cornelian cherry)		cf. *Prunus*	
Corylaceae		*Quercus* deciduous (oak)	*
Corylus sp. (hazel)	*	*Quercus* evergreen (evergreen oak)	*
Fraxinus sp. (ash)	*	*Quercus* sp.	*
cf. *Hippophae rhamnoides* (sea buckthorn)		cf. *Rosa* (rose)	
Juniperus sp. (juniper)		*Salix* sp. (willow)	*
Leguminosae		*Ulmus* sp. (elm)	
Maloideae			

Table 2: Taxa identified in the charcoal assemblages from the Boila Late Glacial rock shelter. The presence of a taxon in all stratigraphic units is indicated with an asterisk.

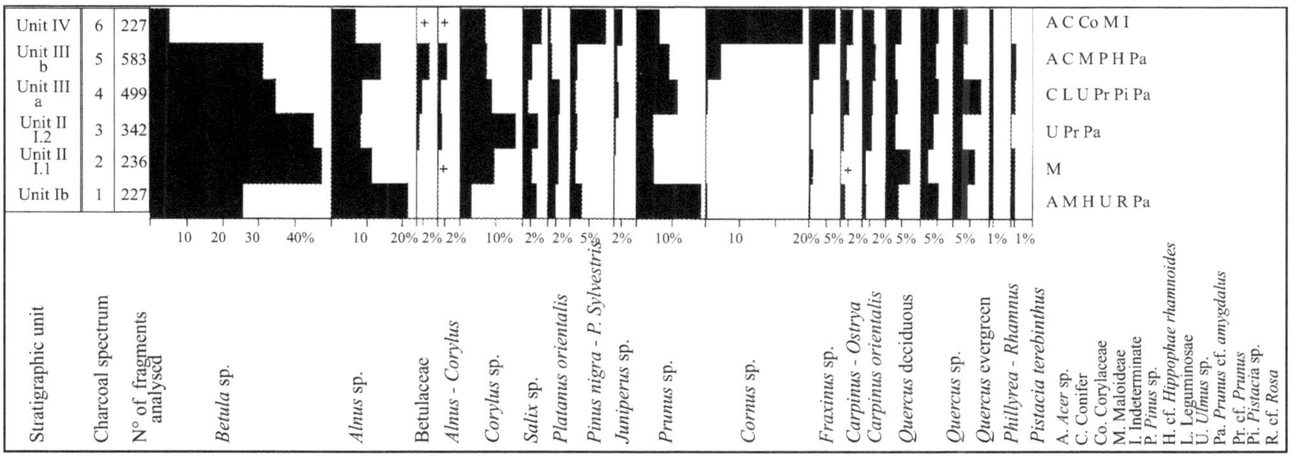

Figure 2: Charcoal diagram from the Boila Late Glacial rock shelter showing relative frequencies of taxa in charcoal assemblages for successive stratigraphic units. Relative frequencies of taxa are calculated on the basis of the fragments identified.

A notable characteristic of the Boila charcoal sequence is the coexistence of ecologically very diverse taxa. On the basis of the ecological characteristics of the taxa identified at Boila, we were able to distinguish between various plant formations.

Riverside vegetation is mainly comprised of *Alnus*, but with a significant component of *Salix* and with an occasional presence of *Platanus orientalis*. These taxa could have formed gallery forests along the riverbanks. *Platanus orientalis* requires special mention. The pollen record in Greece documents the presence of this taxon beginning in *c*. 3500 BP (Bottema, 1982: 274) and its expansion during the Holocene could be associated to geomorphologic, climatic and plant cover changes (Bottema, *ibid*). At Boila, it is present in low frequencies all along the sequence. This could indicate a complementary role of *Platanus* in the vegetation as early as the Late Glacial, assuming that we are not faced with recent infiltrations. Direct AMS dating of this taxon's charcoal fragments could further clarify the stratigraphic provenance at the site context or the dating of its appearance in the Boila sequence and the vegetation record of Greece at large (Bernabeu *et al*. 1999).

Pioneer vegetation is primarily represented at Boila by *Betula*. This is a heliophilous taxon that colonizes open habitats and resists adverse climatic conditions well. We associate this taxon with open forest formations, humid soil types (possibly close to the riverbanks), or even poor and thin ones. Today, *Betula* is very scarce in Greece and it extends southward only as far as the Greek-Bulgarian border (Ntafis *et al*. 1997).

Open parkland vegetation is represented at Boila by heliophilous shrubs like *Prunus* sp., *Prunus* cf. *amygdalus*, *Juniperus* sp. and *Hippophae rhamnoïdes*. These taxa tend to colonize open areas, poor soil and precipitous stony terrain. At Boila, *Prunus* is among the best-represented taxa in the whole sequence. The presence of this taxon is frequently revealed by charcoal analysis of European Late Glacial sites (Heinz and Barbaza, 1998). In Greece, *Prunus* is the most heavily represented taxon all along the Theopetra Cave sequence from *c*. >40,000 BP to *c*. 25,000 BP (Ntinou, 2000b, N.D.). In the assemblages from Boila, various species are present, and anatomical characteristics allowed us to identify *Prunus spinosa*/*P. mahaleb*, *Prunus avium*/*P. prostrata* and *Prunus* cf. *amygdalus*. With regards to *Prunus amygdalus*, it is important to note that this taxon is also a component of present-day steppe forest formations in Irano-Turanian continental territories (Rossignol-Strick, 1995; Zohary, 1973). *Juniperus* sp. in charcoal analysis of the circummediterranean belt is associated with open forest steppe formations, along with *Pinus nigra* or other pioneer sun-seeking taxa (Badal, 1995; Badal, 1997; Heinz and Barbaza, 1998). One of these taxa, *Hippophae rhamnoïdes*, which thrives in sandy, unstable soils and riverbanks, is not encountered in Greece today. However, this taxon is sporadically present at Boila.

Mountain vegetation is represented by *Pinus nigra-P. sylvestris*. Both species are resistant to cold conditions. At present, they form extensive forests within which other plant species scarcely grow due to the high competitiveness of these conifers.

Temperate deciduous vegetation is represented by *Corylus*, *Quercus* deciduous, *Carpinus*/*Ostrya*, *Carpinus orientalis*, *Fraxinus* sp., *Acer* sp., *Ulmus* sp. and *Cornus* sp. *Corylus* sp. is the mesophilous taxon that stands out in the charcoal diagram. Owing to its sun-seeking character, it could have grown in open spaces in the surrounding areas of Boila, as well as among other trees on humid soils close to the riverside. It could be related to different vegetation formations such as those that grow along riverbanks, birch-dominated formations and deciduous thickets. *Cornus* sp. stands out because of its peculiar evolution in the charcoal diagram. The presence of *Cornus* in the Pleistocene pollen record is rare. The taxon is under-represented in the pollen samples (Bottema, 1974: 41). In charcoal analysis, however, it is documented in the French Pyrenees during the Late Glacial interstadial and the Youngest Dryas and has been related to humid, open biotopes under mountain or subalpine dry conditions (Heinz and Barbaza, 1998). The remaining deciduous taxa at Boila were probably growing in low numbers and limited to favourable niches in the vicinity of the site.

Mediterranean thermophilous vegetation is represented in the sequence of Boila by *Quercus* evergreen, *Phillyrea-Rhamnus* and *Pistacia terebinthus*. The overall low frequencies of these taxa could imply that they would have grown in low numbers at favourable places (*e.g.* southerly oriented slopes).

On the basis of the assertions made above, we could attempt to place the plant formations (albeit in a heuristic way) within the Voidomatis catchment. Riverside taxa could have been components of gallery forests along the river. Similarly, birches could have grown in the vicinity of riverbanks or on the neighbouring slopes. Open spaces and rocky sides at the lower gorge walls, which are not particularly steep, could have supported *Prunus*, *Hippophae*, *Juniperus* and other sun-seeking and resistant species. Mediterranean and temperate taxa would have had delimited growing niches some place in the sinuous gorge or in the open plain. Finally, *Pinus* species could have grown on the upper levels of the gorge walls.

At this point, we can focus the discussion on the vegetation evolution as reflected in the Boila charcoal diagram (fig. 2). In the oldest, spectrum 1, riverside vegetation is strongly represented, along with pioneer forest and open parkland elements. According to the radiometric readings, this picture reflects conditions around 15,000 to 14,000 BP. By this time, the qualitative vegetation image pertinent to the whole sequence of Boila can already be detected.

The time span between 14,000 and 13,000 BP corresponds to spectra 2 and 3 and testifies the first well-documented occupation of the rock shelter. The main feature in the vegetation is a notable expansion of birch. In parallel, hazel is becoming more abundant. We could thus postulate an expansion of *Corylus* among sparsely growing birches as a result of progressively better climatic conditions. Temperate and thermophilous taxa that hardly respond to climatic changes are also present in these spectra. Their presence may be viewed as a result of the particularity of the gorge environment. With the exception of a few favourable niches, unstable conditions would have still prevailed.

The first three spectra of the charcoal diagram at Boila seem to correlate well with the postulated improvement in climate of the Late Glacial in Epirus sometime after 15,000 BP (see Turner and Sanchez-Goñi 1997). Before then, steppe formations constituted the main feature in the landscape. The presence of *Prunus* in the charcoal record is reminiscent of Pleniglacial parkland formations (Ntinou, 2000b, N.D.). Nonetheless, it should be noted that the expansion of pioneer forest species, *e.g. Betula*, is documented in this early stage. Spectra 4 and 5 distinguish themselves from the preceding spectra in regards to the values of the prevailing taxa. In spectrum 4, *Betula* continues to be the dominant taxon, but its abundance is reduced by 10%, and *Corylus* shows the same decreasing tendency. On the other hand, *Alnus* and *Prunus* increase, especially the latter. *Juniperus* and some temperate taxa increase slightly. *Quercus* evergreen and *Quercus* sp. are well represented. In spectrum 5, birch keeps descending, as do also some temperate and evergreen elements. *Prunus* exhibits a similar tendency. *Alnus* increases and *Cornus* is clearly present for the first time. Other low frequency mesophilous elements show increasing trends.

Two hypotheses rising from the data of spectra 4 and 5 can be discussed. On the one hand, conditions similar to the earliest spectrum (1) of the sequence may be reflected. If the parkland vegetation represented by *Prunus* is increasing and *Betula*, though still dominant, is decreasing, there should be some kind of change in the prevailing conditions that triggered these modifications, *e.g.* the Younger Dryas episode. However, only the dates from spectrum 5 (Unit IIIa) correspond partly to the Younger Dryas episode where only a slight increase in some temperate and mesophilous taxa is also documented. This in turn makes the hypothesis of a climatic change unfavourable for tree vegetation improbable. On the other hand, the increase of *Prunus* in spectrum 4, dated to the Late Glacial interstadial, could be associated with increasing aridity caused by continuously rising temperatures not compensated by higher precipitation (van Zeist and Bottema, 1982). In spectrum 5, the charcoal diagram detects no changes that could reflect climatic deterioration due to glacial advance. The regression of birch could be seen as a result of a slight expansion of other taxa out of their original growing space. In all, we can assert that this part of the Boila charcoal profile identifies a particular behaviour of plant formations that is also expressed in pollen diagrams of the eastern Mediterranean. The hypothesis proposed is that during that period, a slight decrease in temperatures caused slightly higher precipitation that, in turn, favoured tree expansion (van Zeist and Bottema, 1982).

Spectrum 6 represents the uppermost part of the sequence and is characterized by the dominance of *Cornus* sp. followed by *Pinus nigra-P. sylvestris, Fraxinus* sp., *Corylus, Alnus* and *Prunus*. For the first time, the representation of birches is scarce. A primary analysis of these data could point to the documentation of early Holocene conditions. The low birch percentage seems indicative of more favourable conditions and/or changes in the density of vegetation. The temperate tree taxa could be those that expand as a response to climatic amelioration. The conifers would grow somewhere in the surroundings and their increased representation could indicate persisting continental conditions and open spaces that were progressively occupied by those heliophilous trees.

High *Cornus* percentages are difficult to explain in pure ecological terms, since an analogous situation is absent from present-day conditions in the area. The marked increase of *Cornus* in spectrum 6 could be closely linked to some particular human selection. *Cornus mas* wood is strong, dense, resistant, and durable, while *Cornus sanguinea* branches are flexible and used in basketry. Those properties of the wood could, for instance, be appreciated for the manufacture of composite hunting equipment. Lithic armatures are the most abundant tool type in the industries of Stratigraphic Unit IV (Kotjabopoulou *et al.* 1999). As things stand at present, the interpretation of the spectrum 6 charcoal composition has to remain an open issue.

In summary, the charcoal diagram from Boila covers part of the Late Glacial with a possible representation of former and later periods. The data from charcoal analysis mainly reflect two major phases in the evolution of the vegetation in the vicinity of the site. The first phase is comprised by spectra 1 through 5, which seem to form part of the same Late Glacial framework without any significant differences in the vegetation composition in proximity to the site. The second distinct phase identified corresponds to spectrum 6, where the changes in the vegetation composition, as evidenced in charcoal remains, can be read in association with early Holocene conditions.

Boila in its local and regional context

The above analysis entails diachronic aspects of the vegetation history in Epirus. In this section, we will consider the interplay between climatically induced changes and competition of interspecific variables, especially in light of this area's sustained long-term tree taxa refugia. Following this line of reasoning, the Boila charcoal profile will be juxtaposed with the pollen records west of the Pindus mountain range.

The charcoal data from the Boila rock shelter do not document any neat division in biozones of Older Dryas-Late Glacial interstadial-Younger Dryas. This is in agreement with pollen diagrams from various areas in Greece that lack the elements to justify such subdivisions. The basic feature of the Greek pollen record is the pronounced regional diversity. Particularly, trends of this nature are documented in the circumscribed Epirus region. Variable vegetation succession schemes are recorded. The vegetation is widely varied, as a result of climatic conditions in evolution, of competition among species and of persistence of trees in favourable niches, probably associated with vegetation refugia. To date, there is a prolific pollen database recording the Late Glacial vegetation structure(s). There are five such records in this area (fig. 3), namely Ioannina I (Bottema, 1974), Ioannina 249 (Tzedakis, 1993, 1994), Gramousti (Willis, 1992a, 1997), Tseravinas and Ziros (Turner and Sanchez-Goñi, 1997). In addition, a summary diagram has recently been published from Ioannina, I284, (Galanidou *et al.* 2000), while the Rezina sequence (Willis, 1992b, 1997), although it does not preserve a record of the Late Glacial, strictly speaking, is of relevance to the issues under consideration. Against such a rich background, we will discuss in the next paragraphs how the Boila evidence falls into proposed models of vegetation succession.

In all, the pollen record is not conclusive as to the characteristics of the vegetation in Epirus during the Late Glacial (after 15,000 BP) (Turner and Sanchez-Goñi 1997: 580). Pollen diagrams from Ioannina I (Bottema, 1974) and Ioannina 249 (Tzedakis, 1993, 1994), the larger basin in the region (480 m a.s.l.), and Gramousti (Willis, 1992a) (*c.* 400 m a.s.l.) share common features. In Ioannina I, the Late Glacial documents a persistence of steppe-forest vegetation at low altitudes. As far as tree taxa are concerned, from 14,000 BP the diagrams detect the expansion of *Pinus* and secondarily of *Abies* in the mountains. *Quercus* has a continuous curve, and some deciduous species are also present. As far as the tree taxa are concerned, the Late Glacial is characterized by the alternating increase of conifers and temperate species. The decline of the conifers, according to van Zeist and Bottema (1982: 285), was the result of a rise in temperature not compensated by higher precipitation. The Younger Dryas is thought to be identifiable in the expansion of *Pinus*, which suggests a rise in humidity, in part due to a decrease in temperature (van Zeist and Bottema, 1982: 285). Finally, the end of the Late Glacial is characterized by a rise of deciduous taxa. The Gramousti sequence seems to support this assertion, as it derives from a medium-sized tectonically controlled basin in the hinterland of Epirus, close to the Voidomatis watershed. The Late Glacial vegetation is characterized by steppe-type communities (Willis 1992a). The Ioannina 249 core shows a condensed sequence for the Late Glacial mainly characterized by conditions typical for steppe formations. Also, the summary pollen diagram from Ioannina, I284, documents a series of fluctuations in terms of intermediate expansions of woodland within an overall open landscape for the Late Glacial (Galanidou *et al.* 2000: 353). The picture for the vegetation derived from the above diagrams points to the prevalence of "steppic" conditions (except perhaps for the summary pollen Ioannina I284). Tree species are limited to favourable locations. The forest expansion is confined to the Late Glacial/Postglacial transition. This interpretation is in agreement with other pollen sequences from sites east of the Pindus massif, namely the Xinias sequence in east-central Greece (Bottema, 1979), the Khimaditis and Edessa sequences in western Macedonia (Bottema, 1974) (fig. 3).

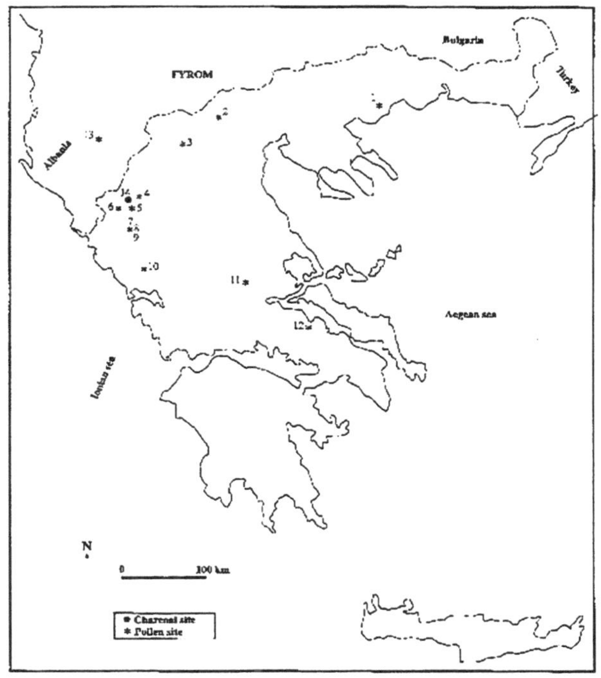

Figure 3: Map of Greece showing location of pollen and charcoal sites discussed in the text. 1: Tenaghi Philippon, 2: Edessa, 3: Khimaditis, 4: Rezina, 5: Gramousti, 6: Tseravinas, 7: Ioannina I, 8: Ioannina 249, 9: Ioannina I284, 10: Ziros, 11: Xinias, 12: Kopais, 13: Maliq, 14: Boila.

Now, other recently published sequences recovered from small basins in Epirus, Tseravinas (*c.* 400 m a.s.l.) and Ziros (50 m a.s.l.) testify to a more complicated picture of the vegetation evolution during the Late Glacial. Furthermore, the dating evidence (AMS) and the vegetation schemes registered in these pollen cores, according to Turner and Sanchez-Goñi (1997: 580) pre-date the conventionally accepted climatic improvement. While Tseravinas is another inland intermountain site, the Ziros Lake is the only lowland site in the southern part of the region and closer to the present coast studied thus far. From 15,000 BP onwards, forest expansion is observed in both diagrams. At Ziros, oak woodland expands more rapidly while at Tseravinas plant formations become dominated by *Quercus* and *Corylus* after the initial expansion of *Betula*. Also, Mediterranean taxa are present especially at Ziros. Some time during the Late Glacial, especially at Tseravinas, tree vegetation declines, *Pinus* becomes dominant and open habitats increase. The dating of this oscillation is attributed to either the Younger Dryas or a former cold episode (Turner and Sanchez-Goñi, 1997: 581). In all, these diagrams manifest a mosaic of vegetation that combines forest and steppe elements. The specific geographical/topographical settings of these sites emerge as decisive factors of variability in regional terms. The information showing the evolution of the vegetation in Epirus derived from those new sequences is not isolated. Other pollen diagrams from the rest of continental Greece already register a similar process of forest expansion as early as the Late Glacial, which culminates with the beginning of the Holocene. This is the case of the Tenaghi Philippon (Wjimstra, 1969) and the two Kopais diagrams (Allen, 1986 cited in Willis, 1994; Allen, 1990; Turner and Greig, 1975) (fig. 3).

In the brief overview above, we see that the pollen record of Epirus (and of Greece on the whole) has given rise to a debate concerning the timing of forest expansion. The question is whether this took place within the Late Glacial or at the Late Glacial/Postglacial transition. The problem rises partly from the correlation of dates and those pollen zones, which show a marked increase of Arboreal Pollen. An in-depth discussion of this complicated issue, raised already during the 70's (Bottema, 1974), is beyond the scope of this paper. However, it remains a fact that the vegetation evolution in Greece covering the Late Glacial and the beginning of the Holocene has particularities that do not conform to the "biozonation" established by North European pollen records. If we narrow our focus to the Epirus region alone, we can say that the proposed models of vegetation succession show differences. No researchers deny that steppe and forest formations contracted and expanded in relation to each other, and that fluctuations occurred probably as a response to climatic change. The discrepancy in interpretation perhaps arises partly due to the different catchment areas of the pollen sites or to the integrity and/or resolution of certain cores (see discussion in Turner and Sanchez-Goñi, 1997).

As we will try to show in the next paragraphs, the Boila charcoal diagram is in tune with the broader framework of vegetation structure(s) derived from the pollen sequences and testifies to the varied character (*i.e.* geographic differentiation) of the vegetation profile in Epirus.

Affinities between the charcoal diagram and those pollen diagrams that detect a clear expansion of tree taxa during the Late Glacial can be traced. The sequences of Tseravinas and Ziros manifest this evolution. The marker of such expansion is notably expressed by the birch and oak curves. Oak, however, is not significant at Boila, where we see woodland expansion manifested by the dominant role of birch from the very beginning of the record. It should be mentioned that at Tseravinas, the pollen site most closely situated to Boila, there is an initial expansion of birch, followed by hazel, before the marked oak progression (at *c.* 15,000 BP, Turner and Sanchez-Goñi, 1997: table 29.1). Those characteristics correspond to the trends of birch and hazel in spectra 2 and 3 of the Boila charcoal diagram. As far as the oak representation at our site is concerned, we are perhaps seeing a very local character of the vegetation conditioned by the physiographically diverse catchment of the Voidomatis. This emphasizes that diachronic patterns of the landscape configuration (bedrock categories, soil properties, relief, altitude etc.) have significant influence on the characteristics and composition of various plant communities.

An added point that is inextricable from the above discussion and brought forward by the Boila charcoal diagram is the importance of endogenous vegetation variables (*e.g.* competition). In this sense, we also have evidence at Boila of the constant presence of tree taxa that appear in all pollen sequences (of Epirus and Greece), regardless of differing interpretations advanced as to the vegetation succession during the Late Glacial. Hornbeam, evergreen oak, willow, alder, ash are examples of this common feature between the Boila charcoal and the pollen records. These temperate, mesic and Mediterranean taxa have been linked with long-term vegetation refugia in the Balkan Mountains, among them the Pindus massif in Epirus (Bennett *et al.* 1991; Tzedakis 1993, 1994). The presence of these taxa at Boila lends further support to the idea of refugia. Locating these areas is difficult. It has been postulated that such niches would be found in high altitudes. The Rezina pollen diagram offers information to this affect (Willis, 1992b). The site is located at a close distance, at 1800 m a.s.l, on the Tymfi range bordering the gorge. The pollen sequence starts at approximately 9,800 BP and from the very beginning the pollen zones are rich in tree taxa, including *Pistacia* and *Phillyrea*. The early expansion of tree taxa at such high altitudes reflects their persistence at least during the Late Glacial in favourable niches in the local setting (Willis, 1997). Recent pollen investigations at Lake Maliq, Albania, document a great diversity of temperate forest taxa as early as 12,000 BP at an altitude of 818 m. The authors (Denèfle *et al.* 2000) consider these middle altitudes more favourable for the maintenance of forest elements and the location of forest refugia, and place the Late Glacial altitudinal boundary between steppe and forest ecosystems at the level of Lake Maliq. In the case of Boila, if the dispersed charcoal is the result of wood procurement in the region surrounding the site but not exceeding distances of a few kilometres, such refugia could have been situated within the drainage basin of the Voidomatis. It is not the objective of this study to assess precisely the mobility of human groups, but we could propose that everyday needs for wood would have been met around the site. Even if the river acted as an agent of transport of dead wood, those taxa still would have grown in the valley/plain surroundings.

Finally, in light of the pollen data in Epirus and particularly the Early Holocene Rezina data, we could view the upper part of the Boila charcoal diagram in conjunction, in case this is an early Holocene depositional record. Charcoal spectrum 6 shows a substitution of birch by other deciduous taxa and a marked rise of *Pinus*. The latter is a common element between our charcoal and the pollen data for the Late

Glacial/Holocene transition. According to the interpretations of the pollen record, such a feature is attributed to the repercussions of the Younger Dryas. This could mean that conditions at Boila during the deposition of Stratigraphic Unit IV manifest continentality. Alternatively, the decrease in the birch frequency could be viewed as a response to increased competition from other tree species. However, a combination of the above processes may have given rise to a still poorly understood Late Glacial/Holocene transition.

Conclusions

The analysis of the dispersed charcoal at Boila rock shelter, in north-western Greece, is a corpus of ecological information that sheds light upon the composition of plant formations in and around the site. This kind of evidence, seen in association with long established pollen research in Epirus, leads us to draw the following conclusions:

- 1. the vegetation around the site was rich in tree taxa representing ecological diversity. Gallery forests along the riverbanks, stands of birch, mountain conifer expanses, open parkland formations, and temperate and Mediterranean thickets are the main aspects of the plant cover.
- 2. The charcoal sequence is characterized by qualitative homogeneity. Nearly half of the taxa identified are constantly present in all spectra.
- 3. The main feature of the charcoal diagram is the sustained forest component from the very beginning of the sequence. Birch is the dominant taxon from 14,500 to 10,000 BP. Minor fluctuations in the relative abundance of the principal taxa are documented from spectrum 1 through 5. Nevertheless, they do not justify any significant modification of the surrounding vegetation.
- 4. The only part of the record that documents a marked change is spectrum 6. *Pinus* and temperate taxa, principally Cornus, dominate. It is possible this change indicates a climatic event (*e.g.* the Younger Dryas) or a specific human selective behaviour.
- 5. The clear-cut subdivision of the Late Glacial in biozones-chronozones is not recorded. This is in agreement with the results from pollen sequences and other palaeogeographical evidence in the region.
- 6. The expansion of woodland is a shared feature between some pollen sequences and the Boila charcoal diagram. Differences concerning the respective dominant taxa could be attributed to macro- and microclimatic conditions, landscape variables and competition between species. The chronological scheme established for the Boila sequence adds weight to an "early" timing for the woodland expansion.
- 7. The coexistence of ecologically diverse tree taxa in an overall homogeneous record contributes to the understanding of the concept of vegetation refugia in the southern Balkans. The nature of the material analysed, *i.e.* fuel wood procured by humans in their immediate surroundings, gives further information as to the location of tree refugia. The lower Vikos gorge, in particular, emerges as a good example of such a potential location.

ACKNOWLEDGEMENTS
We extend our thanks to Dr. Ernestina Badal and Dr. Charles Turner for comments on an early draft of this paper. Any responsibility regarding interpretation however, remains with the authors.

References

ALLEN H. 1990 - A Postglacial record from the Kopais Basin, Greece. In Bottema S., Entjes-Nieborg and van Zeist (eds): *Man's role in the Shaping of the Eastern Mediterranean Landscape*, Balkema, Rotterdam, p. 173-182.

BADAL E. 1992 - L'anthracologie préhistorique : à propos de certains problèmes méthodologiques. *Bull. Soc. bot. Fr.*, 139, Actual. bot. 1992-2/3/4. Les charbons de bois, les anciens écosystèmes et le rôle de l'homme, p. 167-189.

BADAL E. 1995 - La vegetación carbonizada. Resultados antracológicos del País Valenciano. *El Cuaternario del País Valenciano*, p. 217-226.

BADAL E. 1997- El paisatge vegetal de la Marina, a partir dels carbons prehistórics. *Aguaits*, 13-14, p. 23-37.

BAILEY G.N. (ed). 1997 - *Klithi: Palaeolithic settlement and Quaternary landscapes in northwest Greece*. Vol. 1: Excavation and intra-site analysis at Klithi. Vol. 2: Klithi in its local and regional setting. Cambridge: Mc Donald Institute for Archaeological Research.

BAILEY G., TURNER C., WOODWARD J. C., MACKLIN M. G. and LEWIN J. 1997 - The Voidomatis Basin: an Introduction. In Bailey G.N (ed): *Klithi: Palaeolithic settlement and Quaternary landscapes in northwest Greece*, Vol. 2. McDonald Institute Monographs. p. 321-345.

BENNETT K.D., TZEDAKIS P.C., WILLIS K.J - 1991. Quaternary refugia of north European trees. *Journal of Biogeography*, 18, p. 101-115.

BERNABEU AUBAN J., PÉREZ RIPOLL M., MARTÍNEZ VALLE R. 1999 - Huesos, Neolitización y Contextos Arqueológicos Aparentes. II Congrés del Neolitic a la Peninsula Ibérica, *Saguntum*-PLAV Extra-2, p. 589-596.

BOTTEMA S. 1974 - *Late Quaternary vegetational history of northwest Greece*. Thesis, University of Groningen.

BOTTEMA S. 1979 - Pollen investigations in Thessaly (Greece). *Palaeohistoria*. 21, p. 19-40.

BOTTEMA S. 1982 - Palynological investigations in Greece with special reference to pollen as an indicator of human activity. *Palaeohistoria*. 24, p. 257-289.

CHABAL L. 1988 - Pourquoi et comment prélever les charbons de bois pour la période antique : les méthodes utilisées sur le site de Lattes (Hérault). *Lattara* 1. p. 187-222.

DENÈFLE M., LÉZINE A.-M., FOUACHE E., DUFAURE J.-J. 2000 - A 12,00-Year Pollen Record from Lake Maliq, Albania. *Quaternary Research*. 54, p. 423-432.

GALANIDOU N., TZEDAKIS P. C., LAWSON I. T., FROGLEY M.R. 2000 - A revised chronological and palaeoenvironmental framework for the Kastritsa rockshelter, northwest Greece. *Antiquity*. 74 p. 349-355.

HEINZ C., BARBAZA M. 1998 - Environmental changes during the Late Glacial and Post-Glacial in the central Pyrenees (France): new charcoal analysis and archaeological data. *Review of Palaeobotany and Palynology*. 104, p. 1-17.

HIGGS E. S., WEBLEY D. 1971 - Further information concerning the environment of Palaeolithic man in Epirus. *Proceedings of Prehistoric Society*. 37, p. 367-380.

KOTJABOPOULOU E., PANAGOPOULOU E., ADAM E. 1997 - The Boila Rockshelter: a preliminary report. In Bailey G.N. (ed): *Klithi: Palaeolithic settlement and Quaternary landscapes in northwest Greece*, Vol. 2. McDonald Institute Monographs: p. 427-437.

KOTJABOPOULOU E., PANAGOPOULOU E., ADAM E. 1999 - The Boila rockshelter: further evidence of human activity in the Voidomatis Gorge. In Bailey G.N., Adam E., Panagopoulou E., Perlès C. and Zachos K. (eds): *The Palaeolithic Archaeology of Greece and Adjacent Areas*. British School at Athens Studies, 3 p. 197-210.

LEWIN J., MACKLIN M.G., WOODWARD J.C. 1991 - Late Quaternary fluvial sedimentation in the Voidomatis Basin, Epirus, Northwest Greece. *Quaternary Research*, 35, p. 103-115.

MACKLIN M.G., LEWIN J., WOODWARD J.C. 1997 - Quaternary river sedimentary sequences of the Voidomatis basin. In Bailey G. N (ed): *Klithi: Palaeolithic settlement and Quaternary landscapes in northwest Greece*, Vol. 2. McDonald Institute Monographs, p. 347-359.

NTAFIS S., PAPASTERGIADOU E., GEORGIOU K., MPAMPALONAS D., GEORGIADIS TH., PAPAGEORGIOU M., LAZARIDOU TH., TSIAOUSI V. 1997 - Odigia 92/43/EOK. To Ergo Oikotopon stin Ellada: Diktyo FYSI. Symvolaio arithmos B4-3200/84/756, Gen. *Dieuthinsi Epitropi Europaikon Koinotiton, mouseio goulandri Fysikis Istorias* - Elliniko Kentro Biotopon Ygrotopon, 932 p.

NTINOU M. 2000a - *El paleoambiente en el norte de Grecia desde el Tardiglaciar hasta el Atlántico: vegetación, recursos y usos*. Unpublished Ph.D., University of Valencia.

NTINOU M. 2000b - Preliminary results of charcoal analysis from Theopetra Cave. (In Greek). *Theopetra Cave. Twelve years of excavation and research 1987-1998*. Proceedings of the International Conference, Trikala, 6-7 November 1998, p. 69-80.

NTINOU M.N.D - *Charcoal analysis at Theopetra Cave*. Final Report for the Geoarchaeology Fellowship, Wiener Laboratory, American School of Classical Studies at Athens.

POLUNIN O. 1980 - *Flowers of Greece and the Balkans: a field guide*. Oxford University Press.

ROSSIGNOL-STRICK M. 1995 - Sea-land correlation of pollen records in the eastern Mediterranean for the Glacial-Interglacial transition: biostratigraphy versus radiometric time-scale. *Quaternary Science Reviews*, 14, p. 893-915.

TURNER J., GREIG J.A.R. 1975 - Some holocene pollen diagrams from Greece. *Review of Palaeobotany and Palynology* , 20, p. 171-204.

TURNER C., SANCHEZ-GOÑI M-F. 1997 - Late Glacial Landscape and vegetation in Epirus. In Bailey G.N. (ed): *Klithi: Palaeolithic settlement and Quaternary landscapes in nortwest Greece*, Vol. 2. McDonald Institute Monographs, p. 559-585.

TZEDAKIS P.C. 1993 - Long-term tree populations in northwest Greece through multiple Quaternary climatic cycles. *Nature*, 364, p. 437-440.

TZEDAKIS P.C. 1994 - *Vegetation change through glacial-interglacial cycles: a long pollen sequence perspective*. Philosophical Transactions of Royal Society of London B 345, p. 403-432.

WIJMSTRA T.A. 1969 - Palynology of the first 30 metres of a 120 m deep section in northern Greece. *Acta Botanica Neerlandica*, 18, p. 511-527.

WILLIS K.J. 1992a - The late Quaternary vegetational history of northwest Greece. I. Lake Gramousti. *New Phytologist*, 121, p. 101-117.

WILLIS K.J. 1992b - The late Quaternary vegetational history of northwest Greece. II. Rezina Marsh. New Phytologist, 121, p. 119-138.

WILLIS K.J. 1994 - The vegetational history of the Balkans. *Quaternary Science Reviews*, 13, p. 769-788.

WILLIS K.J. 1997 - Vegetational history of the Klithi environment: a palaeoecological viewpoint. In G.N. Bailey (ed) *Klithi: Palaeolithic settlement and Quaternary landscapes in northwest Greece*, Vol.2. McDonald Institute Monographs: p. 395-413.

WOODWARD J.C., HAMLIN R.H.B., MACKLIN M.G., KARKANAS P., KOTJABOPOULOU E. 2001 - Quantitative Sourcing of Slackwater Deposits at Boila Rockshelter: A Record of Late-Glacial Flooding and Palaeolithic Settlement in the Pindus Mountains, Northwest Greece. *Geoarchaeology*, 16, 5, p. 501-536.

ZEIST VAN W., BOTTEMA S. 1982 - Vegetational history of the Eastern Mediterranean and the Near East during the Last 20,000 years. In: J. L. Bintliff and W. van Zeist (eds): Palaeoclimates, Palaeoenvironments and Human Communities in the Eastern Mediterranean Region in Later Prehistory. *British Archaeological Reports International Series*, 133, p. 277-321.

ZOHARY D. 1973 - *Geobotanical foundations of the Middle East*. Gustav Fischer Verlag, Stuttgart, 2 vol., 738 p.

Late-Glacial and Holocene Vegetation and Prehistoric Wood-Use in the Istrian Karst, Croatia

William John Fletcher

Quaternary Palaeoenvironments Group, Department of Geography,
University of Cambridge, Cambridge CB2 3EN, United Kingdom
E-mail: wjf20@cam.ac.uk

Abstract

Recent archaeological excavations in the Istrian peninsula (Croatia) provide evidence for the occupation of karstic caves during the late-Glacial and Holocene. Within the highly erosional karstic landscape, caves are important depositional locations, containing records of both human activity and palaeoenvironmental conditions. Macroscopic charcoal has been identified from four caves located in or near a single gorge in northeastern Istria. The identification of woody taxa within the charcoal assemblages provides a record of wood use at the sites, and important evidence of vegetation changes in the local landscape. Three major vegetation phases are identified. This sequence is interpreted primarily in terms of climate change since the late-Glacial and micro-climatic effects of the karst topography, with consideration given to the possible influence of post-Glacial tree migration and human activity.

Keywords: *charcoal analysis, vegetation history, climate change, human impact, karst, Adriatic, Istria*

Résumé

Des fouilles récentes dans la péninsule d'istrie (Croatie) ont mis en évidence l'occupation des grottes au Tardiglaciaire et à l'Holocène. Ces grottes constituent d'importants dépôts témoins des activités humaines et des conditions paléoenvironnementales. Quatre cavités, situées à proximité ou dans une gorge localisée au nord de l'Istrie, ont fourni des charbons de bois qui ont été identifiés. Les résultats de ces identifications apportent des données sur l'utilisation du bois dans ces gisements et montrent d'importants changements de végétation dans le paysage local. Trois phases sont identifiées. Cette séquence est d'abord interprétée en terme de changements climatiques depuis le Tardiglaciaire et les effets micro-climatique de la topographie karstique, en prenant en considération la possible migration des arbres au Tardiglaciaire ainsi que les activités humaines.

Mots-clés: *Anthracologie, végétation, changement climatique, impact humain, kerst, Adriatique, Istrie.*

Introduction

This paper presents the results of introductory charcoal analyses from four archaeological cave sites in the Istrian karst. The study formed the basis of a Masters degree in Quaternary Science at Cambridge University and is part of a wider project of archaeological investigation in Istria, the Pupićina Cave Project, under the direction of Dr Preston Miracle of Cambridge.

The study region is located in Istria, in the northern Adriatic basin (fig. 1a), within the broad band of Dinaric karst which stretches from Trieste and the Italian border to Albania and northwestern Greece (Herak, 1972). The distinctive landforms of the karst such as dolines, poljes, gorges and caves are highly evident in the region. Climatic conditions in the region are Mediterranean, though moderated by cooler temperatures and higher precipitation than more southerly regions (Palutikof *et al.* 1996). The regional flora (fig. 1b) follows a general pattern of altitudinal variation similar to that of many parts of the Mediterranean region (Polunin and Walters, 1985). At the scale of the study region, local vegetation patterns are also structured by the rugged topography of the karst landscape (Polunin, 1980), which provides a great variety of conditions in terms of slope, micro-climate, and soil stability. The rugged karst landscape also moderates the influence of human impact, with terrain features such as gorges providing refuges for plants from agricultural activity (Polunin, 1980).

The Vela Draga gorge is situated between 200 and 300 m a.s.l., measures c. 4 km in length, and drops 100 m in altitude below the surrounding landscape (fig. 1c). Within the gorge, three limestone caves (Pupićina, Vela and Vešanska caves) have been recently excavated. Excavation of the sites has produced archaeological evidence for periodic occupation of the gorge since the end of the last glacial period (Miracle, 1997). A fourth site, Šebrn, located c. 2.5 km from the gorge, at 800 m a.s.l. provides comparative deposits of early Holocene age (Miracle *et al.* 2000). The collection of charcoal material from these sites represents a potential source of information regarding both the vegetation history of the region and firewood collection in prehistoric times. New environmental information is especially important in the region because previous palaeoecological research has been of limited extent and focused on the Istrian coastline (Culiberg and Šercelj, 1995; Beug, 1977) and polliniferous sediments are rare. The close proximity of the sites and the location of three of the sites within a single topographic feature means that the identified charcoal assemblages should be complementary and should provide a good indication of vegetation changes in spite of taphonomic and cultural factors which may have influenced the individual assemblages.

Site

As excavation details are not yet published for all the sites, only stratigraphic summaries are provided here.

Pupićina Cave

Pupićina cave measures 25m wide by 30m deep, and is situated 18m above the floor of the Vela Draga Gorge (Miracle, 1997). Excavation has exposed a stratigraphy of 4m maximum depth, containing late Glacial and Holocene deposits, and has not reached bedrock. In the lower part of the stratigraphy (Horizons V-P), stony occupation layers alternate with archaeologically sterile, massive silty-clay deposits. The earliest archaeological level (Horizon V) is dated to the Allerød chronozone, c. 13,000 cal BP. With the onset of the Holocene, the stratigraphy is dominated by thick Mesolithic midden deposits of bone, shell, ash and charcoal (Horizons N,M). The top of the midden forms a hard-packed crust (Horizon J), interpreted as an erosional surface. Radiocarbon

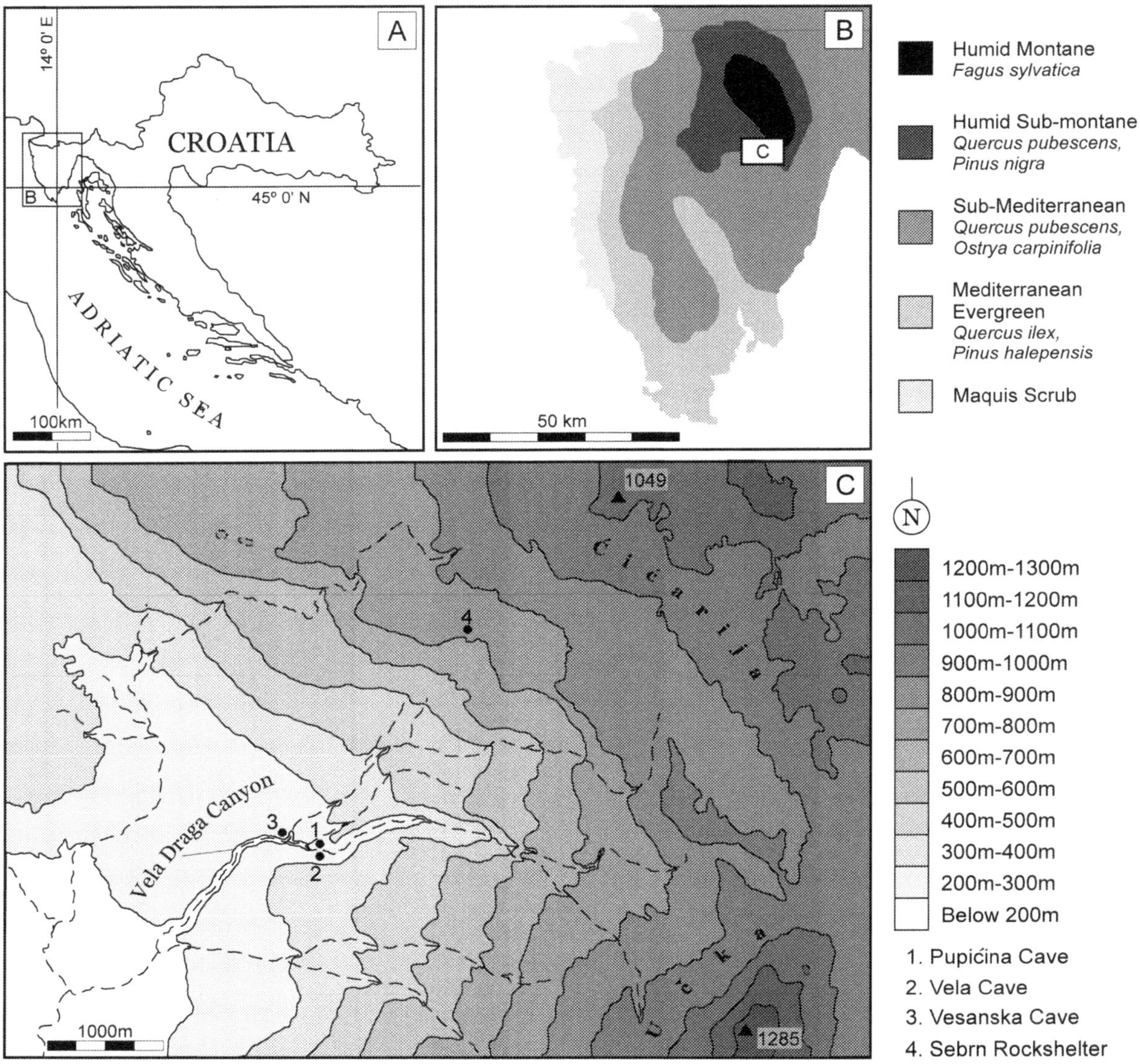

Figure 1: Geography of the study region. **a:** Croatia; **b:** Istria, with major vegetation zones indicated (adapted from UNESCO 1970); **c**: Topography of the Vela Draga Gorge, with archaeological sites indicated.

dates from above and below the crust suggest missing deposits representing perhaps as much as two thousand years. Above the crust, there are middle and late Neolithic strata (Horizon I-G). Into these deposits intrudes a large Bronze Age pit (Horizons F,E). Finally, there are post-Bronze Age levels (Horizons C,B) containing Iron Age and Roman period ceramics, and above these, mixed surface deposits (Horizon A). Abundant charcoal fragments were collected from all but the post-Bronze Age levels.

Vela Cave

Vela cave is 28m wide by 20m deep, and is located only 50m from Pupićina on the opposing face of the gorge. To date, excavation has exposed a Holocene stratigraphy of greater than 5m depth. Unlike Pupićina, Vela contains a deep historic sequence of deposits, which have been disturbed by burrowing. The lower, prehistoric levels are undisturbed. A radiocarbon date of c.11,000 cal BP on charcoal from level 141 confirms that the excavated sequence spans almost the entire Holocene. Charcoal samples were identified from the lower levels (141-118).

Vešanska Cave

Vešanska cave (14m wide by 14m deep) is located less than 1km from Pupićina and Vela, within the Vela Draga gorge. Excavation of a 1x2m trench has exposed a complex stratigraphy which suggests several phases of erosion and deposition. It appears likely that sediments entered the cave by an internal passage as well as via the present cave entrance. Excavation yielded a late Upper Palaeolithic lithic assemblage and a large faunal assemblage. Radiocarbon dates on charcoal from stratigraphic unit II.3, which is the primary cultural deposit, and unit II.3A (a lense of humic clay within II.3) indicate during the Allerød chronozone between 14,000 and 13, 000 cal BP. Little charcoal was recovered from the site and all available fragments from the dated deposits II.3 and II.3A were identified.

Šebrn Rockshelter

Šebrn is a small, open rockshelter, 6m deep by 13m wide and 7m high, located in Ćićarija above the Vela Draga Gorge. The site is at the foot of a cliff where early Holocene deposits have been preserved by fallen limestone blocks. Excavation of these deposits revealed a simple stratigraphy of 90cm depth (Miracle *et al.* 2000). Anthropogenic deposits (levels 3-6) contained assemblages of Mesolithic faunal and lithic remains. Radiocarbon dates on charcoal from levels 3 and 6 indicate that the main occupation of the site occurred within a brief period of at most c.800 calendar years around 10,000 calendar years BP. Abundant charcoal was recovered during excavation. Charcoal fragments from a single level (3B) were identified.

Method

Charcoal samples, collected during the on-site sieving of excavated sediment, were selected for identification with respect to the stratigraphy and chronology of the sites. Prompted by the observation that archaeological features generally provide samples of lower taxonomic diversity than occupation levels (Badal-García, 1992; Heinz, 1991), only samples from occupation levels were identified. From Pupićina cave, which contains a long stratigraphic sequence, samples were selected with the aim of identifying vegetational changes over the chronological span of the site. To gain a comparative sequence, samples were identified from the intact, lower prehistoric sequence at Vela. From Vešanska and Šebrn, which present shorter stratigraphic sequences, charcoal fragments were identified from radiocarbon dated levels to provide a "snapshot" comparison for a particular date. Fragments greater than 2mm were considered identifiable. Fractured surfaces were observed under both low-power (6X-50X) and high-power (100X-1000X) magnification. Identifications were made to the highest confident taxonomic level with reference to printed material (Schweingruber, 1990), reference charcoal from the region, and wood in thin section.

Results

The results of the charcoal analysis are presented for each site (tabls. 1-4). Four charcoal phases are identified, on the basis of the changing presence of taxa in the samples, and will aid the discussion of the results. It is stressed that the interpretation that follows is based primarily on the positive identification, or presence, of taxa in the samples, as the relative frequencies of taxa (particularly when overall sample size is small) may not necessarily be closely representative of the state of past vegetation (Figueiral and Mosbrugger, 2000; Cartwright and Parkington, 1997).

Geological period	PLEISTOCENE							HOLOCENE								
Archaeological period	Upper Pal.				Mesolithic				Neolithic		Bronze Age		Post-Bronze Age			
Horizon & Level	V 373	S 35	O 207	N 206	M 203A	M 203	M 202	I 431	H 424	G 412	E 411	E413	B 408	B 407	B 405	B 404
Cal BP (1 σ)	13,200-12,980		12,150-11,150	11,170-10,690		10,150-9500	10,640-10,190	7690-7250				3640-3210, 3700-3410			2710-2350	
Charcoal Phase				1				2			3				4	
Abies alba							1									
Pinus sp.	6	1	1			1	1									
Pinus sylvestris/nigra	230	89	90	39	35	47	92									
Juniperus sp.											1					
Taxus baccata											5	2				
Clematis vitalba												6				
Ulmus sp.					3			1	4	3	4	2				
Juglans regia										2						
Fagus sylvatica								3	12	5	10	1	1			
Quercus sp.								1	1							
Quercus deciduous			5				2	13	6	12	22	4	6		3	
Quercus deciduous/ *Castanea sativa*								1								
Quercus ilex/coccifera									1		5	23				
cf. *Quercus* sp.										1	2	3	2			
Carpinus sp.											5	11			1	1
Ostrya carpinifolia											2	4				
Ostrya/Carpinus											6	5				
Corylus cf. *Avellana*									1							
Erica type *arborea*											2					
Prunus cf. *avium*													1	1		
Prunus type *avium/padus*													1			
Prunus cf. *spinosa*	3															
Pomoideae									1	1	1	3				
Leguminosae													4			
Cornus sp.									9	11	33	12	5		1	
cf. *Cornus* sp.											2					
Euonymus sp.															2	
Rhamnus alaternus/Phillyrea										24	17					
Vitis vinifera												12				
Aesculus hippocastanum															1	
Acer sp.									2							
Acer type *campestre*										6	1					
Acer platanoides/pseudoplatanus							1		12	3		2				
Acer non-*platanoides/pseudoplatanus*		1							8							
cf. *Acer* sp.										1	2					
Fraxinus sp.	1		1					87	36	24	89	22			13	1
Fraxinus cf. Ornus								10	4	3	6	2	3		2	
Indeterminate Type 1											1					
Unidentified Angiosperm								1	2	1	5	2				
Unidentified Gymnosperm	5	5	5			1	2									
Unidentifiable	8	2	3	1	2	1	1		1	3	1	2	1			
Total	253	98	105	40	40	50	100	117	100	100	222	118	24	1	23	2

Table 1: Absolute numbers of charcoal fragments per level by taxon, for Pupićina Cave.

Pupićina Cave

The charcoal assemblage from Pupićina (tabl. 1) is marked by a strong contrast between the late-glacial and early Holocene samples, which are dominated by *Pinus sylvestris/nigra*, and the mid and late Holocene samples, which are composed of a range of deciduous and evergreen taxa. This contrast is made particularly apparent by the hiatus in deposition between the early Holocene Mesolithic levels and the mid-Holocene Neolithic levels. In the older samples, (Charcoal Phase 1), pine is accompanied by *Abies alba* in one level, and by deciduous taxa *Ulmus* sp., *Fagus sylvatica*, deciduous *Quercus*, *Rosaceae*, *Acer* sp. and *Fraxinus* sp. In Phase 2, pine is absent, and has been replaced by *Fagus sylvatica*, deciduous *Quercus*, and by *Fraxinus* sp., which is strongly represented.

Phase 3 is identified on the basis of the appearance of several new taxa in the assemblage. These include Mediterranean taxa such as *Quercus ilex/coccifera*, *Rhamnus alaternus/Phillyrea* and *Erica* type *arborea*; conifers *Juniperus* sp. and *Taxus baccata*; deciduous taxa *Cornus* sp., *Juglans regia*, and *Ostrya carpinifolia* and *Carpinus* sp.; and climbers *Clematis vitalba* and *Vitis vinifera*. A separate Phase 4 is identified on the basis of apparent changes in the representation of taxa; these may relate to the overall reduced number of available fragments from the Post-Bronze Age levels at the site. Taxa which are observed for the first time in Phase 4 include an unidentified member of the *Leguminosae* family, *Euonymus* sp. and *Aesculus hippocastanum*.

Vela Cave

Assemblage phases similar to those identified at Pupićina are identified (tabl. 2). Phase 1, comprising the oldest sample from the site, is composed of *Pinus sylvestris/nigra*, accompanied by *Fraxinus* sp. and *Corylus* cf. *avellana*. Phase 2 is identified on the basis of the decline of pine and the expansion of deciduous taxa, particularly *Quercus* and *Fraxinus* sp., accompanied by *Ulmus* sp., *Fagus sylvatica*, *Rosaceae* and *Acer* sp.

In phase 3a, several taxa appear for the first time, including Mediterranean taxa such as *Rhamnus alaternus/ Phillyrea*, *Buxus sempervirens*, and *Quercus ilex/coccifera*; conifers *Juniperus* sp. and *Taxus baccata*; and deciduous taxa *Cornus* sp. and *Ostrya carpinifolia*. On the basis of these first appearances, phase 3a appears to correlate with Phase 3 at Pupicina, although correlation between the sequences will require further radiometric dating. Phase 3b is characterized by the consistent presence of *Ostrya carpinifolia*, deciduous *Quercus*, *Fraxinus* sp. and *Rosaceae*. *Ulmus* sp., *Cornus* sp., *Rhamnus alaternus/ Phillyrea* sp., and *Acer* sp. are also present. Phase 4 at Vela, as at Pupićina, is distinguished on the basis of apparent changes in taxon representation. Again, small sample size for these uppermost levels accounts for the decreased diversity of the assemblage.

Geological period	HOLOCENE															
Archaeological period	Mesolithic	Neolithic							Bronze Age							
Level	141	140	139	137	135	133	131	130	129	128	127	125	124	120	119	118
Cal BP (1s)	11,230-10,740															
Charcoal Phase	1	2			3a				3b					4		
Pinus sylvestris/nigra	27	2		1												
Juniperus sp.						6										
Taxus baccata								4								
Ulmus sp.		1						1		1			3			
Fagus sylvatica		2				7	2	3								
Quercus sp.									2							
Quercus deciduous		5	3	5	5	1	1		10	2		13	4	1		1
Quercus ilex/coccifera								2								
cf. *Quercus* sp.											1					
Carpinus sp.										3		2				
Ostrya carpinifolia									7	8	1	2	9			2
Ostrya/Carpinus								4	2	5	2		3			
Corylus cf. *Avellana*	1															
Prunus sp.													1			
Prunus cf. *avium*				1						1		1	2			
Pomoideae					1		2		1		1	1	1			
Pomoideae type *Sorbus* sp.												1				
Cornus sp.					5	3	13	3		1				1		
cf. *Cornus* sp.						1				1						
cf. *Buxus sempervirens*							1									
Rhamnus alaternus/Phillyrea						1				2			1			
Acer platanoides/pseudoplatanus			1		5	2							1			
Acer non-*platanoides/pseudoplatanus*					1		1						1			
Fraxinus sp.	2	16	22	20	13	7	6	11	7	4	1	9	4	1	1	5
Fraxinus cf. Ornus		3	3	3		2	4	1		2		1				1
Unidentified Angiosperm		1	1						1							
Unidentified Gymnosperm																
Unidentifiable								1								
Total	30	30	30	30	30	30	30	30	30	30	6	30	30	3	1	9

Tabl. 2. Absolute numbers of charcoal fragments per level by taxon, for Vela Cave.

Vešanska Cave

The assemblage from *Vešanska* (tabl. 3) consists of *Pinus sylvestris/nigra* accompanied by *Fagus sylvatica*, and corresponds with Phase 1 at Pupićina and Vela. Although the charcoal was collected as dispersed material within an occupation layer, the small size and low diversity of the assemblage suggests that the charcoal may be the product of small-scale and low frequency fires.

Geological period	PLEISTOCENE	
Archaeological period	Upper Palaeolithic	
Level	II.3A	II.3
Cal BP (1s)	13,810 - 13,330	13,460 - 13,180
Charcoal Phase	1	
Pinus sp.	1	
Pinus sylvestris/nigra	15	54
Fagus sylvatica	5	6
Unidentified Gymnosperm	1	2
Unidentifiable		1
Total	22	63

Tabl. 3. Absolute numbers of charcoal fragments per level by taxon, for Vešanska Cave

Šebrn Rockshelter

The charcoal assemblage from Šebrn (tabl. 4) is dominated by *Pinus sylvestris/nigra* accompanied by deciduous oak and *Prunus* cf. *spinosa*, and corresponds with Phase 1 at Pupićina and Vela. Again, the low diversity of the assemblage suggests the possibility that the charcoal is attributable to a restricted number of burning events. The relatively high proportion of unidentifiable fragments is due to a slag-like modification of the wood which has left only the most localised rudiments of the original cellular structure, and which was probably caused by rapid combustion at high temperatures (Schweingruber, 1978).

Geological period	HOLOCENE
Archaeological period	Mesolithic
Level	3B
Cal BP (1s)	10,150 - 9690
Charcoal Phase	1
Pinus sp.	1
Pinus sylvestris/nigra	165
Quercus deciduous	8
cf. *Quercus* sp.	1
Prunus sp.	1
Prunus cf. *spinosa*	3
Unidentified Angiosperm	1
Unidentified Gymnosperm	5
Unidentifiable	17
Total	202

Tabl. 4. Absolute numbers of charcoal fragments per level by taxon, for Šebrn Rockshelter.

Discussion

Before an ecological interpretation can be put forward to account for the composition of a charcoal assemblage, the potential influence of cultural selection must be considered. Several factors suggest that the assemblages primarily reflect local abundances of different tree taxa. First, Pupićina, Vela and Vešanska are located in a steep-sided gorge, a naturally defined catchment area, whose physical characteristics constrain movement and the transport of material. For this reason, it is likely that wood was collected within the gorge during all archaeological periods. Ecological interpretation is therefore supported because it is unlikely that the assemblage represents a sampling of several ecological zones or different ecological zones between periods. Second, although the sites were occupied both by foragers and farmers, the use of the sites as temporary shelters and hunting bases appears to have been similar in all periods. There is no evidence of more complex activities such as construction or craft production which might have demanded specific choices in wood selection (Western, 1971), nor is there evidence for use of fire other than for domestic purposes. Although activity in the wider landscape is likely to have changed with the regional transition to farming (Montagnari-Kokelj, 1993; Chapman and Müller, 1990), activities in the gorge are likely to have remained more constant, with herding and clearance greatly restricted by the steep and rugged topography of the gorge. Third, the high taxonomic diversity of the Pupićina and Vela assemblages relative to their size suggests that wood selection was not conditioned by strong preferences for particular wood types.

With this support for a palaeoecological interpretation in mind, a tentative history of the vegetation of the region may be put forward. The Phase 1 assemblage, identified at all four sites, suggests montane-type forests of *Pinus sylvestris* or *P. nigra* within the gorge from at least the Allerød chronozone, and at higher altitude from at least 10,000 cal BP. Although pine was probably the most common tree, another characteristically montane taxon, *Abies alba*, was present in the gorge. In addition to these, several deciduous thermophilous trees including *Quercus* sp., *Ulmus* sp., *Fraxinus* sp. and *Fagus sylvatica* were present both in the gorge and on the surrounding slopes. The success of *Pinus sylvestris* or *Pinus nigra* throughout the late glacial and early Holocene may reflect tolerance of climatic variability including both wet and dry conditions and the ability to grow on nutrient poor soils depleted through weathering processes associated with cold stage conditions (Barbéro *et al.* 1998; Willis *et al.* 1998). The sporadic presence of thermophilous taxa in the assemblage suggests their presence in small numbers or in clustered pockets in the landscape. It is likely that the gorge provided micro-climatic opportunities for such populations. Their presence alongside pine since at least the Allerød suggests the local presence or regional proximity of glacial refugia. Istria has been recognized as a potential refuge area (Hazler-Pilepic *et al.*, 1999; Culiberg and Šercelj, 1995), and it will be interesting to determine the history of these deciduous taxa in the gorge if older deposits are excavated at the sites.

The transition from Phase 1 to Phase 2 probably reflects a significant change in the gorge vegetation, with the decline of pine and the expansion of broadleaf taxa. The transition occurred after 10,000 calBP, i.e. somewhat later than more southerly parts of the Balkans, but corresponding with pollen sequences from neighbouring Slovenia (Gardner, 1997; Culiberg and Šercelj, 1996). The expansion of broadleaf taxa with a longstanding presence in the gorge, such as deciduous *Quercus* and *Fraxinus* sp., suggests a response to climatic changes, as migration from other areas would not have been a factor in the timing of the expansion. Climatic changes accompanying the onset of the Holocene are generally recognized to underlie the expansion of mixed deciduous forest in Europe (Huntley and Prentice, 1993). In the Mediterranean, especially at low elevations, increased precipitation may have been particularly important, and it is interesting to note the growing recognition of a pluvial phase, with conditions significantly wetter than today, in the Adriatic between c. 9500 and 6800 calBP (Ariztegui *et al.* 2000). Warmer and especially wetter climatic conditions likely underlie the transition from Phase 1 to Phase 2, with success of taxa such as deciduous oak, ash, beech and elm.

Phase 3 reflects several important expansions in the gorge vegetation, and represents the beginning of the modern sub-Mediterranean vegetation. The appearance of several evergreen taxa, the expansion of small trees and shrubs such as *Cornus* sp. and *Rosaceae*, and the appearance of an important tree in the region, *Ostrya carpinifolia*, would have significantly changed the character of the mid- to late-Holocene forest. These changes occurred after 7000 cal BP and prior to 3500 cal BP, and suggest important ecological changes in the gorge.

The changes in forest composition at this period may have several underlying causes, including the late arrival of immigrant taxa, macro-climatic change and anthropic influence. The arrival of some new taxa, for example *Ostrya carpinifolia*, may reflect migration from more southerly refuge areas, although migrational lags on the whole are not characteristic of Balkan vegetation sequences (Willis, 1994). The expansion of Mediterranean sclerophyllous and evergreen elements appears to represent a shift in the competitive balance between taxa which are favoured by regular precipitation throughout the year and those which can withstand prolonged summer water shortage (Zavala *et al.* 2000). This shift may have be related to the onset of Mediterranean climatic conditions characterised by summer drought and uneven precipitation during autumn, winter and spring—a climatic transition recorded in limnological sequences at c. 6500 cal BP in Dalmatia (Wunsam *et al.* 1999).

While a climatic interpretation concurs with those who consider climatic change responsible for widespread vegetation changes since c. 6800 cal BP (e.g. Prentice *et al.* 1996; Kelly and Huntley, 1991), the influence of a late-Holocene climatic transition on the development of the Mediterranean vegetation remains under discussion. Recent research has aimed to demonstrate a predictable relationship between the development of evergreen vegetation and Mediterranean climatic conditions, leading to the development of a model of the time-transgressive installation of the Mediterranean climate during the late-Holocene (Jalut *et al.* 2000). This view may underestimate, however, the effects of human activity. Many argue that the expansion of evergreen elements and the establishment of the modern Mediterranean vegetation was the direct result of human impact on the landscape (Blondel and Arondsen, 1999; Quezel, 1999; Vernet and Thiébault, 1987; Vernet, 1973). This view highlights the role of clearance, grazing and anthropogenic fire in the development of the semi-natural or degraded vegetation communities of the Mediterranean such as the maquis (Polunin and Walters, 1985).

While interpretations of mid- to late-Holocene vegetation change are typically phrased in terms of either climatic change or anthropogenic activity, it may be necessary to explore a conjunction of factors, and to resolve the relative influence of these factors in terms of the sensitivity of the vegetation cover at a local scale. A high-resolution approach is probably especially important as one approaches the boundaries of the Mediterranean climate area (e.g. northern Croatia) and the topographic limits of human activity during the past (e.g. the Vela Draga gorge). At the micro-scale of the Vela Draga gorge, it is possible that changes in forest composition represent a sensitive response to regional climatic variation. Although human pressure in terms of the cutting, burning, grazing and uprooting of forest for pasture, agriculture and settlement had significant impact on vegetation in the wider landscape, these effects may have been more limited in the gorge itself, where local topography is likely to have discouraged extensive agricultural activity. The interpretation put forward here is that the decline of the mature broadleaf forest of Phase 2 and the development of sub-Mediterranean woodland in Phase 3 primarily reflects increased summer dryness during the late-Holocene. However, climatic changes would have been concomitant with small-scale herding in the area which exerted a secondary, and not insignificant, impact on the vegetation cover. The overall variety of taxa recorded within Phase 3 probably also reflects the micro-climatic effects of the gorge.

Although Phase 4 in the charcoal assemblage records the presence of some new taxa, such as *Aesculus hippocastanum*, the overall character of the woodland was probably similar to that of Phase 3. Further research will be required to confirm the similarity of the woodland composition during this period to that of Phase 3.

Conclusion

Through charcoal analysis of the Vela Draga sites, three vegetation phases are identified: 1) montane-type pine forest during the Late-Glacial and early Holocene 2) humid deciduous forest during the Mid-Holocene and 3) sub-Mediterranean woodland from the mid- to late-Holocene. The identification of distinct vegetation phases since the end of the last glacial period has many parallels in both charcoal and pollen records from the Balkans and the Mediterranean (e.g. Heinz & Thiébault, 1998; Willis, 1994; Vernet and Thiébault, 1987). Macro-climatic conditions, specifically the increased warmth and wetness of the early to mid-Holocene and the subsequent onset of summer aridity in the Adriatic basin may underlie the vegetation phases observed in the gorge. Throughout the different phases, taxa of varied ecological preferences were present in the gorge, a feature consistent with the range of micro-climatic niches which are recognised to exist within a karstic gorge (Polunin, 1980). Therefore, while macro-climatic changes probably structured the sequence of vegetation phases, the composition and character of the forest within vegetation phases reflects the influence of the local landscape. Anthropic pressure, which is recognised to be one of the key factors in the evolution of the modern Mediterranean landscape, may be of secondary importance at the micro-scale of the Vela Draga gorge. This interpretation may adapt as archaeological understanding of the region develops. Nevertheless, changes in the woodland landscape would have meant significant changes in forest resources, and provided a dynamic setting for major cultural changes such as the spread of new pastoral and agricultural economies in the region. It is hoped that this outline of the vegetation history of a local area of the Istrian peninsula, and its interpretation in terms of macro- and micro-climatic conditions, will be explored and adapted through further research.

Acknowledgements
The support and comments of Dr Marco Madella, Dr Lydia Zapata and Professor Tjeerd van Andel, and the financial support of Trinity College towards fieldwork in Croatia are all gratefully acknowledged.

References

ARIZTEGUI, D., ASIOLI, A., LOWE, J.J., TRINCARDI, F., VIGLIOTTI, L., TAMBURINI, F., CHONDROGIANNI, C., ACCORSI, C.A., BANDINI MAZZANTI, M., MERCURI, A.M., VAN DER KAARS, S., MCKENZIE J.A., OLDFIELD, F. 2000 - Palaeoclimate and the formation of sapropel S1: inferences from late Quaternary lacustrine sequences in the central Mediterranean region. *Palaeogeography, Palaeoclimatology, Palaeoecology,* 158, p. 215-240.

BADAL-GARCÍA, E. 1992 - L'anthracologie préhistorique: à propos de certains problèmes methodologiques, *Bulletin de la Société Botanique de France, Actualités Botaniques*, 139(2/3/4), p. 167-189.

BARBERO, M., LOISEL, R., QUEZEL, P., RICHARDSON D.M., ROMANE, F. 1998 - Pines of the Mediterranean basin. In D.M. RICHARDSON (ed.) *Ecology and Biogeography of Pines*. Cambridge, Cambridge University Press, p. 153-170.

BEUG, H.J. 1977 - Vegetationsgeschichtliche Untersuchunger im Küstenbereich von Istrien (Jugoslawien). *Flora,* 166, p. 357-381.

BLONDEL, J., ARONSON, J. - 1999. *Biology and Wildlife of the Mediterranean*. Oxford, Oxford University Press.

CARTWRIGHT, C., PARKINGTON, J. 1997 - The wood charcoal assemblages from Elands Bay Cave, southwestern Cape: Principles, procedures and preliminary interpretation. *South African Archaeological Bulletin,* 52, p. 59-72.

CHAPMAN, J., MÜLLER, J. 1990 - Early farmers in the Mediterranean basin: the Dalmation evidence. *Antiquity,* 64, p. 127-134.

CULIBERG, M., ŠERCELJ, A. 1995 - Anthracotomical and palynological research in the Palaeolithic site Šandalja II (Istria, Croatia). *Razprave iv. Razreda Sazu,* 36, p. 49-57.

CULIBERG, M., ŠERCELJ, A. 1996 - Slovenia. In BERGLUND, B.E., BIRKS, H.J.B., RALSKA-JASIEWICZOWA, M., WRIGHT H.E. (ed.), *Palaeoecological Events During the Last 15000 years: Regional Syntheses of Palaeoecological Studies of Lakes and Mires in Europe*. New York & London, John Wiley & Sons, p. 687-700.

FIGUEIRAL, I., MOSBRUGGER, V. 2000. - A review of charcoal analysis as a tool for assessing Quaternary and Tertiary environments: achievements and limits. *Palaeogeography, Palaeoclimatology, Palaeoecology,* 164, p. 397-407.

GARDNER, A. 1997 - Biotic response to Early Holocene human activity: results from palaeoenvironmental analyses of sediments from Podpeπko jezero. *Porocilo o raziskovanju paleolita, neolita in eneolita v Sloveniji,* 24, p. 63-77.

HAZLER-PILEPIC, K., COMPS, B., ŠUKOR, I., ŠOLIC, E. 1999 - Isozyme polymorphism of Croatian beech populations (*Fagus sylvatica* L.). *Periodicum Biologorum,* 101(2), p. 165-170.

HEINZ, C. 1991 - Upper Pleistocene and Holocene vegetation in the south of France and Andorra. Adaptations and first ruptures: New charcoal analysis data. *Review of Palaeobotany and Palynology,* 69, p. 299-324.

HEINZ, C., THIEBAULT, S. 1998 - Characterization and palaeoecological significance of archaeological charcoal assemblages during Late and Post-Glacial phases in southern France. *Quaternary Research,* 50, p. 56-68.

HERAK, M. 1972 - Karst of Yugoslavia. In M. HERAK & V.T. STRINGFIELD (ed.), *Karst: Important Karst Regions of the Northern Hemisphere*. Amsterdam, Elsevier Publishing Company, p. 25-83.

HUNTLEY, B., PRENTICE, I.C. 1993 - Holocene Vegetation and Climates of Europe. In WRIGHT, H.E. JR., KUTZBACH, J.E., WEBB, T. III, RUDDIMAN, W.F., STREET-PERROTT, F.A., BARTLEIN, P.J. (ed.), *Global Climates*. Minneapolis & London, University of Minnesota Press, p. 136-168.

JALUT, G., AMAT, A.E., BONNET, L., GAUQUELIN, T., FONTUGNE, M. 2000 -Holocene climatic changes in the western Mediterranean, from south-east France to south-east Spain. *Palaeogeography, Palaeoclimatology, Palaeoecology,* 160, p. 255-290.

KELLY, M.G., HUNTLEY, B. 1991 - An 11,000-year record of vegetation and environment from Lago di Martignano, Latium, Italy. *Journal of Quaternary Science,* 6(3), p. 209-224.

MIRACLE, P.T. 1997. Early Holocene foragers in the karst of northern Istria. *Porocilo o raziskovanju paleolita, neolita in eneolita v Sloveniji,* 24, p.43-61.

MIRACLE, P.T., GALANIDOU, N., FORENBAHER, S. 2000. Pioneers in the hills: Early mesolithic foragers at Šebrn Abri (Istria, Croatia). *European Journal of Archaeology,* 3(2), p.219-255.

MONTAGNARI-KOKELJ, E. 1993. The transition from Mesolithic to Neolithic in the Trieste Karst. *Porocilo o*

raziskovanju paleolita, neolita in eneolita v Sloveniji, 21, p.69-83.
PALUTIKOF, J.P., CONTE, M., CASIMIRO MENDES, J., GOODESS, C.M., ESPIRITO SANTO, F. 1996. Climate and Climatic Change. In BRANDT, C.J., THORNES, J.B. (ed.), *Mediterranean Desertification and Land Use*. Chichester, John Wiley & Sons Ltd., p.43-86.

POLUNIN, O. 1980. *Flowers of Greece and the Balkans: a field guide*. Oxford, Oxford University Press.

POLUNIN, O., WALTERS, M. 1985. *A Guide to the Vegetation of Britain and Europe*. Oxford, Oxford University Press.

PRENTICE, I.C., GUIOT, J., HUNTLEY, B., JOLLY, D., CHEDDADI, R. 1996. Reconstructing biomes from palaeoecological data: a general method and its application to European pollen data: a general method and its application to European pollen data at 0 and 6 ka. *Climate Dynamics,* 12, p.185-194.

QUEZEL, P. 1999. Les grandes structures de végétation en région Méditerranéene: facteurs déterminants dans leur mise en post-Glaciare. *Geobios,* 32(1), p.19-32.

SCHWEINGRUBER, F.H. 1978. *Mikroskopische Holzanatomie*. Birmensdorf, Swiss Federal Institute of Forestry Research.

SCHWEINGRUBER, F.H. 1990. *Anatomie Europäischer Hölzer*. Bern/Stuttgart, Haupt.

UNESCO 1970. *Vegetation Map of the Mediterranean Zone*. Paris, UNESCO.

VERNET, J. 1973. Etude sur l'histoire de la végétation du sud-est de la France au Quaternaire d'après les charbons de bois principalement. *Paléobiologie continentale,* 4(1), p.1-90.

VERNET, J., THIEBAULT, S. 1987. An approach to northwestern Mediterranean recent prehistoric vegetation and ecological interpretations. *Journal of Biogeography,* 14, p.117-127.

WESTERN, A.C. 1971. The ecological interpretation of ancient charcoals from Jericho. *Levant,* 3, p.31-40.

WILLIS, K.J. 1994. The vegetational history of the Balkans. *Quaternary Science Review,* 13, p.769-788.

WILLIS, K.J., BENNETT, K.D., BIRKS, H.J.B. 1998. The late Quaternary dynamics of pines in Europe. In RICHARDSON, D.M. (ed.) *Ecology and Biogeography of Pines*. Cambridge, Cambridge University Press, p.107-121.

WUNSAM, S., SCHMIDT, R., MÜLLER, J. 1999. Holocene lake development of two Dalmatian lagoons (Malo and Veliko Jezero, Isle of Mljet) in respect to changes in Adriatic sea level and climate. *Palaeogeography, Palaeoclimatology, Palaeoecology,* 146, p.251-281.

ZAVALA, M.A., ESPELTA, J.M., RETANA, J. 2000. Constraints and trade-offs in Mediterranean plant communities: The case of holm-oak—Aleppo pine forests. *Botanical Review,* 66, p.119-149.

Evidence from charcoal analysis for palaeoenvironmental change during the Late Glacial and Post-Glacial in the Central Pyrenees (France)

Christine Heinz

CNRS-UMR 5059, Laboratoire de Paléoenvironnnements, Anthracologie, Action de l'Homme,
Institut de Botanique, 163, rue Auguste Broussonet, 34090 Montpellier, France.
E-mail: heinz@crit.univ-montp2.fr

Abstract

The archaeological sites of Troubat (department of Hautes-Pyrénées) and Buholoup (department of Haute-Garonne) are situated very close to each other, and have provided extensive charcoal horizons. The archaeological work carried out at the cave-shelter of Troubat and Buholoup has provided an interesting stratigraphical sequences concerning the Late Glacial and the Post-Glacial periods. The results of the charcoal analysis provided important information concerning the climatically induced transformations in the vegetation cover from the Late Glacial Bölling-Alleröd interstadial to the middle of the Post-Glacial. At Troubat, between about 13,000 BP and 11,320 ± 410 BP (phase T1) a vegetation with *Juniperus*, some Rosaceae (*Prunus* types 1, 2, 3), *Rhamnus alpina*, *R. cathartica/saxatilis*, *R. pumila*, *Hippophae rhamnoïdes* and *Betula* cf. *pendula* is evidence of an open environment with colonising shrub cover established in a climate of dry mountain or even subalpine type. This change is correlated to the Late Glacial interstadial. At about 10,770 ± 100 BP, a reduction of the *Juniperus* pioneer stage cover, which developed thanks to the Bölling/Alleröd warming, followed by a *Rhamnus* and Rosaceae phase, might correspond already to the vegetation changes characterising the very beginning of the Younger Dryas (phase T2). Between 9700 ± 80 BP and 8880 ± 75 BP, a phase of forest colonisation by *Corylus* and the spread of deciduous oak forests which corresponds to the Post-Glacial climatic amelioration is noted (phase T3). Then, at about 8625 ± 80 BP, there is affirmation of an increase in deciduous oak forests of "collinean" type. *Pinus sylvestris*, whose role is minor at Troubat, as well as *Corylus*, see their frequency diminish in response to an increasingly dense forest environment (phase T4). These two last phases are also identified at Buholoup (B1 and B2). Another archaeological site, the cave of Gazel (Aude) is also discussed. It is located in the meso-Mediterranean vegetation zone and offers an interesting point of comparison.

Key-Words: *Charcoal analysis, Late Glacial, Post-Glacial, Central Pyrenees, Palaeoecology.*

Résumé

Les sites archéologiques de Troubat (Hautes Pyrénées) et de Buholoup (Haute Garonne), très proches l'un de l'autre, ont livré des horizons très riches en matériel anthracologique. Les recherches archéologiques de ces deux abris sous-roche ont mis au jour d'importantes séquences stratigraphiques allant du Tardiglaciaire au Postglaciaire. Les données anthracologiques apportent des informations précises sur les transfomations de la végétation en relation avec les changements climatiques au cours de l'interstade tardiglaciaire et du Postglaciaire. A Troubat, quatre grandes phases illustrent la cinétique de végétation enregistrée par l'anthracologie. Entre 13 000 BP environ et 11320 ± 410 BP (phase T1) une végétation forestière avec *Juniperus*, des Rosaceae (*Prunus* type 1, 2, 3), *Rhamnus alpina*, *R.cathartica/saxatilis*, *R. pumila*, *Hippophae rhamnoïdes* et *Betula* cf. *pendula*, traduit un milieu ouvert et une formation de type fruticée mise en place sous un climat de type montagnard sec voire subalpin. Cette dynamique est corrélée à l'interstade tardiglaciaire. Vers 10 770 ± 100 BP, une rétraction de la fruticée à *Juniperus*, qui s'est développée à la faveur du réchauffement Bölling/Alleröd, suivie d'une phase à *Rhamnus* et Rosaceae serait une réponse de la végétation à l'amorce du Dryas récent (phase T2). Entre 9700 ± 80 BP et 8880 ± 75 BP, une phase de colonisation forestière par *Corylus* et l'installation d'une chênaie caducifoliée en relation avec l'amélioration climatique postglaciaire est notée (phase T3). Puis, vers 8625 ± 80 BP, on enregistre l'affirmation de la chênaie caducifoliée de type collinéenne. *Pinus sylvestris*, dont le rôle est mineur à Troubat, ainsi que *Corylus*, voient leurs occurrences diminuées ceci en relation avec un milieu forestier de plus en plus dense (phase T4). Ces deux dernières phases, sont également enregistrées à Buholoup (B1 et B2). A titre de comparaison, les données de la grotte de Gazel (Aude) en contexte mésoméditerranéen sont discutées.

Mots-clés : *Anthracologie, Tardiglaciaire, Postglaciaire, Pyrénées centrales, paléoécologie.*

Introduction

The cave shelter of Troubat (Hautes-Pyrénées) was formed in the Mesozoic limestone which makes up the northern border of the Pyrenees (fig. 1). It is situated at an altitude of 541 m near the Ourse basin in the region of the Garonnais piedmont. The site itself is located on the flank of the Gouadère massif which rises to 800 m at the conjunction of the supra-Mediterranean vegetation zone and the Atlantic "collinean" zone (terminology after Ozenda, 1975; Quezel, 1976). The former is characterised by the seral stage of *Quercus pubescens*, the latter by the seral stage of *Quercus robur* (Dupias, 1985). The shelter of Buholoup is situated at an altitude of 310 m in the Volp valley (Haute Garonne), in the same biogeographical context, except perhaps with a stronger mediterranean influence.

The abundance of charcoal as well as systematic sampling have provided the first charcoal sequences for the central Pyrenees; almost 6000 fragments of carbonised wood have been studied at Troubat, 2000 fragments at Buholoup. They both present a species list of 37 trees, shrubs or bushes representing all the major forest species (tabl. 1). Thus using this charcoal data we can trace the evolution of the plant cover and the changes in the vegetation over about 6000 years, from about 13,000 BP to 7000 BP.

Site sampling

Systematic sampling of charcoal and other material was carried out by square metre and artificial layer; sediments extracted during the archaeological excavation were systematically sieved with running water, with a mesh of 2 mm, and charcoal recovered. The charcoal concentrated in the archaeological structures was treated separately from that scattered in the occupation levels. We will recall the conditions necessary for a palaeoecological study (Chabal, 1991, 1997; Heinz, 1990, 1991; Badal, 1990):

- charcoal should come from domestic firewood, building wood and worked wood were subjected to strong selection and are therefore not considered.
- only charcoal from firewood, dispersed in the occupation levels, is taken into consideration in the palaeoecological

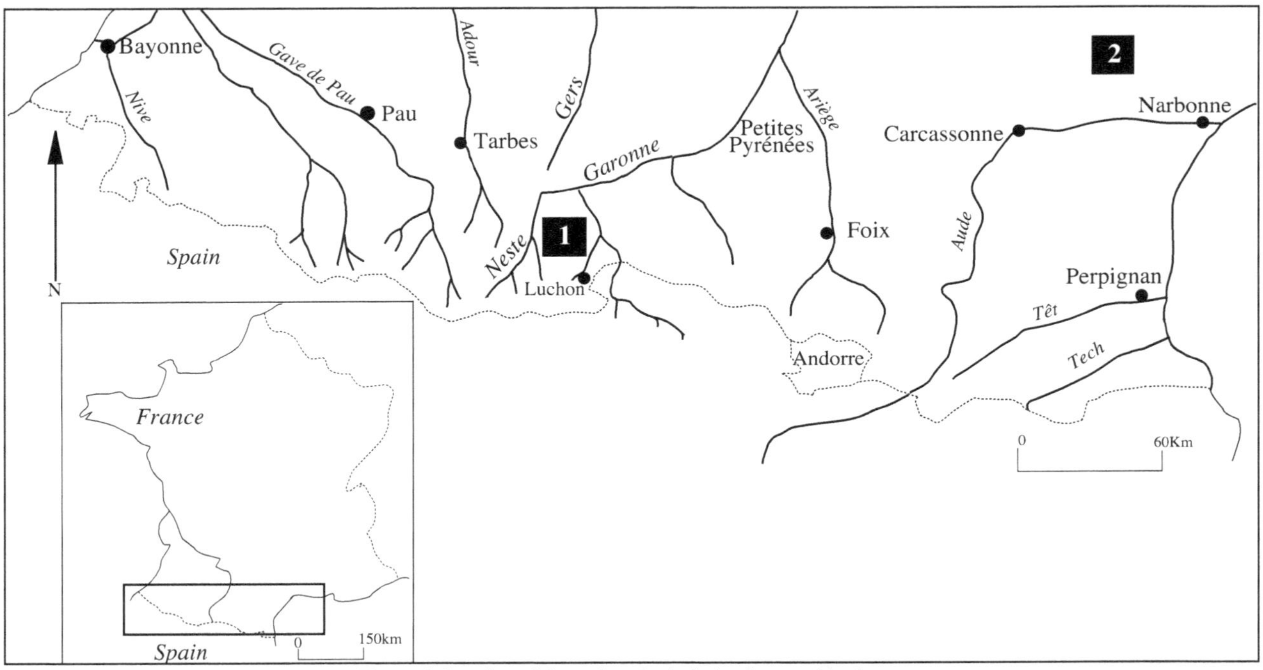

Figure 1: location of the archaeological site discussed in the text.

study. Charcoal from archaeological structures (hearths, ditches...), representing brief events which provide low taxonomic diversity (last fire, last use) should be considered from a more archaeobotanical viewpoint and are not taken into account here. In fact, the charcoal from each period would represent a long period of activity. This is true for the charcoal scattered in the occupation levels. In fact, there is a direct correlation between the duration of wood collection and the surface area sampled.

The minimum number of charcoal fragments to be counted is defined by the use of taxonomic curves; these present the number of taxa encountered during identification as a function of the number of charcoal fragments. It is a convenient criterion to determine at which point the optimal representation of taxa is obtained (Chabal, 1991; Heinz, 1990; Badal, 1990).

Identification of charcoal

The anatomical identification of the charcoal was carried out using a reflected light microscope. The charcoal was fractured by hand and the three anatomical planes of wood observed: transverse, longitudinal-tangential and longitudinal-radial. The observed structures were compared to those described in the atlas of wood anatomy (Boureau, 1956; Metcalfe and Chalk, 1950, 1983; Greguss, 1955, 1959; Jacquiot, 1955; Jacquiot *et al.* 1973; Schweingruber, 1978, 1990) and to a reference collection of modern charcoal.

Some anatomical descriptions are necessary in order to justify certain taxonomic identifications. Concerning the genus *Prunus*, 3 types are distinguished at the sites: type 1 corresponds to fragments for which the rays are no wider than 2 cells; type 2 is between 3 and 5 cells and type 3 more than 5 cells. Taxonomically, type 1 might correspond to *Prunus avium/padus*, type 2 to *Prunus spinosa/mahaleb*, and type 3 to *Prunus spinosa/amygdalus* (Heinz and Barbaza, 1998).

At least three species of *Rhamnus* were identified at Troubat: *Rhamnus alpina, R. pumila*, and *R. cathartica/saxatilis*, the last one present also at Buholoup. This is the first appearance of *Rhamnus alpina* and *R. pumila* in prehistoric charcoal (Heinz and Barbaza, 1998). The width of the rays is the most reliable taxonomic criterion for distinguishing between them as well as the density of the pores in the early wood. *Rhamnus pumila* has rays which are extremely narrow (1 to 2 cells), *Rhamnus cathartica/saxatilis* has rays which are 3 to 5 cells wide with a pore density which is high in the early wood, and *Rhamnus alpina* has very large rays (5/6 to 8 cells). Finally the taxon *Rhamnus* sp. refers to fragments too small or too badly preserved so as the ray width cannot be observed.

The junipers identified here have very low rays (1 to 4 cells maximum), which characterizes species such as *Juniperus communis, J. thurifera, J. nana* or even J. *sabina*. The more thermophilous species *J. oxycedrus* and *J. phoenicea* have rays 1 to 12 cells high, with a mean of 6 cells.

Results

The results concerning Troubat are presented in a charcoal diagram (fig. 2). The diversity of taxa identified and the succession observed in the charcoal assemblages suggest that human populations at Troubat collected wood from a wide range of taxa available in their surroundings; therefore the succession observed may be considered representative of the vegetation dynamics during the Late Glacial–early Holocene.

The changes in plant cover revealed at Troubat can be divi-

	Troubat	Buholoup
Acer type 2	X	
Acer type 1	X	X
Alnus sp.	X	
Betula cf. *pendula*	X	
cf. *Betula* sp.	X	
cf. *Buxus sempervirens*	X	X
Cornus sanguinea	X	X
Corylus avellana	X	X
Corylus/Alnus	X	X
cf. *Cotoneaster* sp.	X	
Crataegus sp.	X	X
Crataegus/Sorbus	X	X
cf. *Daphne* sp.	X	
Hippophae rhamnoides	X	
Juniperus sp.	X	X
Pinus type *sylvestris*	X	
Populus sp.	X	
Prunus sp.	X	
Prunus type 1	X	X
Prunus type 2	X	X
Prunus type 3	X	X
Prunus type *mahaleb*	X	
Quercus (deciduous)	X	X
Rhamnus alpina	X	
Rhamnus catharticalsaxatilis	X	X
Rhamnus pumila	X	
Rhamnus sp.	X	
Rosaceae Maloideae	X	X
Rosaceae Prunoideae	X	
Salix / Populus	X	
Salix sp.	X	X
cf. *Sambucus racemosa*	X	X
Sorbus sp.	X	X
Ulmus minor	X	X
Viburnum cf. *lantana*	X	X
Viburnum cf. *opulus*	X	
cf. *Viburnum* sp.	X	

Table 1: taxa identified by charcoal analysis at Troubat (Hautes-Pyrénées) and Buholoup (Haute-Garonne). Nomenclature used follows *Flora Europaea*, except concerning family Rosaceae; this large family is not sub-divided in *Flora Europaea*; therefore we have followed the division proposed by Mabberley (1997).

ded into 4 phases, respectively called T1, T2, T3 and T4.

Phase T1 contains the archaeological levels C10, C8c, C8a and C7b (Ly-5272: 11,320 ± 410 BP). During T1, a forest vegetation of *Juniperus* type became established in a dry mountain/subalpine climate during the general climatic amelioration. These changes are correlated with the Late Glacial interstadial, and C7b would even correspond to the interstadial forest optimum.

Juniperus clearly predominates with *Hippophae rhamnoïdes*, *Betula* cf. *pendula*, *Rhamnus alpina*, *Rhamnus pumila*, *Rhamnus catharticalsaxatilis*, *Prunus* (type 1 and 2) and *Salix* sp. *Pinus sylvestris* appears in C8b and C7b, but always with very low frequencies. Concerning *Juniperus*, four different species are possible: *J. communis*, *J. thurifera*, *J. nana* or *J. sabina*; if all these pioneering species indicate a plant cover which is undergoing change, specific anatomical determinations would be very interesting from an ecological and biogeographical viewpoint. *Hippophae rhamnoïdes,* a heliophyte, is a coloniser thanks to its adventitious suckering roots. It is found in open scrub pioneer associations on the edges of mountain streams. To be noted is the significant presence of *Rhamnus pumila*, a low spreading shrub which is frequent in rocky habitats and in pioneer associations in the higher mountain and subalpine zone. *Rhamnus alpina* also develops in these contexts as well as in the mountain zone and even in the supra-Mediterranean zone. Moreover, concerning *Rhamnus catharticalsaxatilis*, *R. saxatilis* is a highly branched, prostrate shrub, being a xerophilous species, growing in half-shade or full sun, found up to the beginning of the mountain zone. *Rhamnus cathartica* is a heliophyte and "calcicole" and grows in open scrub pioneer associations but also in hygrophilous habitats. The four species of *Prunus* concerned are frequently associated with open environments (*Prunus spinosa*), with hygrophilous habitats (*Prunus padus*, *Prunus avium*) or again with the pioneer associations (*Prunus padus*, *Prunus mahaleb*). Thus, the abundance of *Juniperus*, of the Rosaceae (*Prunus* type 1, 2, 3), of *Rhamnus alpina*, *R. catharticalsaxatilis*, *R. pumila*, with *Hippophae*, *Betula*, *Salix* indicates an open environment and evokes a pioneer association in a climate of dry mountain or even subalpine type. A hygrophilous habitat was also in evidence with *Cornus*, *Corylus*, *Viburnum* and *Sambucus*. Two interrelated biotopes might have been exploited by man, *i.e.* the open hygrophilous habitat and most specially the hillsides covered with *Juniperus* pioneer associations. These two habitats provided the catchment area for the wood gathered for fuel.

Phase T2 includes levels C7a, C6b (Ly-5275: 10,770 ± 100 BP) and C6a. It is characterised by a very clear decrease in *Juniperus*. The occurrence of *Hippophae* and *Rhamnus pumila* is diminished in C7 and disappears in C6b. *Rhamnus alpina* subsists and *Betula* is better represented than in T1. On the other hand, a rise in Rosaceae (*Prunus* type 1, *Sorbus*) and *Rhamnus catharticalsaxatilis* is clear. The curve of *Pinus sylvestris* although discreet is in slight progression from C6b onwards. The river bank plants become more abundant: *Alnus*, *Daphne* sp., as well as *Viburnum*, *Sambucus* and *Cornus*.

From the classic Azilian levels C6b and C6a onwards the first deciduous oaks are in evidence, accompanied by *Acer* type 1 (*Acer* cf. *campestre*) and by *Corylus*, already present in T1. The presence of a few fragments of deciduous *Quercus* correlates with the rapid development of this taxon at the beginning of the Holocene. As Beug (1975) has pointed out, the refugia of mesophilic trees were probably located in middle altitudes in "collinean" situations less dry than the plains; then at the very beginning of the Holocene, the plains were the first to be colonised. The existence of Late Glacial refugia in the Pyrenean context has been demonstrated elsewhere (Jalut and Vernet, 1989; Andrieu, 1991; Reille and Andrieu, 1995). The appearance of deciduous *Quercus* in levels C6b and C6a at Troubat confirms the existence of these refugia.

Thus, a reduction of the *Juniperus* pioneer association, which developed thanks to the Bölling/Alleröd warming, is observed; a transition phase to *Rhamnus* and *Prunus* follows. Some broad-leaved trees subsist; the rareness of *Juniperus* in C7a suggests a colder episode which would have taken place after 11,320 ± 410 BP. This seems to appear at the beginning of the Younger Dryas. Thus the reduction of the *Juniperus* association together with an increase of *Rhamnus* and Rosaceae might correspond alrea-

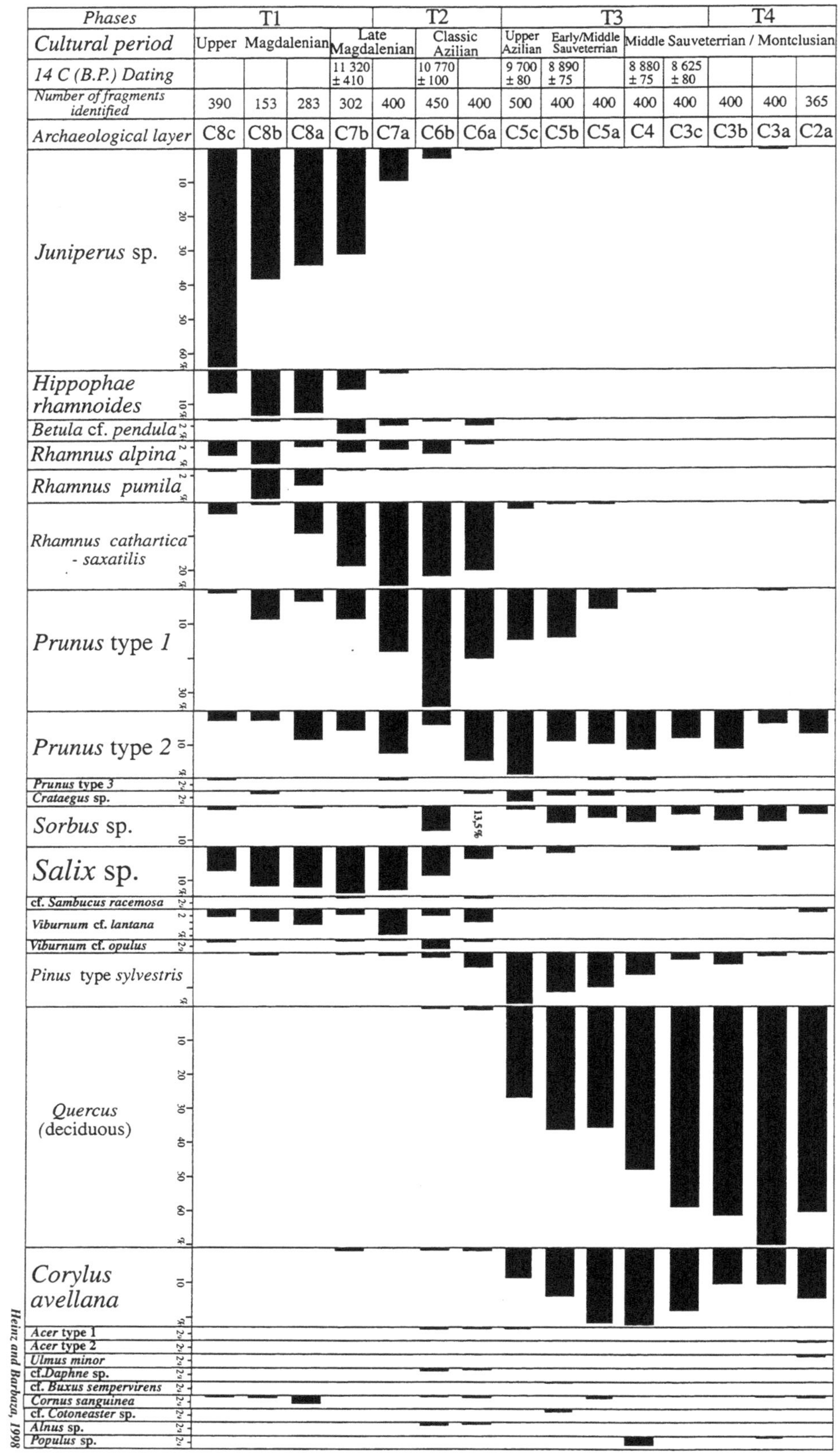

Figure 2: charcoal diagram analysis of Troubat (Hautes-Pyrénées, France).

dy to the vegetation changes characterizing the very beginning of the Younger Dryas. We believe that the remaining moments of the Younger Dryas are not represented at this site. In fact, between C6a (phase T2), and C5c (beginning of phase T3, Ly-6405: 9700±80 BP), in which the charcoal spectrum is typical of the beginning of the Post-Glacial, it is probable that there is a gap in the occupation of the shelter. At the beginning of the Post-Glacial, the curve of *Quercus* and *Corylus* in C5c indicates that the broad-leaved species that developed slightly in T1 survived during the Younger Dryas at a middle altitude. This is in accordance with the regional palynological data (Reille and Andrieu, 1995). The details of the stratigraphical sequence as well as the information it contains agree with this view which includes a gap in the occupation of the site (Heinz and Barbaza, 1998).

The changes in the vegetation due to climate as seen in the charcoal diagram can be summarised thus: from phase T1 to T2, the change is from a period corresponding to the Late Glacial interstadial (Bölling/Alleröd) to the first cooling period (Younger Dryas), with a reduction of the *Juniperus* pioneer stage cover. The Azilian levels C6b (Ly-5275: 10,770 ± 100 BP) and C6a would relate to the beginning of this last cold period. From a chrono-cultural point of view, the date of 10,770 ± 100 BP is acceptable for the classic Azilian which several studies situate wholly or partly in the Younger Dryas (Beaulieu *et al.* 1982). Only the very start of the Younger Dryas is represented at Troubat, as shown by the gap between C6a and C5c.

Phase T3 consists of levels C5c (Ly-6405: 9700 ± 80 BP), C5b (Ly-5274: 8890 ± 75 BP), C5a, C4 (Ly-5271: 8880 ± 75 BP). From 9700 ± 80 BP, the occurrence of *Pinus sylvestris* and above all that of deciduous *Quercus* is in progression, as well as that of *Corylus*. *Rhamnus catharticalsaxatilis* is in regression; *Juniperus* is very rare in C5c, and disappears in C5b; *Sambucus* is absent, *Salix* and *Betula* occasional. The hygrophilous forest includes *Populus* and *Cornus*. *Viburnum* disappears. It is possible that this heliophyte no longer found its optimal conditions in a gallery forest which was perhaps more closed. *Pinus sylvestris* (*c*.10%) is present nearby, at a somewhat higher altitude or as part of the deciduous oak forest.

This forest vegetation is characteristic of the beginning of the Post-Glacial. The dominant plant association is then of "collinean" type with some mountain influence, indicated by the presence of *Pinus sylvestris*. Then for levels C3c (Ly-5273: 8625 ± 80 BP), C3b, C3a, C3c and C2a, the deciduous oak forest is established; this is phase T4. From level C3c onward, *Pinus sylvestris* and *Corylus*, both heliophytes, diminish in frequency, and this may indicate a progressively more dense forest environment.

The preliminary study of Buholoup is presented (fig. 3); the results display two major phases during the Holocene. Phase B1 comprises the levels of habitats C7 (Azilian) and C6 (early Sauveterrian, about 8500 BP). From a palaeoecological point of view, the charcoal analysis suggests a mosaic landscape with three biotops: the first characterised essentially by pioneer shrub species with dry mountain affinities (*Juniperus, Rhamnus, Prunus*), tardiglacial survivals; the second one defined by the spread of the deciduous oak

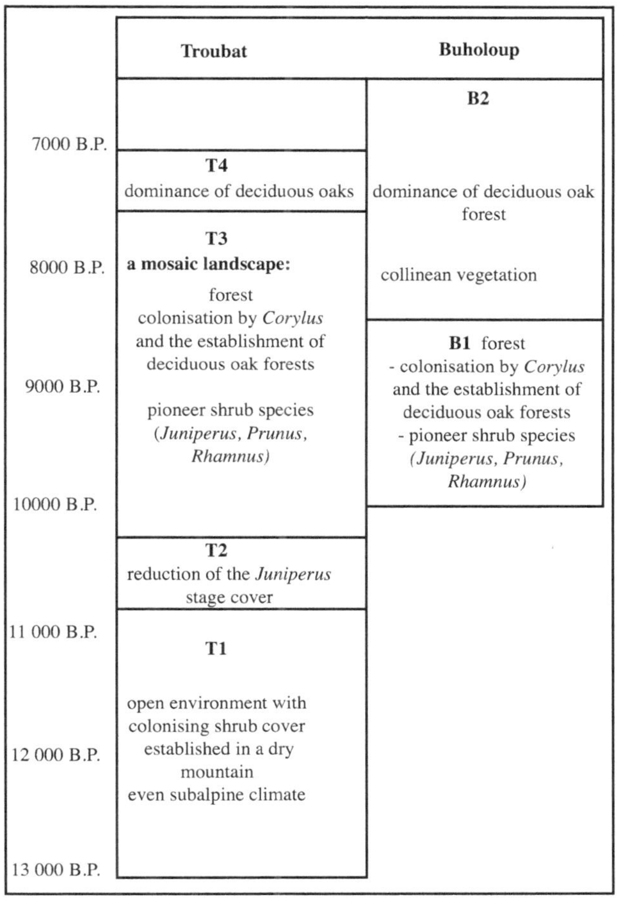

Figure 3: synchronism of the phases between Troubat (Hautes-Pyrénées) and Buholoup (Haute-Garonne).

forest (*Quercus, Acer campestre, Corylus avellana*) and the last one with riverbank species (*Ulmus minor, Fraxinus excelsior, Cornus sanguinea, Sambucus*). These three biotopes - the deciduous oak forest, the pioneer *Juniperus* and the hygrophilous formations - might have been exploited by man and provided the catchment area for the wood gathered for fuel.

Phase B2 comprises levels of habitat C5 (middle Sauveterrian), C4 (Montclusian, about 7500 BP) and C3-C2 (early Neolithic). It is characterized by a preponderance of collinean vegetation with deciduous *Quercus* and associated species. Thus a modification of environmental conditions, associated with higher precipitations and temperatures, took place as indicated by this phase of forest colonisation by *Corylus* and the establishment of deciduous oak forests with *Acer, Corylus* and *Buxus*.

Discussion

The new results obtained at the Cauna de Gazel (240 m a.s.l., Aude) following those first obtained by J.-L. Vernet (1980) for the epi-Magdalenian C7-C6, C6 (10,760 ± 190 BP) and Azilian (C4, C3) (Heinz, 1994, 1999) offers an interesting point of comparison with Troubat. The Epi-Magdalenian layers (Sacchi, 1986) reveal a *Pinus sylvestris* open forest with Fabaceae and *Juniperus*. *Betula* and *Salix* are scarce and deciduous *Quercus* appears. Thus the Late Glacial interstadial favours the establishment of an open vegetation with *Juniperus*, Rhamnaceae and Rosaceae at Troubat (T1) and a scrub with *Juniperus* and Fabaceae at Gazel. About 10,700 BP, at the first cold period of the Younger Dryas, *Juniperus* regresses and *Pinus* and Fabaceae characterize the landscape. At Troubat, we see the retraction of the *Juniperus* scrub association along with a phase with *Rhamnus* and Rosaceae. At Gazel, the Azilian levels of the Post-Glacial (after 10,080 BP) reveal the juxtaposition of two forest types: a *Pinus sylvestris/Juniperus* open forest and a deciduous oak forest (*Quercus pubescens, Acer campestre, A. monspessulanum, Hedera, Corylus...*). At Troubat, phase T3 (from 9700 ± 80 BP) and at Buholoup, phase B1 (about 9000 BP), clearly correspond to the spread of collinean deciduous oak forest.

From the palynological viewpoint, the glacio-lacustrine fill at Barbazan (450 m, Central Pyrenees) has provided very precise data for the palaeoenvironment of the mountain valley of the Garonne from the end of the glacial period to the present (Andrieu, 1991; Reille and Andrieu, 1995). This site is very close to the cave-shelter and in a context comparable to that of Troubat (541 m). This interdisciplinary work involving palynology, sedimentology and isotope data offers material for comparison. The tree species are particularly taken into account.

Palynologically, the Late Glacial interstadial begins with the expansion of *Juniperus* and the rapid development of Poaceae. The progression of *Juniperus*, associated with that of *Hippophae* and *Betula*, indicates the establishment of pioneer shrub associations near the site (Andrieu, 1991). Then a *Juniperus* optimum and the beginning of the expansion of *Betula* is noted at 12,700 ± 270 BP. Finally, the strong progression of *Betula* and the modest success of *Pinus* which characterize, at Barbazan, the end of the Late Glacial interstadial. The optimum of the interstadial forest at Barbazan is dated to 11,590 ± 270 BP (Andrieu, 1991). At Troubat, the charcoal phase T1 would correspond to this optimum. The modest place of *Pinus sylvestris* in the pollen assemblages is surprising for a species which at the same period appears to characterize the pollen sequences of the northern Pyrenees (Jalut *et al.* 1982; Jalut *et al.* 1988; Andrieu, 1991). The convergence of the pollen values with the charcoal data is thus clear: *Pinus sylvestris* probably played a reduced role in the Garonnais piedmont, in contrast to what has been recorded, for example in Andorra (Heinz, 1991).

Phase T2 is to be correlated with the climatic change at the very beginning of the Younger Dryas. At Barbazan, the palynology indicates an expansion of steppic plants. *Juniperus, Hippophae* and *Rhamnus* are also present. The trees *Pinus* and *Betula* are less well represented and indicate a more open environment. The isotope measurements carried out on the sediments coincide with a drop in temperature; the climatic characteristics of the Younger Dryas might have been analogous to those of the Older Dryas (Guiot, 1987; Guiot *et al.* 1989; Andrieu, 1991).

Then at the beginning of the Post-Glacial, at about 10,300 BP if we consider the data for southern Europe (Beaulieu *et al.* 1982; Beaulieu *et al.* 1988), a phase of expansion of deciduous *Quercus* on a regional scale is noted (Andrieu, 1991). The numerous palynological data obtained in the Pyrenees (Jalut, 1977; Jalut *et al.* 1982; Jalut *et al.* 1988; Andrieu, 1991; Reille and Andrieu, 1995) testify to the simultaneous expansion of *Quercus*, at about 10,300 BP, throughout the northern slopes of the Pyrenees, from the West to the East. The start of this expansion marks the beginning of the Post-Glacial period. Phase T3 (Troubat) and B1 (Buholoup) is characterised by the very clear progression of *Quercus* and to a lesser extent by that of *Corylus* from 9700 ± 80 BP onwards. Phase B1 at Buholoup presents the same characteristics. At about 8625 ± 80 BP, the deciduous oak forest is dominant (phase T4 at Troubat and phase B2 at Buholoup).

Conclusion

This charcoal analysis data (Troubat and Buholoup) which has provided an original succession may be considered as representative of the vegetation dynamics from the Late-Glacial Bölling-Alleröd interstadial to the middle of the Post-Glacial. It offers information concerning the transformation in the vegetation cover due primarily to climate. The palaeoecological results concerning the Holocene are recurrent both at Troubat and Buholoup and validate, in this way, the methods used by charcoal analysis.

References

ANDRIEU V. 1991 - *Dynamique du paléoenvironnement de la vallée montagnarde de la Garonne (Pyrénées centrales, France) de la fin des temps glaciaires à l'actuel.* Unpublished Ph.D dissertation, Université de Toulouse 2, 330 p.

BADAL E. 1990 - *Aportaciones de la antracologia al estudio des paisaje vegetal y su evolucion en el cuaternario reciente, en la costa mediterranea del pais valenciano y Andalucia (18,000-3000 B.P.).* Unpublished Ph.D dissertation, Universitat de Valencia, Facultad de Geografia e Historia, 321 p.

BARBAZA M. - 1988. La grotte-abri du Moulin à Troubat (Hautes Pyrénées), Premiers travaux et dispositions initiales de recherches. *Travaux de l'Institut d'Art Préhistorique*, Université de Toulouse-le-Mirail. vol. XXX. p.7-38.

BARBAZA M. 1993 - Technologie et cultures du Mésolithique moyen au Néolithique ancien dans les Pyrénées de l'Est. In : J. Guilaine (dir.). *L'abri du Roc de Dourgne.* Centre d'Anthropologie des Sociétés Rurales. Toulouse, p. 425-442, 1 fig.

BARBAZA M., HEINZ C. 1992 - La grotte-abri du Moulin (Troubat, Hautes Pyrénées), Environnement et Archéologie. In: Charcoal, ancient ecosystems and human impact, colloque international Montpellier 10-13 septembre 1991. *Bull. Soc.bot. Fr.*, 139, Actual. bot. (2/3/4), p. 685-695.

BARBAZA M., BRIOIS F., VALDEYRON N., VAQUER J. 1999 – L'Epipaléolithique et le Mésolithique entre Massif central et Pyrénées. Actes du Vème congrès international UISPP. *L'Europe des derniers chasseurs*, Grenoble (Isère, France), p. 125-143.

BAZILE-ROBERT E. 1983 - Flore et végétation du Magdalénien final de la grotte des Eglises (Ussat, Ariège) d'après l'analyse anthracologique. Etudes préhistoriques pyrénéennes, *Bulletin de la Société Préhistorique de l'Ariège*, t. 38, p. 87-90.

BEAULIEU J.-L. de, PONS A., REILLE M. 1982 - Recherches pollenanalytiques sur l'histoire de la végétation de la bordure nord du massif du Cantal (Massif-Central, France). *Pollen et spores*, t. XXI, fasc. 2, p. 251-300.

BEAULIEU J.-L. de, PONS A., REILLE M. 1988 - Histoire de la flore et de la végétation du Massif central (France) depuis la fin de la dernière glaciation. *Cahiers de Micropaléontologie*, t. 3-4, p. 5-36.

BEUG H.J. 1975 - Changes of climate and vegetation belts in the mountains of mediterranean Europe during the Holocene, *Biuletyn Geologiczny*, fasc. 10, p. 101-110.

BOUREAU E. 1956 - *Anatomie végétale*. P.U.F., Paris, 752 p.

CHABAL L. 1991 - *L'Homme et l'évolution de la végétation méditerranéenne, des âges des métaux à la période romaine : recherches anthracologiques théoriques, appliquées principalement à des sites du bas Languedoc*. Unpublished Ph.D dissertation, U.S.T.L.Montpellier II, 435 p.

CHABAL L. 1997 - Forêts et sociétés en Languedoc (Néolithique final, Antiquité tardive), L'anthracologie, méthode et paléoécologie. *Documents d'Archéologie Française*, Paris, 188 p.

DUPIAS G. 1985 - *Végétation des Pyrénées*, Notice détaillée. Editions du CNRS, 209 p.

GREGUSS P. 1955 - *Identification of living gymnosperms on the basis of xylotomy*, Akadémiai Kiado Budapest, 263 p.

GREGUSS P. 1959 - *Holzanatomie der Europaïschen Laubhölzer und Straücher*, Akadémiai Kiado Budapest, 330 p.

GUIOT J. 1987 - Late Quaternary climatic change in France estimated from multivariate pollen time series. *Quaternary Research* 28, p. 100-118.

GUIOT J., PONS A., BEAULIEU J.-L. de, REILLE M. 1989 - A 140 000 year continental climate reconstruction from two European pollen records, *Nature*, fasc. 338, p. 309-313.

HEINZ C. 1990 - Dynamique des végétations holocènes en méditerranée Nord-occidentale d'après l'antracoanalyse de sites préhistoriques : méthodologie et paléoécologie, *Paléobiologie continentale*, vol. XVI, N°2, 212 p., Thèse Université Montpellier II.

HEINZ C. 1991 - Upper Pleistocene and Holocene vegetation in the south of France and Andorra. Adaptations and first ruptures: New charcoal analysis data. *Review of Palaeobotany and Palynology*, 69, p. 299-324. Elsevier Science Publishers, Amsterdam.

HEINZ C. 1994 - Quelques exemples pyrénéens illustrant les transformations du paysage végétal des quinze derniers millénaires. *Actes du Xème colloqui internacional d'Arqueologia de Puigcerda*, Homenatge al Profesor Jean Guilaine, p. 79-86.

HEINZ C. 1999 - Les transformations du paysage végétal pyrénéen au cours de l'Epipaléolithique et du Mésolithique. Actes du Vème congrès international UISPP, *l'Europe des derniers chasseurs, Epipaléolithique et Mésolithique en Europe, Peuplements et Paléoenvironnements*, Editions du CTHS, p. 115-123.

HEINZ C., VERNET J.-L. 1995 - La Balma Margineda, analyses anthracologiques, l'homme et la végétation in "*Les fouilles à la Balma Margineda*", sous la direction de Guilaine J. et Martzluff M., Ministère des affaires sociales et culturelles du gouvernement d'Andorre, III, p. 26-64.

HEINZ C., BARBAZA M. 1998 - Environmental changes during the Late Glacial and Post-Glacial in the central Pyrenees (France): new charcoal analysis and archaeological data. *Review of Palaeobotany and Palynology*, fasc.104, p. 1-17.

JACQUIOT C. 1955 - *Atlas d'anatomie des bois des conifères*. Centre Technique du Bois, Paris, 135 p.

JACQUIOT C., TRENARD Y., DIROL D. 1973 - *Atlas d'anatomie des bois des Angiospermes*. Centre Technique du Bois, Paris, 175 p.

JALUT G. 1977 - Végétation et climat des Pyrénées méditerranéennes depuis quinze mille ans - *Archives d'Ecologie préhistorique*, Ecole des Hautes Etudes en Sciences Sociales Toulouse, 2, 141 p.

JALUT G., VERNET J.-L. 1989 - La végétation du Pays de Sault et de ses marges depuis 15 000 ans : réinterprétation des données palynologiques et apports de l'anthracologie. In: *Pays de Sault, Espaces, Peuplements, Populations*, CNRS. éd., p. 23-35.

JALUT G., DELIBRIAS G., DAGNAC J., MARDONES M., BONHOURS M. 1982 - A palaeoecological approach to the last 21000 years in the Pyrenees: the peat bog of Freychinede (Alt.1350m, Ariège, south France). *Palaeogeography, Palaeoclimatology, Palaeoecology*, 40, p. 321-359.

JALUT G., ANDRIEU V., DELIBRIAS G., FONTUGNE M., PAGES P. 1988 - Paleoenvironment of the valley of Ossau (Western french Pyrenees) during the last 27000 years. *Pollen et spores*, Vol. XXX (1988), n° 3-4, p. 357-394.

MABBERLEY D. J. 1997 - *The plant book, a portable dictionary of the vascular plants*, Cambridge University Press, second edition, 858 p.

METCALFE C.R., CHALK L. 1950 - *Anatomy of the Dicotyledons*. Clarendon Press, Oxford, 1500 p.

METCALFE C.R., CHALK L. 1983 - *Anatomy of the Dicotyledons*. Clarendon Press, Oxford, 297 p.

RAMEAU J.C., MANSION D., DUME G., LECOINTE A., TIMBAL J., DUPONT P., KELLER R. 1993 - *Flore forestière française*, guide écologique illustré, I.D.F., 2, 2421 p.

REILLE M., ANDRIEU V. 1995 - The late Pleistocene and Holocene in the Lourdes Basin, Western Pyrénées, France : new pollen analytical and chronological data. *Veget. Hist. Archaeobot.* Fasc. 4, p. 1-21.

SACCHI D. 1986 - Le Paléolithique supérieur du Languedoc occidental et du Roussillon. XXIe suppl. *Gallia Préhistoire* CNRS éd., 284 p.

SCHWEINGRUBER F.H. 1978 - *Mikroskopische holzanatomie*. Zürcher A.G. Zug, 226 p.

SCHWEINGRUBER F.H. 1990 - *Anatomie europaïcher Hölzer*. Paul Haupt Bern undStuttgart. 800 p.

VERNET J.-L. 1980 - La végétation du bassin de l'Aude, entre Pyrénées et Massif central, au Tardiglaciaire et au Postglaciaire, d'après l'analyse anthracologique. *Review of Palaeobotany and Palynology*, Amsterdam, fasc. 30, p. 33-55.

Charcoal analysis at La Falaguerra rockshelter (Alcoi, Alacant, Spain) from the Mesolithic to the Bronze Age: landscape and the use of plant resources

Yolanda Carrión

University of Valencia, Department of Prehistory and Archaeology,
Av. Blasco Ibáñez, 28, 46010 Valencia, Spain.
E-mail: yolanda.carrion@uv.es

Abstract

Human groups occupied La Falaguera rockshelter during Prehistory and supplied their domestic fires with wood from the local environment. Charcoal analysis of La Falaguera shows some changes in vegetation from a wet forest dominated by some deciduous species to a characteristic evergreen *Quercus* Mediterranean woodland. Ash trees may have been exploited in order to feed cattle during the Neolithic period. Thus, La Falaguera rockshelter displays either a pattern of landscape use by prehistoric groups or an increase of arid conditions noticed in other neolithic sites.

Key-words: *Charcoal analysis, La Falaguera, Neolithic, Ash*

Résumé

Les groupes humains qui ont occupé l'abri de La Falaguera pendant la Préhistoire ont approvisionné leurs feux domestiques avec du bois ramassé dans un environnement proche. L'analyse anthracologique de l'abri de La Falaguera montre le remplacement progressif d'une forêt humide avec des espèces à feuillage caduc, par une forêt typique méditerranéenne à chênes sclérophylles. L'exploitation du frêne a été documentée pour la nourriture du bétail pendant l´époque Néolithique. L'abri de La Falaguera montre un modèle d'utilisation de l'environnement par les groupes préhistoriques, qui provoquent l'augmentation des conditions d'aridité perçues sur d'autres sites néolithiques proches.

Mots-clés : *Anthracologie, La Falaguera, Néolithique, frêne, alimentation du bétail.*

La Falaguera rockshelter: location and current vegetation

La Falaguera rockshelter (Alicante, Spain) is found at 860 m a.s.l., in a deep gorge belonging to a watershed on the right slope of the Serpis river. These areas are typically humid and shady with a highly characteristic vegetation at the bottom, though most of the streams are inactive. This is one of the nothernmost slopes of La Font Roja nature reserve (fig. 1), its summit being 1200 m high, and where a dense supramediterranean gall oak (*Quercus faginea*) woodland develops, as well as other deciduous species (Costa *et al.* 1997). This reserve is considered a vegetation refuge, as it is one of the few deciduous formations that can be found in the area today.

Currently, the gorge is colonised by thick vegetation in which holm-oak (*Quercus rotundifolia*) and Jerusalem pine (*Pinus halepensis*) are predominant. Manna ash (*Fraxinus ornus*), fig trees (*Ficus carica*) and wild cherry trees (*Prunus avium*) are also present. A dense scrub includes kermes oak (*Quercus coccifera*), juniper (*Juniperus phoenicea*), rosemary (*Rosmarinus officinalis*), small-flowered furze (*Ulex parviflorus*), and white cistus (*Cistus albidus*); there are also some lianes and another climbing plants, *e.g.* ivy (*Hedera helix*) or madder (*Rubia peregrina*).

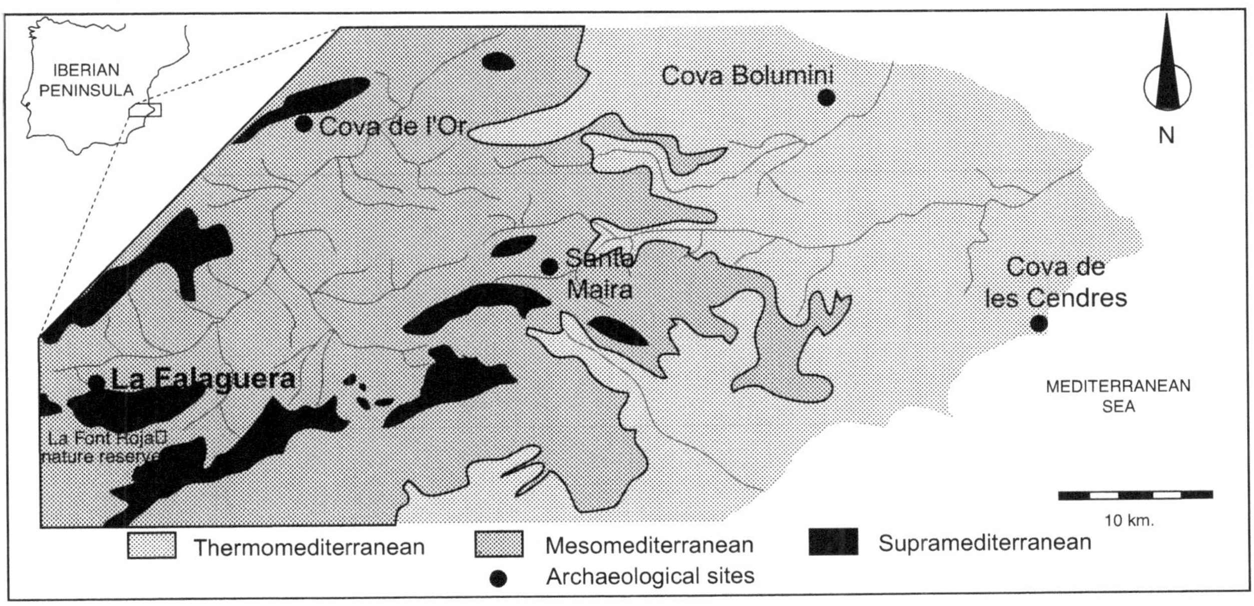

Figure 1: Location of the sites mentioned in the text.

The arboreal vegetation, located only at the bottom of the gorge, contrasts with the rather sparse bushes that grow on the sunny slopes. Some of these slopes are terraced for farming. This formation has benefitted from the humidity and shade prevailing in the narrow ravine.

Currently, the site belongs to a dry mesomediterranean bioclimatic stage included into the vegetation series of the kermes oak *Bupleuro rigidi-Querceto rotundifoliae sigmetum* (Rivas Martínez, 1987).

Fieldwork in the rockshelter and stratigraphy of the site

The rockshelter was discovered after surveying the valley of the Serpis river in 1981. Three excavations have been carried out since then, one by the Museum of Alcoi, and the other two directed by E. Aura Tortosa and O. García Puchol of the University of Valencia. Very thick layers ascribed to neolithic times were documented from the very beginning. There was evidence of significantly dismantled fires, as only some charcoal spots were left (Carrión, 1999). The primary aim in the last excavation was to document the moment of transition from Mesolithic to Neolithic levels, one of the main research questions in regards to the Neolithic in the Iberian peninsula.

Currently, the rockshelter is surrounded by a small, partly dismantled stone wall, which suggests that it was used as a stockyard untill recent times.

The stratigraphic sequence of La Falaguera includes seven phases. Phases IV, V and VI refer to neolithic levels, where fire evidence has been documented. Phase IV would refer to the Neolithic IIB, where plain pottery and a notched and stemmed lithic points have been found. Material from Phase V, particularly plain pottery and some bone awls, define these strata as general Neolithic. The exact chronology cannot be determined because very few pieces have been found. In Phase VI, there is a high number of impressed or incised ceramics, as well as some fragments of cardial ceramics, gradine decorated ceramics with smooth or decorated cords and a great amount of carbonized grains of cereals in the lowest part of the stratum. Phase VII refers to mesolithic levels. Among the only lithic pieces which have been found there is a backed bladelet.

The area where the neolithic and mesolithic levels intersect is especially significant to the documentation of the transition to the Neolithic. However, in La Falaguera it seems to be of an erosive type, with a significant gap in information. Further research of the materials from these layers may help define the chronological difference between the two levels in contact more accurately.

Aims of the charcoal analysis at La Falaguera

One of the key points in neolithic research is the transition from mesolithic hunting and gathering societies to neolithic farming and stockbreeding. The great number of sites and the existence of a concentration of "pure Neolithic" groups with the most ancient chronology in the Iberian peninsula make this area in Alicante exceptionally interesting to explore for possible evidence of contact, acculturation or substitution between both cultures (Martí Oliver and Juan Cabanilles, 1997).

After finding the first mesolithic stratum in La Falaguera rockshelter, researchers hoped that this site could shed some light on these questions and provide a complete sequence from the Epipalaeolithic. The whole set of materials found there is currently being studied to establish possible cultural, economic or ecological variations along the sequence. The research carried out in La Falaguera follows an interdisciplinary research model, in which the studies of lithics, ceramics and bones are supplemented with an environmental approach through different palaeobotanical disciplines, microsedimentology and research on fauna and microfauna.

Charcoal in La Falaguera is abundant and well preserved, which has allowed researchers to obtain a representative sample from each stratum, though some stratum provided very little archaeological material. From this starting point, we established the following methodological, ecological and ethnographic goals:
- To implement the various sampling strategies both when collecting charcoal during the excavation process and when analysing and re-counting data in the laboratory;
-To reconstruct the vegetation around the site, which has been exploited by human groups to collect wood, and to study how this vegetation developed along the whole sequence;
- To understand the use of fire in particular activities through the analysis of concentrated charcoal.

Anthracological data, particularly data from anthropic activities, should be interpreted on the basis of all the archaeological information provided by the whole set of materials found at the site. Also, charcoal analysis results from La Falaguera may help confirm or verify the information gathered from other sources. Accordingly, our main aim in this charcoal analysis is to reconstruct how the environment was used by human groups, within a more complex economic model.

Methodology used in collecting and studying charcoal from La Falaguera

The procedure for collecting charcoal at the site develops in tandem with the excavation process. Both procedures share the same problems concerning the stratigraphic sequence. This procedure allows researchers to implement the best sampling strategies depending on the characteristics of each site (Badal, 1990, 1992; Badal and Heinz, 1991; Chabal, 1988, 1992, 1997).

Charcoal in la Falaguera was collected systematically within an area of 9 m^2, each square metre being divided into four subunits and in layers of 5 to 10 cm, with careful attention to natural strata. All of the charcoal was collected through hand sifting with 2 and 5 mm screens. In addition, a sample of 10 to 12 litres from each stratigraphic unit was taken to be sifted through flotation, which allowed us to recover a great number of charcoal fragments, as well as other carbonized remains (mainly seeds). When the amount of charcoal found in a sample is not sufficient, charcoal obtained through flotation can be added to get a representative quantity.

Charcoal analysis in the laboratory is based on the anatomical differences in the wood of the various plant species, *i.e.* its inner structure, in each of the three sections: —transverse, tangential longitudinal and radial longitudinal—. Charcoal is observed through a reflected light microscope (Vernet, 1973) and is compared with specialized literature (Greguss, 1955, 1959; Jacquiot, 1955; Jacquiot *et al.* 1973; Schweingruber, 1978, 1990) or with an actual comparative collection. The collection of the Department of Prehistory and Archaeology of the University of Valencia includes more than 300 wood species with several examples of each species, sampled in different ecological environments.

Results of the charcoal analysis at la Falaguera

After analysing a sample of 4892 charcoal fragments, the following 34 taxa were identified: *Acer* sp. (maple), *Arbutus unedo* (strawberry tree), Cistaceae, Coniferae, *Erica* sp. (heather), *Fraxinus* sp. (ash tree), *Hedera helix* (ivy), *Juniperus* sp. (juniper or savin), Labiatae, Leguminosae, Maloideae, Monocotyledoneae, *Olea europaea* (olive tree), *Pinus halepensis* (Jerusalem pine), *Pinus* cf. *pinea* (umbrella pine), *Pinus* sp., *Pistacia lentiscus* (mastic tree, lentisc), *Pistacia terebinthus* (turpentine tree, terebinth), *Pistacia* sp., *Prunus* sp. (cherry tree), cf. *Pyrus* sp. (pear tree), deciduous *Quercus* (deciduous oak), evergreen *Quercus* (evergreen oak), *Quercus* sp., *Rhamnus-Phillyrea*, *Rosmarinus officinalis* (rosemary), *Salix-Populus* (willow-poplar/aspen), *Sorbus-Crataegus* (white beam-hawthorn), *Viburnum* sp. (laurustinus), Undetermined I, Undetermined II, Undetermined III and Undetermined IV (fig. 2).

There is a high specific variety in the taxonomic list of La Falaguera. It includes species from different ecological environments. In some cases we could identify them only at a family level; different vegetation strata —arboreal, shrub and even some lianas and another climbing plants— are well represented.

Quercus, both evergreen (which reaches values over 60% in the upper stratrigraphic units) and deciduous (with a percentage higher than 35% in the lower units) is the best represented species in terms of percentage.

Quercus genus is generally spread in various bioclimatic stages, from the thermomediterranean stage to the supramediterranean. In the case of evergreen *Quercus*, it is possibly holm-oak (*Quercus rotundifolia*), considering the other taxa that are found with it and that it is abundant in our area of study today. This species spreads systematically along the whole Mediterranean strip with a certain preference for continental conditions, though it can be found from sea level to 1,500 m above sea level. It can adapt to arid conditions where deciduous *Quercus* cannot survive. This latter species is mainly found in the supramediterranean stage, but it can also be found in a mesomediterranean stage with a humidity level that enables its development. Gall oak (*Quercus faginea*) is the characteristic deciduous species in our area of study (Costa *et al.* 1997: 242-246).

Fraxinus sp. shows the highest percentage after *Quercus*. Ash trees are often found next to rivers or shady areas inside deciduous or mixed forests, which provide them with the permanent level of humidity they need. Manna ash (*Fraxinus ornus*) is the characteristic ash species in the studied area; it can successfully survive without water and can thus be found in the whole inner eastern area.

The documented presence of the ash tree traces a downward curve, reaching the highest value (27%) in unit 2054, and the lowest value (3%) in the unit 2010/2008. Its dynamics

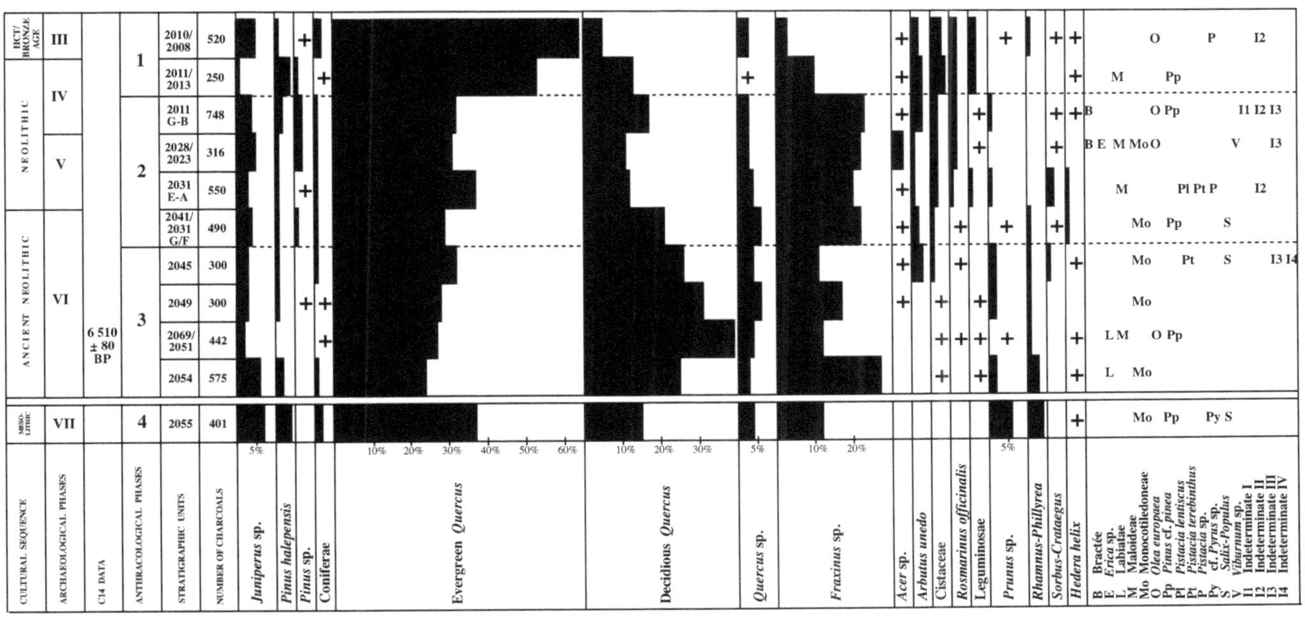

Figure 2: Diagram based on charcoal analysis from La Falaguera.

are somewhat similar to that of deciduous *Quercus*, but with very different features from unit 2041/2031 G-F up to unit 2011 G-B. In these units, the percentage of ash trees rises sharply over 20% and after keeping constant along this interval it falls sharply again. This phenomenon cannot be interpreted in ecological terms from the general dynamics of the other taxa, as it is not consistent with them. For this reason, we conclude that the overrepresentation of ash trees may be provoked by ethnographic reasons, which will be discussed later in this article.

The mixed formation described up to this point includes a forest with a well-developed arboreal stratum that creates humid and shady areas, where certain deciduous species which grow under these conditions may live (Costa *et al.* 1997: 279), *e.g. Acer* sp., *Prunus* sp. or *Sorbus-Crataegus*.

Taxa associated with the characteristic set of species that usually enrich these formations, particulary those of evergreen *Quercus* mediterranean forests have also been identified: *Arbutus unedo*, Cistaceae, *Rosmarinus officinalis*, Leguminosae and *Rhamnus-Phillyrea*.

Conifers are poorly represented. Juniper or savin (*Juniperus* sp.) is the only taxon in this group that reaches a percentage higher than 5% in the lowest units of the diagram. On the whole, junipers are pioneering species that colonise impoverished and degraded soils. They are therefore heliophilous, *i.e.* they grow within open shrub vegetation areas where they receive a high level of sunlight (Riou-Nivert, 1996). For this reason, they can be found on rocky slopes with a poor edaphic development, which cannot be occupied by other more demanding species in these conditions. The percentage of *Pinus halepensis* is always lower than 5%. This pine is resistant to high temperatures and to long periods of drought (Romo Díez, 1997: 72-73). This species often grows in open formations and, like junipers, it can colonise areas with poorly developed soils.

The modest significance of conifer plants in La Falaguera is consistent with the significant development of leafy and particularly deciduous plants whose growth requires certain conditions of edaphic development and humidity. For this reason, the needs of both conifers and deciduous plants are not compatible (Costa *et al.* 1997). In our case, then, conifers may be found in very localised sites, *e.g.* in some very steep and sunny slopes.

Percentages of all the other taxa identified in La Falaguera are very low, generally lower than 1%. While their percentages are not very representative, some important ecological information can be gathered from them. We may highlight, for instance, the existence of Monocotyledonae and *Salix-Populus*, two taxa that are associated with environments with a continuous provision of humidity. *Olea europaea* var. *sylvestris* is a species with a typically Mediterranean nature, as it is sensitive to cold conditions and can spread to higher places only on very sunny slopes or when sheltered from continental cold winds (Romo Díez, 1997: 287). This is also the case of the mastic tree (*Pistacia lentiscus*), which grows mainly in littoral areas with mild winters and no higher than 1000 m (Romo Díez, 1997: 272). The existence of these mild species in La Falaguera can be explained by the thermal conditions at the bottom of valleys, where these species can spread under conditions which are more favourable for them than the thermal conditions of less sheltered slopes.

The anthracological sequence of La Falaguera has been divided into four phases on the basis of species percentages (fig. 2):

- Evergreen *Quercus*, as well as deciduous *Quercus*, *Fraxinus* sp., *Prunus* sp. and *Rhamnus-Phillyrea* figure predominantly in the anthracological Phase 4 which represents the oldest part of the sequence. The first three species amount to 70%, which shows that the formation is mixed and dense, with a certain level of humidity, which allows the growth of other deciduous species. Conifers represent only 12,98 % and they were possibly limited to the steep and sunny slopes, *e.g.* the walls of the gorge.

- The beginning of anthracological Phase 3 coincides with the Neolithic (Archaeological Phase 4). A seed of *Triticum* was found through the use of the AMS method to date to 6510 ± 80 BP, 5620-5320 cal. BC (Beta-142289). The percentages of taxa in this phase are clearly different from those in the previous phase. Evergreen *Quercus* represents between 24% and 32%, and its values here did not reach those in Phase 4. In this phase deciduous *Quercus* shows its highest values, near 40%, and *Fraxinus* sp. presents a value over 25%. There are also *Prunus* sp., *Rhamnus-Phillyrea*, *Hedera helix*, Monocotyledonea and *Salix-Populus*. The development of these deciduous plants shows evidence of high humidity, probably due to a reactivation of a stream and to an only marginal human exploitation of the environment.

On the other hand, taxa such as *Acer* sp., *Arbutus unedo*, Cistaceae, *Rosmarinus officinalis*, Leguminosae, *Olea europaea* var. *sylvestris* and *Pistacia terebinthus* are found for the first time in this phase, resulting in a richer taxonomic variety.

Today, these mixed formations of evergreen and deciduous species in the Mediterranean area are limited to quite humid environments, with fairly mild terrain and regular temperatures with no frosts (Costa *et al.* 1997: 25). *Quercus faginea* and *Quercus rotundifolia* are often found together in a shrub stratum, dominated by *Prunus* and *Rhamnus* species. Given its characteristics, we think that the vegetation in La Falaguera is as clear of an example as La Font Roja —located between 1000 and 1200 m a.s.l. and near our area of study— is today (Costa *et al.* 1997: 251).

- The anthracological Phase 2 shows a similar structure to that at the end of the previous phase, *i.e.* a progressive increase of evergreen *Quercus*, *Acer* sp., *Arbutus unedo*, Cistaceae, *Rosmarinus officinalis* and Leguminosae. *Olea europaea* var. *sylvestris*, *Pistacia* and *Pinus halepensis* become more usual in this phase. Some fragments of pinecone bracts have also been identified. Deciduous *Quercus*, *Prunus* sp. and *Rhamnus-Phillyrea* decrease.

Ash trees are a special case. They increase sharply to steady values between 20% and 23% in this phase, differing from other deciduous taxa that tend to decrease. In our opinion, such a phenomenon cannot be explained on ecological terms, but it is overrepresented by an anthropic factor, *i.e.* an intensive collecting of this species. The stratigraphic units where this phenomenon is observed are those where fire evidence was found. These burnt levels show a characteristic morphology: they show a black or very dark brown sediment, with no archaeological remains, which belongs to the ground where combustion stopped. Another level with plenty of carbonized reddish organic matter appears on the mentioned sediment, where remains of acorns, small fragments of charcoal and coprolites have been identified in La Falaguera. These levels, which have also been documented in other caves in neolithic chronology, *e.g.* Cendres, Bolumini and Santa Maira, have been understood as remains from the burning of stockyard grounds, which were done cyclically to clear them of animal excrement (Badal, 1999: 71). Charcoal in these levels usually comes from branches brought to the site to feed cattle. The use of ash leaves has been documented in ethnographical data for cattle feeding; these leaves were highly appreciated, particularly by goats (Romo Díez, 1997: 284-286). For this reason, we think that the great number of ash charcoal in Phase 2 in La Falaguera results from an intensive use of this species.

The increase of evergreen species shown in Phase 2 intensifies in Phase 1, in which evergreen *Quercus* reaches nearly 65%; Cistaceae, Leguminosae, *Juniperus* sp. and *Pinus halepensis* also increase. *Olea europaea* var. *sylvestris* and *Pistacia* are not very abundant, while deciduous *Quercus* and *Fraxinus* sp. drop to their lowest values.

This suggests a rather arid nature and a formation that is simpler and more open than the conditions which cause deciduous *Quercus* to reach its highest values (Costa *et al.* 1997: 295), although some taxa suggest that a certain degree of humidity is still present.

Discussion and conclusions

A significant development of *Quercus* dominated plant formations is documented along the whole sequence on the basis of charcoal in La Falagera. Broadly speaking, it is a transition process from deciduous formations where deciduous *Quercus* and ash trees are dominant, to another formation where evergreen *Quercus* is dominant. Accordingly, there are two groups of taxa: those representing their highest values at the beginning of the sequence and then decrease, which are associated with the dynamics of deciduous *Quercus* (*Prunus* sp. and *Rhamnus-Phillyrea* are the most obvious); and those whose percentages increase by the end of the sequence, associated with the dynamics of evergreen *Quercus*. These are mainly *Arbutus unedo*, Cistaceae, *Rosmarinus officinalis* and Leguminosae.

Thus a change can be seen in the dominant species in the forest cover, with a development of evergreen *Quercus* and its characteritically shrub-like accompanying species. Conifers, however, do not take up space in the *Quercus* woodland along the whole sequence, and thus forest cover cannot be said to be degraded. These heliophilous species grow very quickly when light increases or in open spaces, but in our case this development was impossible due to the replacement of the deciduous *Quercus* woodland by another well-developed evergreen *Quercus* woodland. Possibly, conifers were located in sunny areas where the *Quercus* woodland was not able to develop.

The observed dynamics in La Falaguera can be compared with anthracological sequences of nearby sites with the same chronology. Forest clearance phenomenon in neolithic levels have also been observed in Cova de les Cendres (Teulada, Alacant) and in Cova de l'Or (Beniarrés, Alacant) (Badal, 1995; Badal *et al.* 1994). Deciduous *Quercus* has no significance in Cova de les Cendres due to its proximity to the coast line and instead, what can be observed is a progressive replacement of evergreen *Quercus* by *Pinus halepensis*. The end of the sequence, however is dominated by a scrub where *Erica multiflora*, *Cistus* sp. and Leguminosae are present. The high percentage of *Olea europaea* shows evidence of the clear thermomediterranean nature of Cova de les Cendres environment. The dynamics in Cova de l'Or are similar, but deciduous *Quercus* is more common here, as should be expected by its location on the mesomediterranean stage. This replacement of *Quercus* woodland by conifers is not present in La Falaguera, but some deciduous species are replaced by evergreen species. This may be due to the following reasons:
- La Falaguera is located in a sheltered area for temperate and humid vegetation that can only grow at the bottom of the gorge and not on the sunnier slopes.
- Human exploitation of the forest is not very intensive; there is not an obvious destruction of the environmental equilibrium as can observed in other sites which are more suitable for farming and where the deforestation process is implemented by human groups (Badal, 1995: 225).

Possibly, La Falaguera is an example of the human exploitation of forests in neolithic times, with an intensive collection of ash branches to feed cattle with their leaves, resulting in an overrepresented presence of this species in charcoal samples. The high percentage of *Olea europaea* in Cova de les Cendres is probably associated with this kind of activity (Badal, 1999). In La Falaguera, *Olea* is not as significant as might be expected due to its location on a mesomediterranean stage; here ash trees play the same role. We can therefore suggest that its exploitation was planned, as the wood gathering efforts were directed towards a particular species. However, the exploitation of the environment is balanced in La Falaguera, since there is no evidence of degradation in the existing woodland.

References

BADAL E. 1990 - *Aportaciones de la antracología al estudio del paisaje vegetal y su evolución en el Cuaternario reciente, en la costa mediterránea del País Valenciano y Andalucía* (18.000-3.000 BP). Tesis Doctoral. Universitat de València.

BADAL E. 1995 - La vegetación carbonizada. Resultados antracológicos del País Valenciano. *El Cuaternario del País Valenciano*, p. 217-226.

BADAL E. 1992 - L'anthracologie préhistorique : à propos de certains problèmes méthodologiques. *Bull. Soc. Bot. Fr.*, 139, Actual. Bot. (2/3/4), p. 167-189.

BADAL E., HEINZ C. 1991 - Méthodes utilisées en Anthracologie pour l'étude de sites préhistoriques. *B.A.R.*, 573, p. 17-47.

BADAL E., BERNABEU J., VERNET J.-L. 1994 - Vegetation changes and human action from the Neolithic to the Bronze Age (7000-4000 BP) in Alicante, Spain, based on charcoal analysis. *Vegetation History and Archaeobotany* 3, p. 155-166.

BADAL E. 1999 - El potencial pecuario de la vegetación mediterránea: las cuevas redil. II Congrés del Neolític a la Península Ibèrica. *Saguntum*-PLAV, Extra-2, p. 69-75.

CARRIÓN Y. 1999 - Datos preliminares del antracoanálisis de l'Abric de La Falaguera (Alcoi, Alacant). Actes del II Congrés del Neolític a la Península Ibèrica. *Saguntum*-PLAV, Extra-2, p. 37-44.

COSTA M., MORLA C., SAINZ H. (Eds.) 1997 - *Los bosques ibéricos. Una interpretación geobotánica*. Ed. Planeta. 572p.

CHABAL L. 1988 - Pourquoi et comment prélever les charbons de bois pour la période antique: les méthodes utilisées sur le site de Lattes (Hérault). *Lattara 1*. A.R.A.L.O. Ed. Caveirac, p. 187-222.

CHABAL L. 1992 - La représentativité paléo-écologique des charbons de bois archéologiques issus du bois de feu. *Bull. Soc. Bot. Fr.*, 139, Actual. Bot. (2/3/4), p. 213-236.

CHABAL L. 1997 - Forêts et sociétés en Languedoc (Néolithique final, Antiquité tardive), L'anthracologie, méthode et paléoécologie. *Documents d'Archéologie Française*, Paris, 188p.

GREGUSS P. 1955 - *Identification of living gymnosperms on the basis of xylotomy*, Budapest, 263p.

GREGUSS P. 1959 - *Holzanatomie der Europaïschen Laubhölzer und Straücher*. Akadémiai Kiado Budapest, 263p.

JACQUIOT C. 1955 - *Atlas d'anatomie des bois des conifères*. Centre Technique du Bois, Paris 135p. 64fig. 2 vol.

JACQUIOT C., TRENARD Y., DIROL D. 1973 - *Atlas d'anatomie des bois des Angiospermes*. Centre Technique du Bois, Paris, 175p. 72fig. 2 vol.

MARTI OLIVER B., JUAN CABANILLES J. 1997 - Epipaleolíticos y Neolíticos: población y territorio en el proceso de neolitización de la Península Ibérica. *Espacio, Tiempo y Forma*. Serie I, 10. Prehistoria y Arqueología, p. 215-264.

RIOU-NIVERT P. 1996 - *Les résineux*. Tome I. Connaissance et reconnaissance. Institut pour le développement forestier.

RIVAS-MARTÍNEZ S. 1987 - *Memoria del mapa de series de vegetación de España 1:400.000*. I.C.O.N.A.. Madrid.

ROMO DÍEZ A.M. 1997 - *Árboles de la Península Ibérica y Baleares*. Guía ilustrada para identificar y conocer todas las especies. Ed. Planeta. 347p.

SCHWEINGRUBER F.H. 1976 - Prahistorisches holz. Die bedeutung von Holzfunden aus mitteleuropa für die Lösung archäologischer und vegetationskundlicher Probleme. *Academica helvetica*. 106 p.

SCHWEINGRUBER F.H. 1978 - *Mikroskopische holzanatomie*. Zürcher A.G. Zug, 226p. 98fig.

SCHWEINGRUBER F.H. 1990 - *Anatomie europaïcher Hölzer* W.S.L.F.N.P. Haupt. 800p.

VERNET J.-L. 1973 - Étude sur l'histoire de la végétation du sud-est de la France au Quaternaire d'après les charbons de bois principalement. *Paléobiologie Continentale* 4 (1), p. 1-90.

Vegetation and firewood management at Cueva de la Vaquera (Segovia, Spain) between 6 and 3.7 kyr. BP: anthracological contribution to the landscape archaeology of the Spanish Central Mountains

Paloma Uzquiano

Departamento de Estudios Clásicos, Facultad de Filosofía y Letras,
Universidad de León and Instituto de Historia CSIC, Duque de Medinaceli, 8. 28014 Madrid, Spain.
E-mail: cehul18@ceh.csic.es

Abstract

The anthracological data obtained at Cueva de la Vaquera shows the woody vegetation and firewood management carried out by humans during the Neolithic, Chalcolithic and early Bronze Ages on the northern slopes of the Spanish Central Mountains. Charcoal analyses from between 6000-4850 BP have indicated the nearly exclusive use of *Pinus sylvestris* as firewood. Towards 4850 BP, the alternate use of *Pinus sylvestris*, *Quercus* deciduous, *Fraxinus*, *Quercus ilex*, *Juniperus* and *Pinus pinea-pinaster* was more characteristic. Chalcolithic and early Bronze occupations have shown the progressive substitution of *Pinus* by *Juniperus* and a firewood management based on the preference of species from open environments such as *Cistus*.

Key-words: *Anthracology, Holocene vegetation, Firewood management, Cueva de la Vaquera, Spanish Central Mountains.*

Résumé

Les résultats anthracologiques obtenus à la Cueva de la Vaquera, située sur le versant nord du système central espagnol, nous permettent de connaître la végétation ligneuse et la gestion du bois de feu réalisée par les humains au cours du Néolithique, Chalcolithique et Bronze ancien. Le contenu anthracologique daté entre 6000-4850 BP a révélé l'usage presque exclusif de *Pinus sylvestris*. A partir de 4850 BP, l'alternance de *Pinus sylvestris*, *Quercus* à feuillage caduc, *Fraxinus, Quercus ilex, Juniperus* et *Pinus pinea-pinaster*, devient plus caractéristique. Les niveaux du Chalcolithique et du Bronze ancien ont révélé le remplacement progressif de *Pinus* par *Juniperus* et une nette préférence pour les essences de milieux plus ouverts (*Cistus*).

Mots-clés : *Anthracologie, végétation holocène, gestion du bois, cueva de la Vaquera, système central espagnol.*

Introduction

The archaeological work carried out at La Vaquera cave has provided an interesting stratigraphy of the middle part of the Holocene showing evidence of human occupation from the Neolithic to the early Bronze Age (6120-3760 BP). The importance of this cave lies in being the first evidence of a Neolithic sequence found in central Spain (Iglesias *et al.* 1995; Estremera, 1999). Systematic sampling has yielded abundant charcoal remains that have been studied, permitting us to establish the first charcoal sequence for the Spanish Central System. Using data obtained from charcoal samples, we can reconstruct the evolution of the woody vegetation around the site and the use of firewood by the occupants. This use is closely related both to changes in the vegetation cover recorded in the palynological record from peatbog sediments (Franco *et al.* 1998) and to livestock movements in this area.

The site: geographic and stratigraphic position and cultural attributions

The Cueva de la Vaquera site opens in the north-western slope of the limestone piedmont at 960 m a.s.l. at the confluence of the Piron and Viejo rivers, about 30 km to the north-east of Segovia (fig. 1). The topography of the area is characterised by the alternation of calcareous and siliceous hills at altitudes between 900 and 1100 m a.s.l., by deep valleys running across these hills forming canyons (Pirón and Viejo) and by ondulating alluvial plains called "campiñas". The climate is continental mediterranean with strong seasonal variations (cold winters and dry summers with 6-8 months of frost) and the existence of local microclimates depending on the altitude, slope orientation and the depth of the valleys (Tejero, 1988; Martín and Rincón, 1990). The flora is influenced by the altitude variations, the differences in temperature and humidity, and human activities.

Juniperus thurifera is located on the limestone piedmont with *J. communis* and a few isolated *Quercus ilex* clumps. These open formations are located on poor soil and they have been modified by livestock activities. On the siliceous piedmont, the transition between *Quercus pyrenaica* and *Q. ilex* woodlands can be found. Shrublands with *Cistus laurifolius, C. ladanifer* and pasture areas are also present. The vegetation of the valleys consists of *Fraxinus angustifolia, Quercus faginea* and several *Salix* and *Populus* species, that alternate with water meadows "navas". *Quercus ilex* wood-

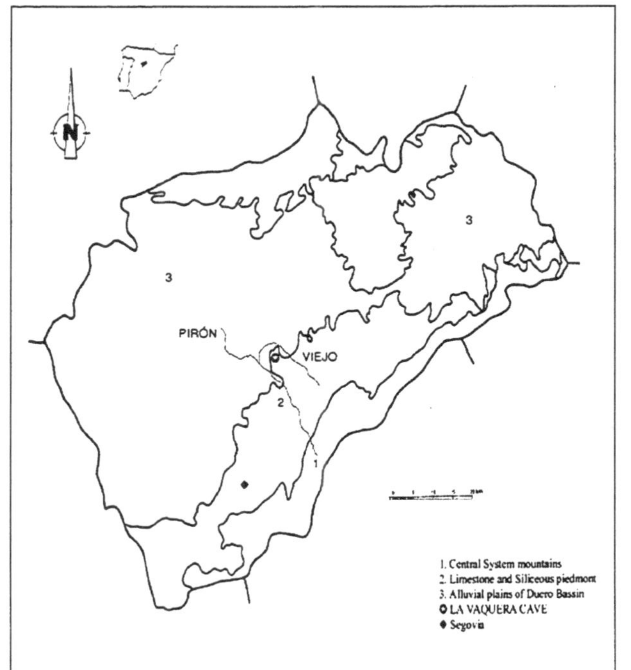

Figure 1: Location of the Cueva de la Vaquera in the province of Segovia (Spain).

lands with some *Juniperus thurifera* and *J. communis* are found on the "campiñas". Small woods of *Pinus pinaster* and *P. pinea* are also present. This vegetation has greatly suffered under human impact (agriculture) and all the communities mentioned are nowadays much reduced.

Traditional use of the land consists mainly of cereal agriculture and livestock activities, with woodland exploitation in some areas. Wheat, barley, rye and oats are the main cultivated cereals. Livestock consists of sheep, cows and goats. The foothills and the valleys are used as grazing areas for sheep and cows respectively. The site is at approximately the same distance from both the "campiñas" and grazing areas (S. Estremera, pers. comm.).

Archaeological studies have detected 8 phases of human occupation based on the evolution of pottery and lithic assemblages that have been attributed to the Neolithic (phases I to III), Chalcolithic (phases IV to VI) and early Bronze (phases VII and VIII) cultures. The radiometric data obtained spans a period of between 6120 and 3760 years BP (Estremera, 1999; O. Alonso, pers. comm.).

Material and Methods

La Vaquera cave yielded 4354 charcoal fragments which were obtained by manual flotation of the entire amount of the excavated sediment. This allows the removal of charcoal from clay sediment without damage. Sieving with water techniques (5 mm and 2 mm) and systematic selection of charcoal from sieving was also employed. All these techniques have been systematically employed in other cave settlements of the Iberian Peninsula (Uzquiano, 1992, 1997). All occupation levels in the cave have produced large quantities of charcoal. Charcoal analyses are carried out by hand-breaking the charcoal in the direction of the three anatomical sections and observing through an optical reflection microscope (Olympus BX60). Wood anatomy atlases were employed for botanical identifications (Schweingruber, 1990; Greguss, 1955, 1959; Jacquiot, 1955; Jacquiot *et al.* 1973) and a reference collection of charcoal from the study area was also very useful in identifying taxa.

Results

Charcoal analyses have yielded almost 30 taxa. Among these *Juniperus* ssp., *Pinus* t. *sylvestris, P. pinea-pinaster, Quercus* deciduous, *Quercus ilex, Fraxinus* t. *angustifolia* and the shrubs *Cistus* t. *laurifolius-ladanifer* are the main woody taxa that show the dynamics of fuel use during the periods of human occupation (fig. 2).

Between 6120 and 5800 BP (Neolithic Phase I and beginning of Phase II) *Pinus* t. *sylvestris* is dominant, followed by deciduous *Quercus*. *Fraxinus* was also present, but to a lesser extent. *Quercus ilex, Juniperus* and shrubs appeared sporadically. Between 5800 and 4690 BP (end of Phase II and Phase III) *Pinus* was still dominant, alternating with greater amount of deciduous *Quercus* and *Fraxinus,* and with a significant presence of *Quercus ilex* and *Juniperus.*

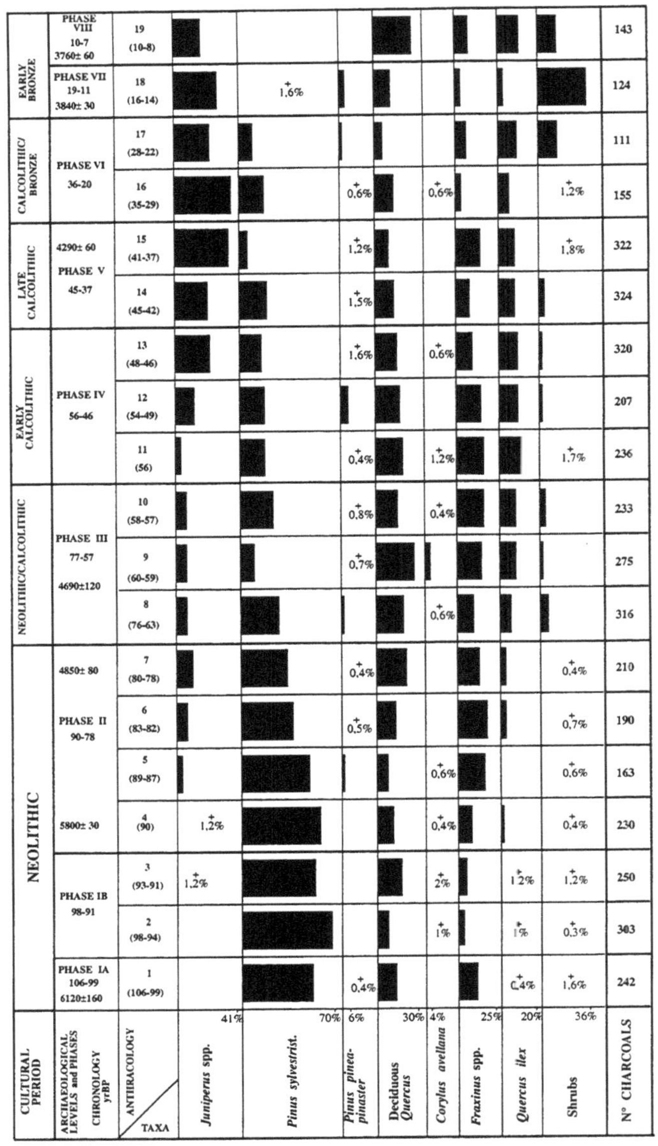

Figure 2: Anthracological diagram with the main woody species exploited by humans at La Vaquera cave.

The presence of *Pinus pinea-pinaster*, although in very small amounts was continuous throughout human occupation.

The results obtained between 4690 and 3760 BP show a progressive increase in *Juniperus* and a gradual decrease in *Pinus* t. *sylvestris*. This implies a change in the areas of firewood supply that occur throughout the Chalcolithic period. A systematic exploitation of deciduous and evergreen *Quercus* communities to a similar extent and a particular exploitation of *Fraxinus*, followed the progressive decrease of *Pinus*. This taxon disappeared from the storage area of the cave at the beginning of the Bronze Age, between 3890-3760 BP. A sharp increase in *Cistus* occured simultaneously with the Chalcolithic-Early Bronze Age transition with a slight regression in the other woody species and with the disappearance of *Pinus*.

These results show the exploitation firewood from different supply areas which are geographically close when the topography of the area mentioned above is taken into account.

Discussion

The relative abundance of *Pinus* from the beginning of the human occupation of the cave, its progressive disappearance from the resources storage area and the increase in *Juniperus* and broadleaved species, lead us to begin by evaluating firstly the palaeoenvironmental factors that have influenced the presence and distribution of all these woody communities as evidenced by anthracology.

Pollen analyses of peat bog sediments from both the Central and Iberian Systems (Franco *et al.* 1998; Ruiz *et al.* 1997; Peñalba, 1989, 1994; García *et al.* 1995) show that *Pinus* was dominant among arboreal vegetation during the first part of the Holocene (8-6 kyr BP) and the broadleaved species were present in small amounts. There were regional variations of vegetation dynamics depending on the geographical position of each deposit considered.

The dominance of *Pinus* in the environment would determine the use of wood resources by the first Neolithic groups (phase I and beginnings of phase II). The range of human movements would be confined to the nearest slopes and valley. Beginning in 5800 BP (end of phase II and phase III), the environment (as recorded by charcoal analyses) was more diverse, and new biotopes were exploited from then onwards. Firewood management is balanced because all the taxa found are present in similar amounts and alternatively exploited. The anthracological results are a reflection of the vegetation dynamics according to the pollen record of the Central System (Franco *et al.* 1998; Ruiz *et al.* 1997), but this particular use of the land is also a consequence of economical activities carried out by human groups during the Neolithic-Chalcolithic transition.

Towards 4 kyr BP pollen data from Rascafría at 1113 m a.s.l. in the Lozoya Valley, relatively near the archaeological site (Franco *et al.* 1998), show a sharp regression in *Pinus* as a consequence of climatic disturbances. The distribution area of this taxon was reduced to small woodlands that turned into an open landscape with the presence of light-demanding taxa. Occuring in 3750 BP, this deposit shows the first evidence of human impact on this area.

The decrease in *Pinus* observed in the pollen record could be related to the progressive substitution of *Pinus* by *Juniperus* observed in the charcoal spectrum from La Vaquera cave (fig. 2). The systematic exploitation of this taxon between phases IV to VIII under unfavourable ecological conditions, would lead to the disappearance of *Pinus* from the catchment area of this settlement. The decrease of *Pinus* would have emphasized the alternative exploitation of deciduous and evergreen *Quercus* communities during the Chalcolithic period. At the beginning of the Bronze Age 3840-3760 BP charcoal showed the first hint of ecological degradation, as indicated by the increase in shrubs and the decrease in arboreal taxa (fig. 2).

According to anthracological data, there was a different use of firewood during the human occupation of the cave. Besides the ecological factors, economical activities have also played a relevant role.

The existence of cereals was recorded from the first neolithic occupation (phase I). However, this is not enough to prove the real existence of agriculture because traces of treshing are still absent. This is recorded only at the end of phase II (Neolithic) and during the phase III (transition to the Chalcolithic) (Gil *et al.* 1999). Seed analyses of the upper levels of human occupation are not complete.

Animal remains are composed of *Suideae* (boar and pig), ovicaprins (mainly sheep and some goat), *Bovideae* (cow) and deer (Gil *et al.* 1999).

The first human ocupation was based on *Pinus* exploitation, which coincides with the arrival of neolithic groups at the cave. The abundance of this species in the environment would have satisfied the need for firewood by the human groups who has arrived there for the first time and began to exploit the surroundings of the site. The archaeological record indicates a high density of human occupation (Estremera, 1999). Cereals were integrated in their diet, but it is not clear whether or not these groups cultivated them. Deer and boar hunting was common, and the presence of ovicaprins took an important place among domestic animals. Phases II and III marked the passage towards a diversification of environments in the anthracological spectrum. According to both archaeological and archaeozoological records, this passage coincided with a lower density of human occupation (periods of alternating abandonment and occupation due to the presence of layers of bat droppings and *Chiroptera* remains). Faunal assemblages are also more diversified in this phase. The use of the site has probably changed due to the livestock activities. This could explain the increase in areas of firewood supply. However, this diversity of environments is also the product of exploration and exploitation of the landscape following the arrival of humans to the site.

From phase III onwards, there was a gradual change in areas of firewood supply and an alternative exploitation of deciduous and evergreen communities that occured simultaneously to a progressive decrease in *Pinus*. Animal remains indicate an increase in *Bovideae* during the Chalcolithic period, with possible cow herding in the phase V (O. Alonso, pers. comm.). This aspect explains the presence of *Fraxinus* in significant amounts during the entire Chalcolithic period. Branches of ash could have been employed for feeding cows and flooring in stables. Anthracological records from the Pre-Alps and the Western Pyrenees have shown this particularity (Thiébault, 1988; Uzquiano, 2000).

During the Early Bronze Age, the exploitation of *Cistus* and *Juniperus*, the disappearance of *Pinus* and the slight decrease in broadleaved species indicate an open environment. The archaeozoological record of this period has established a link between a decrease in hunting, with the disappearance of *Cervus,* and an important presence of free range habitat animals (O. Alonso, pers. comm.).

The progressive decrease in *Pinus* as a consequence of climatic disturbances (Franco *et al.* 1998) combined with human exploitation under non-favourable ecological conditions meant a drastic reduction of firewood biomass that emphasised the exploitation of communities of diverse ecology. This mode of exploitation of the vegetation led to the ecological degradation of the landscape at the beginning of the Bronze Age. The disappearance of cynegetical resources such as deer, could have also lead to the definitive abandoning of the cave.

Conclusions

First anthracological data from an archaeological deposit located in the northern slopes of the Spanish Central System have been obtained. The human occupation of the cave spans the period of 6000-3700 years BP. The proximity of palynological records allows us to establish direct palaecological correlations between vegetation dynamics and firewood management carried out by humans to keep their domestic fires burning.

The abundance of *Pinus* in the surroundings areas provides an important fuel biomass supply, mainly for the human occupations during the Neolithic period. The decrease in *Pinus* recorded from the Chalcolithic age onwards is related to the progressive reduction of the area of distribution of this taxon in the environment as a consequence of a climatic event (Franco *et al.* 1998). This palaeoecological episode would have forced human groups to obtain their firewood from both deciduous and evergreen woody communities regions.

From an economical point of view, the archaeozoological record has yielded ample proofs of important grazing activities in this area. Otherwise, there was a strong correlation between livestock and firewood management during each level of human occupation. The alternative exploitation of piedmont areas and valleys during the Chalcolithic and early Bronze periods, according to the anthracological record, was probably the consequence of grazing activities based on the dual exploitation of ovicaprins and *Bovideae* respectively. This traditional exploitation still remains to the present day.

ACKNOWLEDGEMENTS
This research is part of a project carried out at the Classic Studies Department (University of Leon) and funded by the Spanish Ministery of Culture and Education. Charcoal analysis was conducted at the Laboratory of Archaeobotany of Spanish Council of Reseach (CSIC) in Madrid. I am very grateful to Drs. G. Delibes, S. Estremera and O. Alonso from University of Valladolid for providing anthracological remains and archaeological data. I also thank Dr. J. Maldonado from the U. Politécnica (Madrid) and Opening English School (Talavera de la Reina) for assistance with the English text, and Sabah Walid from Prehistory Department of CSIC (Madrid) for assistance with scanning.

References

ESTREMERA Mª S. 1999 - Sobre la trayectoria del Neolítico interior : precisiones a la secuencia de la Cueva de la Vaquera (Torreiglesias, Segovia). Actes del II Congrés del Neolític a la Península Ibérica. Universitat de Valencia 1999. *Saguntum*, extra-2, p. 245-250.

FRANCO F., GARCIA M., SAINZ H. 1998 - Vegetation dynamics and human impact in the Sierra de Guadarrama, Central System, Spain. *The Holocene*, vol. 8, fasc.1, p. 69-82.

GARCIA M., FRANCO F., MALDONADO J., MORLA, C., SAINZ H. 1995 - Una secuencia polínica en Quintana Redonda (Soria). Evolución holocena del tapiz vegetal en el Sistema Ibérico septentrional. *Anales del Jardín Botánico de Madrid*, t. 52, fasc. 2, p. 187-195.

GIL P., LOPEZ P., MACIAS R., LOPEZ J.A., MARTIN M., UZQUIANO P., ARNANZ A.Mª., MORALES A., MARTINEZ S. 1999 - Paleoambiente y antropización durante el Neolítico en la Meseta norte : análisis arqueobotánicos y arqueozoológicos de la Cueva de la Vaquera (Segovia). *II Congrés del Neolític a la Península Ibérica*. Valencia 7-9 de Abril 1999. Abstract, 3 p.

GREGUSS P. 1955 - *Identification of living Gymnosperms on the basis of xylotomy*. Akademikai Kiado, Budapest. 263 p., 120 pl.

GREGUSS P. 1959 - *Holzanatomie der Europaïschen Laubhölzer und Sträucher*. Akademikai Kiado, Budapest. 330 p., 200 pl.

IGLESIAS J. C., ROJO M.A., ALVAREZ V. 1996 - Estado de la cuestión sobre el Neolítico en la submeseta norte. *Rubricatum*, t. 2, fasc. 1, p. 721-734.

JACQUIOT C. 1955 - *Atlas d'anatomie des bois des conifères*. Paris, Centre Technique du Bois, CNRS. 135 p., 64 pl.

JACQUIOT C., TRENARD Y., DIROL D. 1973 - *Atlas d'anatomie des bois des Angiospermes (essences feuillues)*. Paris, Centre Technique du Bois, CNRS. 175 p., 72 pl.

MARTIN J.F., RINCON A. 1990 - Estudio de ordenación territorial en un sector de la cuenca del río Pirón (Segovia). "Valle del Río Pirón en Covatillas : reserva natural". En Premios del I Concurso sobre Medioambiente, Segovia, *Publicaciones de la Caja de Ahorros de Segovia*, Obra Social, 176 p., 20 fig.

PEÑALBA M.C. 1989 - *Dynamique de végétation tardiglaciaire et holocène du Centre-Nord de l'Espagne d'après l'analyse pollinique*. Thèse Sciences Spécialité : Paléoécologie. Aix-Marseille III, 165 p., 30 fig.

PEÑALBA M.C. 1994 - The history of the Holocene vegetation in northern Spain from pollen analysis. *Journal of Ecology*, t. 82, p. 815-832.

RUIZ B., ANDRADE A., DORADO M., GIL Mª J., FRANCO F., LOPEZ P., ARNANZ A.Mª., LOPEZ J.A., MACIAS R., UZQUIANO P. 1997 - Las transformaciones del ecosistema de la Comunidad de Madrid. En El Paisaje Vegetal de la comunidad de Madrid durante el Holoceno final, p. 95-135. *Arqueología, Paleontología y Etnografía*, Monográfico nº 5. Madrid, Consejería de Educación y Cultura de la Comunidad de Madrid. 201 p., 38 fig.

SCHWEINGRUBER F. 1990 - *Anatomie Europäischer Hölzer*. WSL/FNP Haupt. 800 p., 710 pl.

TEJERO DE LA CUESTA J.M. 1988 - *Análisis del medio físico de Segovia. Delimitación de unidades y estructura territorial*. Valladolid, Consejería de Fomento, Dir. General de Urbanismo, Vivienda y Medioambiente, Junta de Castilla León, 175 p., 30 fig.

THIEBAULT S. 1988 - L'homme et le milieu végétal. Analyse anthracologique de six gisements des Pré-Alpes au Tardiglaciaire et au Postglaciaire. Paris, Maison des Sciences de l'Homme, *Documents d'Archéologie Française*, 15, 110 p., 40 fig.

UZQUIANO P. 1992 - *Recherches anthracologiques dans le secteur pyréneo-cantabrique (Pays Basque, Cantabria et Asturias) : Environnements et relations homme-milieu au Pléistocène supérieur et débuts de l'Holocène*. Thèse Physiol. et Biol. ses Org. et Pop. Spécialité: Environnements et Archéologie. Université de Montpellier II, 401 p., 113 fig.

UZQUIANO P. 1997 - Antracología y métodos : implicaciones en la Economía prehistórica, *Etnoarqueología y Paleoecología. Trabajos de Prehistoria*, t. 54, fasc. 1, p. 145-154.

UZQUIANO P. 2000 - El aprovechamiento del bosque en la cuenca de Arudy (Pirineos Occidentales, Francia) durante el Tardiglaciar y el Holoceno : Antracoanálisis de las cuevas de Espalungue y Malarode. *Complutum*, t. 11, p. 143-156.

Charcoal analysis in the Castle of Ambra (Pego, Alicante, Spain)

Sonia de Haro Pozo

Departament de Prehistòria i Arqueologia- Facultat de Geografía e Història,
Universitat de València. Blasco Ibáñez 28, 46010 Valencia, Spain.
E-mail: JHAROJ@nexo.es

Abstract

The present work is part of a project entitled "Transformations in the medieval population of La Marina Alta". This project seeks to expand upon knowledge of the rules of establishment and exploitation of territory during the XII-XIIIth centuries, developed as much by Islamic societies as by the later feudal societies. In this sense, charcoal remains recovered in the Castle of Ambra were analysed with the objective of evaluating transformations of the vegetable landscape during the time period mentioned above. The castle is a fortification of temporary defence, built in the first quarter of the XIIIth century by the Islamic population of the valley of Pego during the feudal invasion of the territory. After the conquest of the castle by the troops of Jaime I, it was dismantled and abandoned in 1279 (Azuar *et al.* 1997). In the archaeological levels studied in the Castle of Ambra (foundation, construction, occupation and abandonment of structures, both Islamic and feudal) 23 taxa belonging to different ecological groups have been identified: *Arbutus unedo*, *Buxus sempervirens*, *Ceratonia siliqua*, cf. *Chamaerops humilis*, *Erica multiflora*, *Fagus* sp., *Ficus carica*, *Fraxinus* sp., *Juglans regia*, Leguminosae I, Leguminosae II, Monocotyledonae sp., *Olea europaea*, *Pinus halepensis*, *Pinus nigra-sylvestris*, *Pistacia lentiscus*, *Prunus dulcis*, *Punica granatum*, *Quercus ilex-coccifera*, *Rhamnus-Phillyrea*, *Rosmarinus officinalis*, *Sorbus* sp., *Ulmus* sp.

Key-Words: *Charcoal analysis, Castle of Ambra, XII-XIIIth centuries, Islamic society, Exploitation of the territory.*

Résumé

Ce travail fait partie du programme : "Transformations des populations médiévales de La Marina Alta" qui a pour but la connaissance des établissements ruraux et des territoires exploités pendant les XII-XIIIème siècles développés par les sociétés islamique et féodale. C'est dans le but d'évaluer les changements du paysage végétal à cette période que l'analyse anthracologique du château d'Ambra a été entreprise. Ce château est une fortification temporaire construite pendant le premier quart du XIIIème siècle par les population islamiques de la vallée de Pego. Après sa conquête, il est démantelé et abandonné en 1279. Les niveaux archéologiques (fondation, construction, occupation, abandon) ont permis d'identifier 23 taxons appartenant à différents groupes écologiques.

Mots-clés : *Anthracologie, Château d'Ambra, XII-XIIIème siècles, société islamique, exploitation du territoire.*

Localisation

The Castle of Ambra is located in the Marina Alta district (Alicante) on the summit of a hill approximately 300 m south of the city of Pego (fig. 1). Three spaces can be clearly distinguished in the region: 1- the Oliva-Pego marsh, an ancient lagoon that developed due to continuous deposits of river silt and debris from currents, resulting in a marshy area; 2- the coastal plain of Pego, the result of alluvial remains formed by deposits which were swept down ravines from the mountainous area above; and 3- the mountainous configurations of La Marina, which are constituted of calcareous rocks that are part of the oriental foothills of the Béticas Mountains, which surround the coastal plain opening out to the sea where Pego is located. Among the mountains a series of valleys and closed basins appears, such as Vall de Gallinera, Vall d´Alcalà, Vall de Laguar and Vall de Ebo, situated on a perpendicular axis to the coast.

Climate and current vegetation

The climate of Marina Alta is typically Mediterranean, with a mild winter and a warm summer. Pego is located in the rainiest area of Alicante and has an average precipitation of 956 l/m^2. Precipitation varies according to the season, with a maximum in autumn and a very marked minimum in summer, and can be marked by intense periods of rainfall in a short interval of time.

The coastal region is characterised by a typically high level of humidity, with monthly averages between 62% in March and 74% in October. Significantly inferior levels of humidity to those of the coast are recorded in the interior, usually falling below 25%.

The thermomediterranean level is represented in the coastal area near Pego. Thus, the archaeological location studied in this project "the Castle of Ambra" is influenced by the series of potential vegetation *Rubio longifoliae-Quercetum rotundifoliae sigmetum* (Rivas Martínez, 1987) and dense, impenetrable forests of *Quercus rotundifolia*. At the moment, in the thermomediterranean level, serial stages of the potential vegetation such as *Quercus coccifera*, *Rosmarinus officinalis*, and *Thymus vulgaris* formations appear, accompanied by *Pinus halepensis*. In more degraded areas, *Ulex parviflorus* and *Erica multiflora* grow. The formations of *Quercus rotundifolia* have disappeared in most of the territory, appearing alone in very specific areas, for example in the Carrascal d´Alcoi. The thermomediterranean level is very fit for agricultural cultivation and at the moment most of the natural vegetation has been substituted by the cultivation of citric and other varieties of cultivated plants. In the marsh area, the potential vegetation is represented by formations of *Phragmites maximus*, which is included in the association *Typho angustifoliae-Phragmitetum maximi*. Traditionally, this marsh area has

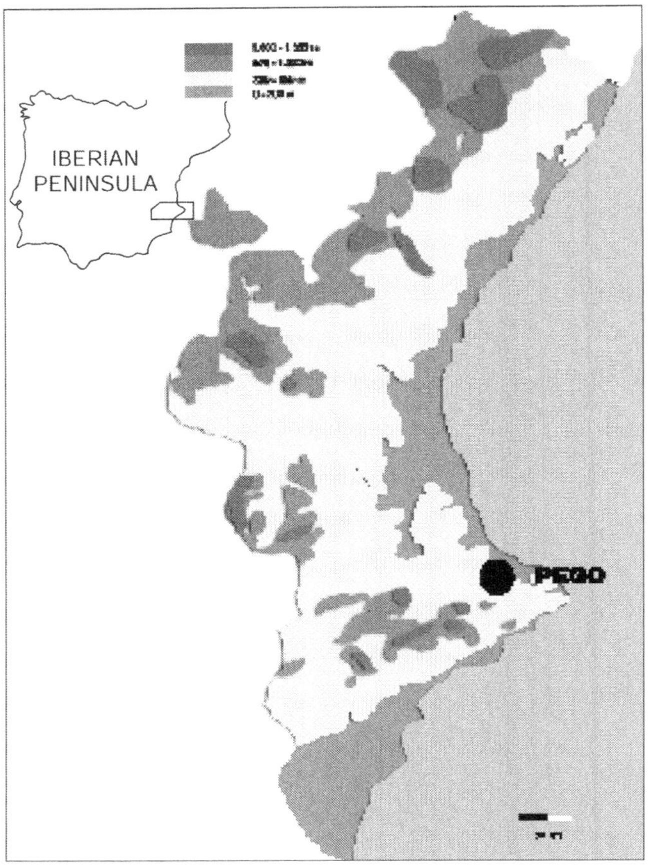

Figure 1: Location of the sites mentioned in the text.

been subjected periodic slash-and-burn practices which have favoured the development of *Cladium mariscus*, allowing formations of this species where the association *Hydrocotylo vulgaris-Cladietum marisci* prevails (Costa, 1999). Currently, a marsh sector is dedicated to the cultivation of rice, while formations of *Tamarix gallica* and *Phragmites maximus* flourish in the interior region.

The mesomediterranean level appears in the mountains and valleys that surround the plain of Pego. The potential vegetation formations of *Quercus rotundifolia* appear in sunny areas and have a floristic composition similar to the formations of *Quercus rotundifolia* thermomediterranean, lacking some clearly thermophilous species. In shady areas above an altitude of 1000 m, the series *Fraxino-Quercetum faginae* appears, which in its ideal biological conditions grows in a thick forest where deciduous trees such as *Quercus faginea* and *Quercus rotundifolia* are found. The strong impact of agricultural development and cattle farming has been the cause of the dramatic reduction of the *Quercus rotundifolia* forest of the mesomediterranean level.

Methods and material

Charcoal samples were recovered by collecting fragments directly from the sediment when they were larger than 1 or 2 cm^3, but such sized fragments do not appear in significant enough quantities to allow complete qualitative and quantitative analysis to be carried out. In order to carry out an acceptable charcoal analysis, it was necessary to supplement the quantity of charcoal by sifting sediment in order to obtain smaller-sized fragments. Sieve openings measured between 2-5 mm. The most effective method for recovering vegetable remains is by water sieving or flotation (Chabal, 1988), pouring a mixture of water and sediment through a sieve. Since it was not possible to use water in this location, fragments were obtained by dry screening. In the Castle of Ambra, two charcoal groups were distinguished: dispersed charcoal and concentrated charcoal. The concentrated charcoal in structures was picked up in its entirety, but separately of what we found dispersed through the sediment. Even though these two types belong to the same *stratum*, the meaning derived from them is different. Dispersed charcoal offers information about the vegetation of the environment, while concentrated charcoal more often provides ethnological data (Grau 1991).

In the group of the levels studied in the Castle of Ambra, 957 fragments of charcoal were analysed. Scarce in number, only 23 taxa appeared, belonging to different ecological means. The fragments of charcoal analysed from the Castle of Ambra were collected during excavations conducted in 1993, 1994, 1995 and 1996. The content of charcoal for different strata was quite scarce. Therefore, I have decided to combine the fragments of charcoal of the different strata in the archaeological level, following the natural strata of the location and respecting the distinction between concentrated and dispersed variations. In this manner, charcoal was collected from the foundation, construction, occupation and abandonment levels of both the Islamic and feudal structures (De Haro, 1998).

Results and discussion

A total of 298 fragments of dispersed charcoal was analysed for the Islamic period. In the foundation level, 60 dispersed

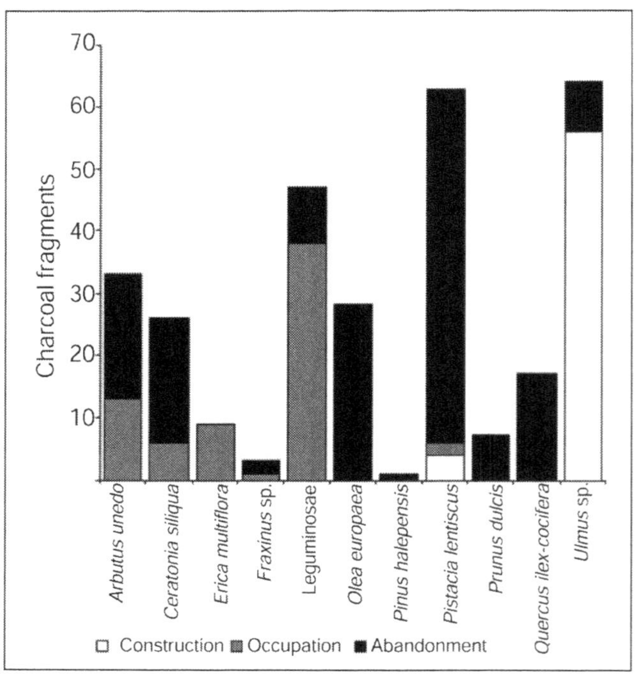

Figure 2: Number charcoal' fragments of Islamic period.

charcoal pieces were analysed, among which two taxa were identified: *Ulmus* sp. (93.3%) belonging to vegetable formations characteristic of waterside areas and *Pistacia lentiscus* (6.7%), characteristic species of shrubby vegetation formations (fig. 2). The charcoal, even those dispersed in the sediment, was not *in situ* but rather were taken out of its original context to be used as filler material for the preparation of the construction. Therefore, it is not evaluated on a quantitative level when carrying out an ecological interpretation (Badal, 1990; Chabal, 1991, 1997; Grau, 1991; Heinz, 1988, 1991; Thiébault, 1988).

In the occupation level, where the remains of charcoal came from firewood used as fuel, 69 fragments recovered in the stratum 363, 320 and 346 were analysed, all of them belonging to Islamic pavements. As for natural vegetation, mostly shrubby species which grow in degraded areas prevailed, such as Leguminosae I, *Erica multiflora* and *Pistacia lentiscus*. A significant percentage of characteristic taxa of forest formations appears, such as *Arbutus unedo*. Another taxon belonging to vegetation formations associated with aquaeous areas such as *Fraxinus* sp. was present, as well as the cultivated species *Ceratonia siliqua* (fig. 2 and 3).

In the abandonment level, 169 fragments of dispersed charcoal were analysed. *Arbutus unedo*, *Fraxinus* sp., *Pistacia lentiscus, Ceratonia siliqua* and Leguminosae I also appear in the occupation level, but only on a qualitative level. Other species appear only in the abandonment level, as is the case with *Pinus halepensis, Quercus ilex-coccifera, Olea europaea* and *Prunus dulcis* (fig. 2 and 3). This qualitative difference could be related to the number of fragments of charcoal analysed for each level, since more than the double the number of charcoal fragments were analysed from the destruction level than from the occupation level. A relationship was established for both the occupation level and the abandonment level between the number of analysed fragments and the number of identified species by means of taxonomic curves. The taxonomic curve of the occupation level seems to be stable, since in the first 25 analysed charcoals, the 6 identified taxa appear. Although, keeping in mind the results of other charcoal analyses which demonstrate the necessity of studying a minimum of 250 fragments of charcoal per layer to obtain a maximum specific wealth (Grau, 1991; Chabal, 1988; Chabal, 1991; Chabal, 1992), the qualitative results of this archaeological level will have to be taken with caution. In the abandonment level the taxonomic curve is not stabilized, and even though more than half of the identified taxa appear among the first 26 analysed fragments of charcoal, three new species are identified between fragments 83 to 131. The increase of identified species seems to be directly related to the increase of analysed fragments of charcoal. Therefore, it seems that the scarce number of taxa identified in the occupation level of the Islamic structures is due to the insufficient anthracological material recovered.

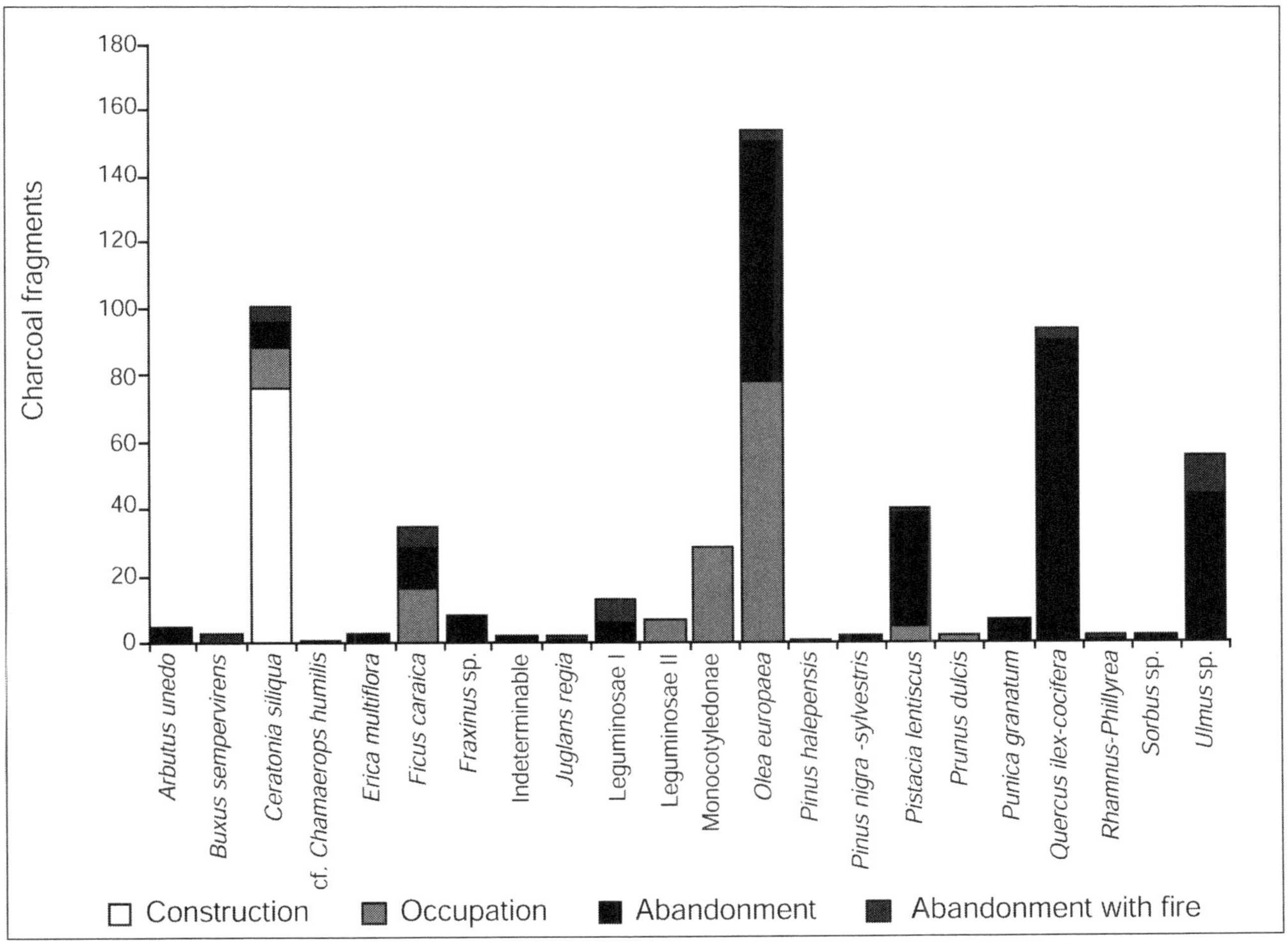

Figure 3: Number of charcoal' fragments from the feudal period.

TAXA	ISLAMIC PERIOD (1228/38-1258)						FEUDAL PERIOD (1258-1280)							
	ARCHAEOLOGICAL LEVELS													
	Construction		Occupation		Abandonment		Construction		Occupation		Abandonment Destruction without fire		Abandonment Destruction with fire	
	N	%	N	%	N	%	N	%	N	%	N	%	N	%
Arbutus unedo			13	18.8	20	11.8					5	1.7		
Buxus sempervirens													3	6.8
Ceratonia siliqua			6	8.7	20	11.8	76	100	12	8	8	2.8	5	11.4
cf. *Chamaerops humilis*											1	0.3		
Erica multiflora			9	13.1							3	1		
Ficus carica									16	10.7	13	4.3	6	13.6
Fraxinus sp.			1	1.4	2	1.2					8	2.8		
Indeterminable											2	0.7		
Juglans regia											1	0.3	1	2.3
Leguminosae I			38	55.1	9	5.3					6	2	7	18.1
Leguminosae II									7	4.7				
Monocotiledoneae									29	19.3				
Olea europaea					28	16.6			78	52	73	24.2	3	6.8
Pinus halepensis					1	0.6					1	0.3		
Pinus nigra-sylvestris											2	0.7		
Pistacia lentiscus	4	6.7	2	2.9	57	33.8			5	3.3	34	11.3	1	2.3
Prunus dulcis					7	4.1			2	1.3				
Punica granatum											7	2.3		
Quercus ilex-coccifera					17	10.1					91	30.2	3	6.8
Rhamnus-Phillyrea											1	0.3	1	2.3
Sorbus sp.											2	0.7		
Ulmus sp.	56	93.3			8	4.7			1	0.7	43	14.3	12	27.2
Total	60		69		169		76		150		301		44	

Table 1: Chart of absolute and relative frequencies of the dispersed charcoals of the different archaeological levels dug in the Castle of Ambra.

On a quantitative level, the percentages of *Arbutus unedo*, *Fraxinus* sp. and Leguminosae I decrease slightly in the abandonment level, except Leguminosae I, which decreases significantly from 55.1% to 5.3%. On the contrary, *Ceratonia siliqua* and *Pistacia lentiscus* increase in percentage in the abandonment level. *Ceratonia siliqua* increases slightly from 8.7% to 11.8%. The increase of the *Pistacia lentiscus* is more dramatic, climbing from 2.9% to 33.8% (fig. 2).

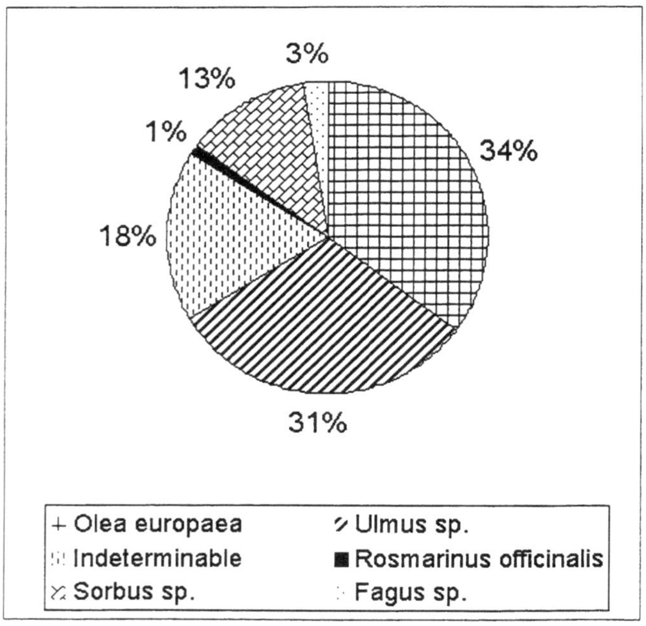

Figure 4: Concentrated charcoal recovered in the Feudal occupation.

For the Feudal period, 571 fragments of dispersed charcoal were analysed. The construction level permitted documentation in the units of preparation of the Christian hermitage, where 76 fragments of charcoal were recovered from silt, all of them belonging to *Ceratonia siliqua* (fig. 2).

In the occupation level, 150 fragments of charcoal from the cleaning of combustion structures were studied. Among the charcoal, cultivated species *Olea europaea*, *Ficus carica*, *Ceratonia siliqua*, and *Prunus dulcis* proved to be abundant. Among the natural vegetation, some taxa that live near river beds, such as Monocotyledonae, *Ulmus* sp., and other shrub-like taxa such as Leguminosae II and *Pistacia lentiscus* appear (fig. 2). These charcoal fragments offer us an aleatory sampling of the vegetation, since they comprise the remains of several fires and reflect diverse firewood collections made by man during a variable period of time. All of the factors cited above allow these results to be interpreted on an ecological level.

A significant abandonment level was also located, inside of which a destruction level without fire and a destruction level with fire were differentiated.

In the first case, the recovered charcoal cannot come from the construction elements, since the structures did not burn. Therefore, an alternative explanation can be offered. The remains could already have been in the occupation level located below the destruction level and mixed with the remains of the other structures when the structure collapsed. Or, the absence of charcoal remains could be attributed to post-depositional causes, such as sediment haulage, etc. Also, some deteriorated wood from man-made structures which was no longer useful as construction material after a collapse was commonly used as fuel. 301 fragments of dispersed charcoal were analysed and 16 taxa were identified. Forest species such as *Arbutus unedo, Pinus halepensis, Pinus nigra-sylvestris, Quercus ilex-coccifera* and *Sorbus* sp. appeared, but none of them surpassed 2%, except *Quercus ilex-coccifera,* which reached 30%. Riverside vegetation is represented by *Fraxinus* sp. and by *Ulmus* sp., which increased in percentage in relation to their presence in the occupation level. Shrubby vegetation also increases in percentage, as illustrated by the appearance of *Pistacia lentiscus* and new taxa like cf. *Chamaerops humilis*, *Erica multiflora*, Leguminosae I and *Rhamnus-Phillyrea*, but none of them surpasses 2%. As for cultivated species, the percentage of *Ceratonia siliqua, Ficus carica* and *Olea europaea* decreases, and two species that were not represented in the occupation level appear (*Juglans regia* and *Punica granatum*) (fig. 2 and 4).

To verify that the qualitative difference between one level and another is a function of the number of analysed charcoal, taxonomic curves were established. The taxonomic curve of the occupation level seems to remain stable for the 78 analysed fragments of charcoal, while the taxonomic curve of the destruction level without fire is not stable in spite of the considerable number of analysed fragments. In this case, starting at fragment 102, 50% of species appear. Starting at fragment 207 the curve seems to stabilise; however, a new taxa appears in fragment 230. From fragment 230 to fragment 301, no new taxon was identified, but we hesitate to confirm that the curve is stable.

In light of these results, it can be established that charcoal remains which are most likely wood charcoal and that appear in the destruction level without fire have some qualities or specific properties that could be attributed to *Arbutus unedo, Pinus nigra-sylvestris, Quercus ilex-coccifera, Sorbus* sp., *Fraxinus* sp., *Ulmus* sp. and *Juglans regia* and could have been used as construction material of excellent quality. None of these taxa appears in the occupation level except *Fraxinus* sp., which appears in an insignificant percentage.

In the destruction level with fire only 44 fragments of charcoal were recovered. Even so, 11 different taxa appeared, coming from stratum 324. This level has an ashen stain which is loosely dispursed and extends over structure 325 and partially on the internal surface of the stay III. This level of fire post-dates the collapse of the walls (Azuar *et al.* 1995). At a qualitative level, the charcoal analysis has revealed 11 taxa. As for the vegetation associated with the forest, *Quercus ilex-coccifera* diminishes in percentage in the destruction level without fire. Neither *Pinus halepensis,* nor *Pinus nigra-sylvestris*, nor *Sorbus* sp. is represented. The presence of *Buxus sempervirens* is surprising, as it is a species which does not appear in any of the previous levels. Riverside vegetation continues to be represented by *Ulmus* sp. and by *Fraxinus* sp., although in superior percentages to the destruction level without fire. As for shrubby vegetation, the increased percentage of Leguminosae I and *Rhamnus-Phillyrea* should be noted. However, the opposite occurs with *Pistacia lentiscus*. As for the cultivated species, a similar trend can be noted, since the percentages of *Ceratonia siliqua, Ficus carica,* and *Juglans regia* increase, while *Olea europaea* decreases (fig. 2 and 4). These charcoal pieces are very difficult to interpret from an ecological point of view, as charred wood coming from the construction remains is very likely to appear conserved as charcoal from a fire, lying among remains of charcoal which could have been concentrated or dispersed through the sediment before the fire.

Concentrated coal was also recovered in an undefined stain of ashes located in the occupation level of the feudal structures. 80 fragments of charcoal were analysed and 5 different taxa were identified: *Olea europaea* (34%), *Ulmus* sp. (31%), followed by *Sorbus* sp. (13%), *Fagus* sp. (3%), and *Rosmarinus officinalis* (1%). Among the 80 analysed charcoal fragments, 18% were not identifiable due to the deformation of their anatomical structure.

A concentration of charcoal was also located in a domestic fire placed on a Feudal pavement, where 100% of the charcoal remains belongs to *Pistacia lentiscus*. As for the possible problem of sufficient representation of the percentage of some taxa in the concentrated charcoal, *Pistacia lentiscus* and *Ulmus* sp. rise dramatically in percentage when dispersed charcoal obtained from the sediment is taken into account. The above-mentioned fire, which appears concentrated inside a fireplace, provides sufficient representation of *Pistacia lentiscus*. As for the fragments of *Ulmus* sp. and *Fagus* sp. they most likely result from a device or structure manufactured totally or partially with these woods.

It is necessary to keep in mind that the changes of vegetation in many occasions are slower than cultural or technological changes. The Castle of Ambra was occupied between 1230-40 and 1280-90, for a maximum of approximately 60 years. This is indeed a short period of time in regards to visible changes in vegetation through the analysis of charcoal.

From an ecological point of view, only the dispersed charcoal used as fuel in domestic fires or domestic ovens is viable for interpretation on an ecological level. Forest species do not appear in the Islamic occupation level nor in the feudal one, although they do appear in the abandonment levels of both social formations. This could indicate that these woods were used for specific purposes and they would be transported from nearby forests which could certainly have been located in the mountainous areas of the region. These forests would have a very important paper of wooden reservation. Riverside species, those located beside riverbeds and ravines, were exploited by both cultures. Primarily the wood of *Fraxinus* sp. and *Ulmus* sp. was used as construction material. As for cultivated vegetation, more species generally appear in the Christian levels than in the late Islamic levels. Only *Ceratonia siliqua, Olea europaea* and *Prunus dulcis* coincide in both moments of the occupation of the Castle of Ambra. *Ficus carica, Juglans regia* and *Punica granatum* only appear in the phase of Christian occupation, a fact that could lead us to conclude that the cultivation and use of these fruit-bearing trees started at that time. However, the appearance of these three species seems to be directly related to the number of coal fragments analysed for each period. Thus, a total of 571 fragments of dispersed charcoal in the sediment were analysed from the feudal period, while only 298 were analysed from the Islamic period.

In light of these results, it seems that the vegetation does not vary in spite of the cultural change that takes place. However, this does not mean that the two groups used the vegetable resources in the same way. The ceramic repertoire recovered in the Castle of Ambra demonstrates that a profound transformation can be verified from one period to another. The kitchen ceramics of the Islamic period disappear and were substituted by new types without local precedent. All evidence points to changes in the way food was prepared and even changes in culinary recipes (Azuar *et al.* 1997).

According to the results gleaned from dispersed coal, a fairly degraded natural vegetation formed in the areas surrounding the Castle of Ambra, which the presence of *Erica multiflora*, Leguminosae I, Leguminosae II, *Pistacia lentiscus,* cf. *Chamaerops humilis* and *Rhamnus-Phillyrea* confirms. This degradation was produced by the creation of cultivation areas located in the fertile and rich alluvial plain of Pego. In the Castle of Ambra, cultivated species such as *Ceratonia siliqua, Ficus carica, Juglans regia, Olea europaea, Prunus dulcis* and *Punica granatum* appeared. This does not imply that each of these species was cultivated in large expanses, since species such as *Ficus carica, Juglans regia* and *Punica granatum* are generally fruit-bearing trees cultivated in isolated locations in field banks or gardens. However, the charcoal remaining in abandonment levels shows that wood from areas next to the location was used for the construction of beams and other architectural elements as well as furniture, tools, weapons, etc. These nearby areas had a less degraded vegetation, as they were situated in the Serra de la Carrasca, Serra del Peñó, and Serra del Migdia mountains and the high parts of Serra Mostalla and Serra de Gallinera that surround the plain. In these mountainous areas, the dominant landscape would be forests of *Quercus rotundifolia* and *Quercus coccifera, Arbutus unedo, Buxus sempervirens*, etc. In valleys such as the Laguar, Ebo, Gallinera and Alcalà and the areas lying near the numerous waterways and ravines of the region, a leafier vegetation including *Fraxinus* sp., *Ulmus* sp., etc. dominated.

The almost total absence of *Pinus halepensis* in occupation levels and in abandonment levels should be noted. *Pinus halepensis* would not have been a very abundant species and it would have been located in more southern and drier mountain areas, where *Quercus rotundifolia* would be limited in fresh and deep valleys. *Pinus nigra-sylvestris* appeared in low percentages in the abandonment level of the feudal structures, where it was probably used as construction material and would have come from the Marina higher mountains, such as the Carrascal de Font Roja, Mariola, Aitana, etc. The fragments of *Fagus* sp. that appear concentrated in the feudal occupation level most likely come from furniture or an instrument manufactured from this superior wood source. It may have been transported by Christians to the castle from more northern areas, since it is a species whose current southern limit is in the Port de Beceite, at the border of the Castellón and Tarragona regions.

Conclusion

In order to give a balanced overview of the results of this work, it is first essential to point out that a good charcoal sampling from the archaeological location is indispensable, since the results of the study depend directly on what, how and how much charcoal has been recovered. In the Castle of Ambra it would seem that more vegetable species would have been exploited in levels of feudal times. However, we found that a larger number of recovered charcoal pieces influences directly on the increase of identified species, therefore this apparent exploitation of vegetable species on the part of the feudal population is not demonstrated in the present work. However, the use of vegetable resources is demonstrated by the exploitable vegetation that was found. The forest provided them with *Quercus ilex* wood, *Arbutus unedo, Buxus sempervirens, Pinus halepensis* and *Pinus nigra-sylvestris,* which was used fundamentally for construction. Riverside vegetation, such as *Fraxinus* sp., *Ulmus* sp. and *Phragmites maximus* was used as much for fuel as for construction material. The low vegetation of heaths located in areas closer to the Castle of Ambra was mostly used as fuel for cooking and heating fires, illustrated by the recovery of species that all belong to a shrubby landscape. The cultivation of some fruit-bearing trees was also demonstrated through charcoal analysis, and besides using fruit, they also took advantage of pruning remains as a fuel source.

References

AZUAR R., MARTÍ J., PASCUAL J. 1995 - *Excavacions arqueológiques al Castell d'Ambra-Pego (Alacant)*. Informe de la IV Campanya d´actuació arqueológica, original mecanoescrito.

AZUAR R., MARTÍ J., PASCUAL J. 1997 - *La cerámica del Castell d'Ambra (Pego). De las producciones andalusies a las cerámicas de la conquista feudal (s. XIII)*. Congreso homenaje a Guillermo Roselló. Jaén 1997. Universidad de Jaén.

BADAL E. 1990 - *Aportaciones de la antracologia al estudio del paisaje vegetal y su evolución en el cuaternario reciente, en la costa mediterránea del País Valenciano y Andalucía (18.000-3.000 BP)*. Tesis Doctoral. Univesrsitat de Valencia.

CHABAL L. 1988 - Pourquoi et comment prélever les charbons de bois pour la période antique: les méthodes utilisées sur le site de Lattes (Hérault). *Lattara* n° 1. p. 187-222.

CHABAL L. 1989 - Perspectives anthracologiques sur le site de Lattes (Hérault). *Lattara* n° 2, p. 53-72.

CHABAL L. 1991 - *L'Homme et l'évolution de la végétation méditerranéenne, des âges des métaux à la période romaine: recherches anthracologiques théoriques, appliquées principalement à des sites du Bas-Languedoc*. Thèse de doctorat, Université Montpellier II, 435 p.

CHABAL L. 1992 - La représentativité paléoécologique des charbons de bois archéologiques issus du bois de feu. *Bulletin de la Société Botanique de France* 139. Actual. bot. (2/3/4) pp. 213-236.

CHABAL L. 1997 - Forêts et societés en Languedoc (Néolithique final, Antiquité tardive). L'antracologie, méthode et paléoécologie, *Documents d'Archéologie Française*, 63, 189 p.

COSTA M. 1999 - *La vegetación y el paisaje en las tierras valencianas*. Ed. Rueda.

DE HARO S. 1998 - *Estudio antracológico del Castell d'Ambra (Pego, Alacant)*. Memoria de licenciatura. Universitat de València.

DE HARO S. (in press) - Paisaje vegetal en la comarca de la Marina Alta durante el siglo XIII a través del antracoanálisis del Castell d'Ambra (Pego, Alacant). *El Medio natural en la España Medieval*. I Congreso sobre ecohistoria e historia medieval, Cáceres, 1999.

GRAU E. 1991 - *El uso de la madera en yacimientos valencianos de la Edad del Bronce a época Visigoda. Datos etnobotánicos y reconstrucción ecológica según la antracología*. Tesis doctoral. Universitat de València.

HEINZ C. 1988 - *Dynamiques des végétations holocènes en Méditerranée Nord-occidentale d'après l'anthracoanalyse de sites préhistoriques : méthodologie et paléoécologie*. Thèse de doctorat, Université Montpellier II, 275 p.

HEINZ C. 1991 - Upper Pleistocene and Holocene vegetation in the South of France and Andorra. *Review of Paleobotany and palynology*, 69, Amsterdam, p. 299-324.

RIVAS-MARTINEZ S. 1987 - *Memoria del mapa de series de vegetación de España* 1:400.000. ICONA. Madrid.

THIEBAULT S. 1988 - L'Homme et le milieu végétal : analyses anthracologiques de six gisements des Préalpes au Tardi- et au Postglaciaire. Paris *Documents d'Archéologie Française*, n° 15, 111p.

Charcoal analysis from Basque archaeological sites: new data to understand the presence of *Quercus ilex* in a damp environment

Lydia Zapata Peña

Department of Archaeology. University of Cambridge, Downing Street CB2 1YE, United Kingdom.
E-mail: lz208@cam.ac.uk and *l.zapata@euskalnet.net*

Abstract

In this paper we present the results of several archaeobotanical analyses which allow us to review traditional interpretations on the origin and chronology of the expansion of evergreen *Quercus* forests on the coast of Biscay. At least in some of the areas where these formations exist, deciduous oaks seem to have been the main arboreal component from the Mesolithic to the Bronze Age (*c.* 6500-1900 cal. BC). The expansion of evergreen oak may have been linked to increasing human action after this period.

Key-words: *Holocene,* Quercus ilex, *human impact, forest, evergreen oak forest, Biscay.*

Résumé

Ce papier présente les résultats de plusieurs analyses archéobotaniques dans le but de réexaminer les interprétations traditionnelles concernant l'origine et la chronologie de l'extension de la chênaie sclérophylle sur la côte de Biscaye. Dans les zones où les formations à chênes caducifoliés existent, il semble qu'elles aient constitué la principale formation du Mésolithique à l'Age du Bronze (6500 - 1900 cal. BC). Le développement des chênes sempervirents doit être mis en relation avec l'accroissement de l'anthropisation après cette période.

Mots-clés : *Holocène,* Quercus ilex, *impact humain, forêt, chêne sempervirent, Biscaye.*

Introduction

Evergreen *Quercus* forests in Iberia are regarded as typical Mediterranean formations. However, *Quercus ilex* (particularly subsp. *ilex*) is present in oceanic areas along the Basque coast and also in inner warm valleys up to the southern exposures of the mountains of the watershed with the Mediterranean Ebro basin.

The coastal *Quercus ilex* forests, *Lauro nobilis-Querceto ilicis*-S (usually known as "encinares cantábricos") thrive in thermoatlantic conditions from Asturias in the West to the western part of the province of Gipuzkoa to the East, an area with a humid ombrotype (800-1400 mm year of precipitation). *Quercus ilex* tends to disappear to the eastern corner of the Bay of Biscay where precipitation is very high (over 1400 mm/year). The area where it grows is characterised by: 1) mild temperatures with little winter frost because of the oceanic influence and certain decrease in summer precipitation and 2) steep mountains, with thin, karstic, A horizon soils which have little ability to retain water. Under these conditions evergreen oaks can compete favourably with other plant species (Meaza, 1991, 1998).

Lauro nobilis-Querceto ilicis-S is dominated by *Quercus ilex* subsp. *ilex* and hybrids along with *Arbutus unedo, Rhamnus alaternus, Phillyrea latifolia, Ruscus aculeatus, Laurus nobilis...* Some elements of Atlantic forest like *Quercus robur* and *Ilex aquifolium* are also present. Understorey vegetation includes xerophyte species like *Osyris alba, Pistacia lentiscus* and *Quercus coccifera* but Atlantic optimum species dominate.

According to the most accepted hypothesis, evergreen oak forests on the coast of the Bay of Biscay would be relict formations that in the past, under warmer and drier conditions, would have had a wider distribution. They might be the remains of a pre-Quaternary vegetation (Blanco *et al.* 1997; Bertrand, 1974 *in* Uzquiano, 1992; Guinea, 1985) or the consequence of a migration from the catalan-provenzal area up the Ebro valley, through the watershed up to the Atlantic coast (Aseginolaza *et al.* 1989; Loidi, 1987; Meaza, 1998; Meaza and Ormaetxea, 1992; Montserrat and Montserrat, 1987; Onaindia, 1986; Vadillo *et al.* 1990). The chronology of this migration is not always specified but the Holocene climatic optimum is sometimes suggested. Here we will review this paradigm in the light of new archaeobotanical data.

Archaeobotanical data

Some of the archaeological sites we have recently studied (Zapata, 1999a) in the province of Biscay, not far from the city of Bilbao, are located in karstic environments where nowadays evergreen oak forests grow. This has provided us with a unique opportunity to study the history of this formation and allows us to review the traditional hypotheses that support that evergreen oak forests are a relict natural formation. Charcoal has been recovered through flotation and fragments > 2 mm have been identified.

Antoliñako Koba (Gautegiz-Arteaga, Biscay): hunter-gatherers of the Upper Palaeolithic

This cave is overlooking the estuary of Gernika in the middle of an important *Quercus ilex* formation. It is being excavated by M.Aguirre and yields Upper Palaeolithic contexts (Aurignacian, Gravettian, Solutrean, Magdalenian) (Aguirre *et al.* 2000). Unexpectedly, in samples that are chronologically close to the Late Glacial Maximum, charcoal analysis shows the dominance of deciduous *Quercus*, probably related to the existence of refuge areas in southern Europe. Other taxa such as *Juniperus, Salix* and *Betula* are important in Aurignacian and Upper Magdalenian contexts (work in progress). *Quercus ilex/Q. coccifera* along with *Laurus nobilis* are extremely scarce but a few fragments are present in Solutrean samples (*c.* 19.000 BP).

Kobeaga II (Ipazter, Biscay): the last hunter-gatherers

Cave located in an area dominated by evergreen *Quercus* formations, with recent plantations of *Pinus radiata*. The

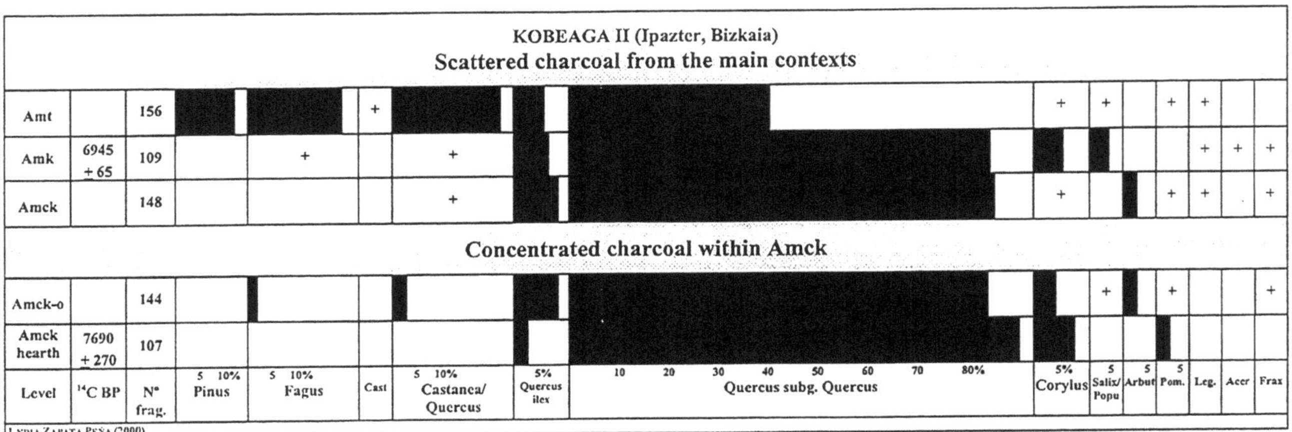

Figure 1: Charcoal analysis diagram from Kobeaga II (Biscay). Amt: superficial/Amk – Amck: Mesolithic.

site was excavated by J.M. Apellániz in 1973 (Apellániz, 1974, 1975) and recently again by J.C. López Quintana (Aguirre *et al.* 2000). Two mesolithic contexts are identified (Amck and Amk) and a superficial level (Amt) that includes modern intrusions of *Pinus*. Charcoal from one hearth and one refuse area has also been analysed. All contexts are dominated by deciduous *Quercus*. Evergreen oak is present in small numbers (fig. 1).

Pico Ramos (Muskiz, Biscay): last hunter-gatherers and Chalcolithic burials

Very small cave located above the Barbadun river, on the coast, 190 m a.s.l. in an open area close to evergreen oak forest elements (Aseginolaza *et al.* 1992). It was excavated by L. Zapata and includes a Late Mesolithic-Early Neolithic habitation level and a Chalcolithic burial level dated *c.* 4900-4540 and 3790-2340 cal. BC, respectively (Zapata, 1995).

The pollen analysis from the site was carried out by M.J. Iriarte (Iriarte, 1994b). The samples from both contexts show an open landscape with values of arboreal pollen under 20%. *Quercus* t. *robur* and *Corylus* are present, but not *Quercus ilex*/*Q. coccifera*. In the charcoal samples, *Quercus* subg. *Quercus* is the main taxon, with percentages over 80%. Evergreen *Quercus* is present on the site only at *c.* 4700 cal. BC. In the Chalcolithic sample there is an increase of *Arbutus unedo* (from 2% to 16%) and *Rhamnus*/*Phillyrea* is also present (fig. 2).

Kobaederra (Oma-Kortezubi, Biscay): first farmers in the Basque Country

This cave was excavated over the last years (Zapata *et al.* 1997), and is located in a karstic area (Oma-Kortezubi, Biscay) covered by one of the largest *Quercus ilex* forests of the province of Biscay (Meaza, 1988). According to fragment number, *Quercus* subg. *Quercus* seems to be the main fuelwood used (fig. 3). Only two fragments of *Quercus ilex*/*Q. coccifera* have been identified, one in Level 3 (*c.* 5250-4230 cal. BC) and the other in Level 1 (*c.* 3310-2900 cal. BC). Along the sequence, the frequency of *Quercus* subg. *Quercus* decreases in favour of shrubby taxa, a possible indication of an opening of the forest and human activity. *Arbutus unedo, Corylus, Fraxinus, Laurus, Rhamnus*/*Phillyrea* and Rosaceae become more common, particularly in the most recent Chalcolithic-Bronze Age level. Agricultural remains have been identified in the neolithic contexts of this cave *c.* 4230 cal. BC (^{14}C date of a grain of *Hordeum*). For the moment pollen analysis has failed to identify *Quercus ilex*/*Q. coccifera* because the conservation of palinomorphs on the site is not good (M.J. Iriarte, pers.com.).

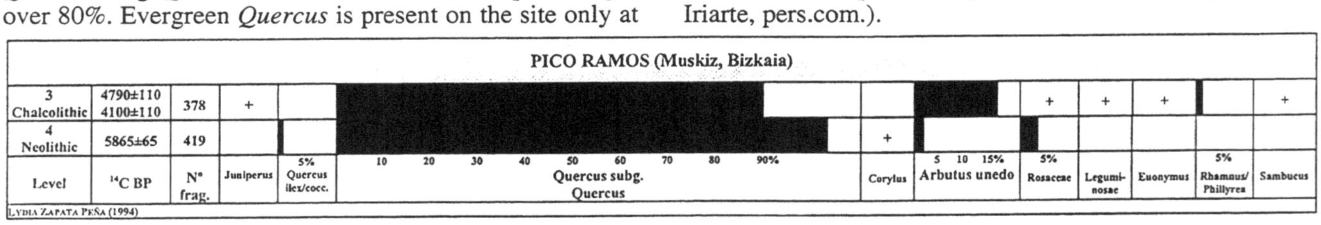

Figure 2: Charcoal analysis diagram from Pico Ramos (Biscay).

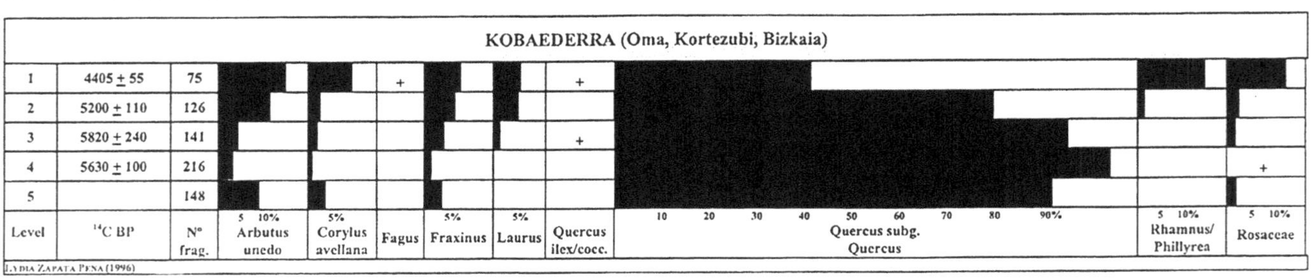

Figure 3: Charcoal analysis diagram from Kobaederra (Biscay).

Dolmens of Hirumugarrieta I y II (Bilbao, Biscay): megalithic tombs of the first farmers

The dolmens of Hirumugarrieta are located on one of the mountain chains that surround the city of Bilbao, at 375 m a.s.l., in a non-forested area that has been considered to be at the limit of a potential vegetation dominated by: 1) *Quercus pyrenaica* and *Quercus robur* and 2) evergreen oaks at lower altitude (Aseginolaza *et al.* 1992). The two dolmens were excavated in 1989-1993 by I. Martín and A. Zubizarreta (Martín and Zubizarreta, 1991; Zubizarreta, 1995a, 1995b).

The scattered charcoal recovered in the megalithic tumuli and chambers (*c.* 4000-3500 cal. BC) shows that *Q. ilex/Q. coccifera* wood is present in very low percentages (< 2%). *Quercus* subg. *Quercus* represents more than 60% of the identified wood in all samples.

We have already discussed (Zapata, 1999a) the difficulty of interpreting wood fragments recovered from megaliths, particularly when other palaeoenvironmental evidence is missing. The mode of arrival is very difficult to evaluate and even the chronology can be difficult to assess, since the charcoal may have been included beforehand in the soil used for building the tomb. However, the available ^{14}C dates were carried out on wood charcoal and they show a coherent Late Neolithic chronology, which shows that at least part of the material might be synchronous with the building of the structures. Because of all these considerations, archaeobotanical interpretation should be cautious but we have suggested the possibility that the high frequencies (even 30%) of *Arbutus*, Leguminosae and Ericaceae in some samples might be an indicator of human activity and deforestation processes simultaneous to the construction or use of the tombs.

Arenaza (Galdames, Bizkaia): Bronze Age farmers

The cave of Arenaza I is located at 180 m a.s.l. in Galdames, Biscay in what is considered an area of potential evergreen *Quercus* forest (Aseginolaza *et al.* 1992). Excavations have been carried out since 1972 (Apellániz, 1974, 1977, 1978; Apellániz and Altuna, 1975a, 1975b, 1975c; Fernández Lombera, 1992, 1994) and have recorded a sequence from the end of the Upper Palaeolithic up to historical times.

M. J. de Isturiz carried out pollen analysis on samples from the Neolithic to the Late Bronze Age (*c.* 4000-750 cal. BC). The most abundant taxon in all the sequence is *Corylus* followed by *Quercus* t. *pubescens/pedunculata, Alnus, Betula* and *Ulmus. Castanea, Pinus, Tilia, Fraxinus, Acer* and *Juglans* are present sporadically. No *Quercus ilex/Q. coccifera* pollen has been identified (M.J. de Isturiz, pers. com.).

P. Uzquiano has carried out the charcoal analysis of this site for samples dated to *c.* 2300-1900 cal. BC (Early and Middle Bronze Age). It shows that deciduous type *Quercus* dominates the samples along with other deciduous taxa (*Corylus, Fraxinus, Tilia, Ulmus, Acer, Castanea*). *Quercus ilex/Q. coccifera* and *Arbutus unedo* are present in percentages < 6% (Uzquiano, pers. com.). The analysis of other plant macro-remains shows the presence of cereals: *Triticum aestivum/durum, T. dicoccum, Setaria italica* *c.* 1900 cal. BC (Zapata, 1999b).

In summary, charcoal analyses identify *Quercus ilex/Q.coccifera* in the archaeological samples but in very small numbers (fig. 4). It could be argued that evergreen oak wood was not used by the people who occupied the sites even if today it is one of the most appreciated fuels in the area (Zapata and Peña-Chocarro, in press) but its scarcity is confirmed by the pollen analyses—when available—. According to archaeobotanical data for these sites, *Quercus ilex-Q. coccifera* seems to be better represented in charcoal than in pollen analyses.

Discussion

Evergreen oak formations in southern Europe seem to have at least two different origins (Neumann, 1992): 1) climatic, as in the Spanish Levant, where these forests spread before the Neolithic (Badal *et al.* 1994), and 2) anthropogenic, as in southern France where they expand from *c.* 5500 cal. BC to the expense of deciduous forests (Chabal, 1997); this may be considered evidence of increasing climatic aridity (Neumann, 1992).

Quercus ilex/coccifera is present at Lezetxiki (Gipuzkoa, Basque Country) at least from Oxygen isotope stage 4, around 70.000 BP (Sánchez Goñi, 1993). During the Upper Palaeolithic and during the Late Glacial it has also been identified in small proportions at the sites of Antoliñako Koba (Zapata, 1999a), Santa Catalina and Laminak II in Biscay and La Peña del Perro in Cantabria (Uzquiano, 1992 and 1994). Thus, the hypothesis of migration should not be considered to explain its presence in this area during the postglacial.

During the Holocene, anthracological data from the caves of La Llana, Mazaculos and Los Canes (Uzquiano, 1992, 1994), along with the evidence we provide here, show again that evergreen *Quercus ilex/coccifera* is present in north Atlantic Iberia from the Mesolithic to the Bronze Age. However, it does not seem to form important extensions. In pollen diagrams often *Quercus* t. *ilex* is not identified. Thus, the spread of large stands of *Lauro nobilis-Querceto ilicis-S* may have been very recent. The reasons behind this pro-

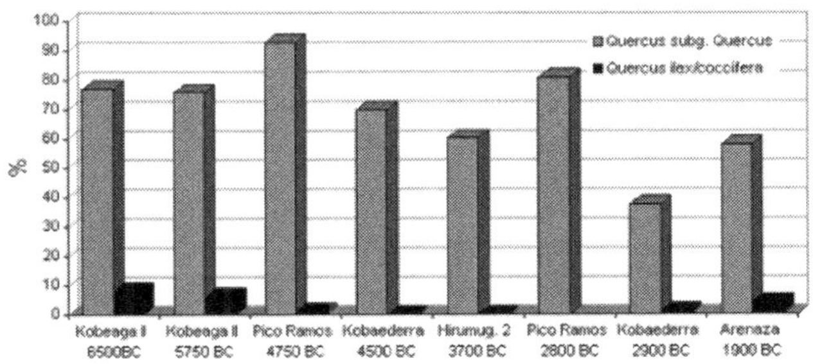

Figure 4: Frequencies of deciduous and evergreen oak identified in different sites located on the coast of Biscay or in the vicinity. Data from Arenaza provided by P. Uzquiano. The dates are approximate and calibrated, derived from the available ^{14}C dates. *Quercus* subg. *Quercus* is the main taxon in all the sites whereas *Q. ilex/coccifera* is present but in very small numbers.

cess are probably not climatic since this vegetation series is not particularly important in the warmest period of the Holocene. In fact, the warmest period is also a wet one (Bell and Walker, 1992) and deciduous forests expand in northern Iberia and nearby areas at that time (among others, Peñalba, 1989, 1992a, 1992b; Sánchez Goñi, 1993; Iriarte, 1994a; Reille and Andrieu, 1995).

If we discard climatic reasons, human action remains a strong hypothesis to explain the spread of evergreen *Quercus* in oceanic Iberia. Early human impact has been detected in Basque sites by archaeobotanical analyses, even in areas where –most probably due to a lack of sampling- evidence of neolithic agricultural and animal husbandry is not available. The rock-shelter of Aizpea at the western Pyrenees has provided a charcoal sequence (fig. 5) where human impact is detected from the Neolithic, something that can also be seen on the pollen record from this site (M. J. Iriarte, pers. com.). However, archaeological work in the area –concentrated on caves and rock-shelters- has failed so far to provide evidence of such early domesticates. On the coastal Basque Country neolithic agriculture has been dated *c.* 4200 cal. BC at the cave of Kobaederra along with some evidence of human impact (fig. 3). But we have already mentioned that *Q. ilex/coccifera* was probably not abundant for the last hunter-gatherers and the first farmers (fig. 2, 3 and 4).

Conclusion

During the middle Holocene, deeper soils with higher nutrient availability along with mild climatic conditions favoured the spread of deciduous woodland (Peñalba, 1989, 1992a, 1992b; Sánchez Goñi, 1993; Iriarte, 1994a; Reille and Andrieu, 1995; Zapata 1999a, among others). On warmer and more protected slopes from the coastal areas deciduous *Quercus* formations might include small numbers of thermophilous species which had been present there continuously at least from the Upper Palaeolithic.

The chronology of the spread of coastal *Lauro nobilis-Querceto ilicis*-S as the extensive formations we know today remains an open question. We suggest that this might have been a very recent dynamic, rooted in previous anthropogenic processes that are little known. For recent periods palaeobotanical data are very limited in Biscay. The most recent samples from the Bronze Age levels of the cave of Arenaza (first millennium cal. BC) yielded no remains of *Quercus* tp. *ilex* (M. J. de Isturiz, pers. com.).

Most probably after the Bronze Age, human action on these slopes may have favoured the extension of evergreen forests through processes we poorly understand: intensification of agropastoral practices, slash-and-burn agriculture, grazing by domestic animals, periodical pruning for fuel, tree fodder or industrial production of charcoal... As we can see in Mediterranean areas (among others, Chabal, 1997; Fabre, 1996 in Chabal, 1997; Vernet, 1999), these activities, on their own or interacting, interfere with the dynamics of *Quercus* forests and may have favoured the spread of *Quercus ilex* also in oceanic environments. The degradation and erosion of the soils (Martínez Cortizas and Moares, 1995) would also help the spread of less demanding species. Because of the location of many archaeological sites on these karstic areas, we are likely to have in the future more raw archaeobotanical data on this subject. However, a greater effort will have to be made in order to interpret the vegetation changes and to assess the chronology and reasons behind them.

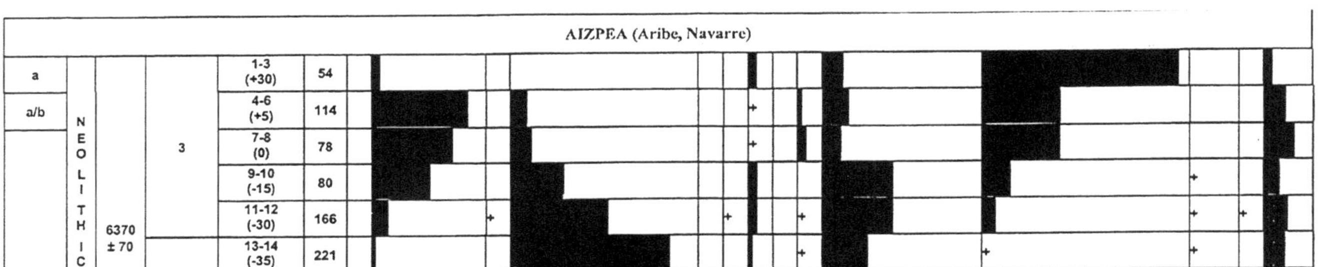

Figure 5: Charcoal analysis diagram from Aizpea (Navarre). 1. *Pinus* tp. *sylvestris* / 2. *Ulmus* sp. / 3. *Alnus* sp. / 4. *Betula* sp. / 5. *Corylus avellana* / 6. *Populus* sp. / 7. Rosaceae Pomoideae / 8. *Acer* cf. *pseudoplatanus*. *Prunus* sps: 90% is *Prunus* cf. *spinosa*; 1% is *Prunus* cf. *avium*. *Rhamnus*: *Rhamnus saxatilis*/*R. catharticus*. *Fraxinus*: 28% is *Fraxinus excelsior*; 72% is *Fraxinus* sp. Samples 30 and 1-2 should be interpreted with caution since the number of fragments available is small. However, they have been included because they are coherent with other samples next to them and with the present vegetation in the vicinity of the site dominated by *Buxus sempervirens*. Level a is superficial and includes very little archaeological material.

ACKNOWLEDGEMENTS
I thank María José Iriarte, María José de Isturiz and Paloma Uzquiano for having provided with unpublished archaeobotanical data. Part of this work has been done under the project PU97/7 funded by the Basque Government: Estudio paleoambiental y disponibilidad de recursos a lo largo de la Prehistoria reciente. Respuesta antrópica a los cambios ambientales e influencia sobre el medio de la actividad humana. Part of the laboratory work has been carried out under a Basque Government postdoctoral grant in the University of Cambridge. Many thanks to William Fletcher for correcting the English version and to Isabel Figueiral and the members of G. Pitt-Rivers Lab (U. Cambridge) for suggestions and comments.

References

AGUIRRE M., LÓPEZ QUINTANA J.C., SÁENZ DE BURUAGA A. 2000 - Medio ambiente, industrias y poblamiento prehistórico en Urdaibai (Gernika, Bizkaia) del Würm reciente al Holoceno medio. *Illunzar*, 4, p. 13-38.

APELLANIZ J.M. 1974 - Excavaciones sobre el Mesolítico de Vizcaya en los años de 1972 y 1973, y el arte rupestre de Arenaza I. Cuevas de Arenaza I (Galdames) y abrigo de Kobeaga II (Ispaster). *Kobie*, 5, p. 31-33.

APPELLANIZ J. M. 1975 - El campamento mesolítico de pescadores de Kobeaga II. Ispáster. Noticiario Arqueológico Hispánico. *Prehistoria*, 4, p. 231-240.

APPELLANIZ J. M. 1977 - Las campañas (IV y V) de excavaciones en la cueva de Arenaza I (S. Pedro de Galdames, Vizcaya), años de 1975 y 1976. *Kobie*, 7, p. 43-44.

APPELLANIZ J. M. 1978 - Avance a la Memoria de la VI Campaña de excavaciones arqueológicas en la cueva de Arenaza I (S. Pedro de Galdames, Vizcaya), año 1977. *Kobie*, 8, p. 113-114.

APELLANIZ J. M., ALTUNA J. 1975(a) - Excavaciones en la Cueva de Arenaza I (San Pedro de Galdames, Vizcaya). Primera campaña, 1972. Neolítico y mesolítico final. Noticiario Arqueológico Hispánico. *Prehistoria*, 4, p. 123-154.

APELLANIZ J. M., ALTUNA J. 1975(b) - Memoria de la II campaña de excavaciones arqueológicas en la cueva de Arenaza I (San Pedro de Galdames, Vizcaya). Noticiario Arqueológico Hispánico, *Prehistoria*, 4, p. 157-181.

APELLANIZ J. M., ALTUNA J. 1975(c) - Memoria de la III campaña de excavaciones arqueológicas en la cueva de Arenaza I (San Pedro de Galdames, Vizcaya). *Noticiario Arqueológico Hispánico, Prehistoria*, 4, p. 185-197.

ASEGINOLAZA C., GOMEZ D., LIZAUR X., MONTSERRAT G., MORANTE G., SALAVERRIA M.R., URIBE-ECHEVARRIA P.M. 1989 - *Vegetación de la Comunidad Autónoma del País Vasco*. Vitoria-Gasteiz, Gobierno Vasco.

ASEGINOLAZA C., GOMEZ D., LIZAUR X., MONTSERRAT G., MORANTE G., SALAVERRIA M.R., URIBE-ECHEVARRIA P.M. 1992 - *Mapa de Vegetación de la Comunidad Autónoma del País Vasco*. Vitoria-Gasteiz, Gobierno Vasco.

BADAL E., BERNABEU J., VERNET J.-L. 1994 - Vegetation changes and human action from the Neolithic to the Bronze Age (7000-4000 B.P.) in Alicante, Spain, based on charcoal analysis. *Vegetation History and Archaeobotany*, 3, p. 155-166.

BELL M., WALKER M.J.C. 1992 - *Late Quaternary Environmental Change*. Physical and Human Perspectives. Essex, Longman Group Ltd.

BLANCO E., CASADO M. A., COSTA M., ESCRIBANO R., GARCIA M., GENOVA M., GOMEZ A., GOMEZ F., MORENO J. C., MORLA C., REGATO P., SAINZ H. 1997 - *Los Bosques Ibéricos*. Una interpretación Geobotánica. Barcelona. Editorial Planeta.

CHABAL L. 1997 - Forêts et sociétés en Languedoc (Néolithique final, Antiquité tardive). L'anthracologie, méthode et paléoécologie. *Document d'Archéologie française*, Paris, Éditions de la Maison des Sciences de l'Homme, 189 p.

FERNANDEZ LOMBERA J. A. 1992 - Cueva de Arenaza I. *Arkeoikuska*, 1992, p. 128-132.

FERNANDEZ LOMBERA J. A. 1994 - Informe de la XXI campaña de excavaciones en Arenaza I (Galdames). 1993. *Kobie* (Serie Paleoantropología), XXI, p. 323-333.

GUINEA E. 1985 - *Vizcaya y su paisaje vegetal*. Bilbao. Diputación Foral de Vizcaya, 432 p.

IRIARTE M. J. 1994(a) - *El paisaje vegetal de la Prehistoria reciente en el Alto Valle del Ebro y sus estribaciones atlánticas. Datos polínicos. Antropización del paisaje y primeros estadios de la economía de producción*. Unpublished PhD. Vitoria-Gasteiz, University of the Basque Country.

IRIARTE M. J. 1994(b) - Estudio palinológico del nivel sepulcral del yacimiento arqueológico de Pico Ramos (Muskiz, Bizkaia). Cuadernos de Sección de la Sociedad de Estudios Vascos / Eusko Ikaskuntza. *Prehistoria-Arqueología*, 5, p. 161-179.

LOIDI J. 1987 - El Pais Vasco. In: Peinado, M. and Rivas-Martínez, S. (Eds.) *La vegetación de España*. Madrid, Universidad de Alcalá de Henares, p. 47-55.

MARTIN I., ZUBIZARRETA A. 1991 - Actuaciones arqueológicas en el municipio de Bilbao, 1989. Cuadernos de Sección. Prehistoria-*Arqueología*, 4, p. 157-195.

MARTINEZ CORTIZAS A., MOARES C. 1995 - *Edafología y Arqueología: Estudios de yacimientos arqueológicos al aire libre en Galicia*. Santiago de Compostela, Xunta de Galicia, 199 p.

MEAZA G. 1988 - Fitogeografía del encinar cantábrico de la Ría de Gernika. *Lurralde*, 11, p. 103-121.

MEAZA G. 1991 - La impronta mediterránea en la vegetación del País Vasco. Carrascales y encinares. *Boletín de la A.G.E.*, p. 45-74.

MEAZA G. 1998 - Geografía de Euskal Herria. Suelos, vegetación y fauna. In: Meaza, G. & Ruiz Urrestarazu, E. (Eds.) *Geografía de Euskal Herria*. El Entorno natural biótico. Lasarte-Oria, Ostoa S.A.

MEAZA G., ORMAETXEA O. 1992 - Propuesta metodológica de valoración fitogeográfica de unidades de paisaje vegetal. Cuadernos de Sección. *Historia*, 20, p. 369-389.

MONTSERRAT G., MONTSERRAT J.M. 1987 - Historical interpretation of the distribution of several floristic elements of Vizcaya (Basque Country, N. Spain). *Pirineos*, 130, p. 65-74.

NEUMANN K. 1992 - The contribution of anthracology to the study of the late Quaternary vegetation history of the Mediterranean region and Africa. *Bulletin de la Société Botanique de France*, 139 (Actual. bot. 2/3/4), p. 421-440.

ONAINDIA M. 1986 - *Ecología vegetal de las Encartaciones y Macizo del Gorbea* -Vizcaya-. Bilbao, UPV/EHU, 271 p.

PEÑALBA C. 1989 - *Dynamique de végétation tardiglaciaire et Holocène du centre-nord de l'Espagne d'après l'analyse pollinique*. Unpublished PhD. Univ. d'Aix. Marseille III.

PEÑALBA C. 1992(a) - La vegetación y el clima en los Montes Vascos durante el Pleistoceno superior y el Holoceno según los análisis polínicos. In: Cearreta A., Ugarte F.M. (Eds.) *The Late Quaternary in the Western Pyrenean Region*, Bilbao, Universidad del País Vasco, p. 171-182.

PEÑALBA C. 1992(b) - Biogeografía holocena de las principales especies forestales del norte de la Península Ibérica. Cuadernos de Sección. *Historia*, 20, p. 391-409.

REILLE M., ANDRIEU V. 1995 - The late Pleistocene and Holocene in the Lourdes Basin, Western Pyrenees, France: new pollen analytical and chronological data. *Vegetation History and Archaeobotany*, 4, 1, p. 1-21.

SANCHEZ GOÑI M.F. 1993 - De la taphonomie pollinique à la reconstitution de l'environnement. L'exemple de la région cantabrique. Oxford, *BAR International Series* 586.

UZQUIANO P. 1992 - *Recherches anthracologiques dans le secteur pyrénéo-cantabrique (Pays Basque, Cantabria et Asturias) : Environnements et relations homme-milieu au Pléistocène Supérieur et debut de l'Holocène*. Unpublished PhD. Montpellier, Université Montpellier II - Sciences et techniques du Languedoc.

UZQUIANO P. 1994 - Estudio antracológico de Laminak II (Berriatua, Bizkaia). *Kobie* (Serie Paleoantropología), XXI, p. 167-172.

VADILLO J.M., ONAINDIA M., BENITO I. 1990 - *Estructura de los encinares de la asociación Lauro-Quercetum illicis en distintos niveles altitudinales del Duranguesado (Bizkaia)*. In: Villar, L. (Ed) Botánica pirenaico-cantábrica, Huesca, CSIC, p. 477-486.

VERNET J.-L. 1999 - Reconstructing Vegetation and Landscapes in the Mediterranean: the Contribution of Anthracology. In: Leveau, P., Trément, F., Walsh, K., Barker, G. (Eds.) *Environmental Reconstruction in Mediterranean Landscape Archaeology*, Oxford, Oxbow Books, p. 25-36.

ZAPATA L. 1995 - La excavación del depósito sepulcral calcolítico de la cueva Pico Ramos (Muskiz, Bizkaia). La industria ósea y los elementos de adorno. *Munibe* (Antropologia-Arkeologia), 47, p. 35-90.

ZAPATA L. 1999(a) - *La explotación de los recursos vegetales y el origen de la agricultura en el País Vasco: análisis arqueobotánico de macrorrestos vegetales*. Unpublished PhD. Vitoria-Gasteiz, University of the Basque Country.

ZAPATA L. 1999(b) - *El combustible y la agricultura prehistórica. Estudio arqueobotánico de los yacimientos de Arenaza, Kanpanoste Goikoa y Kobaederra*. Isturitz, 10, p. 311-343.

ZAPATA L., IBAÑEZ J.J., GONZALEZ URQUIJO J.E. 1997 - El yacimiento de la cueva de Kobaederra (Oma, Kortezubi, Bizkaia). Resultados preliminares de las campañas de excavación 1995-97. *Munibe* (Antropologia-Arkeologia), 49, p. 51-63.

ZAPATA L., PEÑA-CHOCARRO L. In press - *Uso y gestión del bosque en la Euskal Herria atlántica: Aprovechamiento tradicional de los recursos forestales en Encartaciones y Gorbea*. Zainak.

ZUBIZARRETA A. 1995(a) - Excavación arqueológica en el dolmen de Hirumugarrieta 2 (Bilbao, Bizkaia). *Kobie* (Serie Paleoantropología), XXII, p. 63-76.

ZUBIZARRETA A. 1995(b) - La estación megalítica de Artxanda (Bilbao, Bizkaia). Excavación del dolmen de Hirumugarrieta 2. Cuadernos de Sección. *Prehistoria-Arqueología*, 6, p. 259-276.

Late Holocene vegetation dynamics in the Cilento region (southern Italy) preliminary data

Gaetano Di Pasquale and Stefano Mazzoleni

ARBOPAVE Dpt., University "Federico II" of Naples,
Via Università 100, I-80055 Portici, Italy.
E-mail: dipasqua@unina.it

Abstract
The major vegetational changes that have occurred in the Mediterranean region are relatively recent, but, with regards to historical times, little data is available. The aim of this work is to evaluate the age of "maquis" and "garrigue" vegetation at Punta Tresino (Southern Italy), and to investigate the possible presence of deciduous woods in the past, by means of pedoanthracology. Although ^{14}C data was lacking, pedoanthracological analysis suggests that current vegetation is recent and replace the pre-existent deciduous woods where deciduous *Quercus* (*Quercus pubescens* probably) and *Castanea sativa* were predominant. Data obtained seems to confirm that *Quercus ilex* woods probably did not exist in this region.

Key-words: *Mediterranean area, Punta Tresino, Historical Period, Maquis, Garrigue, Pedoanthracology.*

Résumé :Les transformations les plus importantes du couvert végétal en région méditerranéenne sont relativement récentes, mais très peu de données sont disponibles pour la période historique. Le but de ce travail est d'évaluer l'âge de la mise en place des formations sclérophylles qui caractérisent la végétation actuelle à Punta Tresino (Italie du sud), et de vérifier la présence, dans un passé récent, d'une végétation caducifoliée grâce à l'analyse pédoanthracologique. Même en l'absence de datations absolues, les données semblent indiquer que la végétation sclérophylle est récente et a remplacé des formations dominées par *Quercus* (*Quercus pubescens* probablement) et *Castanea sativa*. Les données ne semblent pas confirmer l'existence de *Quercus ilex* pour cette région.

Mots-clés : *Région méditerranéenne, Punta Tresino, période historique, maquis, garrigue, pédoanthracologie.*

Introduction

There is an increased awareness that major vegetational changes occurred in historical times and in periods immediately prior to them within the Mediterranean region (Grove and Rackham, 1998).

The lack of suitable sites for pollen studies in the Mediterranean area and the need to obtain data of high spatial definition led to a working protocol that was defined by pedoanthracological methodology. Pedoanthracology (Thinon, 1978, 1992) is based on the identification and dating of charcoal found in soils; such charcoal, which originates from fires, is present in all soil types. This scientific discipline arose in France in 1978, and has so far been applied in Europe, chiefly in Alpine environments (Carcaillet and Thinon, 1996; Carcaillet and Talon, 1996; Carcaillet and Brun, 2000), while there have been limited applications in the Mediterranean (Carcaillet *et al.* 1997; Di Pasquale, 1998). Our study aims to highlight any recent changes in the forest landscape, and especially to verify, for this area of the Tyrrhenian coast, the hypothesis of a potential presence of deciduous woods rather than evergreen sclerophyllous, similar to findings in the French Mediterranean region (Barbero and Quezel, 1990; Barbero *et al.* 1990a; Barbero *et al.* 1990b; Barbero *et al.* 1992).

The study was carried out at Punta Tresino (40°16'53" N; 14°16'30" E) (fig.1), a small "flysch" promontory less than 360 m high with an overall area of about 1100 ha. Annual rainfall reaches 700 mm; the current vegetation consists of "garrigue" with *Cistus monspeliensis*, *Pistacia lentiscus* and *Myrtus communis* and maquis with, in addition to the above species, *Erica arborea*, *Crataegus monogyna*, *Spartium junceum* and *Calicotome spinosa*. Woodland is restricted to the northern side, basically consisting of small stands of *Quercus ilex*, *Q. pubescens*, *Fraxinus ornus* and, locally, *Ostrya carpinifolia*.

The area is currently included within the National Park of Cilento and Vallo di Diano, and is used as seasonal pasture. The data presented concern the analysis of three soils, and no absolute dating is attempted.

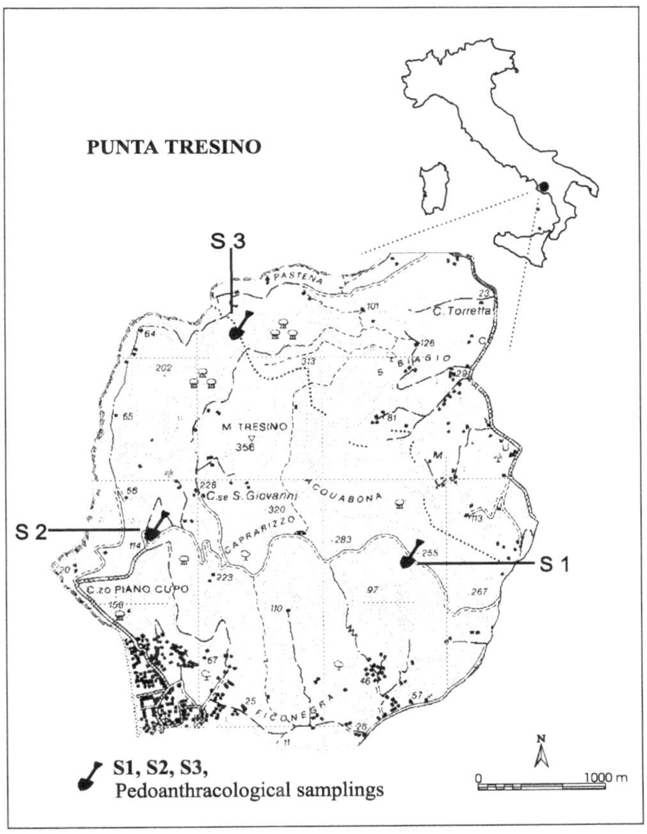

Figure 1: Study area.

Materials and methods

Soil sampling and pedoanthracological analysis

In the area of Tresino, 12 soils were chosen for sampling; the physical characteristics found at the sampling points of the three soils studied are reported in table 1.

For the soil sampling phase and for the phases of preparation and analysis of the charcoal, we followed the criteria defined by Thinon (1992) and Carcaillet and Thinon (1996). Given the general abundance of charcoal in Mediterranean soils (Di Pasquale, 1998), for each sampling level the minimum quantity of collected material was set at 5 l. In the identification phase, an episcopic microscope (x100, x200, x500) with interferential contrast was used. Wood anatomy books and atlases (Cambini, 1967; Nardi-Berti, 1979; Schweingrüber, 1990), literature specifically covering the observation of charcoal (Susmel, 1950; Couvert, 1967) and the reference collection being established at the Laboratory of Anthracology of Naples University were all consulted. For each soil level, at least 100 charcoal samples were analysed. Results are shown quantitatively (Di Pasquale, 1994; Carcaillet and Thinon, 1996); anthracomass values are presented separately. The graphs show for each taxon a relative anthracomass value which corresponds to the percentage weight of the charcoal of the taxon over the total anthracomass of a given level (Carcaillet and Thinon, 1996). However, due to the lack of quantitative experimental data, the interpretation is exclusively based on qualitative results.

Results

Charcoal pattern

The anthracomass data confirm the wealth of material for all three soils (fig. 2). The quantity of charcoal in the soil decreases markedly as depth increases (fig. 2); this distribution type is a function of time (Di Pasquale, 1998), since charcoal assemblages are ordered chronologically in soils.

S1
The current vegetation is *Ampelodesmos mauritanicus* "garrigue", with occasional *Pistacia lentiscus* and *Myrtus communis*. The entire area is used for pasture and is subject to frequent fires; oral testimony links traces of numerous old terraces seen everywhere to carob farming. The results of charcoal analysis are summarised in fig. 3. The first and second level show similar charcoal assemblages, corresponding to the present-day vegetation, except for the presence of *Erica* and *Arbutus unedo*. The third—and deepest—level is clearly differentiated due to the presence of deciduous *Quercus*, *Castanea sativa* and *Olea europaea*.

Site	Vegetation type	Exposure	Slope (°)	Height (m)	Shrub cover (%)	Vegetation cover (%)
S1	Garrigue with *Ampelodesma* (abandoned farmland)	S-SE	30	200	5	0
S2	maquis (abandoned farmland)	SW	7	150	50	10
S3	high maquis with *Q. pubescens*	N	5	175	80	80

Table 1: Study site characteristics.

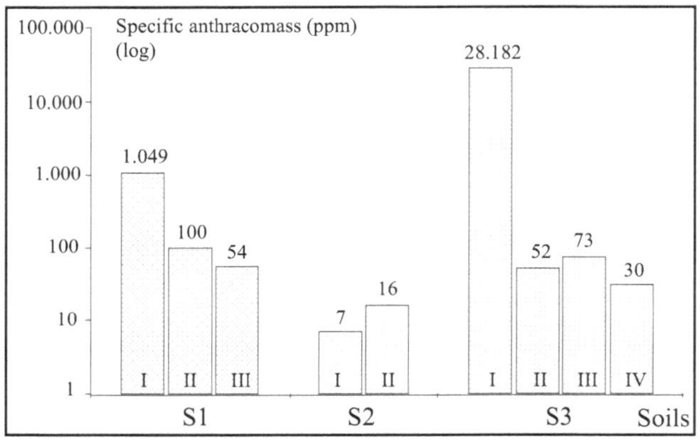

Figure 2: Specific anthracomass per level at 500 μm (ppm).

S2
This soil was sampled at the site of an abandoned olive orchard where current vegetation, besides olive trees, consists of a recolonisation "maquis" dominated by *Pistacia lentiscus*, *Spartium junceum*, *Phillyrea latifolia* and *Crataegus monogyna*. Two sampling depths were defined in the soil profile (fig. 4); the charcoal assemblages identified are fairly similar at both depths and correspond to the present-day vegetation, except for the presence of *Castanea sativa* and deciduous *Quercus*. Currently *Quercus pubescens* is found at a distance of about 500 m as isolated individuals and serving as boundary function.

S3
The current vegetation consists of high "maquis" dominated by *Quercus ilex*, *Phillyrea latifolia*, *Myrtus communis* and *Erica arborea*. Four different sampling levels were identified

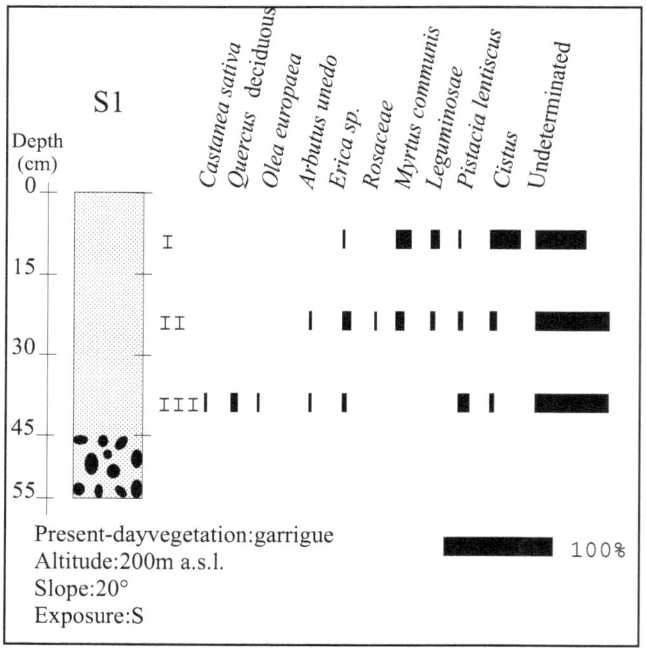

Figure 3: Pedoanthracological diagram S1.

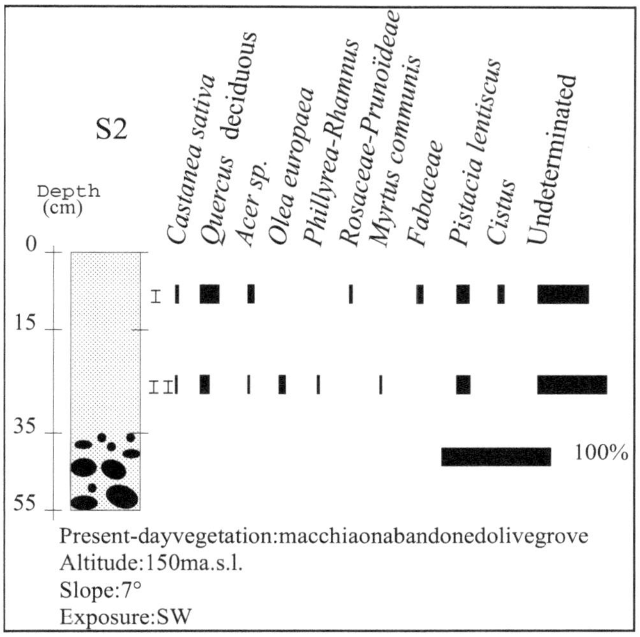

Figure 4: Pedoanthracological diagram S2.

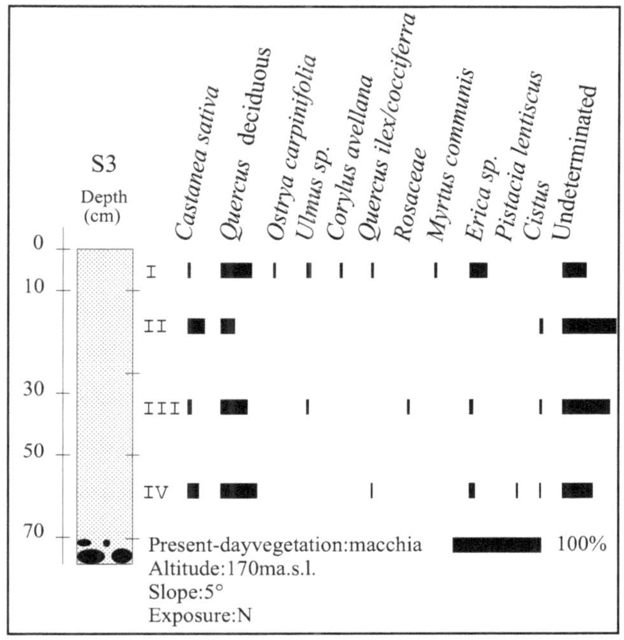

Figure 5: Pedoanthracological diagram S3.

in the soil (fig. 5). The evergreen species that characterise the current vegetation are well represented only in the first level where, in addition to deciduous *Quercus* sp., *Castanea sativa*, *Ostrya carpinifolia* and *Ulmus* also occur. Deciduous *Quercus* and *Castanea sativa* are found in all levels, accompanied by other deciduous species such as *Ulmus* (I and III) and *Ostrya carpinifolia* (I).

Discussion

The lack of chronological data makes it impossible to formulate any hypothesis regarding the age of the charcoal assemblages. The nature of the soils suggests that overall ages do not exceed two millennia (Terribile, pers. comm.). Nevertheless, analysis of the spatial distribution of the charcoal allows us to interpret the position of the charcoal assemblages in each soil with a relative chronology which is a function of the depth at which they were distributed (Di Pasquale, 1998).

Evergreen shrubs and Quercus ilex *woods*

Species in the current "maquis" vegetation are found in all the charcoal communities of the soils in question. In general, the presence of these taxa is more marked in the more recent upper levels, where they prevail over tree species. The presence of taxa such as *Cistus* and *Leguminosae* suggests the previous occurrence of fires and post-cultural recolonisation phases, while other species such as *Erica*, *Myrtus communis*, *Arbutus unedo*, *Phillyrea* are indicators of more or less severely degraded woodland.

The presence of *Quercus ilex-coccifera* requires a particular consideration; holm-oak woods are considered the expression of potential vegetation for the whole Tyrrhenian coast (Tomaselli, 1970; Pedrotti, 1989). On the Tresino promontory, the *Quercus ilex* woods consist presently of small stands on the north-facing slopes. Charcoal analysis has so far not established a greater distribution of such formations in the past. The constant presence of deciduous species, rather than the weak presence of this taxon, which indicates that past forest communities were not made up by holm-oak woods. Stands of *Quercus ilex* must always have existed, probably growing on rocky outcrops, without ever forming extensive woods. Moreover, studies of the recolonisation of abandoned fields, still under way in the Tresino area, show that this species has a very poor capacity to recolonise available space, unlike deciduous *Quercus*.

Deciduous Quercus *and* Castanea sativa

Deciduous *Quercus* charcoal appears in all three soils, especially at middle and deep levels. The oaks currently found in the surroundings are *Quercus pubescens* and *Quercus dalechampii*, which form stands on the north-facing slope and at the summit. Diachronic studies of the variations in vegetation cover carried out on a landscape level throughout the Tresino area for the period 1943-1990 (Mazzoleni *et al.* in press) show a clear increase in mixed deciduous woodland dominated by *Quercus pubescens*. According to anthracological data, such thermophylous deciduous species, especially *Quercus pubescens*, formed woods in Tresino in the past, also on the slopes where the temperature was significantly higher. This would confirm the data already obtained for other areas of inquiry in the Mediterranean, both from pollen (Reille and Pons, 1992) and anthracological studies (Vernet, 1997): during the Holocene, deciduous woods disappear or are severely reduced in area at different moments according to region and, due to human impact. These woods were thus replaced by formations of *Quercus ilex* and by evergreen shrubs, which are more resistant to disturbance.

Castanea sativa is found in the charcoal assemblages of all three soils, hence at altitudes that do not exceed 200 m, and at a few thousand metres from the shoreline. Currently at Tresino, there are still two chestnut trees left on the north-facing slopes. Pedoanthracological data indicate that in the past *Castanea sativa* was much more commonly found, reaching warmer sites where "garrigue" vegetation is currently found. Clearly, its widespread distribution must be attributed to human activity; however, it is worth recalling that the presence of *Castanea sativa* alongside deciduous *Quercus* evidently indicates a forest landscape with meso-thermophilous species rather than sclerophyllous evergreens.

Forest structure

The current vegetation is dominated by evergreen shrubs dating back to various dynamic phases and by stands of coppiced trees. It is no easy matter to attempt to hypothesise the structure of past forest ecosystems by using only pedoanthracological data, but the presence of some taxa rather than others may provide useful indications. We know that this area was essentially used for seasonal pasturage, while tree cropping developed in limited areas only in the 19th century (Bellelli, 1993).

The constant presence of deciduous *Quercus* and *Castanea sativa* always accompanied by heliophilous shrubs suggests a landscape dominated by scattered woodland, functionally similar to woodland pastures. Such landscapes, which are now commonly found in the Iberian peninsula "Dehesas", "Montados", were a feature of European landscapes since early historical times and continuing through the 19th century, and correspond to a variety of land uses which have now disappeared (Di Pasquale and Zanzi Sulli, 1993; Rackham, 1998; Di Pasquale and Mazzoleni, in press).

Conclusion

Due to the absence of absolute dating, no definitive conclusions may be drawn regarding the obtained data. However, the following general conclusions may be formulated both with regard to the method and the vegetation dynamics:
- 1) for a medium-term diachronic study with high spatial detail, pedoanthracology proves to be a very valid and useful method;
- 2) in the case of Tresino, the data set shows that the current dominance of evergreen shrubs is relatively recent and that it must be linked to human land pressure;
- 3) the woods were dominated by thermophilous deciduous species;
- 4) for a certain period, the forest landscape must have been markedly characterised by "savanna" and scattered woodland connected with pasturage.

These initial results still do not allow us to supply clear indications of the primary vegetation at Punta Tresino and the changes it has undergone. However, it appears extremely clear that for this coastal area, as already shown for Mediterranean France, there is a potentiality for deciduous woods rather than holm-oak; prior to the currently occurring evergreen shrubs, the deciduous woodland had given way to a farmed chestnut plantation, which probably lasted for several centuries.

ACKNOWLEDGEMENTS
The study was supported by the ModMED Project (ENV4-CT97-0680). We thank M. Walter for the English translation. Many thanks to L. Chabal and E. Badal-Garcia for their helpful comments and suggestions.

References

BARBERO M., QUEZEL P. 1990 - La déprise rurale et ses effets sur les superficies forestières dans la région Provence-Alpes-Côte d'Azur. *Bulletin de la Societé Linnéenne de Provence* 41, p. 77-88.

BARBERO M., LOISEL R., QUEZEL P. 1992 - Biogeography, ecology and history of Mediterranean *Quercus ilex* ecosystems. *Vegetatio*, vol. 99-100, p. 19-34.

BARBERO M., BONIN G., LOISEL R., QUÉZEL P. 1990a - Changes and disturbances of forest ecosystems caused by human activities in the western part of the mediterranean basin. *Vegetatio*, 87, p. 151-173.

BARBERO M., QUEZEL P., LOISEL R. 1990b - Les apports de la phytoécologie dans l'interprétation des changements et perturbations induits par l'homme sur les écosystèmes forestiers méditerranéens. *Forêt Méditerranéenne*, 12 (3), p. 194-215.

BELLELLI M. 1993 - *Ricerche sull'evoluzione del paesaggio dei promontori di Tresino e Licosa. Indagine storica, cartografia fisionomica e dinamica della vegetazione.* Tesi di laurea, Università "Federico II" di Napoli, 80 p.

CAMBINI A. 1967 - *Riconoscimento microscopico del legno delle querce italiane.* Contributi scientifico-pratici. Roma, C.N.R., 69 p.

CARCAILLET C., BRUN J.J. 2000 - Changes in landscape structure in the northwestern Alps over the last 7000 years. Lessons from soil charcoals. *Journal of Vegetation Science*, 11, p. 705-714.

CARCAILLET C., TALON B. 1996 - Aspects taphonomiques de la stratigraphie et de la datation de charbons de bois dans les sols; exemple de quelques sols des Alpes. *Géographie physique et Quaternaire*, 50, 2, p. 233-244.

CARCAILLET C., Thinon M. 1996 - Pedoanthracological contribution to the evolution of the upper treeline in the Mauriene Valley (North French Alps); Methodology and preliminary data. *Review of Palaeobotany and Palynology*, 9, p. 399-416.

CARCAILLET C., BARAKAT H. N., PANAÏOTIS C., LOISEL R. 1997 - Fire and late-Holocene expansion of *Quercus ilex* and *Pinus pinaster* on Corsica. *Journal of Vegetation Science*, 8, p. 85-94.

COUVERT M. 1967 - Atlas d'anatomie des charbons des foyers préhistoriques. Afrique du Nord Tempérée. *Memoires du centre de recherche anthropologiques préhistoriques et ethnographiques*, XXVI, Alger, 168 p.

DI PASQUALE G. 1994 - *L'apporto dell'analisi pedoantracologica allo studio dell'ecologia del paesaggio mediterraneo: due esempi.* Tesi di Dottorato, Università di Padova, 155 p.

DI PASQUALE G. 1998 - Evidences of past fires by soil charcoals in Mediterranean area. International Journal of Widland Wildfire, in Trabaud L., (Ed.), *Fire Managment and Landscape Ecology*, International Association of Widland Fire, Fairfield, Washington, p. 289-295.

DI PASQUALE G., ZANZI SULLI A. 1993 - *Purposes of standard trees in Coppices according to Silvicoltural Theory during 19th and 20th centuries.* Mitteilungen der Forstlichen Versuchs- und Forschungsanstalt Baden- Würtemberg, 175, p. 95-106.

DI PASQUALE G., MAZZOLENI S. (in press) - The Mediterranean vegetation landscape: overview of its history and dynamics. In Mazzoleni S., Di Pasquale G., Di Martino P. and Rego F. (eds.) *Recent Dynamics of Mediterranean Vegetation Landscape*. Reading, Gordon and Breach Publishers.

GROVE A.T., RACKHAM O. 1998 - History of Mediterranean Land Use. In P. Mairota, J.B. Thornes and N. Geeson (eds.) *Atlas of Mediterranean Environments in Europe*. Chichester, New York, Wiley & Sons, p. 76-78.

MAZZOLENI S., STRUMIA M., BUONANNO M., BELLELLI M. (in press) - Evidences of vegetation landscape changes in the Campania region, southern Italy. In: In Mazzoleni S., Di Pasquale G., Di Martino P. and Rego F. (eds.) *Recent Dynamics of Mediterranean Vegetation Landscape*. Reading, Gordon and Breach Publishers.

NARDI BERTI R. 1979 - *La struttura anatomica del legno ed il riconoscimento dei legnami italiani di più corrente impiego*. Contributi scientifico-pratici. Firenze, C.N.R., 24, 155 p.

PEDROTTI F. 1989 - *Carta della Vegetazione reale d'Italia, 1/1.1000.000*. Ministero dell'Ambiente, Roma.

RACKHAM O. 1998 - Savanna in Europe. in Kirby K. J. and Watckins C. (Eds.) *The Ecological History of European Forest*, p. 1-24.

REILLE M., PONS A. 1992 - The ecological significance of sclerophyllous oak forests in the west part of the Mediterranean Basin: a note on pollen analytical data. *Vegetatio*, 99-100, p. 13-17.

SCHWEINGRÜBER F.H. 1990 - *Anatomie europäischer Hölzer*. Anatomy of European woods. Bern-Stuttgart, Verlag Paul Haupt, 800 p.

SUSMEL L. 1950 - *Caratteri e riconoscimento dei principali carboni di legna*. Italia Forestale e Montana, 13, p. 149-151.

THINON M. 1978 - La pédoanthracologie : une nouvelle méthode d'analyse phytochronologique depuis le Néolithique. *C. R. Acad. Sc. Paris*. 287. série D, p. 1203-1206.

THINON M. 1992 - *L'analyse pedoanthracologique. Aspects métodologiques et applications*. Thèse Doct. ès Sc. Biol. Un. d'Aix-Marseille III. 317 p.

TOMASELLI R. 1970 - *Note illustrative della Carta della vegetazione naturale potenziale d'Italia*. Min. Agri. For., Collana Verde, 27, p. 1-63.

VERNET J. L. 1997 - *L'homme et la Forêt Méditerranéenne de la Préhistoire à nos jours*. Paris, Editions Errance, 248 p.

Vegetation changes in the Sahel of Burkina Faso (West Africa) – Analysis of charcoal from the Iron Age sites Oursi and Oursi-village

Alexa Höhn

Seminar für Vor- und Frühgeschichte Archäologie und Archäobotanik Afrikas
Robert Mayer-Straße, 160325 Frankfurt, Germany.
E-mail: alhoehn@stud.uni-frankfurt.de

Abstract

Charcoal from two settlement mounds in the Sahel of Burkina Faso was studied. While one site provides material from the Early Iron Age the other comprises Middle and Late Iron Age material. The results allow an overview of vegetation changes from the first centuries AD to the XII/XIIth century. This was a period of intensified sedentary cultivation of pearl millet (*Pennisetum glaucum*), which must have affected the woody vegetation. In the charcoal assemblage of the Early Iron Age site, *Acacia raddiana*-type clearly dominates. *Guiera senegalensis* and *Combretum glutinosum* become increasingly important at the Middle/Late Iron Age site. Both species are collected either on fallow fields or in gallery forests. The switch of dominating taxa is interpreted as intensified land use, *Acacia*-savannas having increasingly been replaced by fields. Pearl millet was cultivated in fields in which useful trees also grew. At first these parks were dominated by *Vitellaria paradoxa*, the Sheabutter tree. In the Middle and Late Iron Age *Faidherbia albida* parks were established, allowing an intensification of agriculture and/or cattle-herding. The woody vegetation was more diverse than today, containing elements which cannot be found in the area anymore. This indicates higher precipitation than today as well as less human impact.

Key-words: *Burkina Faso, Sahel, Iron Age, Charcoal.*

Résumé

Les charbons de bois issus de deux monticules archéologiques situés dans la partie sahélienne du Burkina-Faso ont été étudiés. L'un a fourni des vestiges datés de l'Age du Fer ancien, l'autre des Ages du Fer moyen et récent. Les résultats perçoivent des changements dans la végétation, des premiers siècles de notre ère jusqu'aux XII-XIIIème siècles. C'est une époque d'intense culture du millet (*Pennisetum glaucum*), qui a dû affecter la couverture ligneuse. Dans les assemblages anthracologiques du site de l'Age du Fer ancien *Acacia raddiana*-type domine clairement. *Guiera senegalensis* et *Combretum glutinosum* deviennent prépondérants pendant les Ages du Fer moyen et récent. Ces deux espèces sont récoltées à la fois sur les terrains en friche ou dans la ripisylve. Le changement des espèces dominantes est interprété comme une intensification de l'utilisation des terres. Les savanes à acacia sont remplacées par des champs. Le millet était cultivé dans des champs où se trouvaient aussi des arbres de rapport. Au début, ces parcs étaient dominés par *Vitellaria paradoxa*. Plus tard, aux Ages du Fer moyen et récent les parcs à *Faidherbia albida* sont établis, ils permettent une intensification de l'agriculture et/ou de l'élevage. La végétation ligneuse était plus diverse qu'aujourd'hui avec des éléments que l'on n'y retrouve plus. Cela indique à la fois des précipitations plus importantes et un impact de l'Homme sur le milieu plus ténu.

Mots-clés : *Burkina-Faso, Sahel, Age du Fer, charbon de bois.*

Introduction

Research setting

Vegetation changes during the Late Holocene, visible in pollen spectra of the Sahel, have been interpreted as responses to an increasingly arid climate in the whole of West Africa (Lézine, 1989; Salzmann and Waller, 1998). Around 1500 BC (3300 BP), Sudanian and Guinean taxa disappear from the pollen profiles of the Manga grasslands in the Nigerian Sahel (Salzmann and Waller, 1998). In the north of Burkina Faso, the pollen profile from Oursi shows an abrupt change around 1000 BC which has, however, been interpreted as being of anthropogenic origin (Ballouche and Neumann, 1995).

The abandonment of previously inhabited regions in the Sahara and the southward movements of Late Stone Age (LSA) herders in the second millennium BC have to be related to these increasingly difficult climatic conditions. About the same time, cereal cultivation was integrated into pastoralist economies (Neumann, 1999). Domesticated *Pennisetum*, along with wild cereals, is present in Dhar Tichit in 1800 cal. BC (Amblard and Pernès, 1989). In the north of Burkina Faso domesticated pearl millet (*Pennisetum glaucum*), dated to about 1000 BC, was found at sites (Tin Akof, Oursi) that presumably belonged to nomadic groups of the Late Stone Age (Vogelsang *et al.* 1999). Full sedentary farming only emerges in the Iron Age. It is well documented at Jenne-jeno (Mali) in the first millennium BC (McIntosh, 1994) and in north-east Nigeria from 800 BC onwards (Breunig and Neumann, 1999; Klee *et al.* 2000).

In the Sahel of Burkina Faso, intensified sedentary cultivation of plants begins around the first centuries AD. While plant remains are scarce in LSA sites, the settlement mounds of the Iron Age provide comparably large amounts of charred wood, fruits and seeds. Sedentary occupation and agriculture could and should have influenced the woody vegetation around the settlements. If and how these influences are visible in the charcoal assemblages of two Iron Age sites shall be dealt with in this paper.

The region

Dunes of Quaternary origin, a typical geomorphologic element of the Sahel, are crossing the north of Burkina Faso in several long, west-east oriented belts (fig. 1). During the rainy season, temporary lakes or "mares" develop in large basins along these belts. Rivers often run alongside the northern rim of the dunes. The plains between the dunes are mostly eroded to the crystalline base.

The climate is semi-arid. A long dry season alternates with a short rainy season from July to September and the precipitation of about 400 mm per year vary considerably in time and space.

Archaeological sites, as well as modern permanent settlements, are concentrated on the dune belts (fig. 1). The explanation has to be sought in the availability of water. The "mares" and rivers supply drinking water and the rela-

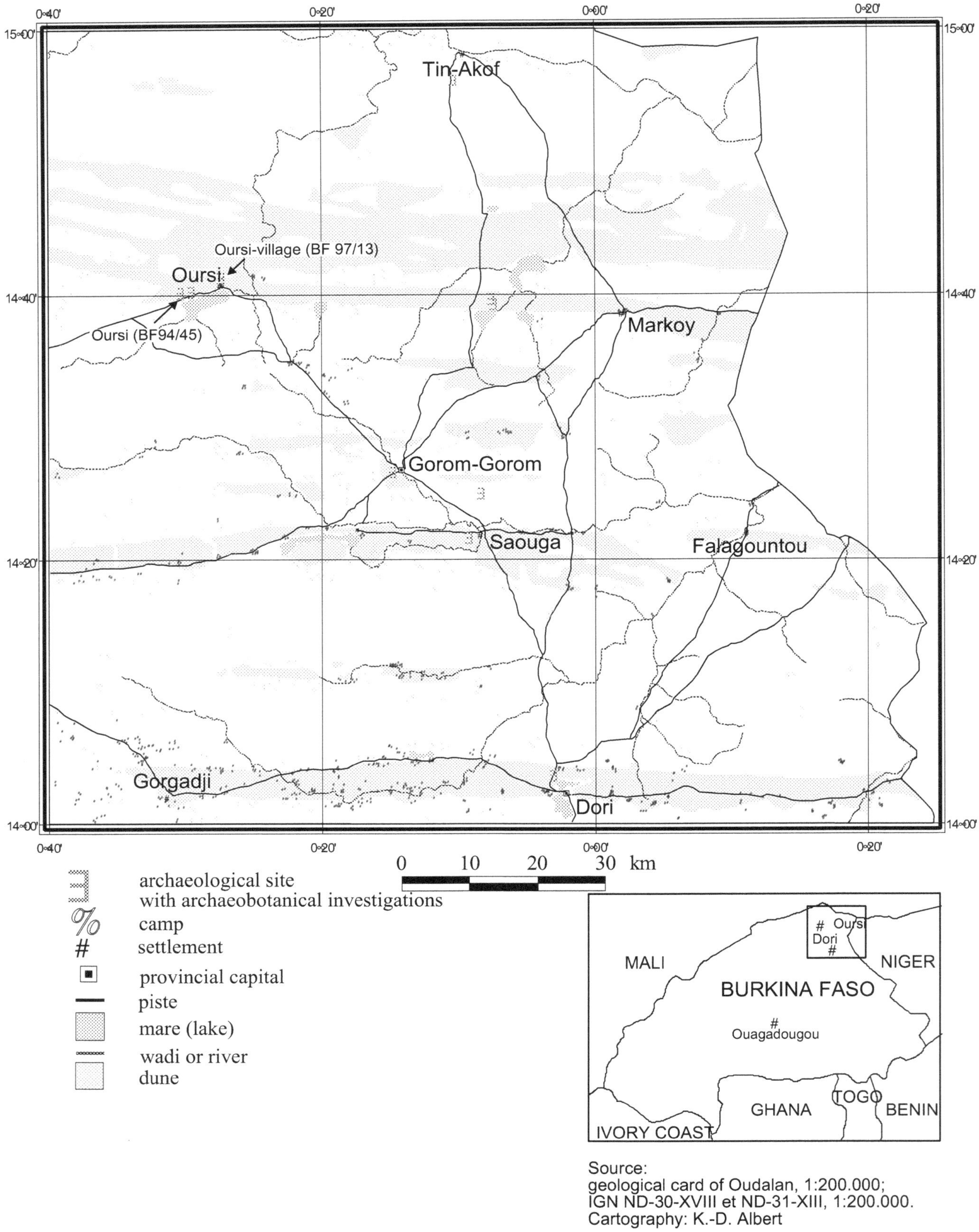

Figure 1: The environment and location of archaeological sites and modern settlements (Vogelsang *et al.* 1999, modified).

tively lush riverine vegetation supplies many resources of both wood and wild fruits. Moreover, cultivation is only possible on the sandy soils of the dunes, which have a relatively high water storage capacity (Vogelsang *et al.* 1999).

Cultivation takes place in agroforestry systems. They are defined as land-use units combining the cultivation of crops and/or livestock breeding with the systematic exploitation of woody plants growing on the same plot (Sturm, 1997). According to its physiognomy, the resulting vegetation can be labelled as park-savanna. In the Sahel of Burkina Faso, the staple crop is pearl millet (*Pennisetum glaucum*), cultivated during the rainy season. The fruit trees, interspersed within the fields, are mainly *Faidherbia albida*, *Balanites aegyptiaca* and *Sclerocarya birrea*. Agroforestry systems are typical for the savanna zone of West Africa. The field crops ensure a supply of carbohydrates, while the fruits of the trees constitute a source of vitamins and minerals (Becker, 1983). Other woody species such as, *Guiera senegalensis* are stimulated by the system of fields and fallow lands. These easily burgeoning species sprout prolifically after cutting and thrive on fallow lands. *Acacia* sp. can also be found on the dunes. *Acacia raddiana*, for instance, is present on the slopes of the dune of Oursi near the "mare".

The tops of the dunes are covered by grasses and only single trees and shrubs are found on them, mainly *Combretum glutinosum* and *Leptadenia pyrotechnica* (Küppers, in press). The elevated parts are not used for cultivation as the soils are poor and often eroded. Instead, they are heavily grazed by livestock. The sites and their vicinities are today covered with wooded grassland just like the plains. Mainly cattle herders live there together with their livestock.

The sites

In the Sahelian North of Burkina Faso, the Iron Age sites are characteristically groups of large settlement mounds. Their development appears to have been related to a sedentary way of life. The lower layers of the settlement mounds date to the first centuries AD. Their occupation ends around 1200 AD, terminating the Iron Age in northern Burkina Faso (Hallier, 1999). The preliminary results presented here are deduced from the charcoal finds from two sites, Oursi (BF 94/45) and Oursi-village (BF 97/13). They are located on the dune of Oursi, only two kilometres away one from another (fig. 1).

The site of Oursi (BF 94/45) is situated at the southern margin of the dune, some hundred meters north of the Mare d'Oursi. After high rainfalls the "mare" holds water the whole year through. The excavation reached a depth of 6 m. Chronologically, two major zones can be distinguished corresponding to the LSA and the Early Iron Age (Hallier, 1998). Radiocarbon dates suggest a chronological hiatus, since charcoal from 5.10 m is dated to about 1100 cal BC and belongs to the LSA, while the samples from above 4.40 m date to about 100 to 300 AD (Hallier, forthcoming). However, the end of the Late Stone Age and the beginning of the Early Iron Age are not distinguishable in the stratigraphy.

The site of Oursi-village (BF 97/13) is located at the northern slope of one of the dune ridges. Within viewing distance there is a "marigot", a small depression filled with water during the rainy season but which dries up soon afterwards. The excavation reached a depth of 8.10 m. The lowest layer dates to the V-VIth centuries AD (Hallier, 1999), thus leaving a gap of 200-300 years between the final occupation of Oursi and the beginning of Oursi-village. The charcoal fragments examined date from the IXth to the XIIth century AD. According to the ceramic chronology, Oursi-village can be ascribed to the Middle and Late Iron Age (Hallier, pers. comm).

Preliminary Results

General differences

Every sample taken from Oursi-village, the younger settlement mound, contains at least 100 fragments, but sometimes there may be as many as 450 fragments per 100 litres of sediment. The Early Iron Age samples from Oursi contain an average of about 30 fragments, with absolute numbers varying from zero to 105. Moreover, the fragments are smaller and their preservation is poor.

At the site of Oursi, 37 samples taken from every stratum (10 cm x 1 m) have been examined, while at Oursi-village only four samples (at 50 cm, 200 cm, 400 cm and 550 cm) have been studied so far. Nonetheless, the results as a whole are comparable as the total number of fragments are 757 and 917 from Oursi and Oursi-village respectively.

Charcoal assemblages

The following woody taxa were identified. They are listed in groups, according to their ecological requirements and to modern occurrences.

Vegetation on the dunes:
Acacia raddiana-type. In the area around the dune of Oursi this group comprises four species (*A. raddiana, A. seyal, A. senegal, A. laeta*). Especially *Acacia raddiana* is typical for the relatively undisturbed vegetation on Quaternary dunes (Trochain, 1940; White, 1983).

Park savanna:
Vitellaria paradoxa, *Faidherbia albida*, *Lannea/Sclerocarya*-type, *Guiera senegalensis*, *Bauhinieae*, *Terminalia* sp., *Combretum glutinosum*, *Ziziphus* sp. *Balanites aegyptiaca* and *Detarium* sp. (only at Oursi-village).

Gallery forest:
Combretum micranthum, *Anogeissus leiocarpus*, *Mitragyna*-type, *Feretia*-type, *Acacia nilotica*-type, cf. *Tamarindus indica*, cf. *Prosopis africana*, *Celtis integrifolia* (only at Oursi), *Diospyros mespiliformis* (only at Oursi), *Khaya senegalensis* (only at Oursi-village), and *Grewia* sp. (only at Oursi-village)

This is not a strict classification since some of the species may belong to two vegetation groups.

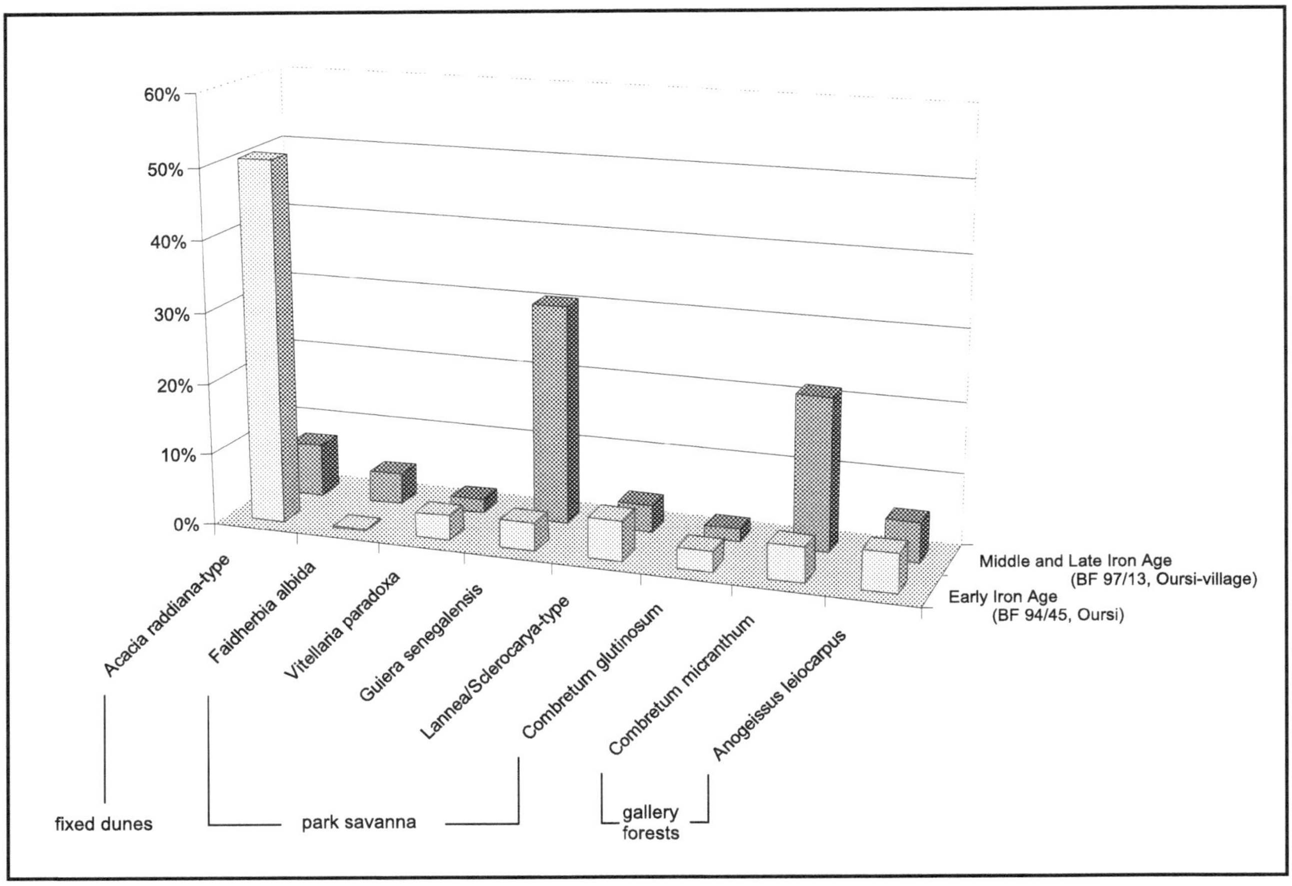

Figure 2: Charcoal from Oursi and Oursi village

In this paper, only main differences between the charcoal assemblages of the sites will be discussed. Correspondingly, figure 2 shows taxa present by more than 40 fragments, considering both sites together. Since no chronological changes can be detected so far within the charcoal assemblages of the site the samples have been summarised.

Acacia raddiana-type dominates all samples of the Early Iron Age site of Oursi. At Oursi-village the percentage of *Acacia raddiana*-type is markedly lower and *Guiera senegalensis* and *Combretum micranthum* dominate. The frequency of *Faidherbia albida* also varies between the sites. While it is present in all the samples at Oursi-village only one fragment has been found so far at Oursi.

Iron Age vegetation changes

Changes in the woody vegetation

The dominance of *Acacia raddiana*-type charcoal in the samples throughout the Early Iron Age can be due both to the importance of *Acacia* sp. in the landscape and to a preferential cutting of acacia wood. The species of this genus are of good burning quality and were thus probably selected for fuel. At the beginning of the Iron Age, *Acacia raddiana* and *A. senegal* might have dominated the vegetation on the dunes. Both species are typical for less disturbed vegetation on Quaternary dunes in West Africa (Trochain, 1940; Benoit, 1988). Today, *A. raddiana* grows in the immediate vicinity of the site, even though the human impact is high.

Good accessibility and easy availability of species that tend to grow in more or less pure stands (Maydell, 1990), combined with excellent burning qualities, might be the reason for high frequencies of *Acacia raddiana* type in the samples. The species is therefore probably over-represented in the assemblage.

The charcoal record from Oursi-village may reflect further replacements of the wooded grasslands by park savannas in the course of the Iron Age since *Acacia raddiana*-type is present in lower percentages. In the vicinity of the settlements, the numbers of *Acacia* trees had apparently decreased due to the persistent anthropogenic pressure. Park savanna had replaced the wooded grassland (fig. 3). People turned to other species with better accessibility and availability. In line with this interpretation, *Guiera senegalensis* and *Combretum micranthum* must have been available near the settlement.

Guiera senegalensis provides very good wood for fuel. The strong presence of this species which is typical for fallow land could point to an intensification of land-use, in the sense that the area around the settlements was fully covered by fields and fallow land, whereas patches of undisturbed vegetation with *Acacia* sp. were previously present closer to

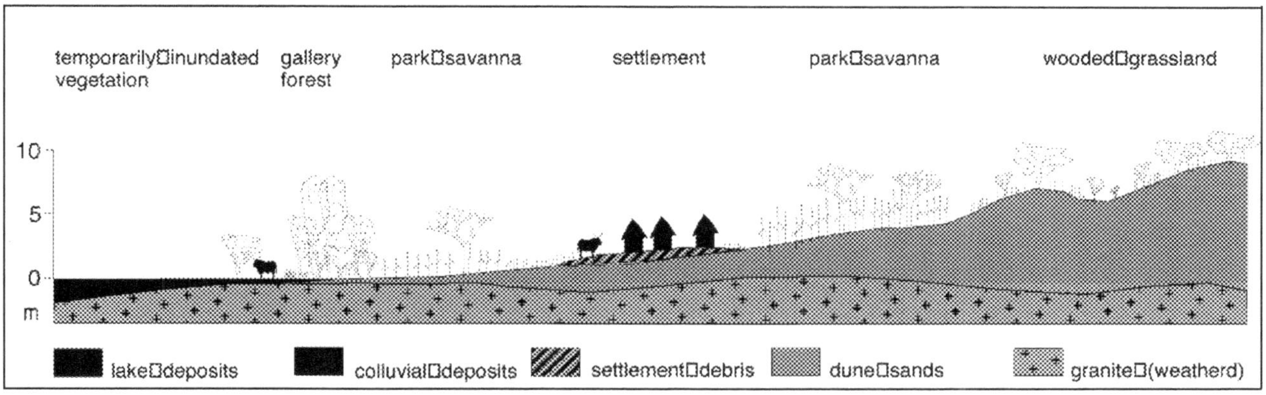

Figure 3: Schematic Iron Age landscape at Oursi (Kahlheber *et al.* forthcoming, modified)

the settlements. *G. senegalensis* could however also have been part of the gallery forest, where it is also still found today.

The high number of *Combretum micranthum* fragments is striking. Neumann *et al.* (1998) found comparable amounts of the species at the medieval site of Saouga. The species is typically growing on the shores of temporary lakes in sandy areas (Maydell, 1990). Like *Guiera senegalensis, C. micranthum* is a shrub rather than a tree which means that the wood is more easily collected as it does not require hard work with an axe. However, the wood of *Combretum micranthum* is not especially well suited as fuel-wood (Maydell, 1990). Although other sources state that the wood is good for fuel and charcoal (Irvine, 1961) the present inhabitants of the Sahel in Burkina Faso do not seem collect the species for this purpose.

Assuming that Iron Age people did not consider *Combretum micranthum* as a suitable fuel - why is the species present in the charcoal assemblage in such high percentages? There are three possible explanations:
- Branches were brought into the villages in order to recover the leaves which are well-known for diuretic, febrifuge and digestive properties. The "useless" branches and twigs were then just discarded into the fire.
- Branches were brought into the settlement as fodder. In northern Nigeria the shrub is reported to be a fodder plant for cattle and of some importance as forage in the dry season (Foster and Mundy, 1962).
- The wood was collected at the rim of the "mare" or "marigot", maybe as a by-product of the daily walk to fetch water. This might then also be true for *Guiera senegalensis*, which can be found in the gallery forests of the Sahel as well.

If the first two explanations are correct, the fragments merely represent the remnants of discarded material. If the last hypothesis is accurate, easy accessibility and availability would have been more important than good burning qualities. However, before any reliable conclusions can be drawn further deeper investigations on the qualities of *Combretum micranthum* wood have to be carried out.

Agroforestry systems

Agroforestry systems with *Faidherbia albida, Vitellaria paradoxa, Balanites aegyptiaca, Parkia biglobosa* and other species are typical of the Sahelian and Sudanian landscapes of today. The results obtained by archaeology and sediment analysis (Kahlheber *et al.* forthcoming) suggest that the beginning of the Early Iron Age was the beginning of a sedentary way of life. It is probable that the cultivation of pearl millet took place in what is today called agroforestry systems (Kahlheber *et a*l. forthcoming). The cultivation of millet does not and did not require cutting of all the trees to clear a field. Therefore, useful trees are and probably were spared and propagated.

The two main constituents of agroforestry systems in West Africa, the sheabutter tree, *Vitellaria paradoxa*, and the winter thorn, *Faidherbia albida*, are also present in the samples from Oursi and Oursi-village. While *Vitellaria paradoxa* is found more or less in all the Iron Age samples, *Faidherbia albida* is only present in the Middle and Late Iron Age samples, except for one fragment found at one metre's depth at Oursi and dated to the Early Iron Age.

The agroforesting system established at the beginning of the Early Iron Age was thus probably characterised by the sheabutter tree, *Vitellaria paradoxa*. It is typical for farming systems where population pressure is low and allows shifting cultivation. Moreover, park-savannas with *Vitellaria paradoxa* occur where people traditionally live without stock breeding. Instead of relying on animal fats, they depend on vegetal fats such as those supplied by the fruits of the sheabutter tree (Pélissier, 1980; Krings, 1991).

The presence of *Faidherbia albida* is related to livestock breeding (Pelissier, 1980; Krings, 1991). Some authors regard it as allochthonous to the area and connect its introduction to the movement of cattle herders (Pelissier, 1980). This fodder tree is valued for its edible leaves and fruits. It is particularly important for the herdsmen because of its inverse phenology: the tree carries leaves during the dry season and thus provides forage when other plants are bare. Moreover, farmers benefit from the fertilising effects of the leaves of this leguminous species as well as from the additional fertilisation caused by the presence of livestock in the shadow of the tree during the dry season. As *F. albida* sheds the leaves before the wet season it does not shade the crops. The presence of *Faidherbia albida* in the Middle and Late Iron Age at Oursi-village (and in the one metre stratum at Oursi) could point to an intensification of land-use, taking into account the fertilising effects of the tree. An intensification of livestock breeding is also probable. The agroforestry system with *F. albida* might have started towards the end of the Early Iron Age, and was fully established during Middle and Late Iron Age. The increasing number of fragments from *Guiera senegalensis* may point in the same direction, since this fallow species is also connected to grazing pressure (Maydell, 1990). *Combretum micranthum*, as mentioned, could have been brought into the settlement as a fodder plant. The interpretation of intensified livestock breeding would also justify the presence of species typical for fallow land, such as *Guiera senegalensis*, in a *Faidherbia albida* park where normally, due to the fertilisation effects, no shifting cultivation is required. At Oursi domesticated sheep, goat and cattle were present (Linseele, pers. comm.). However, since we do not have the results for Oursi-village, no conclusions about the increased importance of stock breeding can yet be drawn.

Comparison with the woody vegetation of today

Today, the human impact on the fixed dunes results in a savanna dominated by *Balanites aegyptiaca* (Trochain, 1940; Benoit, 1988) and/or *Combretum glutinosum* (Albert and Küppers, forthcoming; Trochain, 1940). In the charcoal assemblages of the sites, both species are present with less than 3%. It seems unlikely that these very well-burning species would be underrepresented. As far as we know, both of them are among the favoured woods for fuel as they provide excellent fuel and charcoal (Maydell, 1990). Moreover, they are present in high numbers. So the change to the *Balanites aegyptiaca* or *Combretum glutinosum* dominated savanna did not take place until after the Middle and Late Iron Age.

Taxa found in the Iron Age samples which today are almost or totally absent from the Sahel of Burkina Faso are *Vitellaria paradoxa, Detarium* sp. and *Prosopis africana*. Higher annual rainfall together with less anthropogenic pressure than today are probable reasons for the higher diversity of woody taxa during the Iron Age. *Prosopis africana* must still have been present in the gallery forests of the area as indicated by the "sad" and unique *Prosopis africana* tree still growing in the gallery forest along temporary river north of Oursi. *P. africana* needs an average of 500 mm of rainfall per year (Thies, 1995), *i.e.* slightly more than the Sahel receives today. Moreover, the wood is valued for the production of high quality charcoal for iron smelting and it is therefore probable that human pressure has finished off a species that was already on the brink of extinction. On the contrary, climatic changes are probably mainly responsible for the disappearance of species with a Sudanian distribution such as *Vitellaria paradoxa* and *Detarium* species. *Vitellaria paradoxa* needs at least 200 mm more of annual precipitation, its limit of distribution to the north being associated with the 600 mm isohyet (Thies, 1995); *Detarium senegalense*, the less demanding of two West African *Detarium* species, needs at least 800 mm of annual rainfall (Thies, 1995). *Vitellaria paradoxa* and *Detarium* sp. are only found south of the Sahel today.

Conclusion

The differences between the charcoal assemblages from Oursi and Oursi-village indicate a vegetation change on the dune of Oursi due to human impact. However, the discussion of the results is far from complete. Further investigations of the charcoal assemblages as well as of the uses and properties of the identified species have to be done. Only then, a thorough discussion about the differences between the two sites and the changes of the vegetation during the Iron Age and later can be carried out.

ACKNOWLEDGEMENTS
The Deutsche Forschungsgemeinschaft (German Research Foundation) is funding our studies in Burkina Faso. For the recovery of the charcoal, I thank Maya Hallier, who led the excavations. Thanks also to Dr. Katharina Neumann and Stefanie Kahlheber for critically reading the manuscript, and to the reviewers George Willcox and Margareta Tengberg. However, for any shortcomings that may remain in the paper the author is fully responsible.

References

ALBERT K.-D., KÜPPERS K. forthcoming - Böden und Vegetation der sahelischen Altdünenlandschaft in NE-Burkina Faso. *Berichte des SFB 268.*

AMBLARD S., PERNÈS J. 1989 – The identification of cultivated pearl millet (*Pennisetum*) amongst plant impressions on pottery from Oued Chebbi (Dhar Oualata, Mauritania). *African Archaeological Review*, 7, p. 117-204.

BALLOUCHE A., NEUMANN K. 1995 - A new contribution to the Holocene vegetation history of the West African Sahel: pollen from Oursi, Burkina Faso and charcoal from three sites in northeast Nigeria. *Vegetation History and Archaeobotany*, 4, p. 31-39.

BECKER B. 1983 - The contribution of wild plants to human nutrition in the Ferlo (Northern Senegal). *Agroforestry Systems*, 1, p. 257-267.

BENOIT M. 1988 – La dune de Gonkol. Dégradation d'un biotope ou histoire d'un paysage ? *Cahiers des Sciences Humaines* 24 (4), p. 487-501.

BREUNIG P., NEUMANN K. 1999 - Archäologische und archäobotanische Forschung in Westafrika. *Archäologisches Nachrichtenblatt*, 4, p. 336-357.

FOSTER W.H., MUNDY E.J. 1962 - Forage species of Northern Nigeria. *Tropical Agriculture*, 38, p. 311-318.

HALLIER M. 1998 - Recherches archéologiques en hiver 1997/1998 au nord du Burkina Faso: les collines d'occupation de l'âge du fer. *Nyame Akuma*, 49, p. 2-6.

HALLIER M. 1999 - Recherches archéologiques dans l'époque historique au nord du Burkina Faso: Rapport préliminaire de la campagne de fouille 1998. *Nyame Akuma*, 51, p. 2-5.

HALLIER M. forthcoming – Recherches archéologiques sur le site d'Oursi. *Berichte des SFB 268.*

IRVINE F.R. 1961 - *Woody plants of Ghana*. London, 868 p.

KAHLHEBER S., ALBERT K.-D., HÖHN A. forthcoming - A contribution to the palaeoenvironment of the archaeological site Oursi in North Burkina Faso. *Berichte des SFB 268.*

KLEE M., ZACH B., NEUMANN K. 2000 – Four thousand years of plant exploitation in the Chad basin of northeast Nigeria I: The archaeobotany of Kursakata. *Vegetation History and Archaeobotany*, 9, p. 223-237.

KRINGS T. 1991 - Kulturbaumparke in den Agrarlandschaften Westafrikas – eine Form autochthoner Agroforstwirtschaft. *Die Erde*, 122, p. 117-129.

LÉZINE A.-M. 1989 - Late Quaternary vegetation and climate of the Sahel. – *Quaternary Research,* 2, p. 317-334.

MAYDELL H.-J. VON 1990 - *Trees and shrubs of the Sahel: their characteristics and uses*. Weikersheim, 325 p.

MCINTOSH S.K. 1994 - Changing perceptions of West Africa's past: Archaeological research since 1988. *Journal of Archaeological Research*, 2, p. 165-198.

NEUMANN K. 1999 - Early plant food production in the West African Sahel – new evidence. New York, M. van der Veen (ed.): *Plants and people in Africa: recent archaeobotanical evidence*. p. 73-80.

NEUMANN K., KAHLHEBER S., UEBEL D. 1998 - Remains of woody plants from Saouga, a medieval west African village. *Vegetation History and Archaeobotany*, 7, p. 57-77.

PELISSIER P. 1980 - L'arbre dans les paysages agraires de l'Afrique Noire. *Cah. O.R.S.T.O.M.*, XVII, 3-4, p. 131-136.

SALZMANN U., WALLER M. 1998 - The Holocene vegetation history of the Nigerian Sahel based on multiple pollen profiles. *Review of Palaeobotany and Palynology*, 100, p. 39-72.

STURM H.-J. 1997 - Kulturbaumlandschaften in der Provinz Boulgou (Burkina Faso). *Berichte des SFB 268*, 9, p. 169-188.

THIES E. 1995 - *Principaux ligneux (agro-) forestiers de la Guinée*. Roßdorf, 544 p.

TROCHAIN J. 1940 - *Contribution à l'Étude de la Végétation du Sénégal*. Paris, 433 p.

VOGELSANG R., ALBERT K.-D., KAHLHEBER S. 1999 - Le sable savant : les cordons dunaires sahéliens au Burkina Faso comme archive archéologique et paléoécologique de l'Holocène. *Sahara*, 11, p. 51-68.

WHITE F. 1983 - *The vegetation of Africa*. Paris, 356 p.

Evidence for ancient forest cover and deforestation from charcoal analysis of ten archaeological sites on the Euphrates

George Willcox
CNRS, Jalés, Berrias, 07460 France.
E mail: g.willcox@wanadoo.fr

Abstract
Identifications of charcoal from ten sites along the Euphrates provide evidence for the presence of tree species which are rare or absent in the area today. This difference, between the charcoal evidence and modern day vegetation is seen to result from extensive deforestation. Evidence from forest remnants in isolated areas which provide information about vegetation history is discussed in the light of charcoal analysis. Although up to 20 taxa have been identified from these sites, three characteristic tree taxa, deciduous oak, *Pistacia* and *Amygdalus*, are given special attention because they are considered to be good markers.

Key-words: *Charcoal, Euphrates, Deforestation, Vegetation his-*

Résumé
Les identifications de charbons de dix sites sur les rives de l'Euphrate ont montré la présence d'arbres, rares ou absents actuellement dans la région. Ces différences entre les indices anthracologiques et la végétation moderne sont interprétées comme les résultats d'une déforestation. Les indices de reliquats de forêts dans les régions isolées qui fournissent également des informations sur l'histoire de la végétation sont examinés en relation avec les données anthracologiques. Plus de 20 taxons ont été identifiés, mais seuls trois taxons chêne f. c. *Pistacia* et *Amygdalus* sont traités ici parce qu'ils sont considérés comme de bons marqueurs.

Mots-clés : *Charbons de bois, Euphrate, déforestation, histoire de la végétation, chêne, pistachier, amandier.*

Introduction

Degradation of the vegetation in the Near and Middle East is well known to field botanists who have frequently remarked on how several millennia of over-grazing and wood exploitation has severely diminished the forests. For example Zohary (1973, 653) wrote "thousands of miles of bare rocky hillside testify to the extreme ruination of the landscape....scattered single trees found far from woodland show how radical the destruction of the forest was". He goes on to say that "...man has introduced floristic and distribution changes that in their magnitude exceed by far those caused by Pleistocene climatic changes".

Deforestation is particularly severe in the Near East because regeneration is made difficult due to the aridity and more importantly, constant grazing, which kills off freshly germinated seedlings. Thus in areas where climatic conditions are severe with maximum pressure from herds, regeneration may not occur at all.

Botanists working on the contemporary vegetation were the first to comment on the deforestation. They noted isolated relic stands of trees in inaccessible areas and these were interpreted as the remains of extensive forests which had existed in the past. However the true extent of past forest cover and the chronological sequence of the destruction can only be evaluated by interpreting the results from analysis of fossil material.

While pollen diagrams obtained from lake sediments have provided evidence of vegetation history, sites where anaerobic conditions permit good pollen preservation are rare and limited to the higher rainfall and more mountainous areas such as the Anatolian highlands and the Zagros, or the Mediterranean regions. Further, accurate dates are not always available from these lake deposits. (Despite these areas of south-eastern Anatolia and inland Syria.) Most of these areas are devoid of trees today, but are within the areas of potential forest or forest steppe (Zohary, 1973).

Charcoal from archaeological sites provides a record of wood exploitation within a chronological framework even if sequences are incomplete. Unlike pollen data archaeological wood charcoal does not directly reflect the vegetation cover, it only provides information concerning the taxa used. Most charcoal comes from fuel for which there is less selection than for building timber. Fuel is generally collected from the nearest resources, thus the charcoal assemblages, to a greater extent, reflect the availability within the site catchment area. It is highly probable that for domestic fire wood availability played the most important role in terms of the choice of wood gathered; thus wood will be collected from the trees or shrubs nearest to the site, regardless of other factors such as combustibility.

In regions of the world where wood was readily available such as forested regions of northern Europe or the Mediterranean basin it has been rightly assumed that in the majority of cases charred wood found on archaeological sites represents locally available timber which was gathered not far from the sites and was essentially autochthonous. However in regions where local wood was scarce it may have been collected farther afield, and imported, or driftwood may have been used; thus finds may be allotochonous.

Present day vegetation and forest remnants

The area covered by this study includes two potential vegetation zones of the Irano-Turanian region which have been defined by the presence of isolated relic formations. The distribution of tree species in this area is strongly influenced by altitude. Thus one can distinguish a northerly high altitude zone (above 750 m a.s.l.) dominated by *Quercus brantii*. Here this species is at the extreme west of its distribution which extends east as far as the Zagros mountains. This zone also contains *Q. boisseri* which has a more

Mediterranean distribution. *Quercus brantii* is the most xeric of the deciduous oaks, penetrating further into the arid steppes than any other species of oak. In addition within this zone there are Mediterranean incursions in climatically favourable areas marked by the presence of *Quercus calliprinos*. However pollen analyses and charcoal studies clearly show that the evergreen oak expansion occurred in the mid to late Holocene (Willcox, 1999; Baruch, 1990; Baruch and Bottema, 1999; Jahns, 1990).

The second major zone is found at a lower altitude to the south (less than 900 m a.s.l.). It is dominated by *Pistacia atlantica* and *Amygdalus* spp. This is a forest steppe association with very open woodlands which today only exists in restricted areas. The best preserved example of this zone is to be found on the Jebel Abdul Aziz in north-east Syria between 700 and 850 m a.s.l. Apart from *Amygdalus* other members of the Rosaceae family are frequent components of both these vegetation zones, and include *Crataegus monogyna*, *C. syriaca*, *Pyrus syriaca*, *Prunus spinosa* and *P. microcarpa*. These species are difficult to identify precisely from wood charcoal and are not good markers. Below this zone one finds the true steppe dominated by short-lived annuals, and perennials such as *Artemisia*, various chenopods such as *Noaea mucronata* and grasses such as *Stipa*.

From the point of view of charcoal analysis three taxa, *Pistacia* sp., *Amygdalus* sp. and *Quercus* (deciduous oak) are good markers and are readily identified in the wood charcoal from archaeological sites. The presence/absence of these three taxa allows identification of the vegetation zones for the area concerned.

Archaeological sites

The archaeological sites discussed here range from the Natufian at Abu Hureyra in the extreme south to Middle Ages for the late samples from the Asvan area. Early

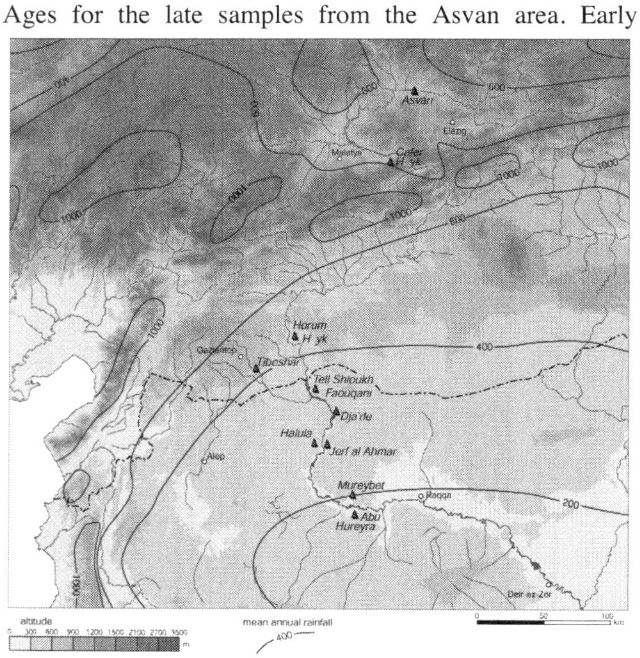

Figure 1: Site location map with contours and isohytes.

Site	Period	Country	Deciduous oak	*Pistacia*	*Amygdalus*
Abu Hureyra	Natufian	Syria	x	x	x
Asvan	Bronze/ Chalco	Turkey	xx	x	x
Cafer Höyük	PPNB	Turkey	xx	x	x
Can Hassan III	Neolithic	Turkey	x	x	x
Dja'dé	Early PPNB	Syria	x	x	x
Halula	Middle PPNB	Syria	x	x	x
Horum Hoyuk	Bronze / Chalco	Turkey	xx	x	x
Jerf al Ahmar	PPNA	Syria	x	x	xx
Mureybet	PPNA	Syria	x	x	x
Tell Shioukh Faouqani	Bronze Age	Syria	x	x	x
Tilbeshar	Bronze /Chalco	Turkey	xx	x	x

Table 1: Three characteristic taxa for ten sites on the Euphrates. Oak is more abundant at sites in Turkey while *Pistacia/Amygdalus* are more abundant on sites in Syria. Abundant = dominant taxa in the majority of samples.

Neolithic sites include Jerf el Ahmar, Mureybet, Dja'de, Halula and Cafer Höyük. Bronze Age sites include Tell Shioukh Faouquâni, Tilbeshar and Horum Höyük. The area defined by these sites has a climate which forms a gradation from the north-west to the south-east. Thus in the extreme south the average annual rainfall is 200 mm at Abu Hureyra, and in the extreme north it is 450 mm at Asvan (fig. 1).

The sites in question are found in areas which today are totally denuded and devoid of the three key species; yet these species represented by equivalent taxa (*Quercus*, *Pistacia* sp. and *Amygdalus*), have been shown to be present in abundance from charcoal identified at all the sites (tabl. 1). The ligneous vegetation in proximity to the sites today consists of a few spiny shrubs, riverine species and plantations of poplars and fruit trees.

Results

By comparing the distribution the presence the three species in the past with their availability in relation to the sites today, it is possible to assess ancient vegetation cover. In order to do this we will start by looking at the more northerly sites, which occur in what botanists consider to be areas of potential oak forest under present-day climatic conditions with a rainfall of more than 400 mm per annum. Then we will look at sites farther south where climatic conditions become progressively more arid as rainfall drops to below 250 mm and summer temperatures increase considerably. Here the vegetation becomes more xeric and more susceptible to over-exploitation.

Starting in the north four sites ranging from Chalcolithic to Medieval in the Asvan region are situated overlooking the flood plain of the Murat, a tributary of the Euphrates. Given the relatively high annual rainfall of between 400 and 450 mm the area could potentially support oak forests. The vegetation in the area today is extremely degraded with only a few spiny shrubs such as *Crataegus monogyna*, *Lycium*, and *Paliurus spina christi* growing locally around the sites. Deciduous oak charcoal was found from all periods at high frequencies (Willcox, 1974). Oak is now very rare in the region. A small damaged stand still existed in 1973 at about nine kilometres from Asvan, but at a considerably higher altitude. Asvan is situated at about 750 m a.s.l. and the oak trees were found at between 1100 and 1200 m a.s.l. This relic formation occurred on a north-facing slope and in 1973 had not been exploited for some time. It covered less than a hectare. Given that there is good evidence for long-term exploitation of oak in the region during several millennia it is inconceivable that this small area of forest could have withstood this kind of long-term exploitation. Thus it appears that oak forests had a much wider distribution in the past and were progressively degraded. In practice it was not possible to trace the degradation over time from the archaeobotanical data. But it was demonstrated that when the site was first occupied the area would have been dominated by deciduous oaks with the assemblage of lesser species such as *Celtis*, *Pistacia*, *Acer* and *Juniperus*. The trees nearest the sites would have been the first to be exploited and then deforestation progressively expanded as the inhabitants were forced to gather their wood farther and farther from the settlement. In the late Medieval period there is evidence of imported wood with the presence of pine, which with willow and poplar began to replace the oak.

Eighty kilometres further south from Asvan charcoal was recovered from the ninth millennium BP (non cal.) neolithic site of Cafer Höyük which is situated on the Euphrates north-east of Malatya (Willcox, 1991). Rainfall is lower than at Asvan and year-round temperature somewhat higher. Only eleven taxa were definitely identified. Willow and poplar combined make up about 50% of the charcoal. Deciduous oak charcoal made up about 35% indicating that oak was a frequently used fuel. Given that oaks almost always are dominants they no doubt formed an important component in the local vegetation. No oaks occur locally in the region today but again this site is well within the potential range of deciduous oak forest. While *Pistacia* and *Amygdalus* were rare at Asvan, these two taxa are more common at Cafer Höyük, which probably reflects its more southerly position and lower altitude compared to the Asvan sites. Riverine species such as poplar, willow and plane would have been locally available on the banks of the Euphrates. Thus the evidence from this site shows that the post-glacial treeless steppe (van Zeist and Bottema, 1982),was colonised by oaks before the ninth millennium BP.

As we move further south we enter a less continental climate, where today Mediterranean species can penetrate from the west, notably the evergreen oak *Quercus calliprinos*. The distribution of the evergreen oak extends from the Mediterranean but does not occur much further east than Gazintape, some 15 km north of which we leave behind the last evergreen oaks. Further south at lower altitudes conditions become hotter and drier. The banks of the Euphrates and the Sajour (a western tributary of the Euphrates) provide a habitat for a rich riverine vegetation. This area is agriculturally rich with extensive dry farming of cereals and pistachio production on a large scale. During the Bronze Age these rich lands were also extensively exploited judging from the high density of archaeological sites of this period in the area.

Charcoal from two Bronze Age sites in this area are in the process of being studied by Hugues Pessin for his PhD thesis: Tilbeshar (Kepinski-Lecompte and Ergec, 2000), situated on the Sajour, and Horum Höyük (Marro *et al.* 2000), situated on the banks of the Euphrates. At both sites Chalcolithic and Bronze Age levels were sampled and preliminary studies indicate that oak charcoal is present at very high frequencies. These sites occur in an area with less than 350 mm of rain per annum and are at the very limit of the present-day distribution of *Quercus brantii*. During the June 2000 campaign at Tilbeshar, I made a number of field excursions in this area in order to discover the most southerly stands of *Q. brantii*. Deciduous oak charcoal had already been found at sites in Syria dated to the Natufian, the Neolithic and the Bronze Age. It was therefore of paramount importance to locate the nearest present-day stands of deciduous oak. In 1995 I had extensively surveyed the Syrian side of the frontier and it was clear that deciduous oaks were only to be found in the extreme north-westerly corner of Syria in the regions of higher altitude and rainfall. Mixed Mediterranean forests of *Q. boisseri* and *Q. calliprinos* can be found starting at an altitude of 1000 m north west of Efrine where a combination of cooler conditions and higher rainfall favours their growth. Thus during the June 2000 campaign at Tilbeshar it was a great surprise to find two stations of *Q. brantii* just eighteen kilometres north of Jerablous on the Turkish-Syrian border. The relic stands were found at between 800 and 900 m a.s.l. (fig. 2). To my knowledge they have not been described before and represent the extreme southern limit of *Quercus brantii* in this region. They consisted essentially of *Q. brantii*, but two species of *Amygdalus*, *Crataegus syriaca*, *Prunus microcarpa*, *Celtis tournifortii*, *Rhamnus* sp, *Jasminum* sp., *Rhus sumac* and *Cotinus cogiggia* were also present.

Preliminary studies have demonstrated that Tilbeshar and Horum Höyük have abundant deciduous oak charcoal, and even some acorns were recorded from Horum Höyük. The small surviving stands of oaks in the area appear to represent the last relics of a much more extensive oak forest which has been progressively degraded since the Bronze Age. Given the density of archaeological sites in this area and the very limited amount of oak available today it appears highly probably that oak forests covered much of this area in the past.

Figure 2: A mature specimen of deciduous oak (*Quercus brantii*) in an isolated stand near the turkish village of Akkuyu, 18 km north of the Syrian border. This region is largely deforested but archaeological sites have produced large quantities of oak charcoal indicating that oak forests were far more widespread in the Bronze.

Moving still farther south we leave behind the present-day habitat of *Q. brantii* and move into the treeless steppe zone of northern Syria, dominated in many areas by *Artemisia herba alba* and *Noaea mucronata* (fig. 3). Here the climate again becomes progressively more arid with higher summer temperatures leading to conditions less favourable to tree growth. Away from the Euphrates valley no tree species are to be found. Within the valley and side valleys favourable micro-habitats consisting of inaccessible north-facing cliffs support occasional shrubs of *Pistacia palestina*, *Celtis tounifortii*, *Ficus carica*, *F. palmata*, *Prunus microcarpa*, and *Rhamnus palestina*. These shrubs are extremely rare and their dispersal into these special ecological niches are probably due to migratory birds.

Charcoal from the site of Tell Shioukh Faouquâni near the Turkish-Syrian border is being studied by H. Pessin (see this volume). Here the frequencies of oak charcoal for the Bronze Age are low compared to the riverine species; higher frequencies in the Iron Age may be due to importation for iron smelting. Otherwise the inhabitants relied on local riverine species for fuel.

Moving again farther south to the more arid steppes of northern Syria (fig. 3), the analysis of wood charcoal, charred seeds and fruits at one late Pleistocene and four early Holocene sites situated on the Euphrates have shown that during these periods the *Amygdalus* and *Pistacia* forest steppe association was present in this area (Hillman, 2000; Roitel, 1997; Roitel and Willcox, 2000; Willcox, 1996; Willcox and Fornite, 1999). Oak charcoal was present at all sites but at low frequencies. A rich riverine assemblage made up by far the most important proportion of the charcoal from these sites and consists of *Populus euphratica*, *Salix* sp., *Tamarix* sp., *Fraxinus* sp., *Platanus* sp., *Alnus* sp., and *Vitis sylvestris*. Fruits of another riverine taxa, *Vitex agnus castus* were found at Jerf el Ahmar. It is striking that all but the first three of these taxa have disappeared from the Syrian Euphrates and neither of the forest steppe species occurs in the area.

Figure 3: The valley of the Euphrates today is totally denuded. Only a few remnants of the riverine species remain and unfortunately these are disappearing under lakes formed behind the recently constructed dams.

Figure 4: The Jebel Abdoul Aziz in north-east Syria has a remnant of a forest/steppe association with a species of wild pistachio (*Pistacia atlantica*) and a species of wild almond (*Amygdalus orientalis*). Fruits and charcoal of these two species have been found in abundance on neolithic sites along the Euphrates some 160 kms to the west. This is the nearest living example of this association.

The corroborative evidence from charred fruits further implies that *Pistacia* and *Amygdalus* were gathered locally. As we have seen these two species form a very open forest/steppe. The nearest example today is found on the Jebel Abdul Aziz 160 kilometres to the east (fig. 4). The low density of wood in this association explains the high frequencies of local riverine species. The finds of oak are more problematic. Hillman has suggested that the oaks may have colonised the side valleys in favourable habitats during the cooler, moister Natufian period (Hillman, 2000). Arguments in favour of local oaks include the finds from Halula some four kilometres up a side valley. Also small charred twigs of oak (Roitel, 1997) have been found which do not easily survive the abrasive action of the river as trees float down. Mountain species such as the conifers, which occur for example in the Taurus, have not been found suggesting that the gathering of driftwood was not a regular activity. However the possibility that oak was collected as driftwood must be considered.

Driftwood exists because as rivers cut into their banks and change their courses they undermine trees which then fall into the river. Flow rates recorded on the Euphrates in Syria show clearly that seasonal floods were particularly strong due to the thawing of winter snows in the mountains of Anatolia coinciding with the spring rains. During August, September and November the average flow is between 200 and 300 cubic meters per second while at the peak season during March, April and May the average flow was around 2000 cubic metres per second and could be even double that (Willcox and Roitel, 1998). Thus the availability of driftwood would have been seasonal. Sites situated at some distance from the river, for example Tilbeshar and Halula, would rely less on the river as a source of fuel and timber. In general the lack of mountain species such as the conifers from the Taurus suggests that systematic collection of driftwood was not the general practice. In summary, oak charcoal found at the sites in Syria may have originated 1) from occasional trees, which occupied favourable habitats with deep soils in the side valleys, during moister periods such as the Younger Dryas, and/or 2) from driftwood brought down from Anatolia. No charred acorns have been found on these sites.

Conclusion

In conclusion, charcoal evidence confirms that both the *Quercus brantii* forests in Turkey and the *Pistacia atlantica/Amygdalus* forests-steppe associations in Syria were much more widespread in the past and that they appear to have extended into lower altitudes. From the archaeological point of view this is important because it shows that when these sites were occupied these areas were more heavily wooded with much richer vegetation compared to the desolate over-grazed steppes that we see today. These habitats may also have been rich in wild cereals and so these results have a bearing on the origins of agriculture. Faunal remains confirm more wooded conditions north of the Syrian frontier during the Bronze Age. At Tilbeshar the principal animals hunted were stag deer and roe deer, whereas further south gazelle and onager, which prefer a more open habitat, were the commonly hunted species for the Neolithic period.

In order to fully understand the vegetation history of this area we need to trace in more detail the chronological advance and retreat of the forests and to determine whether changes are due to climatic or human impact. At present climatic oscillations such as the Younger Dryas are not reflected in the charcoal sequences. Some palaeoenvironmentalists might tempt to interpret these results in terms of climatic change; however, anthropogenic desertification, essentially through over-grazing leads to destruction of the vegetation and increased erosion, which in turn leads to more aeolian activity. In addition, trampling by herds will exaggerate these effects. It is thus cleat that human factors played an important role. The role of climatic change is difficult to assess with the present data.

References

BARUCH U. 1990 - Palynological evidence of human impact on the vegetation as recorded in Late Holocene lake sediments in Israel. In: Bottema S. Entjes-Nieborg, G., and van Zeist W., (eds.), *Man's role in the shaping of the Eastern Mediterranean landscape*. Balkema, Rotterdam, p. 283-293.

BARUCH U., BOTTEMA S. 1999 - A new pollen diagramme from lake Hula in H. Kawanabe, G. W. Coutler et A.C. Roosevelt (eds) *Ancient Lakes : their Cultural and biological Diversity*, p. 75-86 Kenobi, Belgique.

HILLMAN G. 2000 - Plant food economy of Abu Hureyra. In Moore, A., Hillman, G., Legge, T., *Village on the Euphrates, from foraging to farming at Abu Hureyra*. p. 372-392. Oxford.

JAHNS S. 1990 - Preliminary note on human influence and the history of vegetation in southern Dalmatia and southern Greece. In. Bottema S. Entjes-Nieborg; G., and van Zeist W., (eds.). *Man's role in the shaping of the Eastern Mediterranean landscape*. Balkema, Rotterdam, p. 333-340.

KEPINSKI-LECOMTE C., ERGEÇ R. 2000 - Tilbeshar 1999, occupation de la vallée du Sajour de la fin du Chalcolithique au Bronze Moyen. *Anatolica Antiqua*, VIII, p. 215-225.

MARRO C., TIBET A., BULGAN F. 2000 - Fouilles de sauvetage de Horum Höyük (province de Gaziantep) : quatrième rapport préliminaire. *Anatolica Antiqua*, VIII, p. 257-278.

PESSIN H. 2000 - *Analyse anthracologique du site de Tell Shioukh Faouqani*. Mémoire de DEA Paris I.

ROITEL V. 1997 - *Végétation et action de l'homme du Natoufien au Néolithique acéramique dans le Haut-Euphrate syrien*. Ph D Thesis, University of Montpellier II.

ROITEL V., WILLCOX G. 2000 - Analyses of Charcoal from Epipalaeolithic levels at Abu Hureyra. In Moore A., Hillman G., Legge J. *Village on the Euphrates*. Oxford University Press.

WILLCOX G. 1974 - A history of deforestation as indicated by charcoal analysis of four sites in eastern Anatolia. *Anatolian Studies*. Vol. XXIV p. 117-133.

WILLCOX G., 1991 - Cafer Höyük (Turquie) : Les Charbons de bois néolithiques. *Cahiers de l'Euphrate* 5-6, p. 139-150. Editions Recherche sur les Civilisations. Paris.

WILLCOX G. 1996 - Evidence for plant exploitation and vegetation history from three Early Neolithic pre-pottery sites on the Euphrates (Syria) *Archaeobotany and Vegetation History*. 5 p. 143-152.

WILLCOX G. 1999 - Charcoal analysis and Holocene vegetation history in southern Syria. In *Quaternary Science Reviews* Special issue INQA symposium Ankara, the late Quaternary in the Eastern Mediterranean, p. 711-716.

ZEIST VAN W., BOTTEMA S. 1982 - Vegetational history of the eastern Mediterranean and the Near East during the last 20,000 years. in *Palaeoclimates, Palaeoenvironments, and Human Communities in the Eastern Mediterranean Region in Later Prehistory*.(eds.) Bintliff, J.L., and van Zeist, W., B.A.R. International Series, No 133, p. 277-321.

ZOHARY M. 1973 - *Geobotanical Foundations of the Middle East*. 738 p. Fischer, Stuttgart

Preliminary report on the charcoal analysis of Tell Shiukh Fawqani (Middle-Euphrates, Syria), IIIrd-Ist millennia BC.

Hugues Pessin

CNRS-UMR 7041, Archéologies et Sciences de l'Antiquité,
21, allée de l'Université, F- 92023 Nanterre cedex, France.
E-mail: huguespessin@hotmail.com

Abstract

The analysis of charcoal from the site of Tell Shiukh Fawqani, situated on the Euphrates in the extreme north of Syria, has provided evidence for wood exploitation during the Early Bronze Age I-II (3200-2600 BC), the Early Bronze Age IV (2300-2000 BC), the Late Bronze Age II (1300-1200 BC), the Iron Age I (1100-900 BC) and the Iron Age III (VIIth century BC). The inhabitants of the site exploited essentially the gallery forest along the Euphrates and a decidous woodland dominated by deciduous oak (*Quercus* decid.). They also appear to have imported pine timber (*Pinus* sp.). The results from Tell Shiukh Fawqani allow us to characterise the species used for timber and for domestic and industrial firewood.

Key-words: *Syria, Euphrates, wood exploitation, charcoal, Bronze Age, Iron Age.*

Résumé

L'étude anthracologique du site de Tell Shioukh Faouqâni, situé dans la région du Moyen-Euphrate à l'extrême nord de la Syrie, a fourni des informations concernant l'exploitation du bois durant le Bronze Ancien I-II (3200-2600 avant J-C), le Bronze Ancien IV (2300-2000 avant J-C), le Bronze Récent II (1300-1200 avant J-C), l'âge du Fer I (1100-900 avant J-C) et l'âge du Fer III (VIIe siècle avant J-C). Les occupants du site ont essentiellement exploité la ripisylve se trouvant le long de l'Euphrate et une formation forestière décidue dominée par le chêne à feuillage caducifolié (*Quercus* f. c.). Ils avaient également recours à l'importation de pin (*Pinus* sp.). Les résultats obtenus pour Tell Shioukh Faouqâni nous permettent de caractériser les essences utilisées pour le bois de construction, ainsi que pour le combustible domestique et industriel.

Mots-clés : *Syrie, Euphrate, exploitation du bois, anthracologie, Age du Bronze, Age du Fer.*

Introduction

The anthracological research in the region of the middle Euphrates has seen an important development during the last few years. The studies concern essentially the Epipalaeolithic and Neolithic aceramic periods (Natufian, PPNA, PPNB) and show the exploitation of two vegetation communities: the gallery forest along the Euphrates and the steppe-forest consisting of *Pistacia-Amygdalus* and low frequencies of deciduous oak (Roitel, 1997; Van Zeist and Bakker-Heeres, 1984).

No anthracological data are yet available from the middle Euphrates region concerning the Chalcolithic, Bronze Age, Iron Age, Classical and Medieval periods. However, charcoal from these periods has been studied from others nearby regions, for example at the site of Asvan, situated on the Turkish Euphrates, approximately 250 km from Tell Shiukh Fawqani (Willcox, 1974, 1992). The results show the exploitation of major forest species, including deciduous oak, pistachio (*Pistacia* sp.), juniper (*Juniperus* sp.), hawthorn (*Crataegus* sp.) and maple (*Acer* sp.) in the earlier periods (Chalcolithic and EBA). In the later periods (Classical and Medieval periods), the progressive exploitation of spiny, riverine and cultivated species can be seen, showing the deforestation of the area of Asvan.

The study of Tell Shiukh Fawqani is part of a larger program of rescue excavations carried out in the Tishrin dam area (fig. 1). The charcoal analysis carried out at the site allows us to obtain for the first time results concerning the Late Uruk, the Early Bronze Age I-II (EBA I-II), the Early Bronze Age IV (EBA IV), the Late Bronze Age II (LBA II), the Iron Age I (Iron I) and the Iron Age III (Iron III) for the middle Euphrates region.

The main objectives of this article are to identify the principal vegetation communities exploited by the inhabitants of Tell Shiukh Fawqani and to determine for what purposes the wood was used.

The site

The site is a typical Middle Eastern tell covering a surface of approximately 6 ha and rising 35 m above the surrounding plain which in its turn is situated at *c*. 325-350 m a.s.l. The climate of the region is continental Mediterranean with hot and dry summers and cold and harsh winters. Mean average rainfall varies between 250 and 350 mm/year. Both diurnal and annual temperatures show great variation.

From a phytogeographical point of view, the region belongs to the Mesopotamian province of the Irano-Turanian region (Zohary, 1973). The vegetation today can be considered as degraded steppe with perennials such as *Artemisia herba-alba, Noaea mucronata, Phlomis* sp., *Ephedra* sp. and a large variety of small short lived annuals including many grasses. Hawthorns, hackberries (*Celtis* sp.), *Lycium* sp. and *Pistacia palaestina* occur very rarely on the cliffs along the Euphrates. On the flood plain we find remnants of a gallery forest of *Populetum euphraticae* type dominated by *Populus euphratica, Salix* sp., *Tamarix* sp., *Typha* sp. and *Phragmites australis*.

Excavations were carried out at Tell Shiukh Fawqani between 1994 to 1998 by the Groupe International de Recherches Archéologiques (G.I.R.A.), consisting of a joint French-Italian and Syrian team, directed by L. Bachelot (CNRS, France) and F. Mario Fales (University of Udine, Italy). Since 1999 a French-Syrian team directed by L. Bachelot conducts the excavations. Five areas have been opened: D, E, F, G and H (fig. 2). They revealed levels from the Chalcolithic until the Medieval period, with the main levels dating to the Bronze and the Iron Ages (IIIrd-Ist millennia BC) (Bachelot, 1996, 1999). Area D was identified as a residential quarter dated to the EBA I-II, but remains dated to the Late Uruk were also found. Area E contains a collective burial (EBA IV), the remains of one or two buildings destroyed by fire (LBA II) and one part of a residential quarter (Iron I). The Iron III is represented in the areas

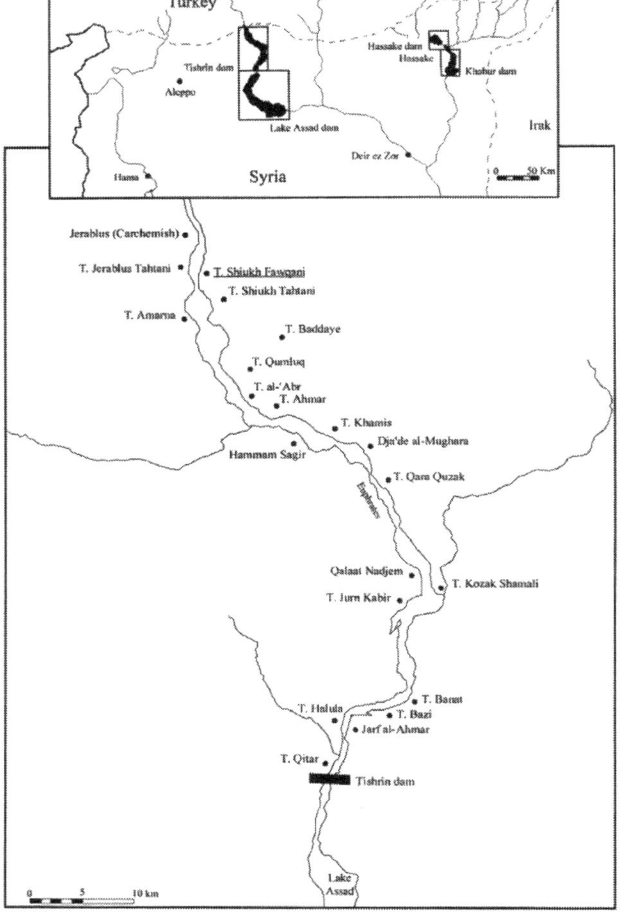

Figure 1: Map of the Tishrin dam area (from Bachelot, 1999).

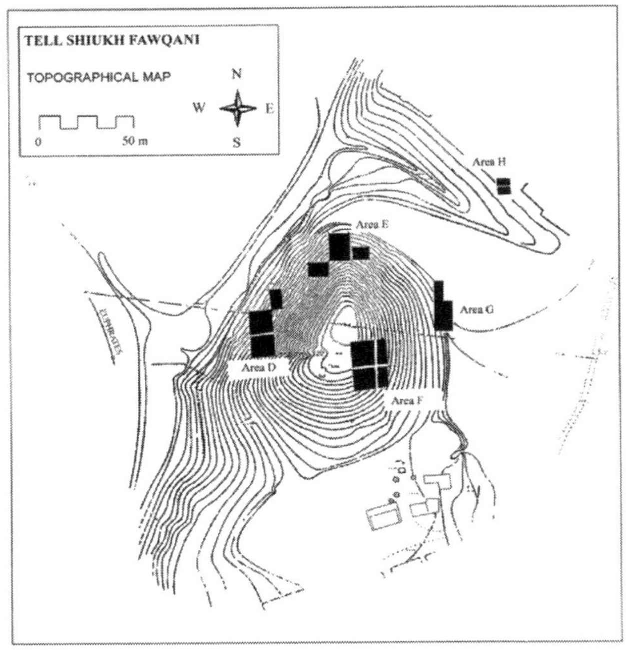

Figure 2: Topographical map of the site (from Bachelot, 1999).

G and H by an industrial quarter and a necropolis respectively. Area F, containing a residential quarter, also presents levels dated to the Iron III as well as to the Roman, Byzantine and Islamic periods.

Material and methods.

Thirty-four samples from thirty different loci situated in the areas D, E and G have been studied. They were recovered either by manual sampling or by the flotation of a little more than 1500 litres of sediment. The charcoal fragments were in most cases (26) dispersed in the archaeological sediments but a few samples (8) came from concentrated contexts such as burnt buildings and ovens.

The charcoal fragments were fractured by hand and observed in three sections (transversal, tangential and radial) using a reflected light microscope with dark and light fields. The identifications were made with the help of wood anatomical atlases (Fahn, *et al.* 1986; Schweingruber, 1990) and a reference collection. Identifications were carried out at the Institut de Préhistoire Orientale at Jalès and at the Maison de l'Archéologie et de l'Ethnologie at Nanterre, under the direction of George Willcox and Margareta Tengberg.

Results

The analysis of 3164 charcoal fragments has allowed the identification of 24 woody taxa (fig. 3). Three different vegetation types, exploited by the inhabitants of Tell Shiukh Fawqani can be distinguished from the charcoal diagram: the gallery forest, the deciduous forest and coniferous woodlands.

Taxa forming the gallery forest are Euphrates poplar (*Populus euphratica*), others poplars and willows (*Populus* sp., *Salix* sp., Salicaceae), tamarisk (*Tamarix* sp.), ash (*Fraxinus* sp.), reeds (cf. *Phragmites*), monocotyledons (Monocotyledonae), elm (*Ulmus* sp.), alder (*Alnus* sp.) and plane tree (*Platanus orientalis*). Only remnants of this type of gallery forest exist today in the region of the middle Euphrates, with the only species being Euphrates poplar, others poplars, willows, tamarisk and reeds.

The deciduous woodland is represented in the charcoal assemblages by deciduous oak, pistachio, almond (*Amygdalus* sp.), juniper, *Prunus*-type species, Pomoideae-type species, Christ-thorn (*Paliurus spina-christi*) and chenopods (Chenopodiaceae). Today, relics of forest-steppe formations dominated by deciduous oak exist about 20 km further north in Turkey (Willcox, in this volume) and in the Djebel Sinjar (Iraq), east of Tell Shiukh Fawqani. A relic of forest-steppe dominated by pistachio and almond also grows on the Djebel Abd al Aziz, North East of the site. However the scarcity of current vegetation in the Near-East does not allow us to ignore the possibility of the presence of another type of deciduous woodland, the deciduous oak forest.

Coniferous taxa are present at Tell Shiukh Fawqani but in low number. They consist essentially of pine, one fragment of yew (*Taxus baccata*) having equally been identified. Pine and yew are not autochtone in the middle Euphrates region and were probably imported from the Taurus or other mountainous regions.

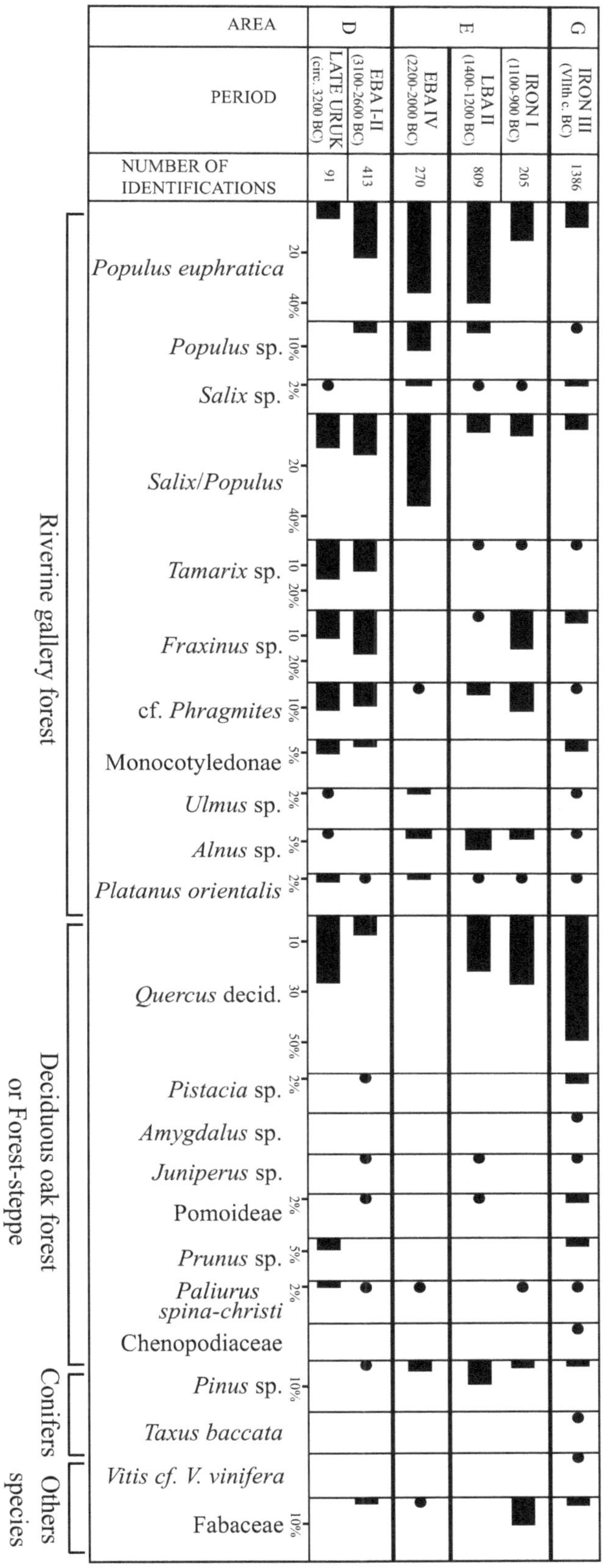

Figure 3: Preliminary anthracological diagram.

Finally, some shrubby species belonging to the Fabaceae family were identified as well as a few monocotyledons, cultivated or wild vine (*Vitis* sp.) and maple.

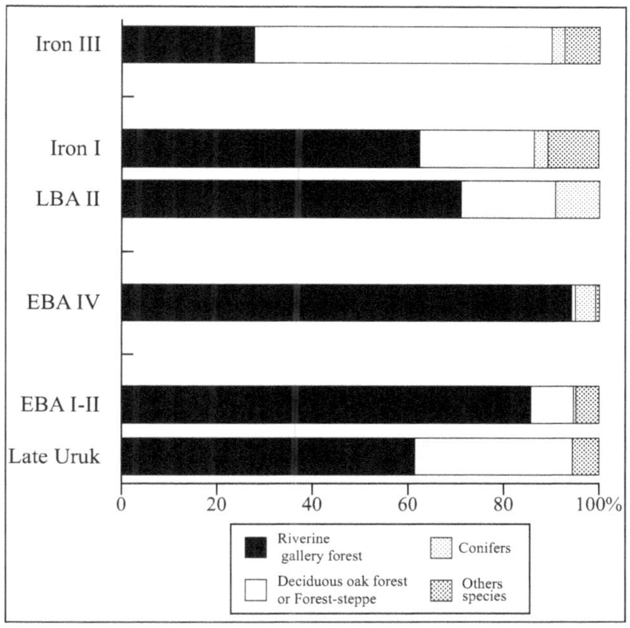

Figure 4: Proportion of vegetation communities for each area and each use. - a: domestic firewood (area D) - b: construction timber (area E) – c: industrial firewood (area G).

Finally, some shrubby species belonging to the Fabaceae family were identified as well as some probably cultivated vine (*Vitis* cf. *V. vinifera*).

Discussion.

The preliminary results allow us to define two main phases of exploitation of the vegetation (fig. 4). Firs, the exploitation of the gallery forest in the immediate proximity to the site during the Late Uruk (61.5 %), Bronze Age (EBA I-II 85.7 %, EBA IV 94.3 %, LBA 71.1 %) and the beginning of the Iron Age (Iron I 62.4 %). During the end of the Iron Age (Iron III) the main area of supply changed, from the gallery forest to the deciduous woodland dominated by deciduous oak (62.2 % for forest-steppe, 27.8 % for gallery forest). This change could be due to the gallery forest being over-exploited or because during the Iron III the needs in term of the quality of wood led the inhabitants to choose high quality combustion woods from the deciduous woodland. Area G (Iron III) revealed vestiges of iron smelting which may have required high calorific woods such as occur in the forest-steppe, for example deciduous oak, Rosaceae and pistachio. Further analysis of samples from area G and samples from area F (Iron III) may bring more light on this problem.

According to the archaeological contexts, different uses of wood could be identified: domestic firewood, industrial firewood and building timber.

In the residential quarter of area D (EBA I-II), the charcoal dispersed in the sediments most likely represent domestic firewood. It consists essentially of gallery forest species (85.7 % of identified fragments) (fig. 4) that were probably exploited in the immediate proximity to the site, *i.e.* along the Euphrates. A few examples of deciduous woodland species were also noted and consist mainly of deciduous oak. The levels of area E dated to the Iron I show also the use of gallery forest species as domestic firewood (62.4 % of identified fragments) (fig. 4). Some deciduous woodland species were used during this period, mainly deciduous oak.

In area G, identified as an industrial quarter specialised in metallurgical activities and dated to the Iron III, the majority (62.2 %) of the identified charcoal fragments belong to the deciduous woodland species (fig. 4), deciduous oak being the most important taxa. Gallery forest species are also present but are less frequent (27.8 %). The predominance of deciduous woodland species could here reflect a choice of wood for specific purposes as discussed above.

The burnt buildings in area E, dated to the LBA II, provide us with information on the wood used for constructions. Most of the fragments analysed (71.1 %) belong to the gallery forest species (Euphrates poplar, others poplars, willows, ash, alder and reeds). The deciduous woodland is also represented in the assemblages of this area (19.7 %) mainly by deciduous oak. Pine wood, represented by 9.3 % of the identified fragments, may have been used for construction purposes (fig. 4).

A collective burial dated to the EBA IV in area D is a good example of biases that can be introduced by specific archaeological contexts. Gallery forest species are over-represented in the assemblages (94.3 % of identified fragments) and deciduous oak is absent in the anthracological diagram (fig. 3, 4). Use of the wood found in the collective burial isn't known, but it seems very specific because Euphrates poplar, others poplars and willows represent the most part of the fragments identified for this assemblage (*c.* 72 %). Such a proportion and absence of deciduous oak certainly indicate a bias introduced in the data by the context of sampling.

Most of the wood used at Tell Shiukh Fawqani seems to have been exploited in the vicinity of the settlement, mainly in the gallery forest along the Euphrates. Even though the deciduous woodland is not present in the surroundings of the site today it is possible that this vegetation type had a much wider distribution in the past, as suggested by the presence of relict formation less than 20 km to the North.

A small proportion of the charcoal fragments represents coniferous wood, essentially pine. This wood could have been floated down the Euphrates from the Taurus or it could have been brought by land. The importation of wood is attested in texts from Mari and also from Assyria dating respectively to the IInd and the Ist millennia BC (Kupper, 1992; Postgate, 1992). Both land and river routes are mentioned for the importation of wood. From the texts, the Euphrates was the main river for the transport of the wood, though others rivers such as the Tigris and the Habur were used. Animal transport was used for land routes, but apparently over much shorter distances (Michel, 1996). Naturally floated drift wood is another possibility for the Tell Shiukh Fawqani finds and this will be the subject of further investigations.

Conclusion.

Preliminary charcoal analysis of Tell Shiukh Fawqani raises some questions about wood identified on the site. Is the deciduous woodland, a deciduous oak forest or a forest-steppe? Can the presence of pine, not currently present in the region, be explained precisely? Is it originate from the moutains *i.e.* allochtonous to the middle Euphrates and imported to the site or is it a reconquest species and present near the site? Is the grape vine the only species cultivated on the site? Are pistachio and almond also cultivated or are they wild? In the last case, these taxa belonged probably to the deciduous woodland or steppe-forest.

To answer to these questions, the charcoal analysis of Tell Shiukh Fawqani and the Middle Euphrates region will be extended, chronologically and spatially.

References

BACHELOT L. 1996 - Burmarina. Découverte d'une cité araméenne. *Archéologia*, 323, p. 34-39.

BACHELOT L. 1999 - Tell Shioukh Faouqâni (1994-1998). In Archaeology of the Upper Syrian Euphrates. *The Tishrin Dam Area* (OLMO LETTE G. del, MONTERO FENOLLOS J. L.). Barcelona, p. 143-162.

FAHN A., WERKER E., BASS P. 1986 - *Wood Anatomy and Identification of Trees and Shrubs of Israel and Adjacent Regions*. Jerusalem, 221 p.

FREY W., KÜRSCHNER H. 1991 - Die Aktuelle und Potentialle natürliche Vegetation im Bereich des Unteren Habur (Nordost-Syrien). In: *Die rezente Umwelt von Tall Seh Hamad und daten zur Umweltrekonstruktion der assyrischen Stadt Dur-Katlimmu (KÜHNE)*. Berlin, p. 87-103.

KUPPER J.-R. 1992 - Le bois à Mari. *Bulletin on Sumerian Agriculture*, VI, p. 163-170.

MICHEL C. 1996 - Le commerce dans les textes de Mari, In: *Amurru 1, Mari, Ebla et les Hourrites, dix ans de travaux* (DURAND J.-M.). Actes du colloque international, Paris (mai 1993), p. 3858-426.

POSTGATE J. N. 1992 - Trees and timber in the assyrian texts. *Bulletin on Sumerian Agriculture*, VI, p. 177-192.

ROITEL V. 1997 - *Végétation et action de l'homme du Natoufien au Néolithique acéramique dans le Haut-Euphrate syrien*. PhD Thesis, University of Montpellier II.

SCHWEINGRUBER F. H. 1990 - *Anatomy of European woods*. Stuttgart, 800 p.

VAN ZEIST W., BAKKER-HEERES J.A. 1984 - Archaeobotanical studies in the Levant 3. Late Plaeolithic Mureybet. *Palaeohistoria*, 26, p. 171-199.

WILLCOX G. 1974 - A history of deforestation as indicated by charcoal analysis of four sites in eastern Anatolia. *Anatolian Studies*, 24, p. 117-133.

WILLCOX G. 1991 - Exploitation des espèces ligneuses au Proche-Orient : données anthracologiques. *Paléorient*, 17, fasc. 2, p. 117-126.

WILLCOX G. 1992 - Timber and trees: ancient exploitation in the Middle East. Evidence from plant remains. *Bulletin on Sumerian Agriculture*, VI, p. 1-31.

ZOHARY M. 1973 - *Geobotanical foundations of the Middle East*. Stuttgart, 2 vol.

Vegetation history and wood exploitation in the Oman peninsula from the Bronze age to the Classical period

Margareta Tengberg

CNRS-UMR 7041, Archéologies et Sciences de l'Antiquité
21, allée de l'Université, F-92023 Nanterre cedex, France.
E-mail: m.tengberg@wanadoo.fr

Abstract

The results of the charcoal analysis carried out at six archaeological sites in the Oman peninsula have allowed a first synthesis on the vegetation history and the wood exploitation in this region, from around 3000 BC until the first centuries AD. While mangrove woodlands constituted the main source of wood along the coasts, the inhabitants of the inland sites exploited essentially open thorn woodlands dominated by *Acacia* spp. Both of these vegetation types have undergone changes in the past with the regression and even the disappearance of certain species and the possible causes of these changes, natural or anthropogenic, are discussed. The charcoal analysis has also allowed the identification of one cultivated species, the date-palm (*Phoenix dactylifera*), as well as several imported taxa.

Key-words: *Charcoal analysis, Oman peninsula, Vegetation change, Mangroves, Thorn woodlands.*

Résumé

Les résultats de l'analyse anthracologique de six sites archéologiques situés dans la péninsule d'Oman ont permis de réaliser une première synthèse sur l'évolution de la végétation et de l'exploitation du bois dans cette région, d'environ 3000 av. notre ère jusqu'aux premiers siècles de notre ère. Tandis que des formations de mangrove constituaient la principale source de bois le long des côtes, les habitants des sites de l'arrière-pays exploitaient essentiellement des formations dominées par des *Acacia* spp. Ces deux types de végétation ont connu des transformations dans le passé avec la régression voire la disparition de certaines espèces et les causes éventuelles de cette évolution, naturelles ou anthropiques, sont discutées. L'analyse anthracologique a également permis l'identification d'une espèce cultivée, le palmier dattier (*Phoenix dactylifera*), et de plusieurs taxons importés.

Mots-clés : *Anthracologie, Péninsule d'Oman, changement de végétation, mangrove, formation d'épineux.*

Introduction

During the last decades archaeological investigations have multiplied in eastern Arabia and particularly in the Oman peninsula, shedding new light on the most ancient history of this part of the Middle East, strategically situated between Mesopotamia, the Indo-Iranian world and southern Arabia. Some of the recent excavations have included the sampling of macrobotanical remains and the analysis of carbonised seeds and fruits as well as mud brick impressions has allowed us to obtain the first palaeoethnobotanical data (Biagi and Nisbet, 1992; Cleuziou, 1992; Cleuziou and Costantini, 1980, 1982; Pena-Chocarro and Barron Lopez, 1999; Willcox, 1995; Tengberg, 1998). Studies of charcoal have, however, been less numerous and dealt, so far, exclusively with shell midden sites situated on the coast of the Gulf of Oman (Biagi and Nisbet, 1992; Cartwright, 1998).

This article presents the results of the charcoal analysis carried out at six archaeological sites, presently situated in the United Arab Emirates and the Sultanate of Oman (fig. 1) and covering a chronological sequence of almost 3400 years, from the beginning of the 3rd millennium BC until the first centuries of our era. Two of these, Tell Abraq and ed-Dur, are located on the flat and low-lying coastal strip that stretches along the Arabian Gulf on the western side of the Oman peninsula. Three settlements, Muweilah, Mleiha and Hili 8, are situated in the hinterland, that is characterized by an alternation of inland salt flats (*sabkhas*), gravel plains (*hamadas* and *regs*) and sand deserts (*nafuds*). Finally, the site of Bat is located in one of the numerous *wadis* intersecting the Oman mountains, or the al Hajar, that stretch like a wide arc, from the cape of Ra's al-Hadd in the South-east to the Musandam in the North.

Several vegetation types are encountered in the studied area, according to the local edaphic and hydrological conditions. The coastal strip is characterised by remnants of mangrove swamps and halophytic communities dominated by various Chenopods. In the sand and gravel deserts of the interior, open xeromorphic dwarf shrublands constitute the most common vegetation type. The foothills and larger wadis of the al Hajar range are characterised by open to very open thorn woodlands with *Acacia* spp. as the main element. Finally, the higher parts of the mountains bear a vegetation of evergreen woodlands dominated by *Juniperus excelsa* subsp. *polycarpos* (Kürschner, 1998).

Most of these formations are today severely degraded, mainly due to the impact of human activities such as wood exploitation and pastoralism, during millennia and the original vegetation can only be guessed at from remnants of natural formations preserved in particularly inaccessible areas. The charcoal analysis constitutes therefore an impor-

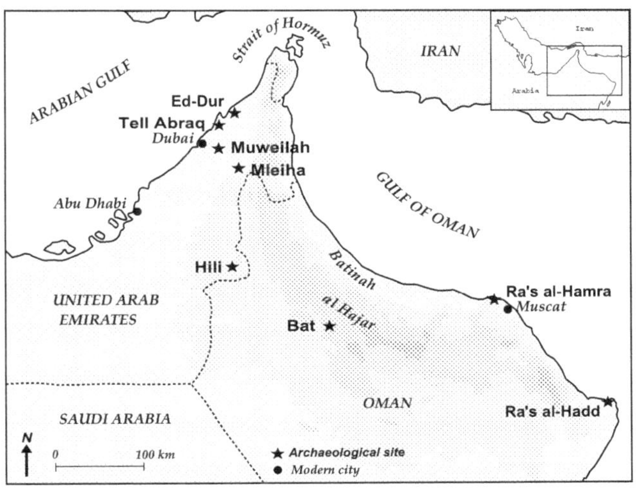

Figure 1: Map of the Oman peninsula with the location of sites mentioned in the text.

tant tool in the appreciation of the past vegetation landscapes.

A primary aim of this paper is to suggest a reconstruction of the vegetation communities that characterized the surroundings of the sites in the past and to determine their evolution through time. A second objective is to investigate, from a palaeoethnobotanical point of view, different aspects of the exploitation of wood resources. From where did ancient man supply his needs for wood? Was the exploitation limited to the immediate surroundings of the sites or was wood also included in the long distance trade that we know (from archaeological and epigraphic sources) existed on the Arabian Gulf and the Gulf of Oman in Antiquity?

Material and methods

The results presented in this article are based on the analysis of almost 7700 charcoal fragments. The number of fragments available for study varied however greatly from one period to another as well as from one site to another (tabl. 1), making direct comparisons difficult. Most of the charcoal pieces were collected manually by the archaeologists whenever concentrations of burnt material appeared during the excavation. They are thus not the fruit of a systematic sampling program, including for example flotation of large volumes of sediment. Moreover, the majority of the charcoal fragments come from specific, confined archaeological contexts such as ovens, hearths or conflagrated buildings and consequently they do not wholly fulfil the requirements for paleoecological representativity as defined by the Montpellier school (Chabal 1992, 1997). For these reasons, I have chosen to represent the various taxa by presence/absence only and not by their relative frequencies in the charcoal assemblages (tabl. 1).

The comparative tools used during the identification work consisted of a reference collection with modern wood samples from Pakistan and the Arabian peninsula as well as wood anatomical atlases (Fahn *et al.* 1986; Jagiella and Kürschner, 1987).

Results

The results of the charcoal analysis are presented in table 1. For the sake of comparison the coastal settlements are grouped together as well as those situated in the hinterland. The majority of the identified taxa are characteristic of the arid subtropical flora of Nubo-Sindian type that still dominates in this part of eastern Arabia (Zohary, 1973; Kürschner, 1998, p. 66-69). On the basis of their ecological requirements they can be divided into three main groups: taxa originate from coastal mangrove communities, taxa characteristic of open wood- and shrublands and taxa growing along intermittent watercourses in wadis and depressions. The charcoal analysis allowed further the identification of one cultivated species, the date palm (*Phoenix dactylifera*) as well as several taxa that are likely to have been imported from elsewhere. Each of these categories will be discussed in some detail below.

The evolution of the mangrove woodlands of the Arabian Gulf

The charcoal assemblages from the coastal sites, Tell Abraq and ed-Dur, are characterized by the presence of two taxa specific to coastal mangrove formations: *Avicennia marina* (fig. 2) and a species belonging to the Rhizophoraceae family (fig. 3) (Willcox and Tengberg, 1996). Due to a lack of reference material, the second has not yet been identified to the species level but can, on anatomical grounds, be ascribed to either the *Rhizophora* or the *Bruguiera* genus. At Tell Abraq these two taxa largely dominate the charcoal record throughout the occupation of the site (*c.* 2500-400 BC), representing between 70% and 97% of the identified taxa, depending on the period. The mangrove woodlands, probably growing in the vicinity of the site, seem thus to have constituted the main source of timber and fuel. The same situation is encountered at other coastal sites such as Ra's al-Hamra (Biagi and Nisbet, 1992) and Ra's al-Hadd 6 (Cartwright, 1998), where *Avicennia marina* is the most frequently identified species. Mangrove formations were also exploited at sites located on the island of Bahrain, further north in the Arabian Gulf. At Qal'at al-Bahrain charcoal fragments from *Avicennia marina* were noted for the Late Dilmun phase (*c.* 1000-500 BC) and at Saar a species belonging to the Rhizophoraceae family was identified from early 2nd millennium levels (Gale 1994).

Today, the mangrove communities of the Arabian Gulf and the Gulf of Oman are reduced to monospecific stands of *Avicennia marina*, distributed in a patchy way along creeks and lagoons (Deil, 1998). In most localities these mangroves, often representing mere remnants of larger communities in

		COASTAL SITES				SITES OF THE HINTERLAND								
	SITE	Tell Abraq			ed-Dur	Hili 8			Bat	Muweilah	Mleiha			
	LOCAL CHRONOLOGY	Umm an-Nar	Wadi Suq	Iron Age		I	II	III	Umm an-Nar	Iron Age II	PIR.A	PIR.B	PIR.C	PIR.D
	DATING	2500-2000 BC	2000-1300 BC	1300-400 BC	100-400 AD	3000-2750 BC	2750-2000 BC	2000-1700 BC	2500-2000 BC	800-600 BC	300-150 BC	150-0 BC	0-200 AD	200-400 AD
	NUMBER	943	1207	729	170	560	1177	50	435	870	130	90	315	1015
MANGROVE WOODLANDS	*Avicennia marina*	X	X	X	X	-	-	-	-	-	-	-	-	-
	Rhizophora/Bruguiera	X	X	X	X	-	-	-	-	-	-	-	-	-
OPEN WOOD- AND SHRUBLANDS	*Acacia* spp.	-	-	-	X	X	X	X	X	X	-	X	X	X
	Prosopis cineraria	-	-	-	X	-	-	-	X	X	X	X	X	X
	Ziziphus spina-christi	X	X	X	-	X	X	-	X	X	-	X	X	X
	Calligonum sp.	-	-	-	-	-	X	-	-	-	-	-	-	-
	Calotropis procera	-	-	-	X	-	-	-	-	-	-	-	-	X
	Chenopodiaceae	-	-	-	-	-	X	-	-	-	-	X	X	X
	Lycium sp.	-	-	-	-	-	-	-	-	-	-	-	-	X
WADI COMMUNITIES	*Acacia nilotica*	-	-	-	-	-	-	-	-	-	-	-	X	-
	Nerium oleander	-	-	-	-	-	-	-	-	-	-	-	X	X
	Tamarix spp.	X	X	X	-	-	X	-	X	X	-	X	X	X
CULTIVATED TAXA	*Phoenix dactylifera*	X	X	X	-	X	X	-	X	X	-	-	X	X
IMPORTED TAXA	*Dalbergia sissoo*	X	-	X	X	-	-	-	-	-	-	X	X	X
	Fraxinus sp.	-	-	-	-	-	-	-	-	-	-	-	-	X
	Pinus halepensis/brutia	-	-	-	X	-	-	-	-	-	-	-	-	-
	Platanus orientalis	-	-	-	-	-	-	-	-	-	-	-	-	X
	Rhizophora/Bruguiera (pro parte)	-	-	-	-	-	-	-	-	-	-	X	-	-

Table 1: Presence/absence of ligneous taxa identified at sites in the Oman peninsula.

the past, are seriously threatened by the rapid process of urbanisation and industrialisation, concentrated since a few decades to the coastal areas of eastern Arabia. Species belonging to the Rhizophoraceae family are altogether absent from these coasts today. Along the shores of the Red Sea, however, both *Rhizophora mucronata* and *Bruguiera gymnorrhiza* are recorded, growing together with *Avicennia marina* (Deil, 1998, p. 216). A distinctive zonation can be observed in these mixed communities with the Rhizophoraceae species occupying the seaward fringe and the border of natural channels where inundations occur daily while the more salt-tolerant *Avicennia marina* thrives on higher soils, affected by seasonal tides only.

The results from the charcoal analysis suggest that the mangrove communities of the Arabian Gulf were both more extensive and floristically richer in the past. This hypothesis is corroborated by the distribution pattern of *Terebralia palustris*, a mangrove-specific marine mollusc, on archaeological sites (Glover, 1998). Indeed, this species seems to have been more widespread and abundant in eastern Arabia, from the mid-Holocene until at least the beginning of our era. While present at several sites on the western coast of the Oman peninsula, for example at Tell Abraq and Shimal, during the Bronze Age *Terebralia palustris* is today entirely absent from the Arabian Gulf (Glover, 1998, p. 67).

According to our data, the Rhizophoraceae would have disappeared from the Arabian Gulf sometime between the first centuries of our era and the present. In fact, a progressive decrease of the relative frequency of the Rhizophoraceae is already noticeable in the charcoal record of Tell Abraq. From having represented almost 60% of the identified taxa for the Umm an-Nar (c. 2500-2000 BC) and Wadi Suq (*c.* 2000-1300 BC) periods respectively, the Rhizophoraceae constitute only 25% of the taxa noted for the Iron Age (*c.* 1300-400 BC). The frequency of *Avicennia marina* increases inversely from 10% in the Umm an-Nar period, to 19% in the Wadi Suq period and 72% in the Iron Age. With all the reserve due to the questionable sampling methods (see Material and methods), these results seem to reflect the progressive regression of Rhizophoraceae species in the Gulf, eventually leading to the reduced monospecific *Avicennia* stands that we find in the region today. Ecological, phytosociological and palynological studies carried out in mangrove systems in India and Pakistan have revealed a similar evolution (Caratini and Blasco, 1973; Tissot, 1973, 1987; Korangi Ecosystem Projects, 1991, 1992a, 1992b). Under ecological stress, due to both natural and anthropogenic factors, the mangrove woodlands undergo several distinctive degradation stages. "Ecologically sensitive" species, such as those belonging to the Rhizophoraceae and Sonneratiaceae families, are the first to disappear and they are substituted by the hardier and more salt-tolerant *Avicennia marina*. At term, the *Avicennia marina* communities are replaced in their turn by a halophytic dwarf scrub, essentially composed by Chenopods such as *Suaeda* and *Salicornia*. The causes of this evolution are multiple and imbricate, involving climatic, geomorphologic as well as biotic factors, as we will see below.

All while growing in the sea, the various mangal species require an input of sweet water and thus they prosper especially at the mouth of large rivers, for example in the deltas of the Indus and the Ganges rivers. In eastern Arabia, where permanent water courses lack altogether, the mangroves rely instead on runoff water from the land, induced by rainfall. A change in the hydrological regime will therefore have direct implications on the size and the species composition of the mangrove communities along the coasts. The mangrove woodlands of the Arabian peninsula were probably favoured by the more humid climate that prevailed in the region until the mid Holocene (McClure, 1988; Lézine, 1998; Sanlaville, 2000, 180-181). Inversely, the subsequent aridification of the climate may have contributed, at least in part, to their decline.

Sea level fluctuations and sedimentation processes would equally have affected the mangroves. The cut-off and filling up of lagoons and creeks would for example have lead to an increase of the salinity and consequently to the regression of certain salt sensitive species.

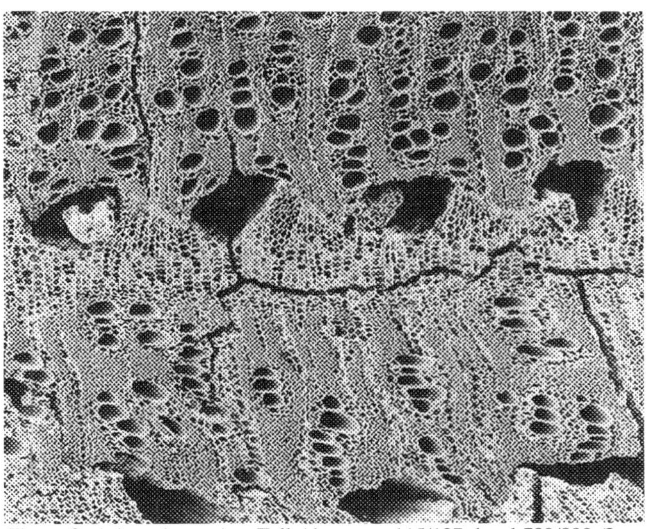

Figure 2: *Avicennia marina*, Tell Abraq, sq. 115/127, level 780/800 (Iron Age), transversal section, x80.

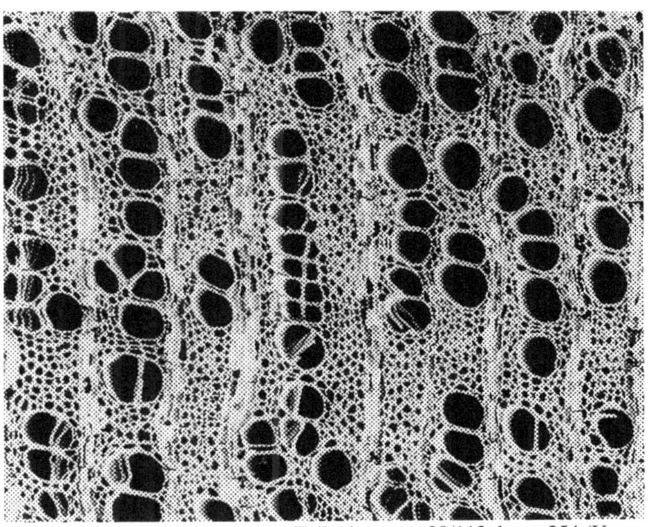

Figure 3: *Rhizophora/Bruguiera*, Tell Abraq, sq. 90/112, locus 254 (Umm an-Nar period), transversal section, x80.

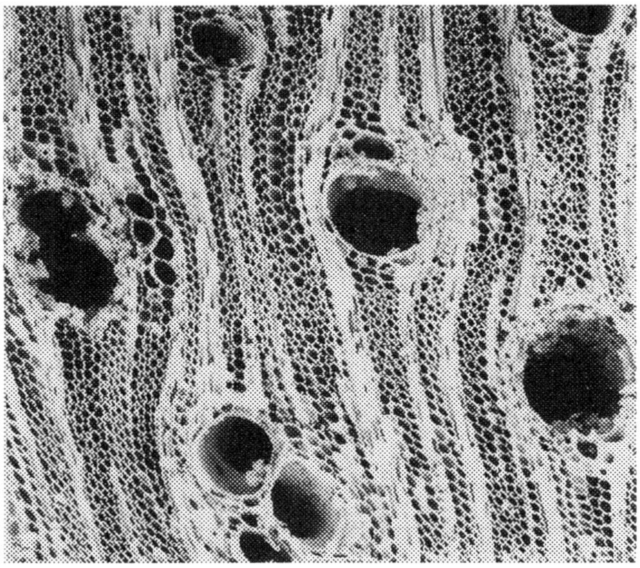

Figure 4: *Acacia* sp., Mleiha, locus 1094 (*c*. 200-400 AD), transversal section, x 80.

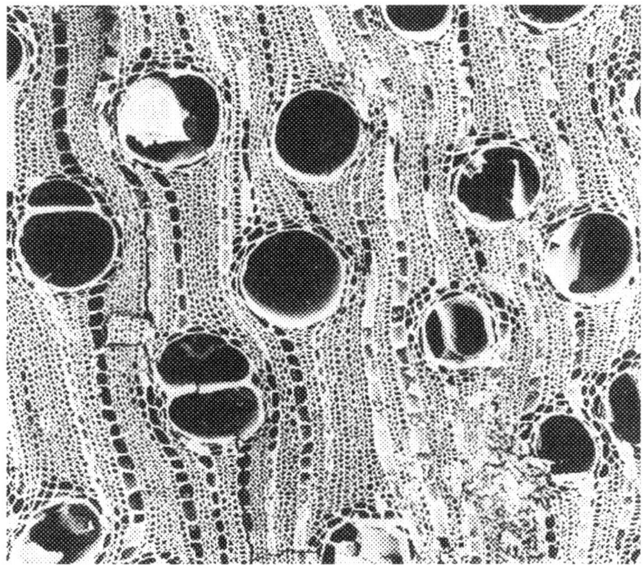

Figure 5: *Ziziphus spina-christi*, Mleiha, locus 617 (*c*. 0-200 AD), transversal section, x80.

Finally, wood-cutting and other human activities did certainly have a major impact on these fragile ecosystems. Along the sparsely vegetated coasts of eastern Arabia, the mangrove woodlands constituted a readily available source of timber and fuel and accordingly they were exploited for this purpose during millennia, as the results of the charcoal analysis testify. Even though other activities, such as the extraction of tanning agents (rich in both *Avicennia marina* and several Rhizophoraceae species) and the grazing of livestock, have not left traces in the archaeological record they are also likely to have taken place.

The vegetation of the interior deserts and plains

At the sites of the hinterland, whether situated on gravel or sandy soils, taxa characteristic of open wood- and shrublands dominate (tabl. 1). The association of *Acacia* spp. (fig. 4) with *Ziziphus spina-christi* (fig. 5) and, in most cases, *Prosopis cineraria* suggest the presence of xeromorphic thorn woodlands, also called pseudo savannas due to their resemblance to the East African "true" savannas, except for the herbaceous cover which is lacking in the more arid Nubo-Sindian counterpart. Depending on the nature of the substrata, these formations included various woody associates such as *Calligonum* spp., *Calotropis procera* and *Lycium* spp.

The results of the charcoal analysis show that this type of vegetation was generalised in the interior of the peninsula during the Bronze and Iron ages. It seems even to have extended until the immediate inland of the coasts as some of its indicator species are also recorded at Tell Abraq (*Ziziphus spina-christi*) and ed-Dur (*Acacia* spp., *Prosopis cineraria*). Today, the open woodlands are in most places reduced to a few scattered trees or they are altogether absent, as for example in the surroundings of Tell Abraq and Muweilah. Nevertheless, we can get an idea of what the past vegetation landscapes would have looked like by studying relict woodland formations, still preserved in the foothill zones and in the larger wadis of the Oman mountains as well as in the Batinah coastal plain (Ghazanfar, 1998, p. 184).

Like for the mangroves, the wider distribution of the *Acacia-Prosopis-Ziziphus* woodlands in the past was probably favoured by the more humid climate that prevailed in the Arabian peninsula during the Holocene optimum. The eastern Sahara seems to have known a similar situation at this period with the advance of the Sahelo-Sudanian savannas by several hundred kilometres to the North (Rithchie and Haynes, 1987; Neumann, 1989). At the subsequent onset of more arid conditions, around the mid Holocene, the limit of these formations progressively moved back southwards to attain its present position. A similar regression is likely to have occurred in the Oman peninsula. However, it is difficult to evaluate to what degree the present situation is the result of natural phenomena or the persistence of wood cutting and grazing during millennia. Indeed, from the archaeological record we know that the extraction of copper, in part destined for export, was a major economic activity in the Oman mountains, at least from the beginning

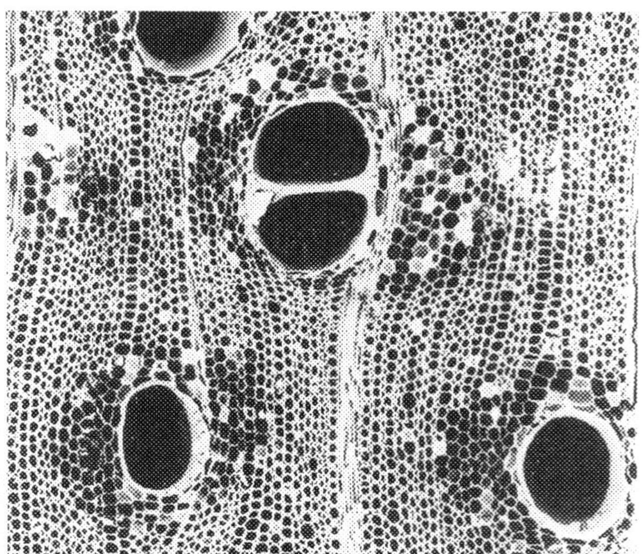

Figure 6: *Acacia nilotica*, Mleiha, locus 617 (*c*. 0-200 AD), transversal section, x80.

of the 3rd millennium BC (Hauptmann, 1985). The melting of copper ore on an industrial scale would have necessitated large quantities of wood, probably exploited in the local vegetation communities.

A few of the identified taxa preferably grow in periodically inundated wadis or depressions. The genus *Tamarix*, represented in the present-day Oman peninsula by several hygro- and halophytes, is recorded at most sites. *Acacia nilotica* (fig. 6) and *Nerium oleander* were identified at Mleiha, where they probably were favoured by the presence of alluvial soils (Dalongeville, 1999). *Acacia nilotica*, a true phreatophyte drawing water from the subsoil by a well-developed root system, is not present in the surroundings of Mleiha today but can be found elsewhere in the United Arab Emirates, for example in the north of Ra's al-Khaimah (Western, 1989, p. 69).

The date palm and the bustan agrosystem

The date palm, *Phoenix dactylifera*, is recorded at practically all the sites (fig. 7). Besides the charcoal fragments, carbonised remains of dates, both seeds (stones) and whole fruits, have frequently been noted by paleoethnobotanical studies (Cleuziou and Costantini, 1980, 1982; Pena-Chocarro and Barron Lopez, 1999; Tengberg, 1998). The ubiquity and the large quantity of date palm remains on the sites of the Oman peninsula, and in eastern Arabia in general, testify to the importance of this fruit tree in the regional economy. Moreover, the analyses of dental remains carried out at several sites indicate a high frequency of caries presumably due to a considerable consumption of sugar-rich dates (Højgaard, 1980).

The date palm did not only supply an important part of the diet, it constituted also the main element of the date palm garden or bustan. This agrosystem, characteristic of the arid zones of the Old World, allows the cultivation of numerous plants on a reduced surface all while economising on precious water resources. The date palms with their large crowns provide shade and shelter to other fruit trees as well

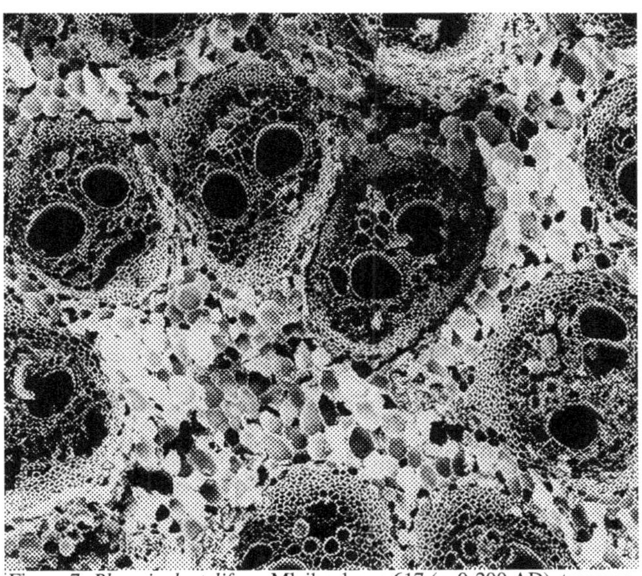

Figure 7: *Phoenix dactylifera*, Mleiha, locus 617 (*c.* 0-200 AD), transversal section, x80.

as various crops grown on ground level. The site of Hili 8 has provided the earliest example of this type of intensive agriculture (Cleuziou and Costantini, 1980, 1982; Cleuziou, 1997) that later became generalized in the region. At the sites of the Oman Peninsula the cultivation of date palms was often combined with that of cereals, mainly barley and naked wheat (Cleuziou and Costantini, 1980; Chocarro and Barron Lopez, 1999; Tengberg 1998).

The importation of wood

A certain number of the identified taxa would not, for ecological reasons, have grown in the surroundings of the sites and must therefore have been brought there, either from other parts of the Oman Peninsula or from more distant regions.

The earliest example of imported wood comes from Tell Abraq where the wood of two dagger handles, found in tombs from the Umm an-Nar period (*c.* 2500-2000 BC) was identified as *Dalbergia sissoo* or Pakistani rosewood (Tengberg and Potts, 1999). Presently, this tree is principally found together with other riparian species along watercourses in continental South Asia: in Afghanistan (Costantini, 1979, p. 104), Iran (Gershevitch, 1957, p. 318) and northern Pakistan and India (Puri, 1960, p. 214-216; Baquar, 1995, p. 282). A few specimens were also recorded from the al Hajar range in Oman (Mandaville, 1977) but it is uncertain if this isolated example represents introduced trees or if *Dalbergia sissoo* is actually indigenous to the Oman peninsula. In case the *Dalbergia* wood noted at Tell Abraq did not originate from the Oman mountains, it may have been imported from across the Arabian Gulf or the Gulf of Oman, i.e. from the Indo-Iranian borderland. We know, from charcoal analyses, that the species was present there in the past, for example in the Makran region of southwestern Pakistan (Tengberg and Thiébault, forthcoming) or at Harappa in the Indus Valley (ongoing research). Moreover, the archaeological record suggest close contacts at this period between the civilisations of the Arabian Gulf and those of ancient Iran and Pakistan.

The *Dalbergia sissoo* from Tell Abraq stands out as an isolated example of "exotic" wood during the Bronze age. In the Classical period, however, the examples of imported wood multiply and at Mleiha for instance this category represents not less than 16% of the identified taxa in the Late Pre-Islamic period D (*c.* 200-400 AD) (Tengberg, 1999). *Dalbergia sissoo* (erroneously identified as *Acacia albida* in Tengberg, 1999) is also present at this site, together with *Rhizophora/Bruguiera*, *Fraxinus* sp. (fig. 8) and *Platanus orientalis* (fig. 9). The Rhizophoraceae species is likely to have been brought from the coast, where it grew at this period as we have seen above. A more northerly origin is to be expected for the mesophilous ash tree (*Fraxinus* sp.) and plane (*Platanus orientalis*), attested in association with two public buildings at the site. Presently these species are found, most often along water courses, in northern Mesopotamia, Turkey, the Levant and the Zagros mountains of Iran (Zohary, 1973, p. 377). The wood of *Pinus halepensis/brutia*, identified at ed-Dur, is likely to have originated in one of the above mentioned regions too. Pine wood has been recorded at a few other sites in the Arabian

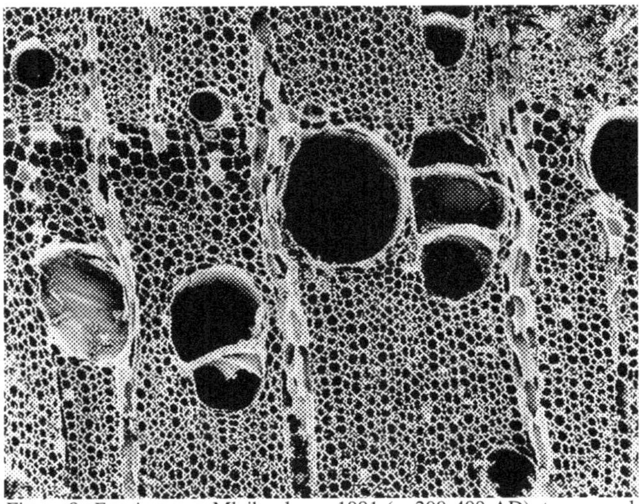

Figure 8: *Fraxinus* sp., Mleiha, locus 1091 (*c*. 200-400 AD), transversal section, x 80.

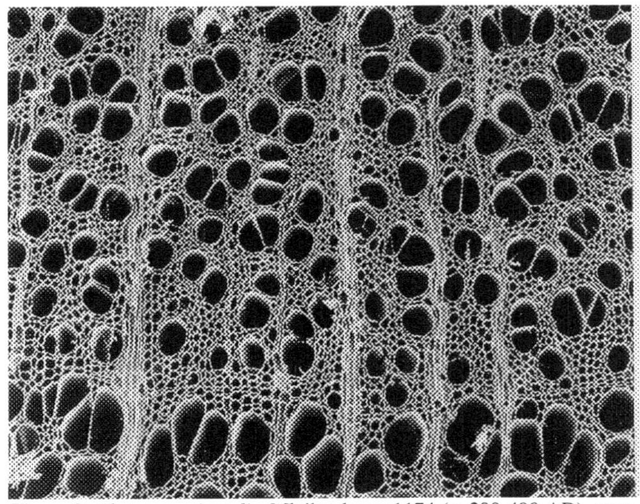

Figure 9: *Platanus orientalis*, Mleiha, locus 1174 (*c*. 200-400 AD), transversal section, x100.

Gulf: at Failaka in Kuwait (Willcox, 1990) and at Qal'at al-Bahrain (Tengberg and Lombard, 2002).

In general the imported wood seems to have been used for specific purposes, as in the case of the Pakistani rosewood at Tell Abraq (dagger handles) or the plane and ash tree at Mleiha (construction material for public buildings). However, it is not possible, on the basis of the charcoal analysis, to determine if the wood was imported to the sites in the form of unworked timber or as already manufactured items or, in the case of construction wood, as planks.

Conclusion

The charcoal analyses presented above have allowed a first synthesis on the evolution of the ligneous vegetation at one hand and on the exploitation and use of wood in the past at the other. The results are, however, based on a limited number of sites where, moreover, the sampling was not always satisfying. Therefore, further research in this region, based on extensive and systematic sampling, is necessary in order to broaden our knowledge on both palaeoecological and palaeoethnobotanical issues. Topics of special interest are the evolution of the mangrove woodlands in the Arabian Gulf, the use of wood resources in relation to specific craft activities, notably copper metallurgy, and the impact of such activities on the vegetation cover.

ACKNOWLEDGEMENTS

The author would like to thank the following foundations for their support: the Royal Swedish Academy of Sciences, the Wenner-Gren Foundations, Åke Wibergs stiftelse, Magnus Bergvalls stiftelse and Stiftelsen Lars Hiertas Minne.

References

BAQUAR S. R. 1995 - *Trees of Pakistan*. Their Natural History, Characteristics and Utilization. Karachi, Royal Book Company.

BIAGI P., NISBET R. 1992 - Environmental history and plant exploitation at the acermaic sites of RH5 and RH6 near the mangrove swamp of Qurm (Muscat - Oman). In Vernet J-L. (ed) Les charbons de bois, les anciens écosystèmes et le rôle de l'Homme. *Bulletin de la Société Botanique Française* 139/2-4, p. 571-578.

CARATINI C., BLASCO F. 1973 - Les mangroves de l'Inde. In Blasco F., Caratini C., Fredou A., Giresse P., Moguedet G., Tissot C., Weiss H. (eds) *Les rivages tropicaux. Mangroves d'Afrique et d'Asie*. Talence, Centre d'études de géographie tropicale (Travaux et documents de Géographie tropicale 8), p. 93-106.

CARTWRIGHT C.R. 1998 - Seasonal aspects of Bronze and Iron Age communities at Ra's al-Hadd, Oman. *Environmental Archaeology* 3, p. 97-102.

CHABAL L. 1992 - Représentativité paléo-écologique des charbons de bois archéologiques issus du bois de feu. In Vernet J-L. (ed) Les charbons de bois, les anciens écosystèmes et le rôle de l'Homme. *Bulletin de la Société Botanique de France* 139/2-4, p. 213-236.

CHABAL L. 1997 - Forêts et sociétés en Languedoc (Néolithique final, Antiquité tardive). L'anthracologie, méthode et paléoécologie. Paris, Editions de la Maison des Sciences de l'Homme. *Documents d'Archéologie Française*, 63, 188 p.

CLEUZIOU S. 1992 - Hili and the beginning of oasis life in Eastern Arabia. *Proceedings of the Seminar of Arabian Studies* 12, p. 15-22.

CLEUZIOU S. 1997 - Construire et protéger son terroir: les oasis d'Oman à l'âge du Bronze. *La dynamique des paysages protohistoriques, antiques, médiévaux et modernes*. Sophia Antipolis, Editions APDCA (XVIIème Rencontres Internationales d'Archéologie et d'Histoire d'Antibes), p. 389-412.

CLEUZIOU S., COSTANTINI L. 1980 - Premiers éléments sur l'agriculture protohistorique de l'Arabie orientale. *Paléorient* 6, p. 245-251.

CLEUZIOU S., COSTANTINI L. 1982 - A l'origine des oasis. *La Recherche* 13 (137), p. 1179-1182.

COSTANTINI L. 1979 - Wood Remains from Shahr-i Sokhta: a Source of Information for Ancient Environment and Technology in Protohistoric Sistan. *South Asian Archaeology* 1977, Naples, p. 87-121.

DALONGEVILLE R. 1999 - Mleiha : étude physique et paléoenvironnement. In Boucharlat, R., Mouton M. (eds), *Environnement, stratégies de subsistance et artisanat*. Lyon (Travaux de la Maison de l'Orient 29), p. 33-54.

DEIL U. 1998 - Coastal and Sabkha Vegetation. In Ghazanfar S., Fisher M. (eds) *Vegetation of the Arabian Peninsula*. Kluwer Academic Publishers (Geobotany 25), p. 209-228.

FAHN A., WERKER E., BAAS P. 1986 - *Wood Anatomy and Identification of Trees and Shrubs from Israel and Adjacent Regions*. Jerusalem, Israel Academy of Sciences and Humanities.

GALE R. 1994 - Charcoal from an Early Dilmoun settlement at Saar, Bahrain. *Arabian Archaeology and Epigraphy* 5, p. 229-235.

GERSHEVITCH I. 1957 - Sissoo at Susa (Opers. Yaka = *Dalbergia sissoo* Roxb.), *Bulletin of the School of Oriental and African Studies* XIX, p.317-320.

GHAZANFAR S.A. 1998 - Vegetation of the Plains. In Ghazanfar S., Fisher M. (eds) *Vegetation of the Arabian Peninsula*, Kluwer Academic Publishers (Geobotany 25), p. 175-190.

GLOVER E. 1998 - Mangroves, molluscs and man. Archaeological evidence for biogeographical changes in mangrove around the Arabian Peninsula. In Phillips C.S., Potts D.T., Searight S. (eds) *Arabia and its Neighbours. Essays on prehistorical and historical developments presented in honour of Beatrice de Cardi*. Brepols (Abiel II. New Research on the Arabian Peninsula), p. 63-78.

HAUPTMANN A. 1985 - *5000 Jahre Kupfer in Oman. Band 1 - Die Entwicklung des Kupfermetallurgie vom 3 Jahrtausend bis zur Neuzeit*. Bochum, Deutsches Bergbau-Museum (Der Anschnitt. Zeitschrift für Kunst und Kultur im Bergbau Beiheft 4).

HØJGAARD K. 1980 - Dentition on Umm an-Nar (Trucial Oman) 2500 BC. *Scandinavian Journal of Dental Research* 88, p. 355-364.

JAGIELLA C., KÜRSCHNER H. 1987 - *Atlas der Hölzer Saudi Arabiens. Die Holzanatomie der wichtigstens Bäume und Sträucher Arabiens mit einem holzanatomischen Bestimmungsschlüssel*. Wiesbaden, Dr Ludwig Reichert Verlag.

KORANGI ECOSYSTEM PROJECT 1991 - Possible effects of the Indus water accord on the Indus delta ecosystem. Meynell P. J. (Program coordinator) *Korangi Ecosystem Project Issues Paper* 1. The World Conservation Union.

KORANGI ECOSYSTEM PROJECT 1992a - Sea level rise - possible impacts on the Indus Delta, Pakistan. Meynell P. J. (Program coordinator) Korangi Ecosystem Project Issues Paper 2. The World Conservation Union.

KORANGI ECOSYSTEM PROJECT 1992b - Sustainable management of mangroves in the Indus Delta. Meynell P. J. (Program coordinator) *Korangi Ecosystem Project Issues Paper* 3. The World Conservation Union.

KÜRSCHNER H. 1998 - Biogeography and Introduction to Vegetation. In Ghazanfar S., Fisher M. (eds) *Vegetation of the Arabian Peninsula*. Kluwer Academic Publishers (Geobotany 25), p. 63-98.

LEZINE A-M. 1998 - Le Yémen désertique : une ancienne région humide. *Pour la Science*, août 1998, p. 54-59.

MANDAVILLE J.P. 1977 - Plants, The Scientific Results of the Oman Flora and Fauna Survey 1975. *Journal of Oman Studies* Special Report, p. 229-267.

Mc CLURE H.A. 1988 - Late Quaternary Palaeogeography and Landscape Evolution of the Rub' al Khali. In Potts D.T. (ed) *Araby the Blest. Studies in Arabian Archaeology*. Copenhagen, CNI Publications, p. 9-13.

NEUMANN K. 1989 - Vegetationsgeschichte der Ostsahara im Holozän, Holzkohlen aus prähistorischen Fundstellen (Mit einem Exkurs über die Holzkohlen von Fachi-Dogonboulo/Niger) In Kuper R. (ed) *Forschungen zur Umweltgeschichte der Ostsahara*. Köln, Institut für Ur-und Frühgeschichte der Universität zu Köln Forschungsstelle Afrika (Africa Praehistorica, Mongraphien zur Archäologie und Umwelt Afrikas), p. 13-181.

PENA-CHOCARRO L., BARRON LOPEZ E. 1999 - Plant remains from the site of Mleiha (U.A.E.). In Boucharlat, R., Mouton M. (eds), *Environnement, stratégies de subsistance et artisanat*. Lyon (Travaux de la Maison de l'Orient 29), p. 63-69.

PURI G. S. 1960 - *Indian Forest Ecology*. New Dehli & Calcutta, Oxford Book and Stationery Co.

RITCHIE J. C., HAYNES C. V. 1987 - Holocene vegetation zonation in the eastern Sahara. *Nature* 330/6149, p. 645-647.

SANLAVILLE P. 2000 - *Le Moyen-Orient arabe. Le milieu et l'homme*. Paris, Armand Colin (Collection U - Série géographie).

TENGBERG M. 1998 - *Paléoenvironnements et économie végétale en milieu aride. Recheerches archéobotaniques dans la région du Golfe arabo-persique et dans le Makran pakistanais (4ème millénaire av. notre ère - 1er millénaire de notre ère)*. Université Montpellier II, Ph.D. dissertation.

TENGBERG M. 1999 - L'exploitation des ligneux à Mleiha - étude anthracologique. In Mouton M. (ed) *Mleiha I - Environnements, stratégies de subsistance et artisanats*. Lyon (Travaux de la Maison de l'Orient Méditerranéen 29), p. 71-82.

TENGBERG M., LOMBARD P. 2002 - L'Homme et les plantes, de Dilmoun à Tylos - recherches archéobotaniques à Qal'at al-Bahreïn. *Paléorient*.

TENGBERG M., POTTS D.T. - 1999 gismes.ma-gan-na (*Dalbergia sissoo* Roxb.) at Tell Abraq. *Arabian Archaeology and Epigraphy*, 10, p. 129-133.

TENGBERG M., THIEBAULT S. (forthcoming) - Vegetation history and wood exploitation in Pakistani Baluchistan from the Neolithic to the Harappan period - the evidence from charcoal analysis. In Weber S., Belcher B. (eds) *Ethnobiology and the Indus Civilization: an interdisciplinary approach to Subsistence, Environment and Change*.

TISSOT C. 1973 - Palynologie et évolution récente de deux mangroves du Tamil Nadu (Inde). In Blasco F., Caratini C., Fredou A., Giresse P., Moguedet G., Tissot C., Weiss H. (eds) *Les rivages tropicaux. Mangroves d'Afrique et d'Asie*. Talence, Centre d'études de géographie tropicale (Travaux et documents de Géographie tropicale 8), p. 111-148.

TISSOT C. 1987 - Recent evolution of mangrove vegetation in the Cauvery delta a palynological study. *Journal of Marine biology*. 29 (1-2), Ass. India, p. 16-22.

WESTERN A.R. 1989 - *The Flora of the United Arab Emirates, an Introduction*. United Arab Emirates University

WILLCOX G. 1990 - The plant remains from hellenistic and Bronze Age levels at Failaka, Kuwait. A preliminary report. In Calvet Y., Gachet J. (eds) *Failaka. Fouilles françaises 1986-1988*. Lyon (Travaux de la Maison de l'Orient 18), p. 43-50.

WILLCOX G. 1995 - Some plant impressions from Umm an-Nar Island. In Frifelt K. (ed), *The Island of Umm an-Nar: the third millenium settlement*. Aarhus, Aarhus University Press (Jutland Archaeological Society Publications XXVI), p. 257-259.

WILLCOX G., TENGBERG M. 1996 - Preliminary report on the archaeobotanical investigations at Tell Abraq with special attention to chaff impressions in mud brick. *Arabian Archaeology and Epigraphy* 6, p. 129-136.

ZOHARY M. 1973 - *Geobotanical Foundations of the Middle East*. Stuttgart, Gustav Fischer Verlag/ Amsterdam, Swets & Zeitlinger, 2 vols.

Late Holocene south-eastern Brazilian fisher-gatherer-hunters: environment, wood exploitation and diet

Rita Scheel-Ybert
CNRS-UMR 5059, Paléoenvironnements, Anthracologie et Action de l'Homme
Institut de Botanique, 163 rue Auguste Broussonnet, 34090 Montpellier, France.
E-mail: scheel@crit.univ-montp2.fr

Abstract
Anthracological analysis of seven shell mounds from the south-eastern Brazilian coast (22°53'-22°57'S, 42°03'-42°33'W) provided palaeoenvironmental information and allowed evaluation of the relationship between human occupation and environment. In spite of some climatic fluctuations that affected the mangrove vegetation, no major changes of the vegetal ecosystem took place during the Late Holocene. This is probably due to the edaphic character of coastal environments. South-eastern Brazilian fisher-gatherer-hunters inhabited the "restinga" environment, close to mangrove and forest vegetation. Environmental stability has certainly been a decisive factor in the permanent settlement of these populations and has contributed to the maintenance of their socio-cultural system for more than 6000 years. Random gathering of dead wood provide most of the firewood for these populations. Nevertheless, *Condalia* sp. was specially selected for cultural reasons still impossible to determine. Gathering vegetables was certainly much more significant in the fisher-gatherer-hunters diet than is usually recognized. All the archaeological sites studied presented carbonized palm fruits, seeds and fragments of tubers, the latter found for the first time in the material from sambaquis.

Key-words: *Palaeoenvironment, palaeoethnology, charcoal analysis, shell mounds, wood exploitation, tubers, vegetation, Brazil.*

Résumé
L'analyse anthracologique de sept amas coquilliers du littoral de l'État de Rio de Janeiro, Brésil (22°53'-22°57'S, 42°03'-42°33'W), a permis de reconstituer l'environnement végétal de cette région et d'évaluer les interrelations entre occupation humaine et milieu. Malgré des oscillations climatiques qui ont affecté la végétation de mangrove, il n'y a eu aucun changement majeur dans l'écosystème végétal pendant toute la deuxième moitié de l'Holocène. Cela est lié, probablement, au caractère édaphique des écosystèmes côtiers. Les pêcheurs-cueilleurs-chasseurs du sud-est brésilien habitaient l'environnement de la "restinga", à proximité de la mangrove et des formations forestières. La stabilité de l'environnement a certainement été un facteur déterminant dans la sédentarisation des populations de pêcheurs-cueilleurs-chasseurs et a probablement contribué à la conservation de ce système socio-culturel pendant plus de 6000 ans. La collecte aléatoire de bois mort fournissait pour ces populations l'essentiel de l'approvisionnement en combustible. Toutefois, la rhamnacée *Condalia* sp. était spécialement recherchée pour des raisons culturelles qu'il est impossible de préciser pour l'instant. La cueillette de produits végétaux était beaucoup plus importante pour l'alimentation des sambaquiens qu'il n'est couramment admis. En effet, tous les sites étudiés contiennent des fragments de fruits de palmiers carbonisés, des graines et des résidus de tubercules de monocotylédones, ces derniers signalés pour la première fois dans le matériel provenant des sambaquis.

Mots-clés : *Paléoenvironnement, paléoethnologie, anthracologie, amas coquilliers, économie du bois, tubercules, végétation, Brésil.*

Introduction

Archaeologists are greatly interested in the life of Brazilian fisher-gatherer-hunter communities, their environment and diet. However, little is known about these subjects. Poor preservation of plant remains has hampered palaeovegetation studies. Desiccated plant remains are seldom recovered in sambaquis, but charcoal fragments are usually abundant and may represent an invaluable source of palaeoenvironmental information. Analysis of charcoal remains recovered in sambaquis allows the reconstruction of local woody vegetation, and provides palaeoethnological information concerning wood exploitation and diet.

The sambaquis, shell mounds constructed by fisher-gatherer-hunter populations that lived along almost all of the Brazilian coast from at least 7000 to 450 years BP (Gaspar, 1996) are the most ancient remains of occupation of the Brazilian coast. Sites are usually established near the sea, on the outer perimeter of lagoons, and in the vicinity of a wide variety of ecological habitats. Archaeological layers are constituted mainly of mollusk shells and fish bones, frequently alternating with sandy layers rich in organic matter and faunal remains (Gaspar, 1998). Archaeological remains associate habitation vestiges, waste deposits and burials.

This paper presents the results of the first anthracological study ever carried out in Brazil. Charcoal analysis of seven sambaquis from the southeastern coast of Rio de Janeiro State provided information on the palaeoenvironment in which Brazilian fisher-gatherer-hunter populations lived, on wood exploitation and diet. This work allowed evaluation of the relationship between human occupation and environment.

Study area

The seven sites studied are situated in a phytogeographical zone known as "The Lake Region" (fig. 1). The vestiges of an intense prehistorical occupation were found in this region, which presents a large diversity of plant associations. The most prominent associations can be divided into four distinct categories: (1) the "restinga" ecosystem, typical of the Brazilian coast and associated to sandy beach ridges. This group presents a mosaic of vegetation types with diverse physiognomies, varying from open communities of sparse herbaceous elements and shrubs, frequently arranged in thickets ("open restinga"), to dense evergreen forest ("restinga forest"); (2) the endemic xeromorphic formations on hillsides near the ocean in the Cabo Frio and Arraial do Cabo region. It includes the rupestrial vegetation typical of the abrupt rocky seashores, shrubby formations and xeromorphic forest; (3) mangrove, along the edges of rivers and lagoons, and (4) the relictual Atlantic Forest inland.

Particular geological, climatic and ecological features cha-

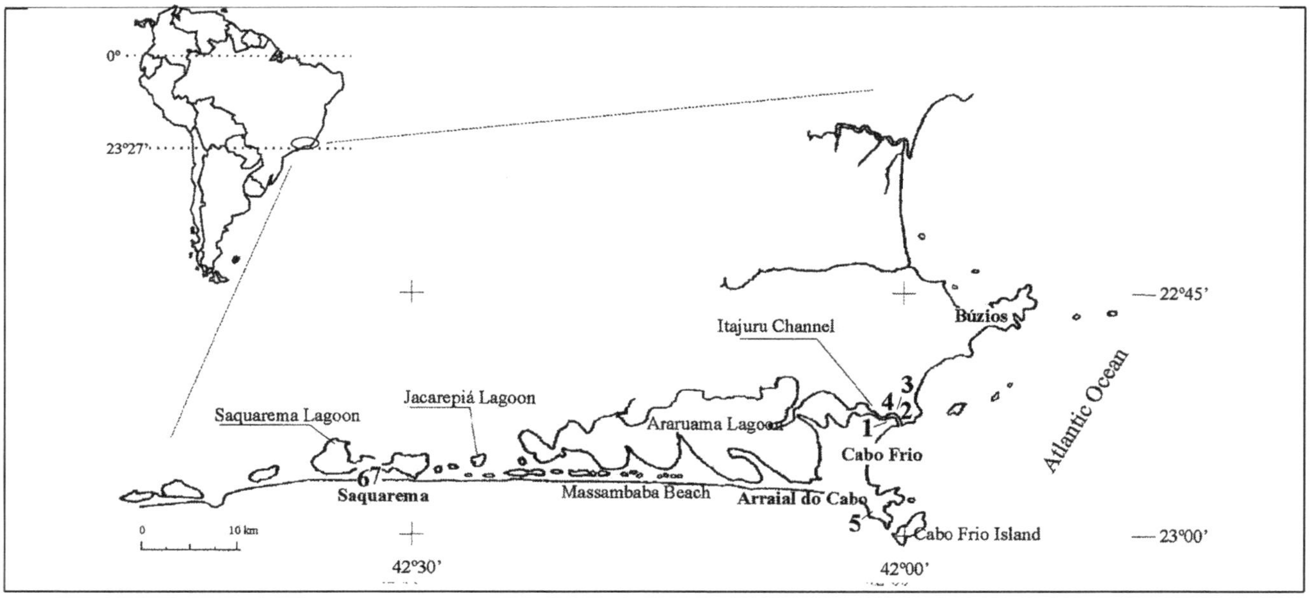

Figure 1: Geographical location of sites (after Scheel-Ybert, 2000). (1) Sambaqui do Forte; (2) Sambaqui Boca da Barra; (3) Sambaqui do Meio; (4) Sambaqui Salinas Peroano; (5) Sambaqui da Ponta da Cabeça; (6) Sambaqui da Beirada; (7) Sambaqui da Pontinha.

racterize this region. It has frequently been pointed out as an important area for palaeoenvironmental studies (Martin *et al.* 1988; Tasayco-Ortega, 1996).

The Sambaquis do Forte, Salinas Peroano, Boca da Barra and do Meio are situated at the Cabo Frio region (22°53'S, 42°03'W), on both sides of the Itajuru channel outlet, which connects the Araruama lagoon to the sea. The Sambaqui Ponta da Cabeça is situated at the Arraial do Cabo peninsula (22°57'S, 42°14'W), on a crystalline hill on the southwestern side of Massambaba beach. The Sambaquis da Beirada and da Pontinha are located at the Saquarema region (22°55'S, 42°33'W), on the back of the Pleistocene beach ridge between the Saquarema lagoon and the sea.

Detailed information concerning the climate, the geomorphology and the present vegetation of this area are presented elsewhere (Scheel-Ybert, 1998, 2000). Information on the archaeological context of each sambaqui may be found in specialized literature (Kneip, 1976, 1977, 1980, 1994; Heredia, 1984, 1986, 1987; Kneip *et al.* 1988, 1989, 1991; Kneip and Pallestrini, 1990; Scaramella *et al.* 1990; Gaspar, 1991; Franco and Gaspar, 1992; Gaspar and Scaramella, 1992; Tenório *et al.*, 1992; Kneip and Machado, 1993), or in the synthesis presented by Scheel-Ybert (1998).

Material and methods

Charcoal samples were collected from vertical profiles in the trenches left by previous archaeological excavation work. Two profiles were sampled at Sambaquis do Forte, Salinas Peroano and Boca da Barra, only one was sampled at Sambaquis do Meio, da Beirada and da Pontinha.

Each sample, consisting of a sediment layer of 2 m width, 50 cm depth and 10 cm thick, was taken with the assistance of a "sampling guide" (Ybert *et al.* 1997). When archaeological strata differed from artificial levels, we always proceeded as not to mix charcoal samples from different strata. Sediment was dry-sieved in the field using a 4-mm mesh. Residual charcoal was later recovered in the laboratory with a flotation machine (Ybert *et al.* 1997).

At Sambaqui Ponta da Cabeça, charcoal fragments were collected during the archaeological excavation by dry sieving of archaeological sediments. Plant remains were sorted by hand.

All the charcoal fragments over 4 mm were analyzed. In tropical regions, smaller fragments are normally impossible to identify, because they generally do not present enough significant characteristics to allow identification.

Charcoal samples were examined under a reflected light brightfield/darkfield microscope. Charcoal pieces were manually cut following the three wood sections. Systematic determination was obtained by comparing the anatomical structure observed with that of extant charred samples and with descriptions and photographs from reference books (Détienne and Jacquet, 1983; Mainieri and Chimelo, 1989; Metcalfe and Chalk, 1950; Record and Hess, 1943 etc.) and from numerous specialized papers on wood anatomy (see Gregory, 1994).

As wood anatomy of tropical plants is very poorly known, we have assembled a reference collection of charred woods containing over 1800 determined samples coming from wood collections donations and from field sampling. Charcoal identification was facilitated by the development of a program for computer-assisted identification specially conceived for charcoal analysis, coupled with a data bank of anatomical features from extant and ancient charcoal (Scheel-Ybert *et al.* 1998).

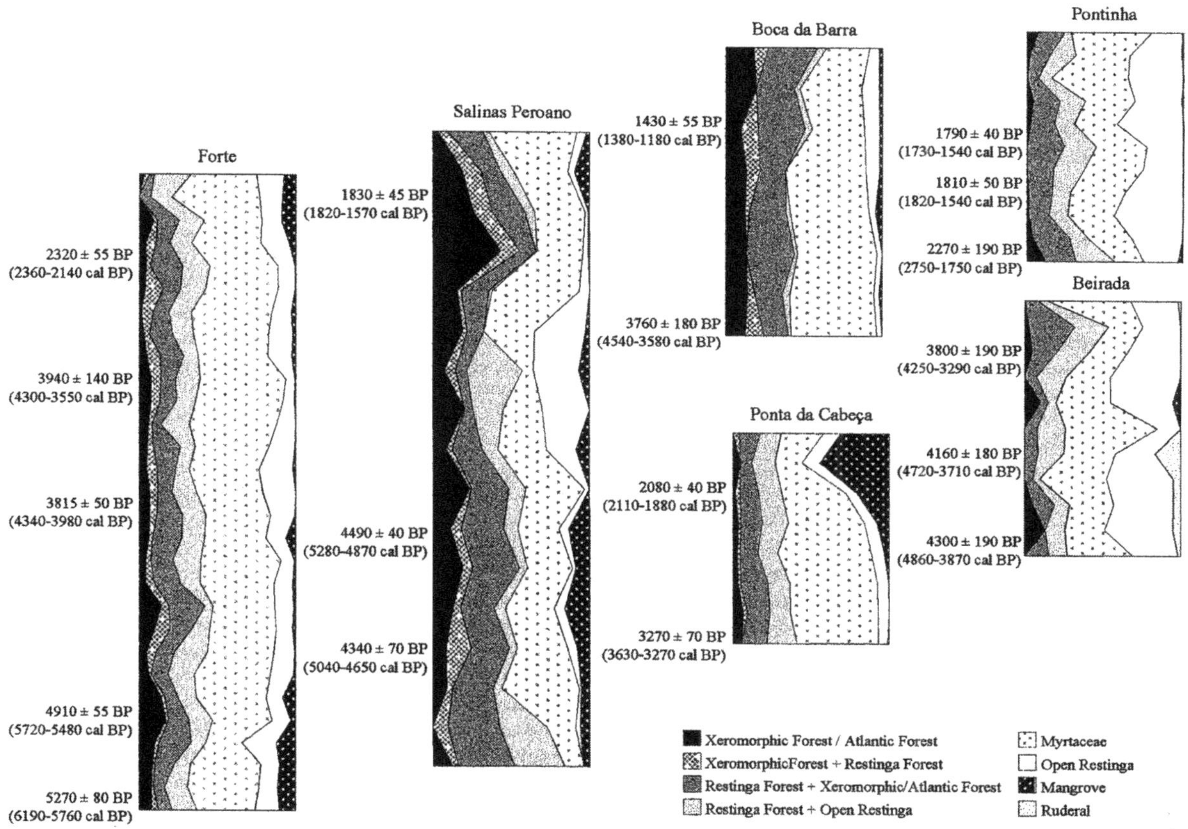

Figure 2: Summary charcoal diagrams from Sambaquis do Forte, Salinas Peroano, Boca da Barra, Ponta da Cabeça, da Beirada and da Pontinha (after Scheel-Ybert, 2000). Sambaqui do Forte - Nt: 8097; Nsp: 102. Sambaqui Salinas Peroano - Nt: 2052; Nsp: 59. Sambaqui Boca da Barra - Nt: 698; Nsp: 47. Sambaqui Ponta da Cabeça - Nt: 1956; Nsp: 58. Sambaqui da Beirada - Nt: 519; Nsp: 39. Sambaqui da Pontinha - Nt: 1621; Nsp: 54. (Nt: number of studied fragments; Nsp: number of identified taxons).

Results and Discussion

Palaeoenvironment

Late Holocene south-eastern Brazilian fisher-gatherer-hunters lived in the confluence of three major plant associations: restinga, mangrove, and forest formations such as xeromorphic and Atlantic forests.

The anthracological results are presented here as summary diagrams (fig. 2), where the different taxa have been grouped by plant formation. Complete anthracological diagrams were published previously (Scheel-Ybert, 2000).

All studied sites are characterized by a high floristic diversity. Each 10 cm artificial level presents, in general, more than 30 taxa. Sambaqui do Forte's entire sample presents more than a hundred taxa. This supports the hypothesis that archaeological charcoal corresponds to sampling (firewood gathering) in a relatively large area surrounding the sites, and to a lengthy temporal activity. These are essential premises to reliable palaeoecological interpretation based on archaeological charcoal (Chabal, 1992).

In the Cabo Frio region (Sambaquis do Forte, Salinas Peroano and Boca da Barra), restinga, mangrove and xeromorphic forest existed for at least 5500 yrs BP (fig. 2). Sambaqui inhabitants might also have access to the inland Atlantic Forest.

All anthracological samples are characterized by a strong predominance of Myrtaceae species. This family is currently present in different communities of Brazilian vegetation, but high percentages are typical for restinga formations. Open restinga is better represented in the Sambaqui do Forte, while forest formations are more important in Sambaquis Salinas Peroano and Boca da Barra. This is a consequence of geographical location. The first sambaqui is situated near the beach, in the phytosociological domain of open restinga, while the later ones are situated on the eastern margin of the channel, on crystalline hills dominated by the xeromorphic forest.

Only nine charcoal micro-fragments were retrieved in the Sambaqui do Meio. They were found at the base of this site, during the excavation work. Only four of these could be identified: two Myrtaceae species and *Condalia* sp. They show that restinga vegetation existed near the site shortly after 5200 BP, but a more extensive palaeoenvironmental reconstruction cannot be established.

At the Arraial do Cabo region (Sambaqui Ponta da Cabeça), open restinga formations were predominant from *c.* 3300 until after 2000 BP (fig. 2), but anthracological analysis shows that mangrove and xeromorphic forest were also present in the region. This agrees with what might be expected from the geographical location of this site, as well. Despite being situated on a crystalline hill at present, it is situated very near the beach, in a region characterized by a particularly dry climate due to an upwelling phenomenon centered in this area.

The identification of mangrove elements in this site is very important, because this vegetation, which was probably established on the south-eastern margin of the Araruama Lagoon, is currently absent from the Arraial do Cabo region. On the other hand, the exploitation of mangrove by Sambaqui Ponta da Cabeça inhabitants is attested for the first time. It corroborates the hypothesis that mangrove was almost always present in the site catchment area of sambaquiens (Tenório, 1996).

Open restinga was also predominant at the Saquarema region, during the occupation period of Sambaquis da Beirada and da Pontinha, *i.e.*, between 4300 and 3800 yrs BP and from *c.* 2300 until after 1800 BP (fig. 2). Low percentages of mangrove and forest taxa do not indicate that these formations were rare during this period. They might have been growing farther away from the sambaquis, where wood exploitation was less significant. The Atlantic Forest was probably situated on the northern side of the Saquarema lagoon, and the mangrove on its margins.

The slight increase of Atlantic Forest elements at the base of Sambaqui da Pontinha, around 2300 years BP, is not significant and cannot be interpreted as a consequence of environmental change. Oscillations in the relative percentages of different plant formation in Sambaqui da Beirada, as well as in some levels of Sambaqui Salinas Peroano (fig. 2) are not significant either. They are a consequence of the very low number of charcoal pieces analyzed for these levels.

The only significant variation on the charcoal diagrams concerns mangrove vegetation. In the Cabo Frio region, these variations might be related to climatic oscillations, essentially related to local raindrop, which provoked lagoon salinity variations (Scheel-Ybert, 2000). This interpretation is corroborated by the analysis of the curve of isotopic variation of carbonates in sediments of the Araruama lagoon (Tasayco-Ortega, 1996). Low values of $\delta^{18}O$ are interpreted to reflect a major influence of rainwater, while high values indicate increased salinity (fig. 3). Climate was more humid before *c.* 5000 yrs BP; after that, a drier period lasted until *c.* 2300 yrs BP. A brief rainy episode is recorded between 2300 and 2000 yrs BP, followed by another dry period that remained at least until the end of the occupation.

In the Arraial do Cabo region, two hypotheses may be proposed to explain the increase of mangrove elements in the upper levels: it may be due to a real increase of this vegetation in the environment, and/or it might be a consequence of a population increase in the sambaqui. Indeed, archaeologists consider the upper 30 cm of this site as the occupation apogee (Tenório *et al.* 1992). This might have provoked an expansion of the wood catchment area.

In spite of the climatic oscillations affecting mangrove vegetation, no other significant evidence of vegetation change is recorded during the Late Holocene.

Holocene palaeoenvironment on the Brazilian coast is poorly known. However, Bissa *et al.* (in press) found similar results in the coastal area of São Paulo State. The palynolo-

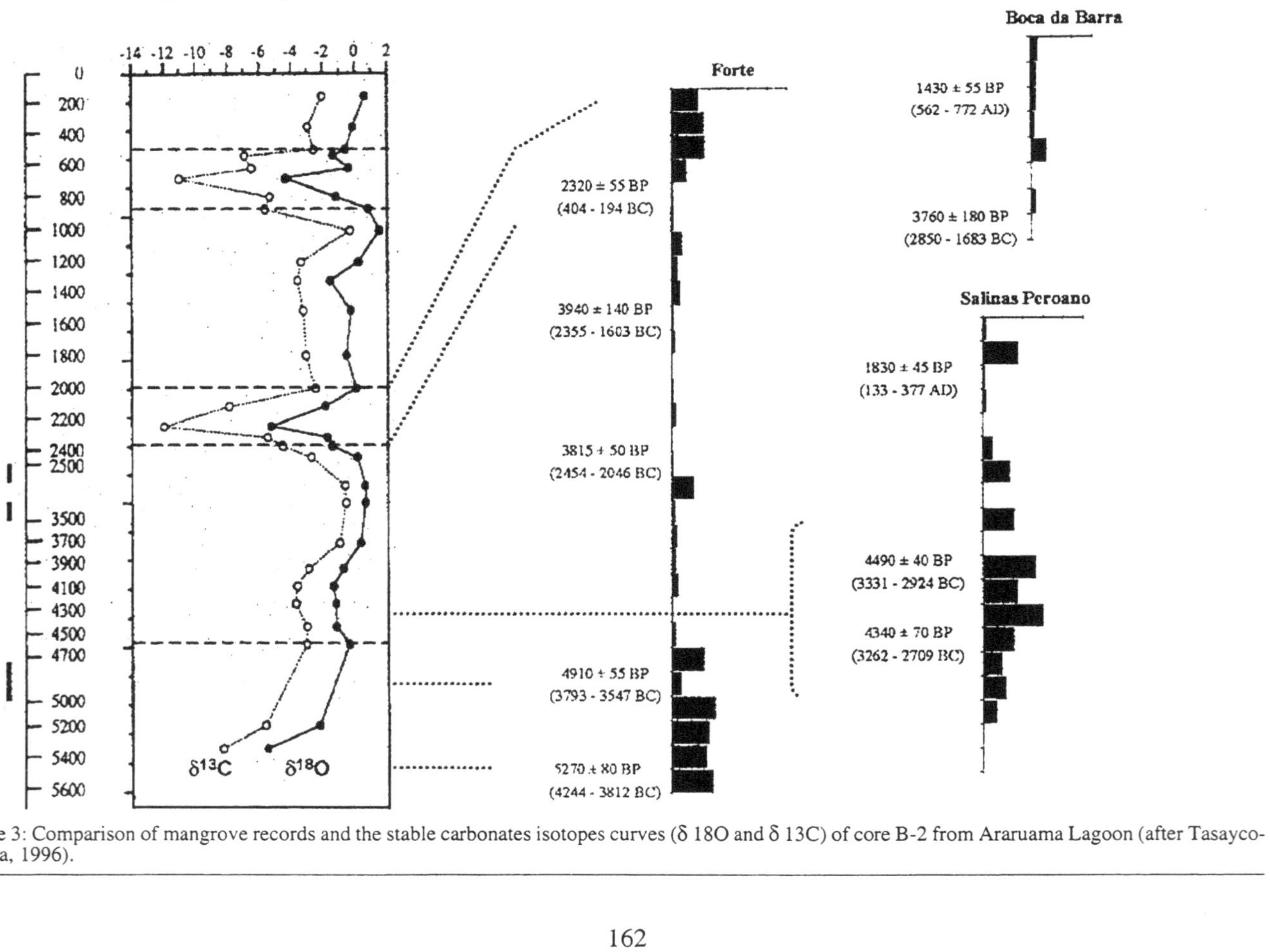

Figure 3: Comparison of mangrove records and the stable carbonates isotopes curves (δ 18O and δ 13C) of core B-2 from Araruama Lagoon (after Tasayco-Ortega, 1996).

gical analysis of a peat-bog core under flooded forest showed no variations in vegetation that could be attributed to climatic change.

The restinga ecosystem is related to sandy soils and to the geomorphologic nature of sandy beach ridges. This implies that it is essentially an edaphic formation. This is also the case of other plant formations typical to the Brazilian coast, such as xeromorphic and flooded forests. We believe that mainland coastal formations, with an important edaphic component, have a marked resilience to climatic change.

South-eastern Brazilian mainland coastal vegetation, specially the restinga and the xeromorphic forest, show surprisingly stability during the Late Holocene. The anthracological assemblage reveals no great variations through time, suggesting that vegetation was not greatly affected by climatic or by anthropogenic perturbations.

The local vegetation, similar to the present vegetation, probably began to suffer significant degradation only after the era of colonization, as a consequence of extractive activities, of the urban development on the coast and, more recently, of tourism.

Environmental stability had undoubtedly important consequences for prehistoric populations. We believe that this has been a decisive factor in the permanence and in the maintenance of a stable sociocultural system by the fisher-gatherer-hunters.

Wood exploitation

Great taxonomic diversity of the charcoal assemblage, either in the dispersed charcoal spectra (Scheel-Ybert, 2000) and in hearth concentrated charcoal samples (Scheel-Ybert, 1999), points out to a random gathering of firewood. Good correspondence between the charcoal assemblages and the present-day vegetation (Scheel-Ybert, 2000) also indicate that wood gathering, most likely for domestic fires, was not selective. Numerous charcoal fragments bearing traces of decay before charring and/or signs of saprophagous insect larvae attack suggest that dead wood was utilized (fig. 4).

However, the frequently high frequencies of *Condalia* sp, specially in the Cabo Frio and Arraial do Cabo regions, are difficult to explain by environmental criteria. This taxon is very important in most of the anthracological diagrams and, along with Myrtaceae, is the only one to be represented in all the studied sambaquis. However, *Condalia buxifolia*, typical to the restinga vegetation and the only species that occurs in the Brazilian coastal vegetation nowadays, is very rare (Silva and Oliveira, 1989).

This species might be especially abundant in the environment; but it implies that the earlier vegetation was quite different from that of the present. This hypothesis is refuted by the anthracological results. It is most probable that the species was not particularly frequent, but was selected for cultural reasons. This hypothesis is supported by several of this taxon's characteristics: its very dense wood is considered as an excellent combustible and is reported to produce a blue pigment; the root bark of some species may be used as soap or medicine; the plant bears edible fruits (Record and Hess, 1943).

A ritual, ceremonial or mystical utilization of this taxon is also possible. Heizer (1963) states that fire related ceremonies are very frequent in indigenous American societies. They frequently imply ritualized gathering of firewood, and species choice cannot be excluded. The fact that all *Condalia* fragments are vitrified might support this hypothesis, because it suggests its wood was always charred green (Scheel-Ybert, 1998).

The use of this species as a pigment source could also lead to charring of green wood. However, better knowledge of this species utilization and of wood treatment techniques, as well as of the vitrification phenomenon, is necessary to support this argument.

Diet

The great number of charred food remains found in all the studied sites suggests that plant gathering was particularly important in the fisher-gatherer-hunter diet. Until now, direct vestiges of this activity were seldom found in sambaquis. Plant consumption was testified to only by the presence of a

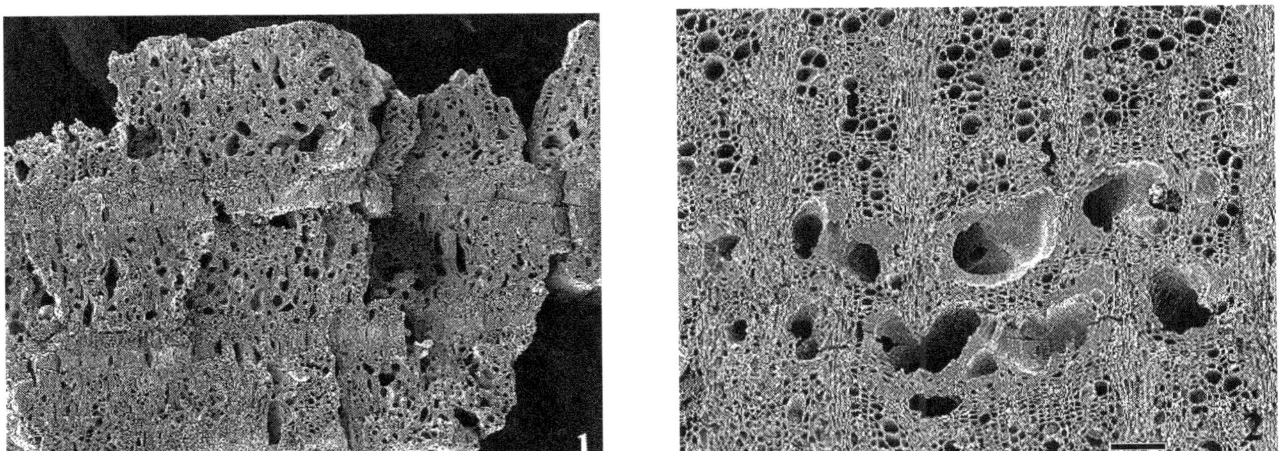

Figure 4: Charcoal fragments presenting signs of decay (1) and of saprophagous insect larvae attack (2) before charring.

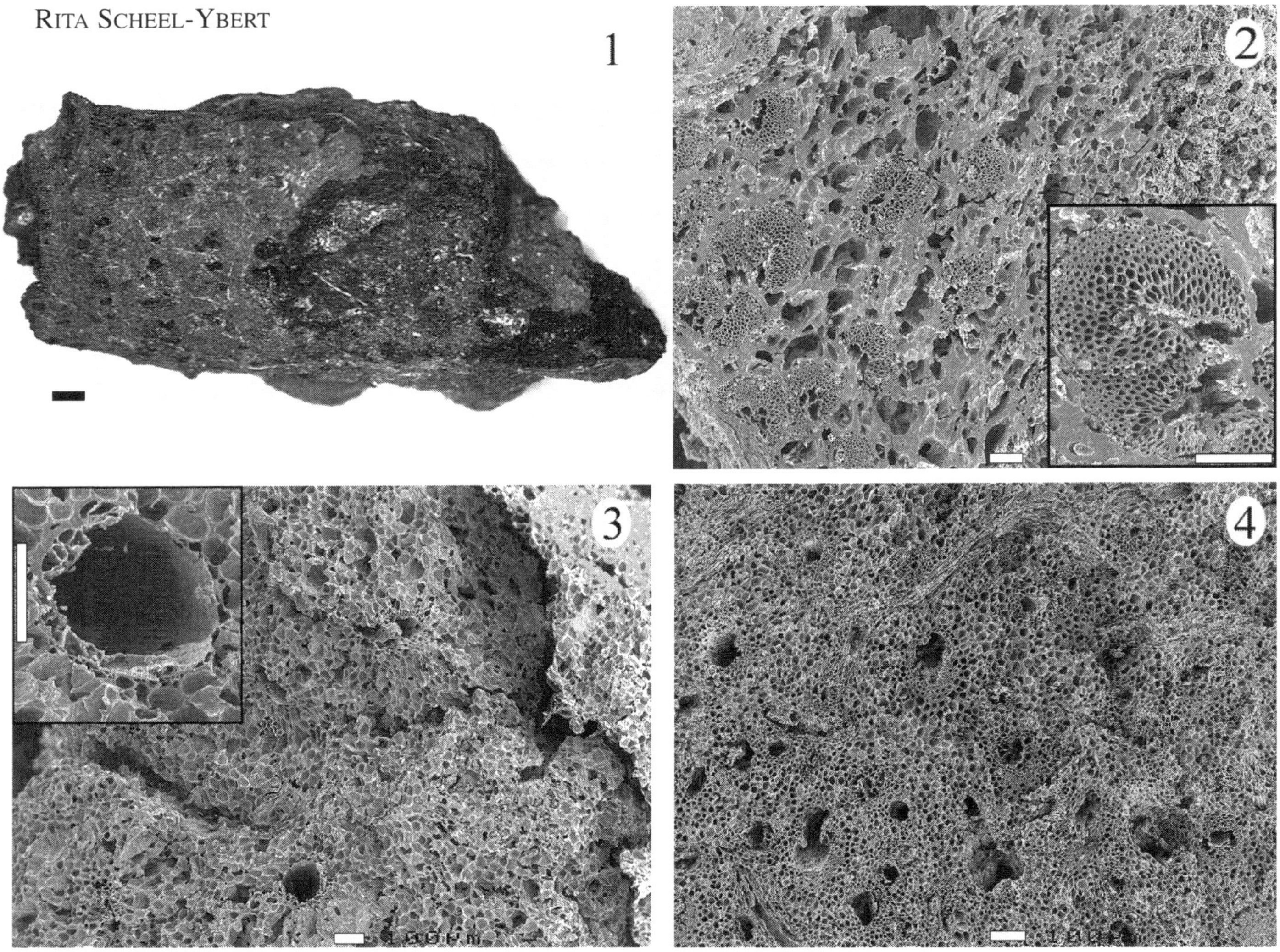

Figure 5: Tubers. 1, 2 - cf. Gramineae/Cyperaceae, macroscopic and SEM views. Sambaqui Salinas Peroano (1830 ± 45 yrs BP). 3 - *Dioscorea*. Sambaqui do Forte (3815 ± 50 yrs BP). 4 - unidentified monocotyledonous tuber. Sambaqui do Forte (*ca*. 4500 yrs BP). Black scale bar = 1 mm; white scale bar = 100 μm.

few seeds, of charred palm fruit nuts and of stone objects probably used in food preparation (Kneip, 1977, 1994; Gaspar, 1995; Tenório, 1991; Tenório *et al*. 1992 etc.).

We analyzed over 15500 charred fragments, 4% of which may be considered as food remains. Fragments of charred palm nuts, seeds and monocotyledon tuber parenchymatous remains are present on most archaeological levels. Although most of them could not yet be specifically identified, a large diversity of species was used. Some tubers were identified as Gramineae/Cyperaceae, others as *Dioscorea* (yams) (fig. 5). Some might belong to *Typha domingensis* (Scheel-Ybert, 2001).

Brazilian fisher-gatherer-hunters have long been considered almost exclusively as "shellfish eaters". Only recently fishing was recognized as predominant over shellfish gathering (Figuti, 1992, 1993, 1996), even when mollusk remains are apparently predominant in the archaeological stratigraphy. Although tacitly admitted, plant gathering is usually considered as a secondary activity, with negligible importance placed on diet.

Diets based on gathering shellfish, fishing or hunting leave many more remains on archaeological sites than plant gathering. As it is usually difficult to prove plant use, this activity is generally under-analysed and diets that leave more visible rests in the archaeological sediments are favored.

Palm nuts are frequently found in shell mounds, but seeds and tubers are seldom retrieved. Palm remains are most frequently from small fruits, seldom identified. Most of them are probably from *Astrocaryum*, *Bactris* and *Syagrus genera*.

Why, under tropical climatic conditions, is conservation of plant remains achieved almost exclusively by carbonization? It depends on whether or not the plant material is directly exposed to fire, intentionally or accidentally, before its preparation or consumption. Palm nuts, besides being much more resistant to degradation, are the non-comestible parts of fruits, which increases the probability of archaeological conservation. They might even be recycled as a supplementary fuel. Seeds, which are commonly parched before consumption or storage, can be accidentally preserved if a few seeds spill into a fireplace, for example. Tubers, on the other hand, are usually eaten fresh or boiled, like leafy greens and pulpy fruits (Munson *et al*. 1971; Miksicek, 1987). They are unlikely to be preserved by carbonization and are therefore rarely identified in archaeological sites.

Indeed, Miksicek (1987) considers their conservation as highly improbable. When preserved, these fragments are fragile and hardly identifiable. Hather (1993, 1994) agrees that tuber remains are seldom identified in archaeological material, but this author points out to the fact that they are sometimes abundant. Since the identification methods are still undeveloped, tuber remains are generally classed as undeterminable fragments.

Although tubers are never abundant in the studied sambaquis, they are present in almost all the analyzed samples. Their conservation suggests that they were widely used by fisher-gatherer-hunters and that plant gathering contributed at a higher extent to their diet than is usually assumed.

There are only few records of plant remains for Rio de Janeiro State sambaquis. The Sambaqui de Sernambetiba exceptionally presented a great number of non-carbonized plant remains, which conservation is due to this site's muddy mangrove sediments. Two kinds of palm fruits (*Acrocomia* sp. and an unidentified one), several Myrtaceae (cf. *Psidium*), Bombacaceae (*Ceiba* or *Bombax*) and Lecythidaceae (cf. *Lecythis pisonis*) fruits were found. The later ones have highly nourishing oleaginous seeds (Heredia and Beltrão, 1980).

Numerous charred fragments of Bactris setosa nuts were found in all archaeological levels of Sambaqui Zé Espinho (Kneip and Pallestrini, 1987). Corondó site presented remains of palm nuts (*Astrocaryum* sp, *Bactris* sp.) and of Myrtaceae fruits (*Psidium* sp., *Eugenia* sp.) in three of its four archaeological levels (Carvalho, 1984). The Sambaqui da Malhada presented seeds and roots remains, as well as plant remains attributed to Menispermaceae family. The later ones were associated with a burial (Machado, 1992).

These few examples attest that palm nuts, as well as seeds and fruits, entered in the composition of sambaquians diet. Frequency of each type of remains is not related to its importance in the diet, because of the differential preservation of plant remains. Assuming that the examples mentioned above represent only a very small fraction of all the plant species really eaten, we deduce that a great diversity of vegetables was used by Brazilian fisher-gatherer-hunters.

The restinga ecosystem is extremely rich in plant food. Legumes that frequently produce edible seeds are very frequent, as well as numerous species bearing edible fruits during all seasons (Maciel, 1984), such as Myrtaceae (*Eugenia, Gomidesia, Myrcia, Myrciaria, Psidium*), Ana-

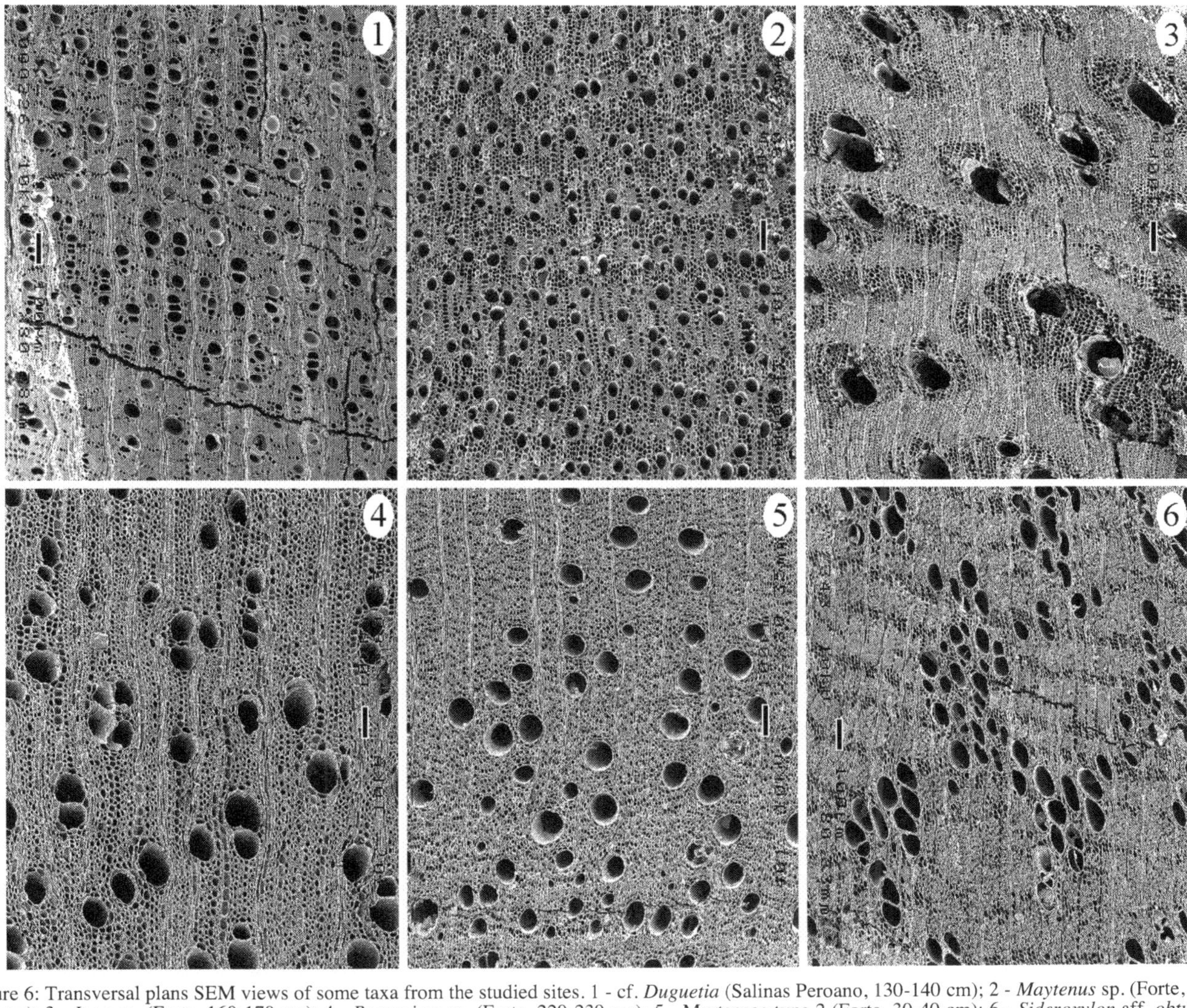

Figure 6: Transversal plans SEM views of some taxa from the studied sites. 1 - cf. *Duguetia* (Salinas Peroano, 130-140 cm); 2 - *Maytenus* sp. (Forte, 280-290 cm); 3 - *Inga* sp. (Forte, 160-170 cm); 4 - *Byrsonima* sp. (Forte, 220-230 cm); 5 - Myrtaceae type 2 (Forte, 30-40 cm); 6 - *Sideroxylon* aff. *obtusifolium* (Forte, 160-170 cm). Scale bar = 100 μm.

cardiaceae (*Spondias, Tapirira*), Annonaceae (*Annona, Duguetia*), Bromeliaceae (*Ananas, Bromelia*), Cactaceae (*Cereus*), Celastraceae (*Maytenus*), Chrysobalanaceae (*Chrysobalanus*), Malpighiaceae (*Byrsonima*), Moraceae (*Ficus*), Palmae (*Astrocaryum, Bactris, Syagrus, Allagoptera*), Passifloraceae (*Passiflora*), Sapotaceae (*Pouteria, Sideroxylon*) etc. It is remarkable that several of these genera figure in the anthracological record (Scheel-Ybert, 2000) (fig. 6).

The case of *Sideroxylon obtusifolium* is particularly interesting. This plant is represented in most of the sambaquis charcoal samples and it is presently very common in the vicinity of sambaquis. This association suggests the hypothesis of a human management of these trees, probably for their edible fruits (Scheel-Ybert, 2001).

An important source of starch was available in the several tubers existing in the restinga environment. Many yam species (*Dioscorea* sp.) grow in the open restinga and in the restinga forest, while *Typha domingensis* and various Gramineae and Cyperaceae tuber-producing species are abundant in the marshy areas.

Conclusions

Charcoal analysis of seven sambaquis from the south-eastern coast of the Rio de Janeiro State provided information on fisher-gatherer-hunters' palaeoenvironment, wood exploitation and diet. Besides being the first anthracological study carried out in Brazil, this work provides an innovative approach to the same set of samples, using both palaeoenvironmental and palaeoethnological analyses.

Our results demonstrate that the analysis of tropical archaeological charcoal supplies reliable information on past environments, but they also show that analysis of isolated sites provides only local information. A regional reconstruction depends on the study of numerous sites distributed over a relatively large area.

South-eastern Brazilian fisher-gatherer-hunters have built the sambaquis in the restinga environment, close to mangrove and forest vegetation. These sites were inhabited for several centuries. However, charcoal taxonomic assemblage is essentially the same throughout the human occupation, showing that the mainland coastal vegetation was not significantly affected by climatic or by anthropogenic perturbations.

Climatic oscillations affected the mangrove vegetation, especially in the Cabo Frio region, but no other significant evidence of vegetation change is recorded during the Late Holocene. This is probably due to the edaphic character of the mainland coastal vegetation ecosystem, which makes coastal formations much less sensitive to climatic change. Vegetation stability has probably been a decisive factor in the permanence and in the maintenance of the fisher-gatherer-hunter sociocultural system.

Palaeoethnological information provided by charcoal analysis particularly concerns wood exploitation and the fisher-gatherer-hunters' diet. Random gathering of dead wood certainly provided most of the wood fuel used. However, *Condalia* sp was probably subject to a cultural choice. Quality of this wood as fuel or its use as a dye or medicine might explain this choice. The hypothesis of a ritual or ceremonial utilization is also advanced. At present, *Condalia* sp is a very rare plant, but it was probably more frequent during the studied period.

Species selection does not invalidate palaeoenvironmental interpretations, provided the choice is limited to a small fraction of the vegetal assemblage. In our sites, most of the species existing in the environment might be represented in the charcoal sample, and the more abundant species are probably represented at higher frequencies. On the other hand, before being subject to a choice, a taxon must be sufficiently frequent in the environment; in other words, in order to be selected, a species must first exist.

The great taxonomic diversity in the sambaquis charcoal samples indicates that domestic firewood was not subject to selection. Associated with the accumulation of vestiges during a large lapse of time, ensuing from the characteristics of these shell mounds, the concept of a lack of selection constitutes an argument that justifies the use of anthracological analysis for the reconstruction of the vegetal palaeoenvironment from this material.

Gathering plant food was certainly much more important to these populations than is often assumed. A great number of charred palm nuts, seeds and monocotyledons tubers (grasses/cyperaceous and yam tubers, among others) were found in all sites. Tubers were found for the first time in Brazilian shell mounds. Tubers are rarely found in a preserved state and testify to the importance of plants in these individuals' diet.

ACKNOWLEDGEMENTS
The author is indebted to many people who contributed to the completion of this work. Special thanks are due to Maria Dulce Gaspar, Lina Kneip, Maria Cristina Tenório, Dorothy Sue Dunn de Araujo and Jean-Pierre Ybert. This work, related to the project "Aproveitamento Ambiental das Populações Pré-Históricas do Estado do Rio de Janeiro" (MN/FINEP/FUJB), was conducted while holding a doctoral scholarship from the CAPES, Brazil.

References

BISSA W.M., YBERT J.-P., CATHARINO E.L.M., KUTNER M. in press. - Evolução paleoambiental na planície costeira do Baixo Ribeira durante a ocupação sambaquieira. *Revista do Museu de Arqueologia e Etnologia*, São Paulo.

CARVALHO E.T. 1984 - Estudo arqueológico do sítio Corondó. Missão de 1978. *Boletim do Instituto de Arqueologia Brasileira*, sér. Monografias, 2, p. 1-243.

CHABAL L. 1992 - La représentativité paléo-écologique des charbons de bois archéologiques issus du bois de feu. *Bulletin de la Société Botanique de France*, Actual. bot., 139 (2/3/4), p. 213-236.

DETIENNE P., JACQUET P. 1983 - *Atlas d'identification des bois de l'Amazonie et des régions voisines*. Centre Technique Forestier Tropical, France. 640 p.

FIGUTI L. 1992 - *Les sambaquis COSIPA (4200 à 1200 ans BP) : étude de la subsistance chez les peuples préhistoriques de pêcheurs-ramasseurs de bivalves de la côte centrale de l'État de São Paulo, Brésil*. Ph.D. Thesis. MNHN, IPH. Paris. 212 p.

FIGUTI L. 1993 - O homem pré-histórico, o molusco e o sambaqui: considerações sobre a subsistência dos povos sambaquieiros. *Revista do Museu de Arqueologia e Etnologia*, São Paulo, 3, p. 67-80.

FIGUTI L. 1996 - Resultados preliminares dos vestígios zooarqueológicos do Sambaqui Espinheiros II (Joinville, SC). *Revista do Museu de Arqueologia e Etnologia*, São Paulo, 6, p. 169-187.

FRANCO T.C.B., GASPAR M.D. 1992 - O sítio arqueológico Salinas Peroano. In: *Anais da VI Reunião Científica da SAB, Rio de Janeiro*. 1, p. 162-171.

GASPAR M.D. 1991 - *Aspectos da organização social de um grupo de pescadores, coletores e caçadores: região compreendida entre a Ilha Grande e o delta do Paraíba do Sul, Estado do Rio de Janeiro*. Ph. D. Thesis. Depto. Antropol., FFLCH, USP. 362 p.

GASPAR M.D. 1992 - Aspectos da organização social de um grupo de pescadores, coletores e caçadores que ocupou o litoral do Estado do Rio de Janeiro. In: Araújo, A.J.G. & Ferreira, L.F. (coords.). *Paleontologia e Paleoepidemologia:* Estudos Multidisciplinares. Ensp, Fund. Oswaldo Cruz. p. 95-109.

GASPAR M.D. 1995 - O "homem" e o ambiente: um estudo de caso. In: III Simpósio de Ecossistemas da Costa Brasileira. *Subsídios a um gerenciamento ambiental*. Academia de Ciências, São Paulo, 1, p. 367-370.

GASPAR M.D. 1996 - Análises das datações radiocarbônicas dos sítios de pescadores, coletores e caçadores. *Boletim do Museu Paranaense* Emilio Goeldi, sér. Ciências da Terra, 8, p. 81-91.

GASPAR M.D. 1998 - Considerations of the sambaquis of the Brazilian coast. *Antiquity* 72 p. 592-615.

GASPAR M.D., SCARAMELLA N.R. 1992 - O sítio do Meio, Canal de Itajuru, Cabo Frio – RJ. In: *Anais da VI Reunião Científica da SAB*, Rio de Janeiro. 1, p. 172-179.

GREGORY M. 1994 - Bibliography of the systematic wood anatomy of dicotyledons. *IAWA Journal*, suppl., 1, p. 1-266.

HATHER J.G. 1993 - *An archaeobotanical guide to root and tuber identification*. vol. 1. Europe and South West Asia. Oxford, Oxbow Monograph 28. 154 p.

HATHER J.G. 1994 - The identification of charred root and tuber crops from archaeological sites in the Pacific. In: Hather, J.G. (ed.). *Tropical Archaeobotany*. Applications and new developments. London, New York, Routledge. p. 51-64.

HEIZER R.F. 1963 - Domestic fuel in primitive society. *The Journal of the Royal Anthropological Institute of Great Britain and Ireland* 93, p. 186-194.

HEREDIA O.R. 1984 - *O aproveitamento ambiental das populações pré-históricas no Estado do Rio de Janeiro*. Research report, June 1983/June 1984. FINEP/MN/FUJB. 120 p.

HEREDIA O.R. 1986 - *O aproveitamento ambiental das populações pré-históricas no Estado do Rio de Janeiro*. Research report, January/August 1985. FINEP/MN/FUJB. 126 p.

HEREDIA O.R. 1987 - *O aproveitamento ambiental das populações pré-históricas no Estado do Rio de Janeiro*. Research report, January/June 1987. FINEP/MN/FUJB. 88 p.

HEREDIA O.R., BELTRÃO M. C. 1980 -. Mariscadores e pescadores pré-históricos do litoral centro-sul brasileiro. Pesquisas, sér. *Antropol.*, 31, p. 101-119.

KNEIP L.M. 1976 - Sambaqui do Forte. Identificação espacial das atividades humanas e suas implicações (Cabo Frio, RJ, Brasil). *Coleção do Museu Paulista*, sér. Arqueol., 2, p. 81-142.

KNEIP L.M. 1977 - Pescadores e coletores pré-históricos do litoral de Cabo Frio, RJ. *Coleção do Museu Paulista*, sér. Arqueol., 5, p. 7-169.

KNEIP L.M. 1980 - A seqüência cultural do sambaqui do Forte - Cabo Frio, Rio de Janeiro. Pesquisas, sér. *Antropol.*, 31, p. 87-100.

KNEIP L.M. (coord.). 1994 - Cultura material e subsistência das populações pré-históricas de Saquarema, RJ. *Documentos de Trabalho*, sér. Arqueol., 2, p. 1-120.

KNEIP L.M., MACHADO L. M. C. 1993 - Os ritos funerários das populações pré-históricas de Saquarema, RJ: sambaquis da Beirada, Moa e Pontinha. *Documentos de Trabalho*, ser. Arqueol., 1, p. 1-76.

KNEIP L.M., PALLESTRINI L. 1987 - Arqueologia: estratigrafia, cronologia e estruturas do Sambaqui Zé Espinho. In: Kneip, L.M. *Coletores e pescadores pré-históricos de Guaratiba*, Rio de Janeiro. Série Livros 5. UFRJ, EDUFF. p. 89-141.

KNEIP L.M., PALLESTRINI L. 1990 - O Sambaqui da Beirada à luz da metodologia por superfícies amplas - Saquarema, Rio de Janeiro. *Dédalo*, São Paulo, 28, p. 171-193.

KNEIP L.M., CRANCIO F., RODRIGUES FRANCISCO B.H. 1988 - O Sambaqui da Beirada (Saquarema – Rio de Janeiro): Aspectos culturais e paleoambientais. *Revista de Arqueologia* 5 (1), p. 41-54.

KNEIP L.M., MAGALHÃES R.M.M., VOGEL M.A.C., MELLO E.M.B., CORRÊA M.M.G. 1989 - O Sambaqui da Beirada (Saquarema - Estado do Rio de Janeiro): Dados culturais, faunísticos e cronológicos. In: *Anais do IX Congresso Brasileiro de Paleontologia*. Curitiba. p. 651-666.

KNEIP L.M., PALLESTRINI L., CRANCIO F., MACHADO L.M.C. 1991 - As estruturas e suas interrelações em sítios de pescadores-coletores pré-históricos do litoral de Saquarema, RJ. *Boletim do Instituto de Arqueologia Brasileira*, sér. Ensaios, 5, p. 1-42.

MACHADO L.C. 1992 - Biologia de grupos indígenas pré-históricos do sudeste do Brasil. As Tradições Itaipu e Una. In: Meggers, B.J. (ed.). *Prehistoria sudamericanca. Nuevas perspectivas*. Washington, Taraxacum. p. 77-103.

MACIEL N.C. 1984 - A fauna da restinga do Estado do Rio de Janeiro: passado, presente e futuro. Proposta de preservação. In: Lacerda, L.D.; Araujo, D.S.D.; Cerqueira, R. & Turcq, B. (orgs.). *Restingas: Origem, Estrutura, Processos*. Niterói, CEUFF. p. 285-304.

MAINIERI C., CHIMELO J.P. 1989 - *Fichas de características das madeiras brasileiras*. 2nd ed. São Paulo, IPT. 418 p.

MARTIN L., FLEXOR J.-M., VALENTIN J.-L. 1988 - Influence du phénomène océanique pacifique "El Niño" sur l'"upwelling" et le climat de la région du Cabo Frio, sur la côte brésilienne de l'État de Rio de Janeiro. *Comptes Rendus de l'Académie des Sciences de Paris*, ser. II, 307, p. 1101-1105.

METCALFE C. R., CHALK C. 1950 - *Anatomy of the dicotyledons*. 2 vol. Oxford, Clarendon Press. 1500 p.

MICKSICEK C.H. 1987 - Formation processes of the archaeobotanical record. *Advances in Archaeological Method and Theory* 10, p. 211-247.

RECORD S.J., HESS R.W. 1943 - *Timbers of the New World*. New Haven, Yale Univ. Press. 640 p.

SCARAMELLA N.R., SCARAMELLA G., BARBOSA D.R, GASPAR M.D. 1990 - Resultados preliminares da análise do material lítico do Sítio do Meio, Cabo Frio, Rio de Janeiro. *Revista do CEPA* 17 (20), p. 175-180.

SCHEEL-YBERT R. 1998 - *Stabilité de l'Écosystème sur le Littoral Sud-Est du Brésil à l'Holocène Supérieur (5500 - 1400 ans BP). Les Pêcheurs-Cueilleurs-Chasseurs et le Milieu Végétal : Apports de l'Anthracologie*. Ph.D. Thesis. Univ. Montpellier II. 3 volumes. Texte, 245 p.; Atlas anthracologique, 114 p. + 1 cdrom; Annexes, 161 p.

SCHEEL-YBERT R. 1999 - Paleoambiente e paleoetnologia de populações sambaquieiras do sudeste do Estado do Rio de Janeiro. *Revista do Museu de Arqueologia e Etnologia,* São Paulo, 9, p. 43-59.

SCHEEL-YBERT R. 2000 - Vegetation Stability in the Southeastern Brazilian coastal area from 5500 to 1400 yr BP deduced from charcoal analysis. *Review of Palaeobotany and Palynology*, 110, p. 111-138.

SCHEEL-YBERT R. 2001 - Man and vegetation in the Southeastern Brazil during the Late Holocene. *Journal of Archaeological Science* (in press)

SCHEEL-YBERT R., SCHEEL M., YBERT J.-P. 1998 - *Atlas Brasil* – Databank for charcoal analysis and computerized key to charcoal determination (in Portuguese, English and French). Version 1.8. CD-ROM. 1500 entries.

SILVA J.G., OLIVEIRA A.S. 1989 - A vegetação de restinga no município de Maricá, RJ. *Acta Botânica Brasílica*, supl., 3 (2), p. 253-272.

TASAYCO-ORTEGA L.A. 1996 - *Variations paléohydrologiques et paléoclimatiques d'une région d'upwelling au cours de l'Holocène : enregistrement dans les lagunes côtières de Cabo Frio (État de Rio de Janeiro, Brésil)*. Ph. D. Thesis. Univ. Paris VI. 321 p.

TENÓRIO M.C. 1991 - *A importância da coleta no advento da agricultura*. M.Sc. Thesis. IFCS, UFRJ. 234 p + annexes.

TENÓRIO M.C., BARBOSA M., PORTELA T. 1992 - Pesquisas arqueológicas no sítio Ponta de Cabeça, Arraial do Cabo, Rio de Janeiro. In: *Anais da IV Reunião da SAB*, Rio de Janeiro. 2, p. 279-291.

TENÓRIO M.C. 1996 - A contribuição da arqueologia na compreensão do desenvolvimento do mangue. *Boletim do Museu Paranaense Emílio Goeldi*, sér. Ciênc. da Terra, 8, p. 123-136.

YBERT J.-P., SCHEEL R., GASPAR M.D. 1997 - Descrição de alguns instrumentos simples utilizados para a coleta e concentração de elementos fósseis de pequenas dimensões de origem arqueológica ou pedológica. *Revista do Museu de Arqueologia e Etnologia*, São Paulo, 7, p. 181-189.

Phytolith and pedoanthracological analysis of a Holocene pedosedimentary sequence from the middle Rhone valley (France)

Claire Delhon

CNRS-UMR 7041, Archéologies et Sciences de l'Antiquité
21, allée de l'Université, F-92023 Nanterre cedex, France.
E-mail: delhon@mae.u-paris10.fr

Abstract

The construction in the middle Rhone valley of the new "TGV-Méditerranée" fast railway has brought to light several archaeological sites and pedosedimentary Holocene sequences. The local vegetation, subject to several climatic influences, is particularly sensitive both to anthropogenic pressure and climatic changes. Unfortunately, the palaeoenvironmental reconstruction has so far been limited by the lack of pollen and by the absence of non woody remains from anthracological assemblages. In this work, phytoliths and microscopical charcoal from the sequence at Espeluche-Lalo are analysed together, in order to test the complementarity and the accuracy of these two methods for the reconstruction of Holocene Mediterranean environments. Off-site sampling allows us to detect local variations and to compare them to the overall regional vegetation pattern. Our results show that pedoanthracology and phytolith analysis provide different but yet convergent images of the past vegetation. Deciduous oak (*Quercus* cf.) was dominant in the vegetation of the middle Rhone valley all through the Atlantic period, the development of oak forests being favoured by stable morphosedimentary phases. Pine is a pioneer species on burnt areas and thus its distribution extended in relation to fire phases. The origin of the fires (lightning or human activity) is unknown, but they seem to have increased when climatic conditions were propitious. Under dry conditions, fire affect all kinds of vegetation, not only inflammable coniferous forests but also riparian and open formations. Fire phases identified at Lalo are synchronous with fire phases recorded from other sequences in the region and this is a further argument for considering climatic conditions as being responsible for the breakout of fires in a vegetation that was also affected by agro-pastoral activities.

Key-words: *Phytolith analysis, pedoanthracology, palaeoenvironment, fire, middle Rhone valley.*

Résumé

La construction du TGV-Méditerranée en moyenne vallée du Rhône a permis de mettre au jour de nombreux sites archéologiques et séquences pédosédimentaires holocènes. Dans cette région, la végétation, soumise à plusieurs influences climatiques, est particulièrement sensible à la pression anthropique et aux changements climatiques. Malheureusement, les reconstitutions paléoenvironnementales sont limitées par le manque de pollen et par l'absence des espèces herbacées des assemblages anthracologiques. Dans ce travail, les phytolithes et les micro-charbons de la séquence d'Espeluche-Lalo sont analysés en parallèle, dans le but de vérifier la complémentarité et l'efficacité de ces deux méthodes pour l'étude des environnements holocènes méditerranéens. Le prélèvement "hors-site" permet de mettre en évidence des variations locales ou stationnelles au sein du modèle de végétation régional. Nos résultats montrent que la pédoanthracologie et l'analyse phytolithique donnent des images différentes mais néanmoins convergentes de la végétation. Les chênes caducifoliés (*Quercus* cf.) ont dominé la végétation de la moyenne vallée du Rhône tout au long de l'Atlantique, mais le développement des chênaies reste favorisé par les phases de stabilité morphosédimentaire. Le pin, pionnier sur les zones brûlées, se développe en corrélation avec des phases d'incendies. La cause première des incendies (foudre ou activités humaines) est inconnue, mais leur recrudescence semble survenir lorsque les conditions climatiques deviennent propices. Dans des conditions plus sèches, n'importe quel type de formation végétale peut brûler, qu'il s'agisse des forêts de conifères, facilement inflammables, ou de formations ouvertes ou ripicoles. Les phases à incendies identifiées à Lalo sont synchrones de phénomènes comparables enregistrés dans d'autres séquences de la région, ce qui est un argument supplémentaire en faveur d'un déterminisme climatique de l'extension des feux, dans un milieu végétal rapidement affaibli par les activités agro-pastorales.

Mots-clés : *phytolithes, pédoanthracologie, paléoenvironnement, incendie, moyenne vallée du Rhône.*

Introduction

In the middle Rhone valley, rescue excavations prior to the construction of the "TGV-Méditerranée" fast railway have brought to light numerous archaeological sites and Holocene pedosedimentary sequences.

The Rhone valley is located in France, in the northern part of the Mediterranean basin. It is situated within the Mediterranean climatic zone as well as in the convergence zone of medio-european and alpine climatic influences (fig. 1). This situation makes the vegetation highly sensitive to both climatic changes and anthropogenic pressure. Even if the regional palaeobotanical background has already been reconstructed by palynological and anthracological syntheses (de Beaulieu, 1977; Triat-Laval, 1978; Jalut, 1974; Vernet and Thiébault, 1987; Argant, 1990; Thiébault, 1999...), traditional palaeobotanical methods are now attaining their limits in the area. Pedosedimentary sequences where pollen is preserved are lacking (Argant, 1990; Argant *in* Berger, 1996). Anthracological studies give an incomplete view of the past vegetation as they only record woody plants and thus do not allow a complete reconstruction of the relative importance and location of grasslands in comparison to forests. Moreover, charcoal from archaeological sites allows a determination of the species collected for fuel but give no indication of the spatial distribution of various plant associations.

To increase our understanding of the dynamics of the middle Rhone valley vegetation since the late Glacial period, we have combined two methods that are both still rarely used: phytolith and pedoanthracological analysis. A first aim of this paper is to discuss the possibilities and the limits of our method. A second objective is to demonstrate the pertinence of this method for treating various aspects of the Mediterranean palaeoenvironment. For this purpose, the first results obtained at Espeluche-Lalo in the middle Rhone valley will be presented.

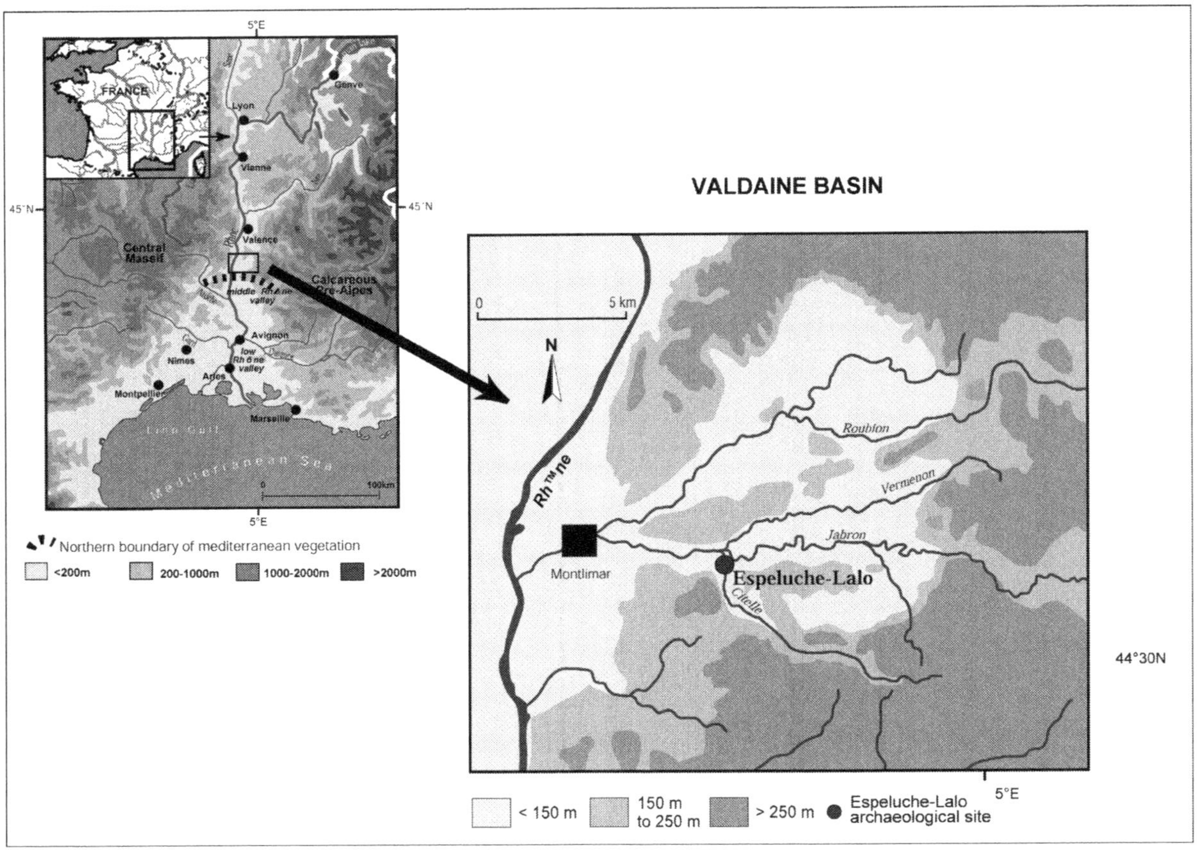

Figure 1: Location of Espeluche-Lalo, the Valdaine Basin and the middle Rhone valley.

Methods

Pedoanthracology

Pedoanthracology, a method essentially developed since a decade (Thinon, 1992; Carcaillet and Thinon, 1996; Thinon and Talon, 1998), consists of the taxonomic determination of fossil charcoal collected off-site in order to reconstruct palaeoenvironments.

It has been shown that pedoanthracological assemblages deposited *in situ* after a forest fire reflect the burnt vegetation (Thinon, 1992; Scott *et al.* 2000). The first studies carried out in the middle Rhone valley further show that the number of identified taxa is high compared to the number of studied fragments (Delhon *et al.* forthcoming; Delhon, 2001; Delhon and Thiébault *in* Beeching and Vital, 2001; Thiébault, 1999). Moreover, not only the dominant forest trees (*Quercus, Fagus, Fraxinus, Pinus, Populus*...), but also less common taxa (*Rhamnus*, Ericaceae, Fabaceae) are represented in the pedoanthracological assemblages. The large taxonomic diversity in some assemblages suggests that the palaeoecological information has been little distorted by taphonomical processes during the primary deposition (Scott *et al.* 2000; Nichols *et al.* 2000). It has been noted that larger fragments (over 125μm; all the charcoal fragments of our study are included in this category) are abundant in the burnt area, and far less numerous or even totally absent in adjacent unburnt areas. Only small particles (such as pieces of leaves and bark) are swept away by convection movements (Clark, 1988). Secondary transport of charcoal fragments is however possible as they are submitted to the same post-depositional processes as other sediment particles, particularly if they are part of the hydrosystem (Berger and Thiébault, in this volume). Thus, they can be found downhill (but never uphill) from the location of the burnt vegetation (Thinon and Talon, 1998).

In order to recover charcoal, 10 kg of sediment from each stratigraphic unit were systematically sifted through sieves with a mesh of 2 mm and 0.4 mm, even when macroscopical charcoal was not visible. Charcoal fragments were small-sized (frequently <2 mm) and often fragile, so the total number collected for each sample was usually low. Fractured charcoal pieces that were kept in connection by the clayey matrix did often not resist the sifting, even when this was carefully done. In the case of very small or badly preserved fragments, it might therefore be preferable not to remove them from the sediment until they have been identified. In the example presented below, flotation was not used to recover charcoal as we wanted to avoid further fragmentation. The small size of the fragments is not a major obstacle to the identification work as long as the anatomical structures are preserved.

Taxonomical identification was done with the help of a reflected light microscope. When possible, the charcoal was broken along three characteristic planes (Western, 1963). For the smallest fragments, only one or two sections could be observed and compared to descriptions in wood anatomical atlases (particularly Schweingruber, 1990) or to charcoal from the laboratory reference collection. The determination usually attained the genus level and could sometimes be more precise. Even in the worst cases (very small and destroyed fragments) it was still often possible to distinguish angiosperms from gymnosperms.

Even though taphonomical processes are now taken into account and better understood (Berger and Thiébault, in this volume), the problem of the sample size still remains. The fragments included in pedoanthracological analyses are seldom numerous enough to allow statistical treatment. The taxa are thus frequently recorded by presence/absence and the results are merely qualitative in nature. Moreover, it must be emphasised that anthracological analysis only deals with ligneous species and despite the occasional presence of charred grass stems, anthracological spectra always remain limited in this respect (Figueiral and Mosbugger, 2000).

Pedoanthracological analysis yields precise qualitative but rarely quantitative data on the composition of the woody vegetation. In this paper, the presence of a taxa has been considered as ecologically significant for the main trends of vegetation dynamics. Thus, regardless of the number of identified fragments, we consider that the occurrence of *Quercus* in the charcoal assemblages indicates the presence of oak forests; the occurrence of heliophilous (*Prunus* sp., *Fraxinus* sp., Rosaceae) and riparian (*Salix* sp./*Populus* sp.) taxa reflect open formations/forest edges and riparian formations respectively. Moreover, the Betulaceae that appears in one assemblage (probably *Betula* sp. according to anatomical characteristics), has been interpreted as a mesophilous taxon, likely to indicate cooler climatic conditions.

Phytolith analysis

Phytoliths are hydrous silica particles formed in living vegetal cells by opal precipitation. Their shape depends on the shape of the cell from which they originate and has a taxonomical significance. Many studies have demonstrated that the analysis of phytolith assemblages is a pertinent palaeoecological proxy (Fredlund and Tieszen, 1994; Carbone, 1977; Rovner, 1988), and more particularly a mean of tracing grassland dynamics and forest-grassland shifts that occurred during the Holocene in America, tropical Africa and Asia (Kurman, 1985; Piperno, 1988; Tsutsuki *et al.* 1993; Inoue and Sase, 1996; Fredlund and Tieszen, 1997a, 1997b; Alexandre *et al.* 1999; Barboni *et al.* 1999) as well as in the Mediterranean area (Delhon, 2001).

Due to their chemical composition, phytoliths are physically and chemically resistant. They are weakly soluble (unless the pH exceeds 9) and are not affected by the oxidation and decarbonatation processes that frequently occur in our study area. They are therefore often abundant in the sediments. Moreover, as for the charcoal, they are subject to limited transport. It has been shown that phytolith spectra from primary positions indeed reflect the station from which samples were taken (Bartoli and Guillet, 1977; Bartoli and Souchier, 1978; Delhon, 2001).

Phytoliths are chemically extracted from the sediments. Around 10 g of dry material is analysed. Phytoliths are extracted by several procedures: removal of calcium carbonate using HCl; oxidation of organic matter using H2O2; granulometric separation; densimetric separation in a ZnBr2 solution with a density of 2.35 (Kelly, 1990; Fredlund and Tieszen, 1994). They are mounted on microscope slides and counted under a petrographic microscope, using a magnification of x400-600.

Phytoliths appear as isotropic opal particles ranging from about ten to a few hundred microns. These particles have a characteristic morphology depending on the morphology of the living plant cell in which they were formed (Twiss *et al.* 1969; Laroche, 1977; Piperno, 1988). As different types of cells can be silicified in the same plant, there is no direct correlation between a phytolith morphotype and a taxon. Fossil phytolith assemblages must therefore be interpreted globally by the comparison with modern soil assemblages. Previous studies in the middle Rhone valley have shown that grass phytoliths are always dominant; that forest vegetation assemblages are characterised by the presence of more than 10% of trees phytoliths; and that broad-leaved tree and pine forests yield distinct phytolith spectras (Delhon, 2001).

Pedoanthracology and phytolith analysis are highly complementary. Phytoliths provide quantitative information on the importance of herbs in the vegetation and the density of arboreal formations. Charcoal analysis yields qualitative information on the ligneous taxa present in the vegetation. Phytoliths and charcoal can both be extracted from pedosedimentary layers that contain no prehistoric remains but which are still close to archaeological sites. Off-site, the accumulation of microbotanical remains is not biased by human activities such as selective exploitation, gathering or storage. However, the vegetation they reflect may have been influenced by man.

Charcoal and phytoliths are normally the subject of very limited transport prior to deposition. Phytoliths and micro-charcoal in primary positions (buried soils) record the vegetation at a small spatial scale, and can be interpreted in terms of local vegetation or vegetation of the studied station (Bartoli and Guillet, 1977; Bartoli and Souchier, 1978; Rovner, 1988; Delhon *et al.* forthcoming; Berger and Thiébault, in this volume). On the contrary, phytoliths and micro-charcoal from alluvial levels are likely to originate from all of the parcels drained by the river (Berger and Thiébault, in this volume; Delhon *et al.* forthcoming). The palaeobotanical assemblages from alluvial levels will therefore be interpreted as reflecting the vegetation of the whole basin while assemblages from palaeosoils will be seen as reflecting the site vegetation. However, when a soil has formed on an alluvium, it will sometimes register a "background noise" relating to the whole basin.

In order to show the complementarity between the two methods which are still insufficiently used in palaeoenvironmental studies, we present here the first results obtained from trench T2 at Espeluche-Lalo, a large pedosedimentary sequence located in the middle Rhone valley.

Presentation of the site of Espeluche-Lalo (middle Rhone valley, France)

The middle Rhone valley is located downstream from the town of Valence in south-eastern France, between the Alps and the Massif Central. The valdaine basin is formed by the joining of the Roubion and Jabron basins (fig. 1). These two rivers join in Montélimar where they form a large alluvial fan before flowing into the Rhone. The studied site, Espeluche-Lalo, is located at the bank of the Citelle river, a tributary to the Jabron. The excavation, carried out by A. Beeching and J. Vital in 1996 (Beeching and Vital, 2001), has provided various structures (mainly pits) attributed to Mesolithic, Neolithic and Bronze Age cultures.

Due to its location, a few kilometres north of the boundary of the Mediterranean area, the region is subject both to the Mediterranean climate (summer drought and mild winters) and to continental and mountain influences. The vegetation, typical of the supra-Mediterranean stage (Ozenda, 1964, 1994), is characterised by *Quercus pubescens* and *Pinus sylvestris*, but locally also by mountainous (*Fagus sylvatica*) or more mesomediterranean species (*Quercus ilex, Pinus halepensis*). Because of the multiple climatic influences, enhanced by winds, edaphic and microtopographic conditions, this type of vegetation is very sensitive to disturbance due to climatic change, fire or anthropogenic activities. Indeed, the vegetation cover in the area is much influenced by human activities.

Besides archaeological structures, the excavation provided a 130 m long pedosedimentary sequence, trench T2, that crosses the low alluvial plain from the Würmian terrace to the river (fig. 2). The sequence included numerous pedosedimentary formations, from the late Glacial to the Holocene, as well as "natural" fire layers, palaeosoils and archaeological remains that allowed several absolute and relative datings.

Due to the high resolution of the pedosedimentary sequence, the early Atlantic period is well documented. The results presented here focus on thirty samples from this period. All of them contained large amounts of phytoliths and microscopic charcoal was present in half of them. As the pedosedimentary processes had already been studied (Berger *in* Beeching and Vital, 2001), the areas from which the samples originated could all be properly evaluated. This multidisciplinary palaeobotanical approach, applied to a well understood pedosedimentary context, constitutes the first step towards a systemic study of Holocene environments in the middle Rhone valley.

Results of the analysis of phytoliths and micro-charcoal (tabl. 1, fig. 3)

All of the processed samples provided large amounts of well preserved phytoliths. On the contrary, only 16 of them yielded identifiable charcoal fragments, even though sediment from every level was sifted. However, this is still a fair

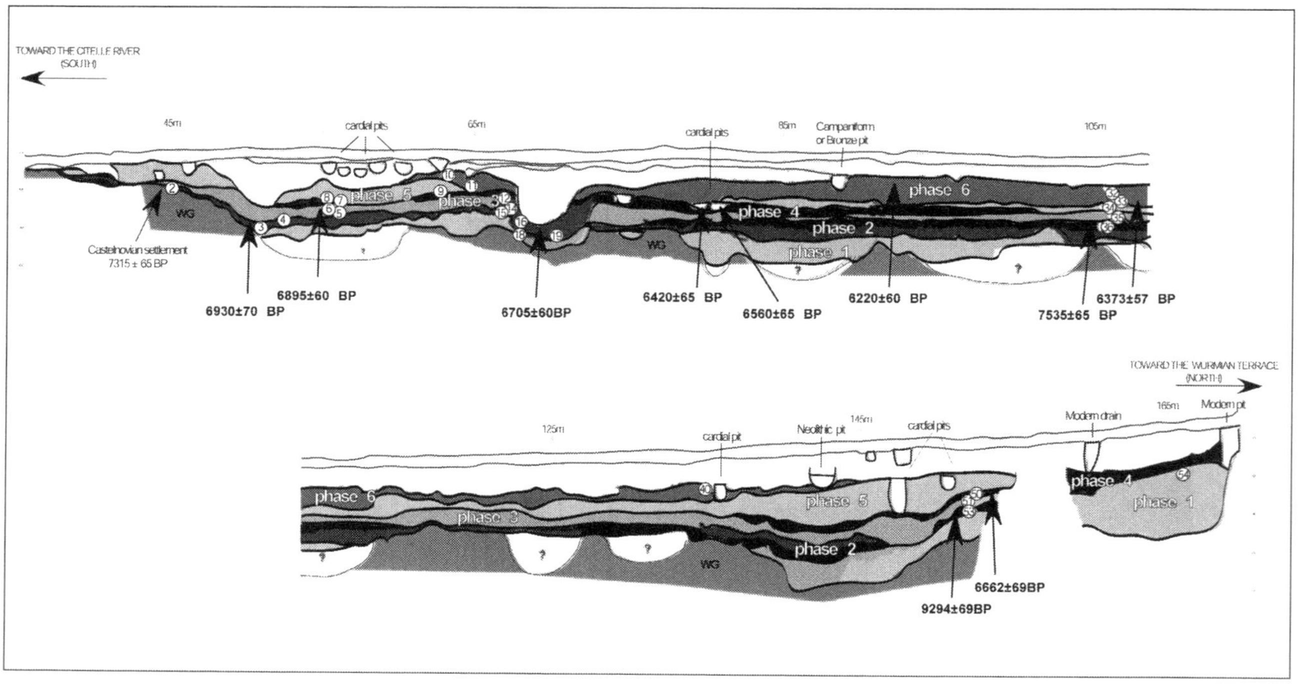

Figure 2: Schematic representation of Trench T2 showing main pedosedimentary events from the late Glacial to the early Atlantic period and the location of sediment samples that provided phytoliths and micro-charcoal:
Phase 1: deposition of alluvial and aeolian elements on Würmian gravel (WG), probably during the late Glacial;
Phase 2: first pedogenetic phase during the early Holocene (9294±69 BP/7535±65 BP);
Phase 3: first hydrological crisis;
Phase 4: second pedogenetic phase, *c.* 6600/6400 BP;
Phase 5: second hydrological crisis;
Phase 6: alternation of alluvial deposits and pedogenesis (cumulic pedogenesis).
(Original drawing and interpretation of the sedimentary section by J-F. Berger).

		Samples																									
		T2-40	T2-32	T2-10	T2-33	T2-11	T2-09	T2-34	T2-08	T2-07	T2-50	T2-12	T2-51	T2-14	T2-19	T2-06	T2-15	T2-16	T2-05	T2-04	T2-18	T2-35	T2-03	T2-02	T2-36	T2-53	T2-54
Phytoliths (%)	grass phytoliths	87,06	92,08	86,14	90,59	93,59	98,03	92,92	88,06	91,54	73,15	97,04	93,33	96,77	72,31	77,31	91,62	84,31	98,09	94,12	91,08	82,96	78,15	83,43	85.85	65,17	83.43
	pine phytoliths	0	1,48	3,46	0,49	0,49	0	0,94	0,99	0,99	0	0	1,96	0	22,32	3,34	2,95	3,92	0	2,26	4,45	8,68	1,45	4	2.44	3,48	7.69
	dicot. phytoliths	9,45	4,45	8,91	7,92	2,95	1,97	5,66	6,96	6,96	22,15	0,49	3,13	2,3	3,77	17,87	5,42	8,33	1,9	1,36	4,45	6,75	16,02	1	6.83	29,35	5.91
	others	3.49	1.99	1.49	1.00	2.97	0	0.48	3.99	0.51	4.70	2.49	1.58	0.93	1.6	1.48	0	3.44	0	2.26	0	1.61	4.38	11.57	4.88	2	2.97
	TOTAL	100%	100%	100%	100%	100%	100%	100%	100%	100%	100%	100%	100%	100%	100%	100%	100%	100%	100%	100%	100%	100%	100%	100%	100%	100%	100%
	D/P(dicot. *versus Poaceae*)	0,11	0,05	0,10	0,09	0,03	0,02	0,06	0,08	0,08	0,30	0,01	0,03	0,02	0,05	0,23	0,06	0,10	0,02	0,01	0,05	0,08	0,20	0,01	0.08	0,45	0.07
	Pi/P (pine *versus Poaceae*)	0,00	0,02	0,04	0,01	0,01	0,00	0,01	0,01	0,01	0,00	0,00	0,02	0,00	0,31	0,04	0,03	0,05	0,00	0,02	0,05	0,10	0,02	0,05	0.02	0,05	0.09
	T/P (Trees *versus Poaceae*)	0,11	0,06	0,14	0,09	0,04	0,02	0,07	0,09	0,09	0,30	0,01	0,05	0,02	0,36	0,27	0,09	0,15	0,02	0,04	0,10	0,19	0,22	0,06	0.11	0,50	0.16
Charcoals (nb. of fragments)	*Quercus sp.* + cf. *Quercus sp.*		8	6		22		3	28	13	6	6	82		8				6	14		6		30		3	
	Fraxinus sp.														3												
	Salix/Populus						3		6	1																	
	Prunus sp.						5		4													3					
	Rosaceae						3																				
	Betulaceae (cf. Betula sp.)								7																		
	Dicotyledon				2	5	4	2	21	2	3		5		1				1	2							
	TOTAL		8	6	2	27	15		66	16	9	6	87		12				7	16		9		30		3	

Table 1: results of the analysis of phytoliths and charcoal: percentages of phytolith morphotypes, values of the phytolith ratios and numbers of fragments for each taxon identified in samples from trench T2.

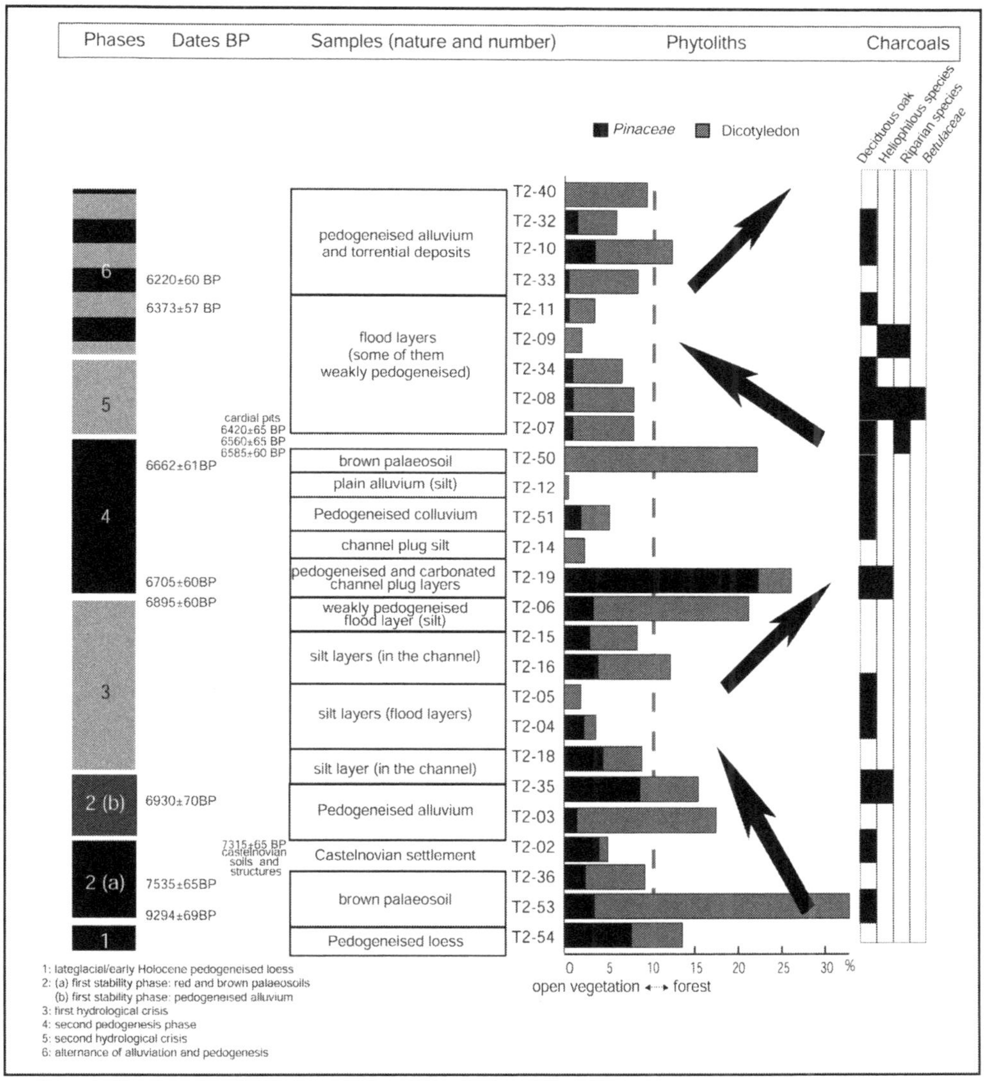

Figure 3: Evolution of the local vegetation around trench T2 during the Atlantic period as reconstructed by phytolith and micro-charcoal analysis (left - pedosedimentary phases; centre - number of tree phytoliths, right - taxa identified by the charcoal analysis).

amount considering that the 16 layers represent fires that occurred during one millennium only (tabl. 1).

The phytolith index, T/P (tabl. 1), indicating the ratio between arboreal (dicotyledons and pine) and Poaceae phytoliths, is a good proxy to the density of the vegetation cover (Alexandre *et al.* 1997; Delhon *et al.* forthcoming). It can be further subdivided into the ratios D/P (Dicotyledonae/Poaceae) and Pi/P (*Pinus*/Poaceae) in order to quantify the relative proportions of pine and broad-leaved tree species in the arboreal vegetation (Delhon, 2001; Delhon *et al.* forthcoming). The charcoal analysis allows us to determine more precisely which species were present in those formations (fig. 3).

At Lalo, the proportion of tree phytoliths hardly exceeds 10%, *i.e.* the critical value used to distinguish between forests and open formations in the study area (Delhon, 2001). This can probably be explained by the location of the trench in the lowest part of the Citelle alluvial plain, where the shifting of the river bed restricts the development of long lasting forests.

Since the beginning of the Holocene, the alluvial plain has known several phases of stability as well as several hydrological crises:

- *First stable phase during the early Holocene (fig. 2, phases 1 and 2)*

The oldest sample (T2-54) probably dates to the late Glacial/early Holocene transition (Berger, *in* Beeching and Vital, 2001). It comes from sandy loess accumulated at the bottom of a Pleniglacial terrace in the northern part of the trench, that was later affected by a beginning pedogenesis. The phytolith analysis shows that trees were quite numerous (T/P=0,16), with pine and broad-leaved trees present in similar proportions. As no charcoal was found in this context, the tree species could not be further identified. An early, brown-coloured pedogenesis phase, associated with decarbonation and moderate alluviation, took place during the Boreal and/or the beginning of the Atlantic (sample T2-53, dated to 9294±69 BP, sample T2-36, dated to 7535±65 BP), in correlation with a period of stability in the alluvial plain. Here the phytolith assemblages are significant mainly at a local scale. Dicotyledon phytoliths are numerous, especially in sample T2-53, and reflect the presence of a forest vegetation (35% tree phytoliths). A few charcoal pieces from sample T2-53 could be identified as deciduous oak. The Citelle riverbed, then located at the foothill of the Würmian terrace, and the plain seem to have been covered by a forest vegetation dominated by oak.

During the stable phase recorded in the plain, settlers of the Mesolithic Castelnovian culture occupied a gravel mound (7315±65 BP, sample T2-02) located in the southern part of the site. The oak dominated forests described above underwent important changes: the frequency of tree phytoliths abruptly falls in sample T2-02 while that of grass phytoliths increases to constitute 90% of the identified particles. The remaining tree phytoliths can mostly be ascribed to a pine morphotype, while all of the identified charcoal pieces belong to deciduous oak. The presence of a human community could have been responsible for a local opening up of the vegetation, an event recorded here by the phytolith analysis only and not by other archaeobotanical methods, usually efficient only on a larger spatial scale. The presence of numerous charcoal fragments from oak suggests that a local clearance affecting this type of formations took place. This level is dated to the same period (around 7300 BP) as many other fire levels known in the Rhone valley (Berger and Brochier, 2000).

The Castelnovian site was recovered by an alluvium that was later subject to pedogenesis (sample T2-03, dated to 6930±70 BP). The phytolith assemblage from that level shows a T/P ratio of 0.22, indicating a vegetation of forest type. The sample probably reflects essentially the station, even though it also presents a strong "background noise" from the basin, as it came from an alluvium that had known a partial pedogenesis. Sample T2-35 comes from an alluvium mainly constituted of transferred material, originally belonging to the previous pedogenetic phase. This alluvium also shows signs of the beginning of a brown pedogenesis, related to a period of stabilisation and fires. The few charcoal pieces remaining after the sifting indicate that the burnt vegetation was an open deciduous oak formation (deciduous *Quercus* and Rosaceae). The increase of tree phytoliths (T/P around 0.20 for samples T2-03 and T2-35) corroborates the pedoanthracological results in showing an extension of the forest cover. The evolution towards a denser vegetation was simultaneous with the decrease of floods and the stabilisation of the alluvial plain. Pine phytoliths account for nearly 9% of sample T2-02 while no pine charcoal has been identified. The absence of pine charcoal all along the Lalo sequence, in spite of the occurrence of pine phytoliths, is an issue that will be discussed below.

- *First hydrological crisis (fig. 2, phase 3)*

After the stable phase described above, began a period of hydrological crisis (from *c.* 6900-6700 BP), characterised by a channel incision of about 70 m, the accumulation of silt layers in the channel (samples T2-18, T2-16 and T2-15), the raising of the alluvial plain (samples T2-04, T2-05 and later T2-06 which was weakly pedogenised between the floods) and torrential deposits in the northern part of the site. The lowest percentages of tree phytoliths for the whole sequence (samples T2-04 and T2-05) are found here and they indicate the dominance of a non arboreal vegetation. The previous oak formations, reflected by sample T2-04 which contained fragments of deciduous oak only, seem to have regressed due to the opening up of the vegetation. While incising a new channel of almost 70 m, the river deposited silt layers. The phytolith spectra from these layers reflect the presence of mostly open formations in the basin (T2-18, T2-16 and T2-15, T/P=0.08 to 0.15), with occasional stands dominated by broad-leaved trees (D/P=0.05 to 0.10) as well as stands of pine (Pi/P=0.03 to 0.05). No charcoal was found in these layers.

Samples T2-06 and T2-07 come from flood layers. The first, dated to 6895±60 BP, indicates rather dense broad-leaved tree forests (T/P=0.27, D/P=0.23). The second sample contained less tree phytoliths (T/P=0.09) but yielded charcoal from deciduous oak and *Salix*/*Populus* (one fragment). It can be assumed that fires affecting the oak forests (and probably also the riparian formations) resulted in an opening up of the landscape, as recorded by sample T2-6.

- Second phase of pedogenesis (fig. 2, phase 4)
This first crisis was followed by a second phase of stabilisation during which the channel progressively filled up (sample T2-19, dated to 6705±60 BP) all while a new riverbed formed (fig. 2: marks 135-140 m). Before drying up, the active part of the earlier watercourse was of considerable width (samples T2-14 and T2-12). The new watercourse was located at 70 m distance (fig 2: mark 70 m), in a context of stabilised slopes. The alluvium of the plain (T2-12) and the colluvial deposits of the northern part of the site (T2-51) were affected by pedogenesis, resulting in the formation of a second brown soil associated with secondary accumulations of carbonates (samples T2-50, dated to 6662±61 BP). The T/P ratio increases in samples T2-19 (with pine being largely responsible for the increase: Pi/P=0.31, T/P=0.36) and T2-50 (dicotyledons most important, D/P=T/P=0.30). Both samples contained charcoal from deciduous oak, sample T2-19 also yielded *Fraxinus*. The difference between the two phytolith assemblages can be explained by the nature of the samples: sample T2-50, originate from a palaeosoil, has a more local significance than silt sample T2-19. The pine phytoliths from the latter probably rather reflect the vegetation of the slopes than that of the plain. Samples T2-12 and T2-14, originate from the silt that filled up the channel, are chronologically intercalated between T2-50 and T2-19. They both contain very few tree phytoliths (T/P=0.01 and 0.02). Sample T2-12 nevertheless provided a few fragments of deciduous oak.

The first Neolithic communities settled in the plain when it dried up. However, the archaeological remains from this period are scarce (Cardial pits, dated to 6420±65 BP, 6560±65 BP and 6585±60 BP) because of the disturbance of the palaeosoils brought about by a later, second hydrological crisis.

- Second hydrological crisis (fig. 2, phase 5)
The flooding of the rivers led to the deposition of thick (30 cm) alluvial layers (samples T2-07, T2-08, and T2-34), subject to pedogenesis between the floods. The abundance of molluscs and charcoal fragments indicate an erosion of the slopes, intervening after fire events. In sample T2-08, the proportion of tree phytoliths starts to decrease (around 8%). This level provided not only oak charcoal, but also fragments of Rosaceae, *Salix/Populus* and Betulaceae, indications of an open vegetation thriving in somewhat cooler and more humid conditions.

- Alternation of alluviation and pedogenesis (fig. 2, phase 6)
During this phase, the flowing of the river changed once more; floods became more frequent and the pedological build-up of the alluvium thus decreased (sample T2-09). While the basin stabilised and the riverbed became more incised (sample T2-11), another Cardial occupation occurred in the sector. This occupation, as the previous one, was disturbed by a later phase of instability and torrential flow (sample T2-33 and T2-32) related to a network of minor channels before 6373±57-6220±60 BP. Then, while the main watercourse progressively retracted into its former bed, a last Neolithic community settled. The amount of tree phytoliths, at first rather low (samples T2-09 and T2-11), now increases up to around 13% (samples T2-10). All of the identified charcoal fragments belong to oak. It is likely that the riverbed moved towards the South, allowing broad-leaved trees to establish at the site. This evolution favoured the pedogenesis (samples T2-40 and T2-10). Cumulic soil samples T2-33, T2-32, and T2-40 also show that relatively open broad-leaved tree formations were present for a considerable span of time. Unfortunately, the cumulic soil and the Neolithic occupation could only be recorded partially as they had been disturbed by a subsequent, brutal return of alluviation.

Discussion

This study, combining the analysis of phytoliths and charcoal, has allowed the characterisation of vegetation dynamics (open versus dense formations), at the level of a single station as well as at the scale of an entire river basin.

The results obtained from pedoanthracological and phytolith analyses show the persistence of deciduous oak and pine all through the Atlantic period (fig. 3). In the middle Rhone valley, deciduous oak (*Quercus pubescens*), was the main forest species all through the early Atlantic, while the presence of pine, a pioneer species on burnt areas, is related to fire phases. The large amount of pine phytoliths in sample T2-19, dated to 6705±60 BP, can be related to the increase of fire events in the middle Rhone valley around 6800 BP, shown by Berger and Thiébault (*in* Berger and Brochier, 2000, and forthcoming). The development of forests, revealed by an increase in tree phytoliths in the samples, is often detected from buried soil levels, also containing *Quercus* charcoal. Samples T2-54, T2-53, T2-06, T2-50 correspond to palaeosoils; T2-03 and T2-35 consist of alluvial sediments including material from earlier palaeosoils that have, in their turn, experienced a beginning of pedogenesis; T2-40 and T2-10 come from cumulic palaeosoil levels. There is a correlation between morphosedimentary stability and the

	% of pine phytoliths in alluvial sediment		**% of pine phytoliths in palaeosoils**
	T2-35: 8,68	T2-12: 0,00	T2-54: 7,69
	T2-18: 4,45	T2-7: 0,99	T2-53: 3,48
	T2-4: 2,26	T2 8: 0,99	T2-36: 2,44
	T2-5: 0,00	T2-34: 0,94	T2-3: 1,45
	T2 -16: 3,92	T2-9: 0,00	T2-6: 3,34
	T2-15: 2,95	T2-11: 0,49	T2-50: 0,00
	T2-19: 22,32	T2-33: 0,49	T2-10: 3,46
	T2-14: 0,00	T2-32: 1,48	T2-40: 0,00
	T2-51: 1.96		
	all samples	excluding ESL-T2 19 (3)	
mean (m)			
all samples (1)	3.05	1.85	2.73
excluding samples without pine phytoliths (2)	3.99	2.46	3.64
standard deviation (sd)			
all samples (1)	5.27	2.22	2.31
excluding samples without pine phytoliths (2)	5.71	2.25	1.95
variation coefficient (sd/m)			
all samples (1)	**1.72**	**1.20**	**0.85**
excluding samples without pine phytoliths (2)	**1.43**	**0.91**	**0.53**

Table 2: Table 2: mean, standard deviation and variation coefficient of the percentages of pine phytoliths in alluvial and palaeosoils assemblages from Lalo:
1. all pedogenised or alluvial samples included;
2. only samples containing pine phytoliths included (samples without pine phytoliths indicate only the absence of pine from the basin, not on the origin of the phytoliths);
3. sample T2-19 (containing an exceptionally high amount of pine phytoliths) excluded.

development of deciduous oak forests during the early Atlantic period.

It is interesting to note that the results obtained by the pedoanthracological and the phytolith studies provide convergent but still different images of the past vegetation. It must be stressed that in spite of the presence of pine phytoliths, this species was never identified at Lalo by the pedoanthracological analysis. Only one fragment of pine (*Pinus* cf. *sylvestris*) has been identified from a Neolithic pit of younger date than sample T2-10 (Thiébault, *in* Beeching and Vital, 2001). This would suggest that pine was absent or at least rare in the burnt vegetation of the Citelle basin. Still, some samples, in particular T2-19, contain a large proportion of pine phytoliths and numerous fragments of oak charcoal. On the other hand, phytoliths from dicotyledons are rare and no pine charcoal has been identified. This discrepancy can be explained in several ways:
- a different spatial distribution of pine and oak in the landscape: oaks were growing in the alluvial plain and pines on higher, marly slopes. The stands of oak may have suffered from fire, while the pines were spared (this is however contrary to what would usually be expected, cf. *infra*).
- a shift between oak and pine forests occurring through time: the same parcel could have been colonised by pines after the destruction of the oak forest by fire.

The replacement of oak forests by pine in the low alluvial plain is nevertheless unlikely as the alluvial soils would not constitute an appropriate substratum for pine. It must be noted though that pine phytoliths are present, even if in limited numbers, almost all through the sequence. If pine did not grow in the studied area, as suggested by the absence of pine charcoal, pine phytoliths must have originated from the surrounding slopes.

Normally, the pedoanthracological and phytolith assemblages present in palaeosoils are supposed to reflect the local conditions of the station, possibly including some weak regional "background noise". At this site, the amount of pine phytoliths in the soil assemblages (mean value of 2.73% or 3.64% depending on which samples were included in calculation, tabl. 2) could indicate the presence of phytoliths coming from both the basin and the actual station. It is probably misleading to consider that only pine phytoliths could come from uphill areas. Dicotyledon and grass phytoliths could equally have arrived from the basin in quantities comparable to those of pine (always >4%, except in sample T2-54) but for these taxonomical groups it has not been possible to distinguish between allochthonous and autochthonous specimens. Sample T2-54, containing 7.69% of pine phytoliths, probably comes from a preboreal palaeosoil (Berger *in* Beeching and Vital, 2001). If so, the higher proportion of pine phytoliths could be explained by a more significant presence of pine in the vegetation cover. This would be in accordance with the vegetation pattern known from this period by other palaeoenvironmental studies carried out in southern France (Heinz and Thiébault, 1998).

The number of pine phytoliths is less variable in palaeosoils than in alluvial sediments, independently of the considered samples (tabl. 2: variation coefficient = 0.85 or 0.53 for palaeosoils assemblages, versus 0.91 to 1.72 for alluvial samples). The low proportion of pine morphotypes in the buried soils of Lalo confirms the local significance of phytolith spectra as it is unlikely that pine would have grown in the low alluvial plain. The differences in the number of pine phytoliths between alluvial sediments and palaeosils suggest that the first reflect the actual extension of pine woods on the slopes while the latter show a general "homogenised" image, which is more difficult to interpret.

Pine and oak formations could successively have replaced each other not only in space but also in time. The pedoanthracological assemblages reflect a vegetation destroyed by one or several fire events (in the latter case by the accumulation of several layers, each corresponding to a minor fire) while phytoliths record the vegetation over a longer period of time.

Palaeobotanical remains which are not subject to much transport record variations in the regional vegetation pattern on a more or less local scale. The Atlantic period is no longer perceived as uniformly forested. Open spaces were still remaining, probably in the most unstable areas. Berger and Brochier (2000) have shown that open areas developed in the middle Rhone valley in response to fire phases occurring in cycles of four to five centuries.

The absence of pine charcoal in the pedoanthracological assemblages, despite its presence in the region (attested by regional palaeobotanical studies such as Argant, 1990; Thiébault *in* Berger, 1996; Argant *in* Berger, 1996) and even in the basin (attested by the phytolith assemblages) suggests that the vegetal formations that burnt were not always those known as to be highly inflammable. Heliophilous (samples T2-35, T2-19, T2-8 and T2-9) and riparian (samples T2-07, T2-08 and T2-09) taxa as well as numerous fragments of *Quercus* were identified from fire layers, showing that any kind of vegetation can burn under propitious climatic conditions or due to anthropogenic factors. In sample T2-08, charcoal belonging to *Quercus*, *Salix/Populus*, *Prunus* and Betulaceae was identified. For the same period (6400-6200 BP), Berger and Brochier (2000) noted carbonated levels, deposited under conditions of high evapotranspiration, as well as fire levels all along the Rhone valley. Climatic factors, together with the weakening of the ecosystems due to agro-pastoral activities (palynological studies bear evidence of such activities from this period on, Argant, 1990) are responsible for the burning of open and hydrophilous vegetation communities. Charcoal fragments are also numerous and belong all to deciduous oak in sample T2-2, dated to the Mesolithic period. Berger and Thiébault (1997 and *in* Berger and Brochier, 2000) have shown that fires frequently affected deciduous oak formations, riparian forests and border species in the middle Rhone valley around 7300 BP. According to these authors, climatic conditions were determinate for these fire events while the activities of Mesolithic populations played only a minor role.

Conclusion

The combined study of phytoliths and micro-charcoal allow us to obtain rich palaeoenvironmental data from pedosedimentary sequences that at first sight seem to contain but few archaeobotanical remains. Due to their chemical composition, phytoliths are resistant and therefore preserved in most sediments. They register variations in the density and composition of the vegetation cover at various levels. Charcoal analysis, that allows a more precise determination of ligneous elements, provides qualitative information completing the quantitative data obtained by phytolith analysis. By this study, the first of its kind in the middle Rhone valley, we have shown that it is possible to characterise the vegetation on a local scale (by taxonomical identification) and to trace the alternation of forests and open formations, due to fire events. Here we have presented the first results obtained in the framework of an ongoing bioarchaeological program in the region. They are promising but need to be elaborated further by accurate statistical tests, when more data will be available. Studies of others sequences from the middle Rhone valley are in progress. They will provide multiple data on local variations of the vegetation cover during the Holocene in a general regional context as well as new information concerning the natural conditions that favour the outbreak of fires.

References

ALEXANDRE A., MEUNIER J.-D., MARIOTTI A., SOUBIES F. 1999 - Late Holocene phytoliths and carbon-isotope record from a latosol at Salitre, south-central Brazil. *Quaternary Research*, 51 (2), p. 187-194.

ARGANT J. 1990 - *Climat et environnement au Quaternaire dans le bassin du Rhône d'après les données palynologiques*. Lyon, Université Lyon 1 (Documents du Laboratoire de Géologie de Lyon 111), 199 p.

BARBONI D., BONNEFILLE R., ALEXANDRE A., MEUNIER J.-D. 1999 - Phytoliths as palaeoenvironmental indicators, West Side Middle Awash Valley, Ethiopia. *Palaeogeography, Palaeoclimatology, Palaeoecology* , 152, p. 87-100.

BARTOLI F., GUILLET B. 1977 - Etude comparée de diagrammes phytolithique et pollinique d'un podzol des Vosges gréseuses. *C.R. Acad. Sc. Paris,* 284, p. 352-356.

BARTOLI F., SOUCHIER B. 1978 - Cycle et rôle du silicium d'origine végétale dans les écosystèmes forestiers tempérés. *Annales des sciences forestières*. p. 189-202.

BEAULIEU J.-L. de 1977 - *Contribution pollenanalytique à l'histoire Tardiglaciaire et Holocène de la végétation des Alpes méridionnales françaises*. Thèse de doctorat de 3ème cycle, Université d'Aix-Marseille III, 358 p.

BEECHING A, VITAL J. 2001 - *Espeluche-Lalo*, Document Final de Synthèse, Ministère de la Culture.

BERGER J.-F. 1996 - *Le cadre paléogéographique des occupations du bassin du valdainais (Drôme) à l'Holocène*. Thèse Université Paris I, 325 p.

BERGER J.-F., THIÉBAULT S. 1997 - *Holocene fire history in south-east France: impact of climatic changes and human activities*. Symposium "Fire and the Palaeoenvironment", Sheffield, march 1997. Unpublished oral communication.

BERGER J.-F., THIEBAULT S. - The study and significance of charcoal as an indicator of ancient fires: an application to the middle Rhone valley (France). in: New methodological approches in anthracology, vegetation history and wood-use since prehistory. - Papers of the 2nd international meeting of anthracology, *British Archaeological Report, international series,* p. 25-42.

BERGER J.-F., BROCHIER J.-L. 2000 - Evolution des paysages et du climat dans la moyenne vallée du Rhône et sa bordure préalpine de 13000 à 5000 BP. in *Les derniers chasseurs-cueilleurs d'Europe occidentale*. Actes du colloque international de Besançon, octobre 1998. Besançon, Presses universitaires Franc-Comtoises, p. 37-57.

CARBONE U. 1977 - Phytoliths as palaeoecological indicators. *Annals of the New York Academy of Science* 288, p. 194-205.

CARCAILLET C., THINON M. 1996 - Pedoanthracological contribution to the evolution of the upper treeline in the Maurienne Valley (North French Alps): methodology and preliminary data. *Review of Palaeobotany and Palynology*, 91, p. 399-416.

CLARK J. 1988 - Particle motion and the theory of charcoal analysis: source area, transport, deposition, and sampling. *Quaternary Research*, 30, p. 67-80.

DELHON C. 2001 - L'analyse phytolithique : un nouvel outil pour l'étude des paléoenvironnements méditerranéens. in press, *Revue d'Archéométrie*.

DELHON C., ALEXANDRE A., BERGER J.-F., MEUNIER J.-D., THIEBAULT S., BROCHIER J.-L. Forthcoming - *Phytoliths assemblages as a promising tool for reconstructing Mediterranean Holocene vegetation*. Submited to Quaternary Research.

FIGUEIRAL I., MOSBUGGER V. 2000 - A review of charcoal analysis as a tool for assessing Quaternary and Tertiary environments: achievments and limits. In: "Fire and the palaeoenvironment", A. Scott, J. Moore and B. Brayshay eds., *Palaeogeography, Palaeoclimatology, Palaeoecology* , 164, n°1-4, p. 397-407.

FREDLUND G., TIESZEN L.T. - 1994. Modern phytolith assemblages from the North American Great Plains. *Journal of Biogeography*, 21, p. 321-335.

FREDLUND G., TIESZEN L.T. 1997 (a) - Calibrating grass phytoliths assemblages in climatic terms: Application to late Pleistocene assemblages from Kansas and Nebraska. *Palaeogeography, Palaeoclimatology, Palaeoecology*, 136, p. 199-211.

FREDLUND G., TIESZEN L.T. 1997 (b). - Phytolith and carbon evidence for late Quaternary vegetation and climate change in the southern Black Hills, South Dakota. *Quaternary Research*, 47, p. 206-217.

HEINZ C., THIEBAULT S. 1998 - Characterization and palaeoecological significance of archaeological charcoal assemblages during late and post-glacial phases in southern France. *Quaternary Research*, 50, p. 56-68.

INOUE K., SASE T. 1996 - Paleoenvironmental history of post-Toya ash tephric deposits and paleosols at Iwate volcano, Japan, using eolian dust content and phytolith composition. *Quaternary International*, 34-36, p. 206-217.

JALUT G. 1974 - *Evolution de la végétation et variations climatiques durant les quinze derniers millénaires dans l'extrémité orientales des Pyrénées*. Thèse Université de Toulouse, 181 p.

KELLY E.F. 1990 - *Method for extracting opal phytoliths from soils and plant material*. Internal Document, Department of Agronomy, Colorado state university, Fort Collins, 10 p.

KURMAN M. H. 1985 - An opal phytoliths and palynomorph study of extant and fossil soils in Kansas (USA). *Palaeogeography, Palaeoclimatology, Palaeoecology*, 49, p. 217-235.

LAROCHE J. 1977 - La silice et les plantes supérieures. *Revue de Cytologie et de Biologie Végétale*, 40, p. 15-45.

NICHOLS G., CRIPPS J., COLLINSON M., SCOTT A. 2000 - Experiments in waterlogging and sedimentology of charcoal: results and implications. In: "Fire and the palaeoenvironment", A. Scott, J. Moore and B. Brayshay eds. *Palaeogeography, Palaeoclimatology, Palaeoecology*, 164, n°1-4, p. 43-56.

OZENDA P. 1964 - *Biogéographie végétale*. Paris, Douin, 374 p.

OZENDA P. 1994 - *Végétation du continent européen*. Paris, Delachaux et Niestlé, 272 p.

PIPERNO D. R. 1988 - *Phytolith Analysis. An Archaeological and Geological Perspective*. New York, Academic press, 280 p.

ROVNER I. 1988 - Macro and micro ecological reconstruction using plant opal phytolith. Data from archaeological sediment. *Geoarchaeology: an International Journal*, 3 (1), p. 155-163.

SCHWEINGRUBER F. H. 1990 - *Anatomie europäischer Hölzer*. Stuttgart, Haupt, 800 p.

SCOTT A., CRIPPS J., COLLINSON M., NICHOLS G. 2000 - The taphonomy of charcoals following a recent hearthland fire and some implication for the interpretation of fossil charcoal deposits. In: "Fire and the palaeoenvironment", A. Scott, J. Moore and B. Brayshay eds. *Palaeogeography, Palaeoclimatology, Palaeoecology*, 164, n°1-4, p. 3-31.

THIEBAULT S. 1999 - *Dynamique des paysages et intervention humaine, du Tardiglaciaire à l'Holocène, de la Méditerranée aux Préalpes sud-occidentales – apport de l'analyse anthracologique*. Rapport de synthèse présenté à l'Université de Paris 1 Panthéon-Sorbonne en vue de l'habilitation à diriger des recherches. 278 p.

THINON M. 1992 - *L'analyse pédoanthracologique, aspect méthodologiques et applications*. Thèse de doctorat ès Sciences, Université Aix-Marseille III, 317 p.

THINON M., TALON B. 1998 - Ampleur de l'anthropisation des étages supérieurs dans les alpes du sud : données pédoanthracologiques. *Ecologie*, t. 29 (1-2), p. 323-328.

TRIAT-LAVAL H. 1978 - *Contribution pollenanalytique à l'histoire Tardiglaciaire et Postglaciaire de la végétation de la basse vallée du Rhone*. Thèse de doctorat d'Etat, Université d'Aix-Marseille III, 343 p.

TSUTSUKI K., KONDO R., SHIRAISHI H., KUWATSUKA S. and Ohnohara Wetland Research group 1993 - Composition of lignin-degradation products, lipids, and opal phytoliths in a peat profile accumulated since 32000 years B.P. in central Japan. *Soil Science Plant Nutrition*, 39 (3), p. 463-474.

TWISS P.C., SUESS E., SMITH R.M. 1969 - Morphological classification of grass phytoliths. *Proc. Soil Sci. Soc. Am.* 33, p. 109-115.

VERNET J.-L., THIEBAULT S. 1987 - An approach to north-western Mediterranean recent prehistoric vegetation and ecologic implications. *Journal of Biogeography*, 14, p. 117-127.

WESTERN A.C. 1963 - Wood and charcoal in archaeology. in: D. Brothwell, E. higgs and G. Clark (eds), *Science in archaeology: a comprehensive survey of Progress and Research*. Thames & Hudson, p. 150-158.

Palaeoenvironmental approach of Ponsonby archaeological site (Riesco Island, Chilean Patagonia)

Maria-Eugenia Solari[1], Vincent Bernard[2], Dominique Legoupil[3], Hervé Richard[4], Paul Roiron[5] and Patrick Schoellammer[6]

1) Universidad Austral de Chile – Instituto de Ciencias Sociales - casilla 567 – Valdivia, Chili.
E-mail : msolari@ uca.uach.cl
2) CNRS-UMR 6566, Université de Rennes I - Campus de Beaulieu – 35042 – Rennes, France.
3) CNRS–UMR 7041, Archéologies et Sciences de l'Antiquité – 21 allée de l'Université –92023 Nanterre cedex, France.
E-mail : legoupil@mae.u-paris10.fr
4) CNRS-UMR 6565, 16, route de Gray – 25030 – Besançon cedex 30, France.
E-mail : herve.richard@univ-fcomte.fr
5) CNRS-UMR 5059, Institut de Botanique - 163 rue A. Broussonet –34090 - Montpellier cedex, France.
E-mail : roiron@crit.univ-montp2.fr
6) Service et Musée d'Archéologie - Espace Paul Vouga – 2068 - Hauterive, Suisse.
E-mail : patrick.schoellammer@ne.ch

Abstract

The site of Ponsonby (Chilean Patagonia) provides evidence of one of the earliest Patagonian coastal occupations. The site is located between the Patagonian steppe "pampa", an area that is currently influenced by human activities, and the Cordillera (Island Riesco), covered with a *Nothofagus* forest. It shows human adaptation to the maritime territory in the middle of the Holocene. In addition, Ponsonby provided palaeobotanical data by means of pluridisciplinary studies (pollen analysis, leaf macroflora, dendrochronology, wood and charcoal) carried out at the main archaeological site (locus 1) and a nearby peat bog (locus 2). The results obtained for these two locus are complementary and provide an accurate record of the local and regional vegetation history for approximately 4000 years. Thus, pollen analysis of levels of peat bog (*c.* 8000-7000 BP) is supplemented by the analysis of charcoal fragments found in levels without pollen from locus 1 (*c.* 7000-4250 BP). Similarly dendrochronological data associated with wood and macroflora analysis give also a more precise view of the local palaeoenvironment at the exact moment of human implantation 7000 years ago. The various palaeobotanical studies provide the first long-term vegetational history for this part of the Riesco Island, which transform an open landscape with thickets of evergreen *Nothofagus* to a shrubby steppic vegetation with *Maytenus* in dry zones, similar to present-day conditions. The implantation of human population in this site corresponds to an important modification of the environment: a marine ingression which lead to the insolation of Riesco Island where the site is located.

Key-words: *Chili, Patagonia, archaeological site, pollen, anthracology, dendrochronology, leaf macroflora.*

Résumé

Le site de Ponsonby qui se trouve à la lisière entre la steppe "pampa" patagonienne actuellement complètement anthropisée et la Cordillère, domaine de la forêt de *Nothofagus* est l'un des plus anciens sites des archipels de Patagonie présentant une stratigraphie continue sur près de 4 millénaires. Il montre l'implantation progressive de l'homme dans le territoire maritime au milieu de l'Holocène. Parallèlement à cette installation, on peut y suivre l'évolution du milieu, grâce au relais stratigraphique de deux fouilles très différentes quoique voisines : une occupation archéologique située dans un cône de colluvions sableuses (le locus 1) et une tourbière ou plus exactement un marais tourbeux (le locus 2). La richesse et la diversité de ce site ont permis une approche pluridisciplinaire particulièrement complète et sans hiatus d'environ 8000 à 4000 ans BP, chaque discipline apportant sa pierre à l'édifice. Ainsi la palynologie, très explicite dans les niveaux inférieurs de la tourbière, est relayée par l'anthracologie du locus 1 pour les niveaux supérieurs, sableux et pauvres en pollen. De leur côté, la dendrochronologie, l'étude des bois et de la macroflore foliaire renforcent et précisent nos connaissances sur l'environnement local à la période-clé de l'installation de l'homme il y a environ 7000 ans. Ces diverses études paléobotaniques montrent que la végétation de l'île Riesco passe d'un milieu ouvert avec des bosquets de *Nothofagus sempervirens* à une steppe buissonnante et arbustive à *Maytenus* dans les zones plus sèches, proche du contexte actuel. L'implantation de l'homme dans cet endroit semble en outre correspondre au moment d'une modification importante du milieu, une ingression marine ayant entraîné l'ouverture du canal Fitz-Roy et l'isolement de l'île Riesco sur laquelle se trouve le site.

Mots-clés : *Chili, Patagonie, site archéologique, pollen, anthracologie, dendrochronologie, macroflore foliaire.*

Introduction

Ponsonby is one of the few sites in the region, that presents a continuous corpus of data that spans several millennia. Moreover, it is the oldest site that retains vestiges of human occupation in archipelagos of the southern part of South America, except in the particular case of the great island of Tierra del Fuego. This site offers the possibility of observing the progressive human adaptation from hunting of terrestrial mammals to a maritime economy. It allows an exceptionally pluridisciplinary investigation of well-preserved fauna and flora remains. The aim of this paper is to give an overview of the different results obtained on plant macro and microremains (wood, charcoal, leaves, pollen) preserved in the site and its close environment.

The study area

Ponsonby is located on the Riesco Island in the Chilean Patagonia (Magallanes zone), about 90 km north-west of Punta Arenas (fig. 1). The archaeological site is situated *c.* 100 m from the Fitz-Roy channel. It was discovered and excavated by J. Emperaire in 1951-52 and 1958, date of his accidental death at the excavation site. From 1994 to 1998, a French expedition team directed by D. Legoupil carried out a new campaign of excavation. Our study is based on these excavations (locus 1 and 2).

According to Koppen, this part of Riesco Island belongs today to the bio-climatic zone of the cold steppe: P. annual average about 400 mm (Zamora *et al.* 1981); annual average

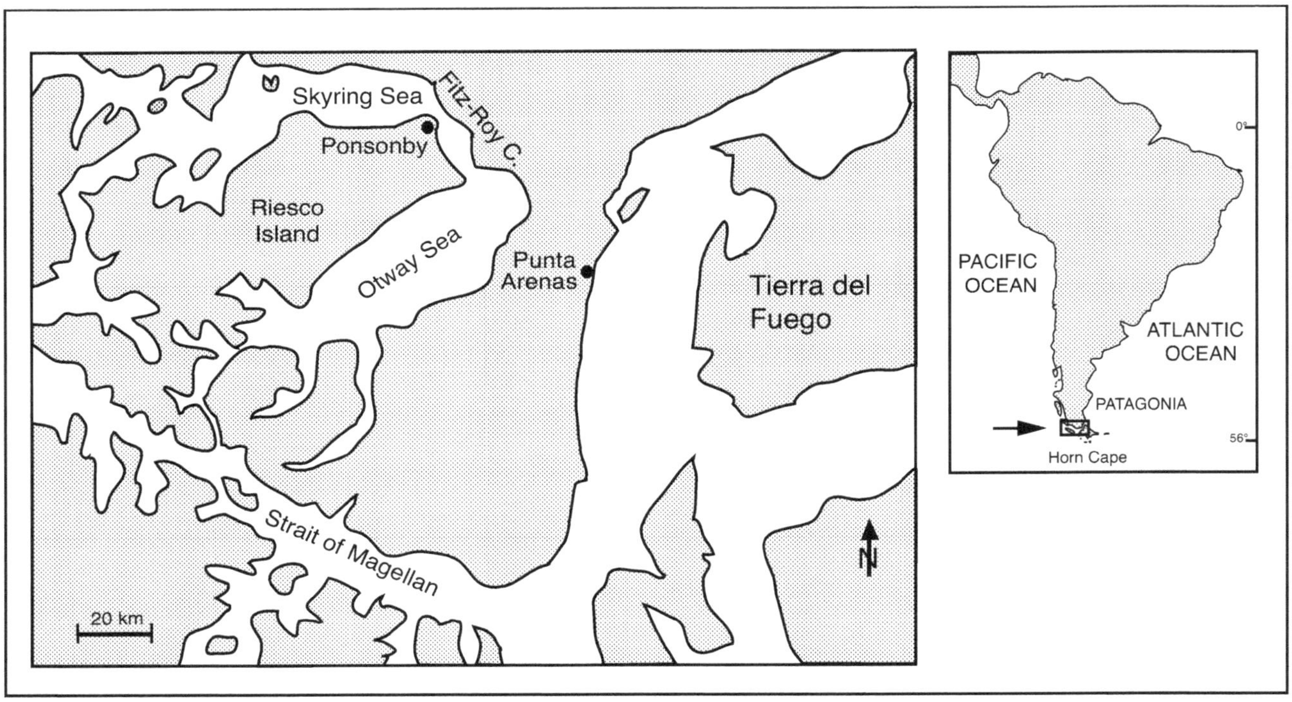

Figure 1: Geographical location of Ponsonby.

temperature between 8 and 9°C (Pisano, 1973). The woody elements of this association are *Nothofagus antarctica* and some shrubs such as *Berberis*, Compositae and *Maytenus magellanica*.

Stratigraphy

Two excavations (fig. 2) were carried out during the second campaign on Ponsonby (cf. Legoupil, in press):

- The main archaeological excavation or locus 1

An area of about 50 m^2 was excavated. The site was occupied from about 7000 (between 7450 ± 80 BP —doubtful— and 6690 ± 130 BP) to 4130 ± 75 BP (for all dates, see Fontugne and Legoupil, in press). The stratigraphy shows three archaeological layers: D-C-B. At the base of the sequence, gleyish clays (layer D) correspond to the marginal facies of the peat bog infilling. This unit is overlain by a thin white clay sediment, which seems in relation to a chalk level observed in the peat bog. Some desiccation cracks filled up by sands disturb the stratification. This can reflect an hygrophilous vegetation adapted to a near-permanent flooded environment. Layer D corresponds to the first human occupation and provided similar results to those obtained from the peat bog.

Upper levels (layers C and B) are sandy slope deposits, slightly disturbed by geomorphological moves; no pollens are conserved.

- The peat bog or locus 2

The bog is about 30 m to the south-east of the main excavation and 2/3 m below. This natural depression, once colonised by swamp vegetation, was formed behind morainic accumulations, parallel to the Fitz-Roy channel. This zone offers a very contrasted stratigraphy that lays between 8150 ± 60 and 4210 ± 75 BP. In the lower part, laminated sandy clays contain two organic layers rich in guanacos bones (*Lama guanicoe*), plant remains and wood, which are in chronological relation with the layer D of the locus 1 (fig. 3). Further up, a carbonated silt level, 20 to 40 cm thick, containing some occasional artefacts, is covered by sands which are poor in pollen, but with two archaeological levels dated from about 4200 BP.

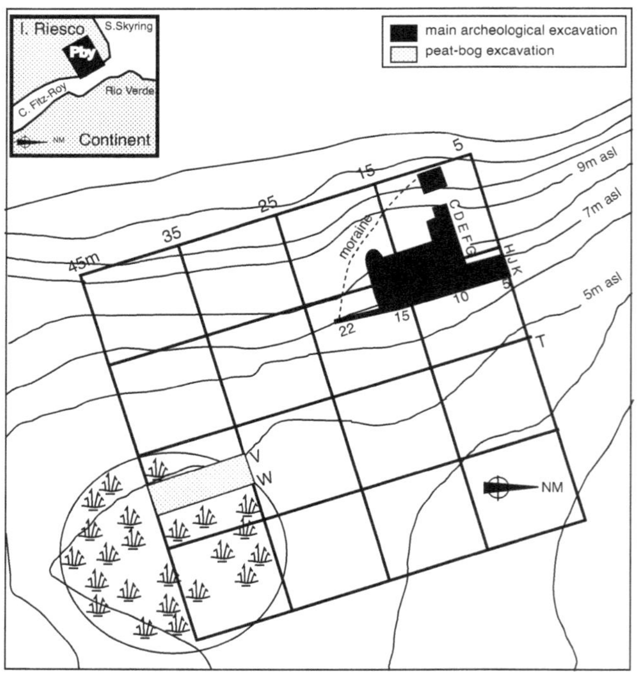

Figure 2: Main excavation (locus1) and peat-bog (locus 2) of Ponsonby.

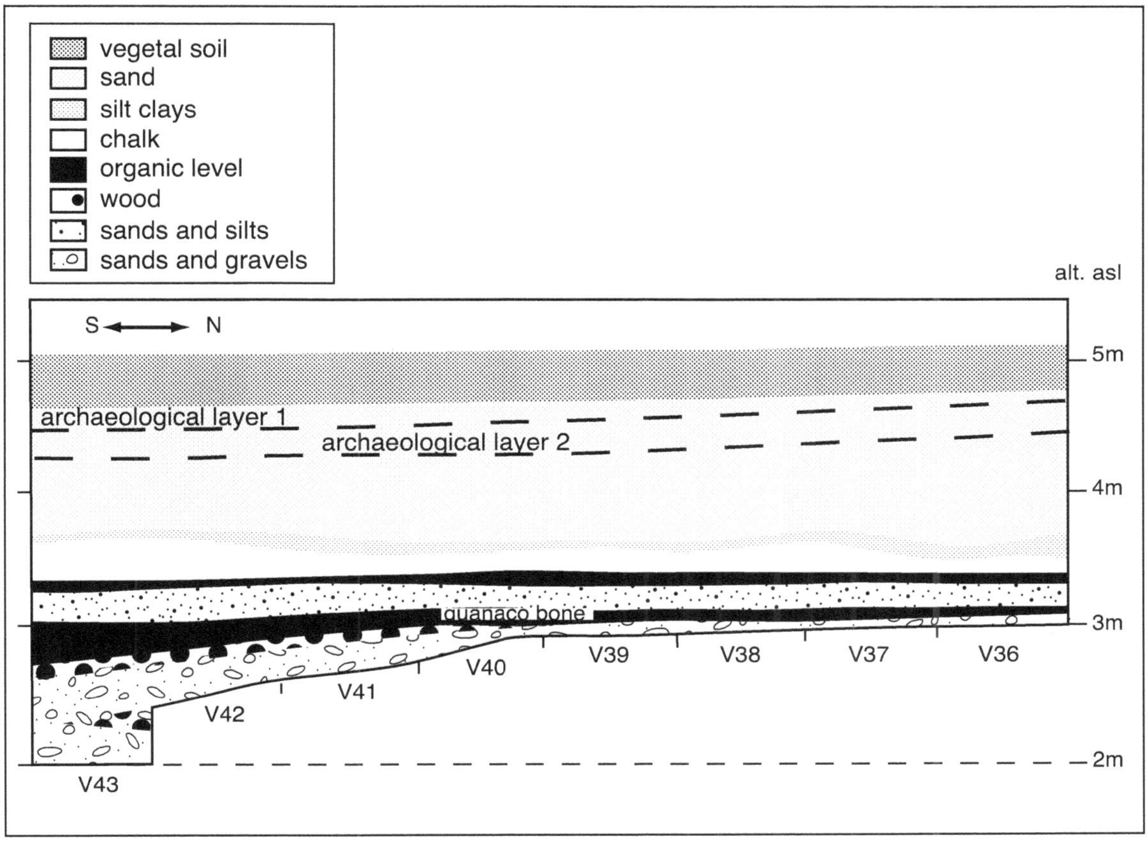

Figure 3: Peat-bog stratigraphy.

Charcoal flora

A charcoal sampling was carried out in a stripe of 2 x 6 m at the main archaeological site. Thus, the reconstruction of the vegetation dynamics is supported by charcoal analysis of the levels D (*c*.7000-6700 BP), C (5400-4500 BP) and B (4500-4200 BP). Results obtained are compared with the study of other plant remains. They offer an image of the vegetation evolution in this zone.

2600 charcoal pieces from the three archaeological layers of locus 1 were identified (tabl. 1). Charcoal from layer D presents ferruginous intrusions in their anatomic structure. It is a wet layer that corresponds to the level of peat bog.

Locus 1	Layer D (from 7450 to 6700 B.P.)	Layer C (4500-5400 B.P.)	Layer B (from 4500 to 4200 B.P.)
Nothofagus (evergreen) type*betuloides*	**104**	**12**	**38**
Nothofagus (decidous) type *pumilio*	22		11
Nothofagus (decidous) type *antarctica*	**3**	**3**	**64**
Nothofagus sp.	63	12	19
Berberis sp.	54	54	176
Compositae type*Chiliotrichium*	1	10	36
Compositae type *Baccharis* sp.	26	12	96
Compositae Baccharis-Chiliotrichum	15	8	175
Maytenus magellanica	**27**	**288**	**1008**
cf. Fuchsia magellanica	1	3	3
cf. Ribes magellanica		3	30
Ericaceae		1	
Drymis winteri	2		
Embothrium coccineum	1		2
Coniferae type*Pilgerodendron uviferae*	2		
Indeterminata	9	6	49
Indeterminable	12	14	112
TOTAL	**342**	**426**	**1819**

Table 1: Charcoal analysis of layers D, C and B from the main archaeological site at Ponsonby .

Nothofagus betuloides, an evergreen tree which grows in wet forests, is the main taxon.

In layers C and B, other taxa beginning continously in the basal layer, have gradual developments at the expense of *Nothofagus betuloides*. In these layers, an opening of the vegetation is testified by the increase of heliophilous species (*Berberis*, Compositae, *Ribes*), and deciduous *Nothofagus,* as well as the predominant species, *Maytenus magellanica*.

At present, an anthropic steppe on the site exists with some heliophilous species, such as *Berberis buxifolia*, *Baccharis patagonica* and *Chiliotrichium diffusum*. *Nothofagus betuloides* grow about 40 km away, near the Skyring sea. Close to the Otway sea, deciduous species of genus *Nothofagus* (*N. pumilio* and *N. antarctica*) grow under steppic conditions, in association or even in hybrid forms.

Pollen flora

Pollen analysis was carried out on the main archaeological site (locus 1), the peat deposit of locus 2, and on a core obtained from a wetland 100 m away from the field excavation (locus Pinguino). Only the results from the peat bog are presented here (fig. 4). 49 samples were taken during the excavation on a 2,50 m high profile. The base of the stratigraphy consists of sandy sediments of lacustrine origin, the middle part is composed of a succession of peat layers and carbonated silts (lacustrine marl), and the upper part is formed by marine sandy deposits.

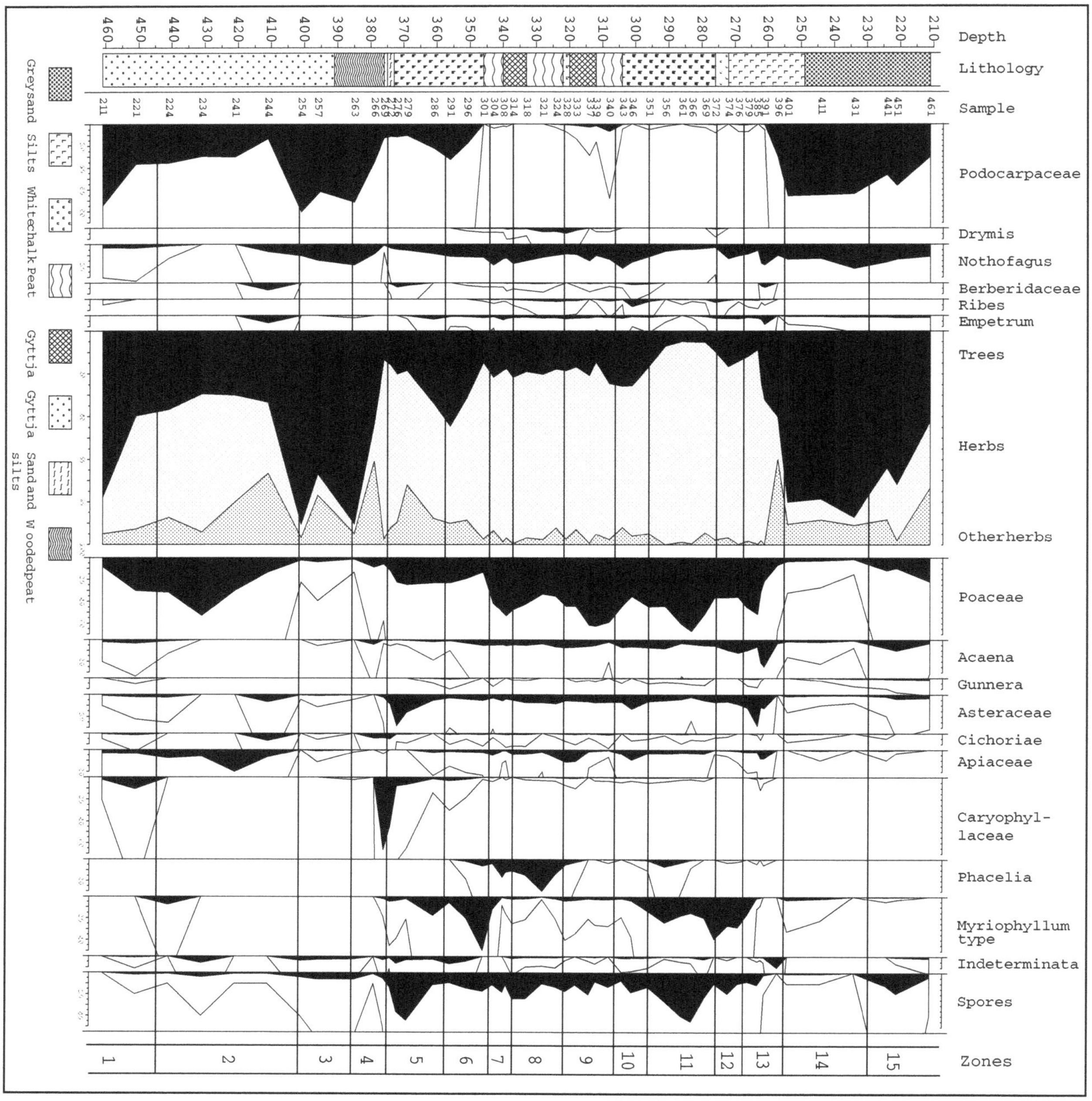

Figure 4: Palynological diagram of peat-bog excavation.

The pollen diagram obtained may be divided into three phases corresponding roughly to the stratigraphic sections: the sandy levels at the base and at the top are dominated by Podocarpaceae, while the peat and carbonated silts are dominated by herbaceous pollen. Relatively large amounts of Podocarpaceae in sandy levels do not reflect surrounding and regional vegetation. They probably result from a long-distance pollen input and a floating concentration effect. Only a few grains should be considered as the "normal" arboreal pollen representation, which can be observed in the median part of the sequence (peat and silt) *i.e.* 330-340 cm a.s.l.

Nothofagus frequencies do not show significant differences; they are always lower than 15 %, even in levels with abundant macro-remains such as trunks and branches of this genus (260-273 cm a.s.l.). Relatively low percentages of *Nothofagus* in the pollen spectrum lead us to consider that, at this time, there is not a local *Nothofagus* forest (see for exemple *Nothofagus* forest pollinic image at the same time in: Heusser, 1989, 1990; Heusser *et al.* 1996; Markgraf, 1980, 1993; Markgraf *et al.* 1992; Pendall *et al.* 2001; Rabassa *et al.* 2000).

Other arboreal and shrub pollens include Berberidaceae, *Ribes*, *Empetrum* and *Drimys*. Their percentages are higher in boggy and silty layers even though low values are observed in silty levels between 343 and 372 cm a.s.l.

This particular stratigraphic section, where peat and silts alternate, is characterised by herbaceous plants such as Poaceae, *Acaena*, *Gunnera* and Apiaceae.

Several levels show high values for one or two species. For example, a peak of Caryophyllaceae is observed between the two principal basic peat levels (sample 269); a significant rise of *Phacelia* (pollen zones 7 and 8) is recognised, and high frequencies of *Myriophyllum* are noted in zones 5-6 and 11-12. Such variations have local origins, showing modifications of the vegetation cover in this area, due to changes in water level.The vegetation indicated by the pollen assemblage is very similar to the present-day vegetation in the area. The ancient pollinic results show only a slightly higher representation of *Nothofagus* and more long-distance elements (Podocarpaceae for example). The increase in the taxonomic diversity and the importance of shrubs in this open landscape suggest logically a less important human impact.

Leaf and fruit flora

Leaf fragments are preserved in the upper peat level, approximately 15 cm thick. This layer is located just below the carbonated silts level. The peat is also rich in pollen, seeds, insect remains and mollusc shells.

The identification of the samples was carried out by comparing their morphology with that of leaves of extant species living in southern Chile.

Nothofagus betuloides, Berberis buxifolia and *Ribes magellanicum* have been identified (fig. 5). Small tough leaves with entire margins indicate the presence of several chamephytes species which are not yet identified. Many small fruits of *Hippuris vulgaris* are also present.

The leaves of *Nothofagus betuloides* confirm the presence of this species in the landscape during this period. Leaves identified as belonging to *Ribes magellanicum*, a shrub growing in wetlands often in the understory of the *Nothofagus* forest also supports this image of an open forest vegetation. However, the presence of *Berberis buxifolia* and the chamephytes also testify to the existence of a shrubby open vegetation where *Ribes magellanicum* may also grow. In addition the many fruits of *Hippuris*, an aquatic plant of shallow water, indicate the presence of wet or marshy zones.

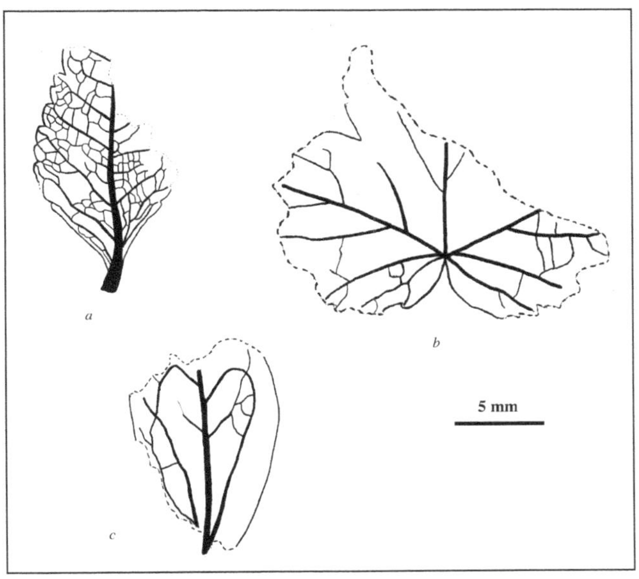

Figure 5: Leaves of *Nothofagus betuloides*, *Ribes magellanicum* and *Berberis buxifolia* in the peat-bog.

Dendrochronology

A level with some 40 floated wood remains, trunks or branches (fig. 6), lays in the lower part of the peat-bog. One was dated from 7275 ± 70 BP. In addition to specific identifications, some are enough in size to make dendrochronologic measurements (fig. 7). However, considering the complexity of the tree-climate response and the limited number of trees represented, the first approach is to try to extract more information from the wood itself.

J.A. Boninsegna, F. Roig and R. Villalba (Dep. Dendrocronologia e Historia Ambiental, Mendoza, Argentina) have worked in Patagonia on many natural sites and on many species of woods for the last decade. They have established 21 chronologies in Fuegian *Nothofagus* forests and a 1400-yr floating chronology based on subfossil woods in the same area (Boninsegna and Villalba, 1996; Roig *et al.* 1996). With 7 samples, our work can only illustrate some of the aspects of the natural landscape around 7200 BP (^{14}C analysis).

Figure 6: The wood level in the peat-bog (7200 BP).

Although ring widths may not be interpretable directly as measurements of the climate, it is clear that narrow rings indicate periods detrimental to growth. Actually, on the 105 years of *Nothofagus* tree-ring chronology constructed from 7 trees from Ponsonby's bog (of floating time scale), we observed five abrupt growth changes, lasting from 10 to 20 years (fig. 7). These changes are characterised by rings less than 1 mm in width and are significant because they affect all the trees. Unusual anatomical characteristics have also been observed: the initial wood presents smaller cells with thinner wall boundaries, while the late wood disappears. The rays are locally quite large (1 mm) and numerous. As it has been well documented elsewhere, some of these events seem to be clearly linked to major variations in the soil water content (Tardif and Bergeron, 1993; Leuschner, 1992; Girardclos *et al.* 1996). A rise in the water table could have limited or prevented root functioning. These *Nothofagus* seem to have lived in an extreme environment, at the limit of tolerable conditions for this species. Therefore, after three or four violent events illustrated by abrupt growth change, some of the trees died. They did not die at the same time; rather they died having produced from 5 to 20 small and disorganised rings.

However, we would stress that other causes for reduced growth, such as drought conditions, physical damage or chemical effects, cannot be ruled out. Fresh samples from living trees around the Skyring sea were repeatedly collected; the study of these reference wood samples constitutes the next step of this work. Comparison with known events will allow us to understand and to interpret a century of growth conditions at the beginning of the Holocene.

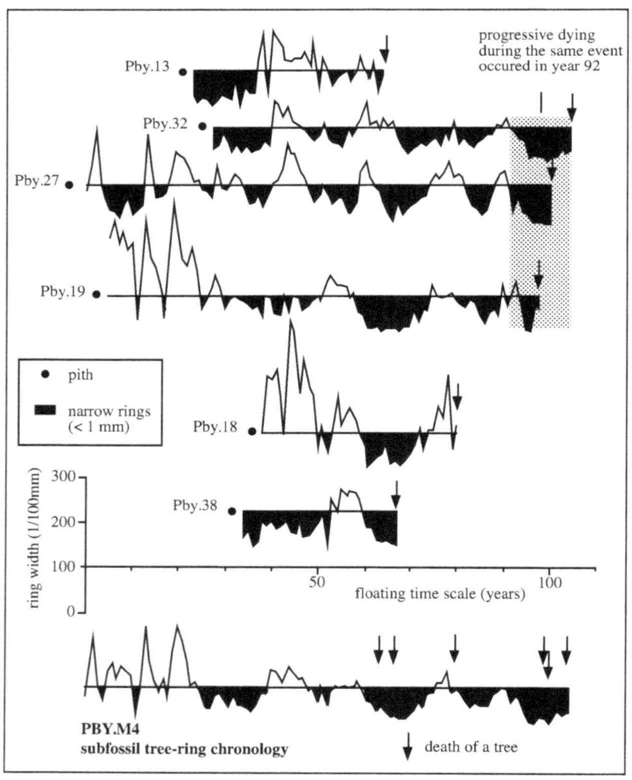

Figure 7: *Nothofagus* series from Ponsonby showing abrupt declines in serial growth.

Discussion

The results obtained are complementary and concordant and unable us to follow the evolution of local and regional environment during near 4000 years.

Palynological results of lower levels of the peat bog show a strong representation of *Podocarpus* pollens coming from distant zones in the north and whose selective conservation is favoured by the sandy sediments. This hypothesis seems to be confirmed by the absence of macroremains (wood or charcoal) of this taxon.

Wood, charcoal and leaves of *Nothofagus* were preserved along the sequence, which seems indicate a local implantation, near the site. But, the presence of this tree in the pollen analysis is not as important as in a *Nothofagus* forest. Two solutions are possible: (1) the local *Nothofagus* trees were not mature (they did not yet produce pollen), or (2) the trees grew only in a few special places in an open landscape.

Several species of *Nothofagus* seem to cohabit in the area during this period. In the lower level of the locus 1 (layer D), the evergreen *Nothofagus betuloides* dominated. This is also the case in the contemporary levels of the peat bog (wood and leaves of *Nothofagus betuloides*). The favorable state of conservation of the external wood rings of the specimens confirms that these trees lived near the deposit location.

Other taxa of drier zones (*Baccharis, Chiliotrichum, Berberis...*) testify to more open formations which could have grown on the morainic bars nearby.

For the upper levels of locus 1, the anthracological flora of layer C is dominated by *Maytenus magellanica,* whereas evergreen *Nothofagus* strongly regresses, which represents a development of dry bushy formations. The pollinic diagram does not contain a lot of pollen of dry entomophilous shrubs as *Maytenus* (Celastraceae), but herbaceous plant pollens indicate steppic landscapes. This phenomena develops in the higher part of the sequence (layer B), whereas the probable opening of the Fitz-Roy channel had caused a marine ingression in the depressions, in particular in the peat bog.

Conclusion

In conclusion, three different phases could be distinguished:
- between 7000-6500 BP, the presence of *Nothofagus betuloides* was noted, particulary around the lake;
- around 5400-4500 BP, the presence of a bushy vegetation with *Maytenus* and small trees and shrubs evolve to a steppic environment, with heliophilous species growing on sandy substrate; deciduous *Nothofagus* (*N. antarctica* et *N. pumilio*) were more present than in the previous period.
- around 4500-4200 BP heliophilous taxa of steppic (bushy) environment such *Berberis* and Compositae were better represented.

This last environment was modified by the arrival of European settlers during the XXth century (sheep husbandry, collection of *Maytenus* for firewood).

On palaeo- and archaeo-botanical levels, the importance of the archaeological site of Ponsonby resides in the presence of numerous plant remains which allow us to observe, in an optimal way, the complementary nature of several disciplines in the analysis of its deposits. Establishing a dialogue between archaeology and four palaeobotanic disciplines (palynology, dendrology, study of fossil leaves and anthracology) is not easy, mainly due to the differences in observation'scales.

In a general sense, the palynology gives an image on three scales: a far scale with the long distance deposits, characterised by the important presence of distant species of gymnosperms; a regional scale and a strictly local one.

During the oldest periods, the pollen analysis are complemented by the deposits, in the peat bog, of wood and fossil leaves (mainly of evergreen *Nothofagus)*, and leaves of *Berberis* and *Ribes*, among others. Dendrochronology attests a series of violent natural events. They can be connected with a predominant natural event: the opening of canal Fitz-Roy.

In the archaeological site, the sequence is relayed by the anthracology that analyses charcoal of the three layers of the deposit and permits us to follow the evolution of the environment beginning in the level D, contemporaneous with peat-bog, and continuing to the most recent level. During that period, the anthracology can give an account of the specific evolution, of the ligneous vegetation. It begins with a mosaic of vegetation, since species of different environments (hygrophilous, heliophilous or others) are already present, but with an abundance of *Nothofagus betuloides*; this will be replaced by bushy species indiginous to a steppic zone, with *Maytenus magellanica* as the main species.

The evolution of Ponsonby environment is determinant for human occupation in this part of Riesco Island: originally, terrestrial hunters, only occupied by guanacos hunting, lived in an open marshy landscape. Then, progressively, during the evolution of the landscape and probably after the opening of the Fitz-Roy Channel and the separation of the Riesco Island, the inhabitants began to exploit the sea products.

References

BONINSEGNA J. A., VILLALBA R. 1996 - Dendroclimatology in the southern hemisphere: review and prospects. Tree Rings, Environnement and Humanity: Proceedings of the International dendrochronological Conference - *Radiocarbon*, p. 127-141.

FONTUGNE M., LEGOUPIL D. en préparation - Les datations du site de Ponsonby. In Les chasseurs-cueilleurs de Ponsonby et leur environnement du VI° au III° millénaire av. J.C. *Anales del Instituto de la Patagonia*, Punta Arenas, Chile.

GIRARDCLOS O., LAMBERT G., LAVIER C. 1996 - Oak tree-ring series from France between 4000 BC and 800 BC. Tree Rings, Environnement and Humanity: Proceedings of the International dendrochronological Conference - *Radiocarbon*, p. 751-768.

HEUSSER C.J. 1989 - Late Quaternary vegetation and climate of southern Tierra del Fuego. *Quaternary Research*. 31, p. 396-405.

HEUSSER C.J. 1990 - Late-glacial and Holocene vegetation and climate of Subantarctic South America. *Review of Palaeobotany and Palynology*. 65, p. 9-15.

HEUSSER C.J., LOWELL T.V., HEUSSER L.E., HAUSER, A., ANDERSEN B.G., DENTON G.H. 1996 - Full-glacial-late-glacial palaeoclimate of the Southern Andes: evidence from pollen, beetle and glacial records. *Journal of Quaternary Science*. 11(3), p. 173-184.

HEUSSER C. J., HEUSSER L. E., LOWELL T. V., MOREIRA A., MOREIRA S. M. 2000 - Deglacial palaeoclimate at Puerto del Hambre, subantarctic Patagonia, Chile. *Journal of Quaternary Science*. 15(2), p. 101-114.

IRIONDO M. 2000 - Patagonian dust in Antarctica. *Quaternary International*. 68-71, p. 83-86.

LEGOUPIL D. ed. (en préparation) - Les chasseurs-cueilleurs de Ponsonby et leur environnement du VI° au III° millénaire av. J.C. *Anales del Instituto de la Patagonia*, Punta Arenas, Chile.

LEUSCHNER H.H. 1992 - *Subfossil trees. Tree-rings and environnement*: Proceedings of the International dendrochronological Symposium - Lundqua Report, fasc. 34, p. 193-197.

MARKGRAF V. 1980 - *New data on the Late and Postglacial vegetational history of "La Mission", Tierra del Fuego, Argentina*. in IV Int. Palynol. Conf. Lucknow.

MARKGRAF V., DODSON J.R., KERSHAW P. A., MACGLONE M. S., NICHOLLS N. 1992 - Evolution of Late Pleistocene and Holocene climates in the circum-South Pacific land areas. *Climate Dynamics*. 6, p. 193-211.

MARKGRAF V. 1993 - Paleoenvironments and paleoclimates in Tierra del Fuego and southernost Patagonia, South America. *Palaeogeography, Paleaoclimatology, Palaeoecology*. 102, p. 53-68.

PENDALL E., MARKGRAF V., HUBER U., WHITE J. W. C., DREIER M., KENNY R. 2001 - Multiproxy Record of Late Pleistocene-Holocene Climate and Vegetation Changes from a Peat Bog in Patagonia. *Quaternary Research*. 55, p. 168-178.

PISANO E. 1973 - Fitogeografia de la Peninsula Brunswick, Magallanes. I. Comunidades meso-higromorficas e higromorficas. *Anales del Instituto de la Patagonia*, Punta Arenas, vol. 4, n° 1-3, p. 141-205.

RABASSA J., CORONATO A., BUJALESKY SALEMME, ROIG C., MEGLIOLI A., HEUSSER C., GORDILLO S., ROIG F., BORROMEI A., QUATTROCCHIO M. 2000 - Quaternary of Tierra del Fuego, Southernmost South America: an updated review. *Quaternary International*. 68-7, p. 217-240.

ROIG F., ROIG C., RABASSA J., BONINSEGNA J. 1996 - Fuegian floating tree-ring chronology from subfossil *Nothofagus* woods. *The Holocene*, p. 469-476.

TARDIF J., BERGERON Y. 1993 - Radial growth of *Fraxinus nigra* in a Canadian boreal floodplain in response to climatic and hydrologic fluctuations. *Journal of Vegetation Science*, fasc. 4, p. 751-758.

VILLAGRAN C. 1988 - Late Quaternary vegetation of southern Isla Grande de Chiloé, Chile. *Quaternary Research*. 29, p. 294-306.

ZAMORA M., SANTANA A., SANTANA E. 1981 - Informe final proyecto Caracteristica climáticas del area central de la Provincia de Magallanes, XII Región, *Inf. Inst. Patagonia*, 100p.

Fuel from protohistorical and historical kilns in north-western France

Dominique Marguerie

CNRS-UMR 6566, Civilisations atlantiques et archéosciences,
Laboratoire d'Anthropologie, Université de Rennes 1, Beaulieu, F-35042 Rennes cedex, France.
E-mail: dominique.marguerie@univ-rennes1.fr

Abstract

The fuel from craft and domestic structures at fourteen Iron Age, gallo-roman and medieval sites in north-western France is studied. The craft structures comprise metal, ceramic, tile, glass or lime kilns. Charcoal fragments are identified under a microscope and the ligneous structure is observed with a binocular lens. Samples from kilns or associated refuse pits in craft sites contain sometimes only one species, while up to ten taxa, originate from several different ecosystems, can be found in domestic kilns. The tree ring curves indicate that oak charcoal from craft structures more often come from timber with a large diameter. In domestic kilns, they derive in equal quantities from trunks and branches. This study clearly shows that artisans made their choice of wood according to both the species and the wood diameter. The status of the workshop influenced on the use of wood that was more or less available locally. Charcoal could either be imported or produced locally. Trunks or large branches of oak and beech are suitable for sustained, "small" fires and useful for a slow rise in kiln temperature. Shrubs or small trees of heathlands, when gathered in bundles, are on the contrary used for rapid "big" fires, in order to obtain high temperatures. The state of the wood used as fuel (diameter, dryness or humidity) is also an essential firing parameter.

Key-words: *charcoal analysis, kiln, palaeoethnography, palaeoenvironment, north-western France*

Résumé

Le combustible provenant de structures artisanales et domestiques a été étudié sur quatorze sites de l'Age du Fer, de la période gallo-romaine et du Moyen Age dans le nord-ouest de la France. Les structures artisanales peuvent correspondre tout aussi bien à des bas fourneaux métallurgiques, qu'à des fours de potiers, de tuiliers, de verriers, voire des fours à chaux. Les charbons ont été déterminés sous microscope et leur plan ligneux préalablement observés à faible grossissement. La diversité taxonomique des échantillons provenant des fours ou des fosses de vidanges associées est parfois monospécifique chez les fours artisanaux tandis qu'elle peut atteindre plus de dix taxons provenant de biotopes variés chez les fours domestiques. L'étude de la courbure des cernes de croissance indique que les charbons de chêne proviennent plutôt de grosses pièces de bois dans les structures artisanales. A l'opposé, les structures domestiques sont alimentées de bois provenant de branches ou de troncs. Cette étude montre clairement que les artisans ont fait un choix d'espèces et de calibre particulier. Le statut des ateliers conditionne une utilisation d'essences disponibles plus ou moins localement voire une importation ou une fabrication locale de combustible sous forme de charbons de bois. Les troncs ou les grosses branches de chêne et de hêtre sont propices au "petit feu" et utiles à une montée lente en température du four. Les arbustes ou petits arbres de la lande sont utilisés pour le "grand feu" surtout lorsqu'ils sont réduits en fagots. L'état du bois utilisé comme combustible (de plus ou moins grosse section, sec ou vert) apparaît également fondamental pour la conduite de la cuisson.

Mots-clés : *Anthracologie, four, palethnographie, paléoenvironnement, nord-ouest de la France.*

Introduction

This paper presents data on the working of protohistorical and historical craft and domestic kilns obtained by charcoal analysis. The kind of fuel used plays in fact a fundamental role for the increase and intensity of temperature in a kiln. The remains of the final combustion are used here in order to determine the choice of taxa and wood diameters made by the artisans.

Study area and archaeological sites

The study area is located in north-western France and most of the sites are in Brittany. This area is a vast peninsula, stretching far out into the ocean, with little surface relief (around 100 m) and with a generally Atlantic climate. The vegetation belongs to the dominant series of pedonculate oak (*Quercus robur* L.) with Atlantic heathlands at the lower dynamic level. The region is sparsely wooded and large areas are characterised by generally regressive and sometimes climacic heathlands. Moreover, the region includes much pasture and farmland bocage. The present-day vegetation is to a large extent the result of human influences on the environment (Marguerie, 1992).

The anthracological study of twenty-four structures belonging to fourteen settlements is presented here (fig. 1). The archaeological sites span a time period from the second Iron Age to the late Middle Ages. The combustion structures studied are variable and can be ascribed to five different categories: ceramic or tile kilns, glass kilns, lime kilns, metal kilns and domestic kilns (tabl. 1).

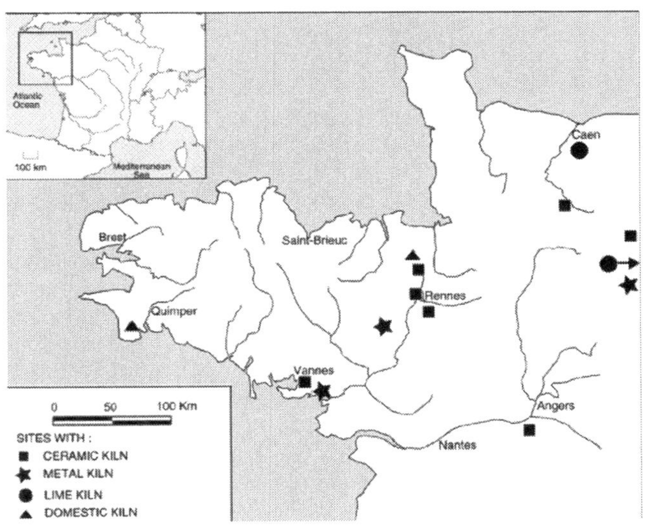

Figure 1: Study area and location of the studied sites.

Site	Type	Period	Origin of charcoals	Archaeologist Excavation dates	References
Tressé, Ille-et-Vilaine Bas Rouault	Tile kiln	Gallo-Roman	Associated pit	F. Le Ny 1987-1989	Le Ny, 1988
Térmentines, Maine et Loire La Fërtellière	Pottery kiln	Carolingian	Kiln	A. Vallais 2000	
Vannes, Morbihan Meudon	Pottery kiln	7th-9th century	Kiln	A. Triste-D.Taquet 1984-1988	
La-Roche-Mabile, Orne La Grande Ouche	Pottery kiln	Late Merovingian	Associated pit	B. Fajal 1990	
Chartres-de-Bretagne Ille-et-Vilaine Four Provost	Pottery kiln	Middle Age	Kiln and Associated pit	L. Beuchet 1992-1994	
La-Haute-Chapelle, Orne La Picaudire	Pottery kiln	14-16th century	Kiln	B. Fajal 1991	
La Bazoge, Sarthre L'Aunay-Truchet	Metal kiln	Iron Age	Kiln and Associated pit	S. Cabboi 1997	
Péllan-le- grand, Ille-et-Vilaine Coudouan	Metal kiln	Gallo-Roman	Kiln	G. Larcher 1991	
Ambon, Morbihan Eglise	Bell kiln	Middle Age	Kiln	P. Andr 1988	
Jouars-Ponchartrain, Yvelines La Ferme d'Ithe	Lime kiln	Gallo-Roman	Kiln and Associated pit	O. Blin 1994-1998	Blin *et al.* 1999
Touffréville, Calvados La Saussaye	Lime kiln	Gallo-Roman	Associated pit	N. Coulthard 1991-2000	
Cesson-Sévigné, Ille-et-Vilaine La Chapelle	Glass kiln	Gallo-Roman	Kiln	D. Pouille 1997	
Pont-L'Abbé, Finistère Kralio	Domestic kiln	Iron Age	Kiln	S. Hingant 1995	Hingant *et al. 1998*
Tinténiac, Ille-et-Vilaine La Cocherais	Domestic kiln	9-10th century	Kiln	A. Provost 1989	Le Boulanger and Provost, 1992.

Table 1: Table of studied sites

Method

The charcoal fragments came from several different contexts (tabl. 1). Many of them were found in refuse pits associated with kilns. Sometimes it was possible to collect carbonised wood from inside the structure; either from the fire chamber, from above or below the hearth or from the mouth of craft kilns.

In order to recover charred wood, sieving (through a 2 mm sieve) and flotation of several litres of sediment from the studied structures was done in the laboratory. The charcoal pieces were then studied under a reflected light microscope with episcopic lenses and identified on the basis of the wood structure. This examination was systematically associated with an observation of the ligneous structure through a binocular lens (magnifications x 8-32). Tree ring curves were noted when the charcoal fragments were large enough. This characteristic enables the identification of the part of the tree that was used: low or smooth curves correspond to the tree trunk while strong or marked curves correspond to the branches.

Finally, some alterations of the wood caused by fusion (vitrification) or radial cracks were noted. These features are either the consequence of specific combustion conditions or reveal the state of the wood before combustion (Marguerie *et al.* forthcoming).

Results

A taxonomic diversity factor of 14 was obtained in the group of the studied sites (fig. 2-5). The wood used in the combustion structures mostly came from heliophilous taxa (*Betula, Corylus, Acer, Genistae*) and from one sciaphilous taxon (*Fagus*). River bank taxa (*Salix, Alnus, Fraxinus*) were used, as well as forest trees such as oak. *Quercus* was the most commonly used taxon and was even the sole genus identified from some metal or lime kilns.

The results show a big difference in the taxonomic diversity between craft and domestic kilns. In ceramic kilns, between 2 and 7 taxa were noted (fig. 2). Even less diversity was recorded for the Cesson-Sévigné glass kiln and the Ambon bell kiln (fig. 3). On the contrary, the domestic kilns at Pont-L'Abbé and Tinténiac contained up to ten taxa belonging to various ecosystems (fig. 5).

There was no obvious correlation between the taxonomic diversity of a sample and its precise location, for example from the inside of a kiln or from associated pits (tabl. 1, fig. 2-5).

The systematically observed tree ring curves indicate that the oak charcoal used in craft structures very often or sometimes exclusively came from timber having a large diameter (fig. 6). The fragments of charred beech found in pottery

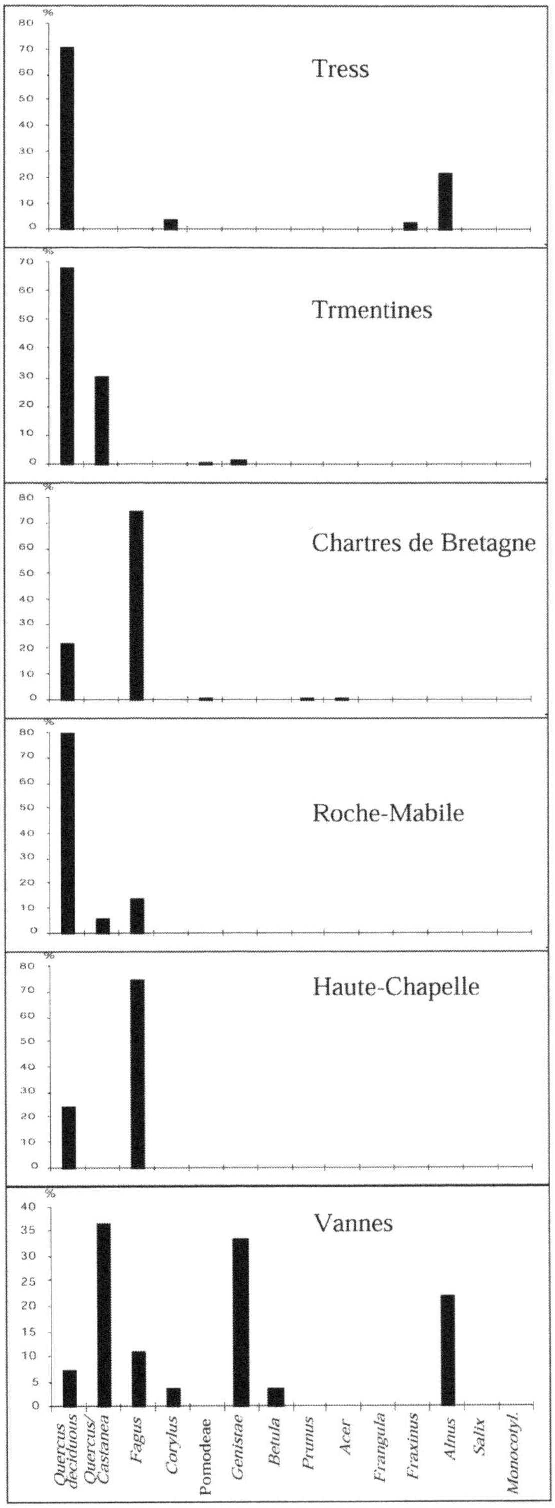

Figure 2: Graphs showing lists of taxa from ceramic and tile kilns.

kilns, also most often came from trunks, except in the case of some samples from La Roche-Mabile (fig. 7). The oak charcoal from the domestic kilns at Pont-L'Abbé and Tinténiac equally came from trunks or branches (fig. 6).

Finally, the examination of charcoal under less magnification brought out some special features. Some charcoal fragments from oak and beech from La Haute-Chapelle presented radial cracks (13,5 %). On the contrary, such features

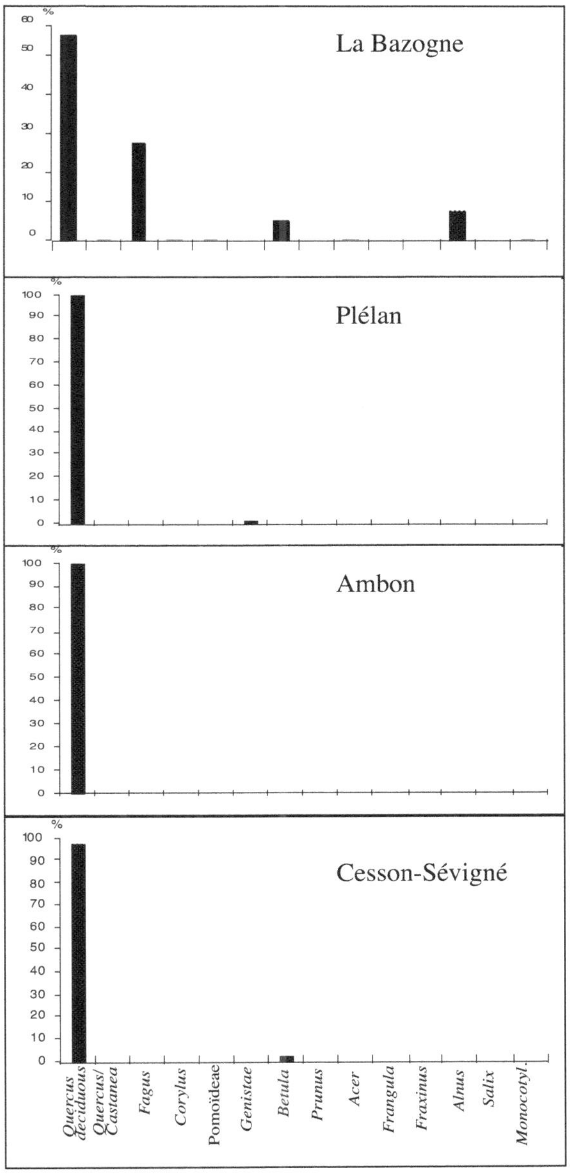

Figure 3: Graphs showing lists of taxa from metal kilns.

were totally absent in the oak charcoal from the lime kilns of Jouars. The cellular tissues of some oak (15 %) and beech (23 %) charcoal found in metal kilns of La Bazoge were molten (vitrified). In the bell kiln of Ambon, more than 50 % of the oak charcoal was "vitrified".

Discussion

This study clearly shows that artisans chose their fuel according to the wood's technical characteristics.

In specialized kilns, a lesser variety of wood was used. The preferential use of first-rate wood types for firing, such as oak and beech, was noted. The molten or vitrified character of some oak and beech charcoal suggests that charcoal and not exclusively wood was used in certain metal kilns. This kind of fuel had to be imported or specially produced locally. For the metalworking craft industry, the heat had to be stable and intense. The choice of fuel used in these kilns

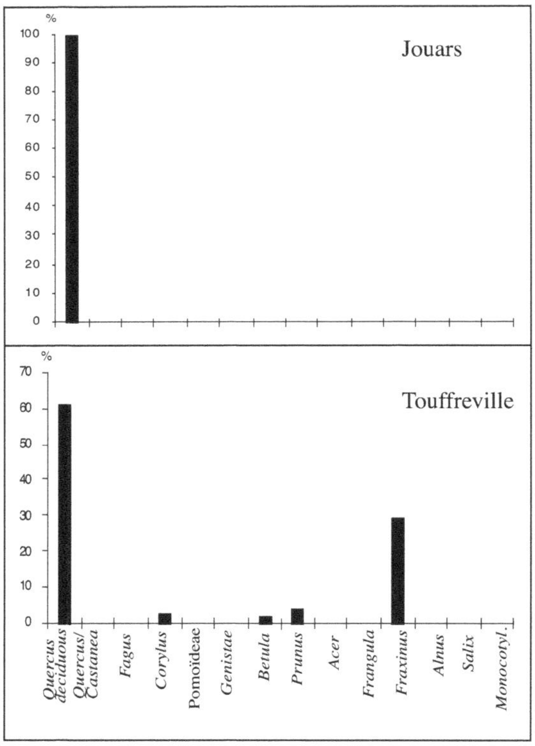

Figure 4: Graphs showing lists of taxa from lime kilns.

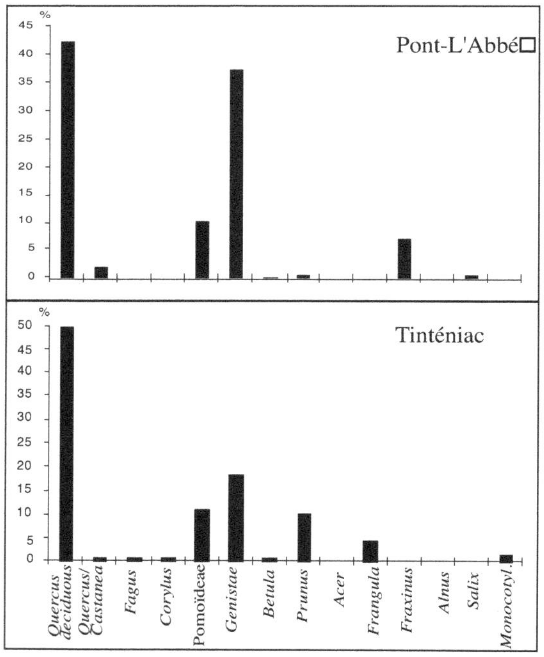

Figure 5: Graphs showing lists of taxa from domestic kilns.

was therefore particularly important. In craft kilns, oak and beech charcoal came mostly from trunks or large branches (fig. 6 and 7). The firing temperature was obtained by selected pieces of chopped wood.

In domestic structures, a wide spectrum of local fuel wood was used. During the Iron Age and the Middle Ages many different species were exploited, including numerous heliophilous taxa from the regressive heathlands. This vegetation had known an important development in Brittany since the

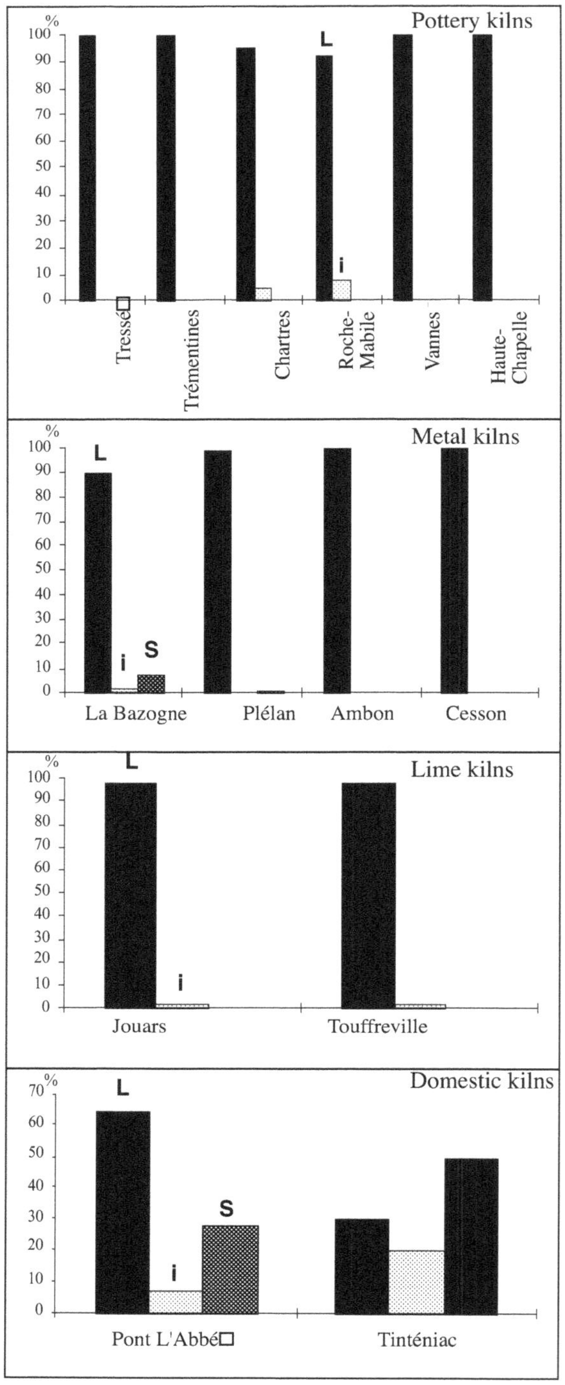

Figure 6: Tree ring curves of oak charcoal (L: low curve; i: intermediate; S: strong).

Protohistoric period. Furthermore, the charcoal came from various parts of the tree. The diameter of the used wood was variable (fig. 6). The artisans of Iron Age Pont-L'Abbé and medieval Tinténiac used wood from both trunks and branches.

Oak and beech produce excellent fire wood characterised by slow combustion and short flames. They are particularly suitable for sustained "small" fires and are used to obtain a slow rise in kiln temperature. This is even more true when the fuel consists of logs from large branches or trunks, instead of bundles of branches (Brongniart, 1977; Pillet, 1982; Sell and Kropf, 1990; Venet, 1974).

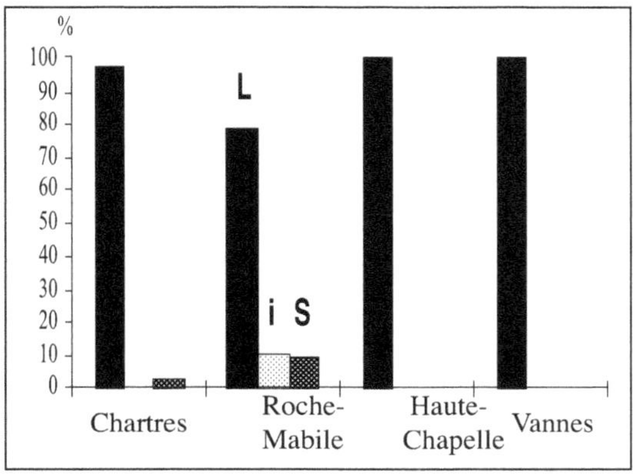

Figure 7: Tree ring curves of beech charcoal (L: low curve; i: intermediate; S: strong).

The soft wood of shrubs or small trees, such as hazel (*Corylus*), Pomoïdeae, birch (*Betula*), maple (*Acer*), broom (*Cytisus*), gorse (*Ulex*) and alder (*Alnus*) provide high flames over a short period of time when gathered in bundles. The "big" fires obtained in this way account for high temperatures. Gallo-roman potters used some plant elements with fast combustion, like straw, reed, heather, bark or pine cones to obtain a high and very hot flame as well as a rapid rise in temperature. When added to the "small" fire, this additional fuel produced the "big" fire needed for the main firing process (Le Ny, 1988).

In the XIIIth century, wood from beech was known as being "calorific", especially when it was "worm-eaten". At the end of the XIVth century, beech was used for many different purposes: as fuel and as a building material. In the XIIIth century, ash (*Fraxinus*) was described as "good wood to burn" (Mille, 1993).

Last, F. Gueriff (1973) mentions the use, at the end of the XIXth century, of bundles of heather, gorse and broom for creating "big" fires in potters'workshops at Herbignac (Loire-Atlantique). In the gallo-roman glass kiln of Cesson-Sévigné, the fire chamber is partially filled with a thick layer of amorphous organic charred matter.

In the lime kilns at Jouars-Pontchartrain, dry oakwood (charcoal without cracks) was used. On the contrary, the oak and beech wood used in the kiln of Haute-Chapelle was humid. This suggests that the use of humid wood in the kiln was deliberate. Green wood releases a lot of smoke and thus the pottery fired under such conditions is a so-called "smoked pottery". The excess carbon deposited on the surface of the pots both served aesthetic purposes and made the ceramics more waterproof.

As the trees growing near a site are often those exploited by the local population, palaeoecological information on the environment of that site can normally be obtained by charcoal analysis. However, for similar types of kilns (for example pottery kilns), the fuel chosen does not seem to have varied significantly, at least not for the short period of time studied here.

In addition, the analysis of this kind of burnt macroremains provides palaeoethnographical information. This paper clearly shows that the species and the size of the wood were carefully selected in some sites. The composition of the samples reveals the choices of wood made for different purposes.

ACKNOWLEDGMENTS
The author is grateful to Jean-Yves Hunot and Françoise Le Ny for enrichening discussions and to Loïc Gaudin, Sylvain Renaudin and Yannick Garcia for their assistance in the fieldwork and the laboratory analysis.

References

BRONGNIART A. 1977 - *Traité des arts céramiques ou des poteries considérées dans leur histoire, leur pratique et leur théorie.* Fac-similé de l'édition d'origine 1877, Paris, Dessain et Tolra, 2 vol., 71 pl., 824 p.

GUERIFF F. 1973 - *Les potiers d'Herbignac : l'art du feu dans un coin de Bretagne.* La Baule, Ed. les Paludiers, 55 p.

BLIN O., BERNARD V., COUTURIER D., DIETRICH A., GEBHARDT A., MARGUERIE D., MORIN J.-M., LEROYER C., G. ALLENET de RIBEMONT, MATTERNE V., SEYRIESSOL K. , ZWIERZINSKI E. 1999 - Impact anthropique et gestion du milieu durant l'Antiquité. Un exemple d'approche paléo-environnementale pluridisciplinaire sur le site d'agglomération secondaire antique de "La Ferme d'Ithe" à Jouars-Pontchartrain (Yvelines, France), *Les Nouvelles de l'Archéologie*, 78, Paris, p. 45-56.

HINGANT S., LE GOFF E., GEBHARDT A., GRALL B., MAGUER P., MARGUERIE D. 1998 - Un site de l'Age du Fer stratifié en milieu rural : l'établissement de Kéralio à Pont-l'Abbé (Finistère). *Revue archéologique de l'Ouest*, 15, p. 59-114.

LE BOULANGER F., PROVOST A. 1992 - Un "village" carolingien sur la déviation de la RN 137 à la Cocherais en Tinténiac (Ille-et-Vilaine). *Les Dossiers du Ce.R.A.A.*, 20, p. 87-117.

LE NY F. 1988 - L'atelier de tuilier gallo-romain au Bas-Rouault en Tressé (Ille-et-Vilaine). *Les Dossiers du Ce.R.A.A.*, 16, p. 91-111.

LE NY F. 1988 - Les fours de tuiliers gallo-romains : méthodologie, étude technologique, typologique et statistique, chronologie. *Documents d'Archéologie Française*, n°12, 142 p.

MARGUERIE D. 1992 - Evolution de la végétation sous l'impact humain en Armorique du Néolithique aux périodes historiques. *Travaux Laboratoire d'Anthropologie*, Université de Rennes 1, n°40, 313 p.

MARGUERIE D., HUNOT J.-Y., BERNARD V. In press - *Charcoal analysis and dendrology : data from archaeological sites in North-western France.*

MILLE P. 1993 - Le choix des essences opéré par les artisans du bois à la fin du Moyen Age. *Revue Forestière Française*, n°2, p. 165-177.

PILLET M. 1982 - *Potiers et poteries populaires en France.* Paris, Ed. Dessain et Tolra, 150 p.

SELL J., KROPF F. 1990 - Propriétés et caractéristiques des essences de bois. Le Mont, Suisse, *Lignum*, 93 p.

VENET J. 1974 - *Identification et classement des bois français.* Nancy, E.N.G.R.E.F., 2e édition, 310 p.

An anthracological method for the study of charcoal kilns in relation to historical forestry management

Laurent Fabre[1] and Jean-Christophe Auffray[2]
1) CNRS-UMR5059, Laboratoire de Paléoenvironnement, Anthracologie et Action de l'Homme, Institut de Botanique, 163 rue A. Brousonnet, 34000 Montpellier, France.
2) CNRS-UMR5554, Institut des Sciences de l'Evolution CC064, Université Montpellier 2, 34095 Montpellier Cedex 05, France.
E-mail: laurfaber@hotmail.com

Abstract
The remains of numerous charcoal burning sites are found in the French Mediterranean forest (garrigue). This once widespread economic activity has now been abandoned in this region. Charcoal recovered from these sites may provide information on the historical dynamics of these forests under human influence. However, there has been no evidence that these charcoal remains reflect more specifically the plant associations exploited or modified by charcoal burners. Our study, based on the analysis of charcoal remains from an experimental kiln, provides the first evidence that these remains clearly reflect the plant association growing in the exploited forest plot. We have carried out a controlled experiment performed by two retired charcoal burners. The results obtained show that the list of taxa identified by the charcoal analysis is very close to the plant composition of the exploited forest. In general, this approach may provide a clear image of the vegetation dynamics in southern France, in relation to this activity and its abandonment.

Key-words: *Charcoal analysis, Garrigue, craft charcoal burning, Human influence,* Quercus ilex.

Résumé
En Languedoc, de très nombreux sites archéologiques de charbonnage parsèment l'espace couvert, aujourd'hui, par la garrigue. Cette activité charbonnière, autrefois très répandue, est abandonnée, ce depuis le début du XXe siècle. Les charbons de bois issus de ces sites contiennent une information susceptible de nous éclairer au sujet des dynamiques forestières induites et maintenues par les activités humaines. Cependant, rien ne prouve que les charbons de bois issus des sites archéologiques de carbonisation reflètent véritablement les associations spécifiques des parcelles forestières exploitées par les charbonniers. Notre étude, fondée sur l'analyse anthracologique des charbons de bois produits par un charbonnage expérimental actuel et de type traditionnel, montre, pour la première fois, que les restes charbonneux traduisent correctement la biomasse ligneuse poussant dans la parcelle contiguë, associée au site de charbonnage. L'expérimentation rigoureusement contrôlée a été menée par deux charbonniers traditionnels, alors à la retraite. Les résultats montrent que la liste de taxons produite par l'anthraco-analyse est très proche de la composition floristique de l'assemblage forestier réellement exploité. Cette nouvelle approche permet alors d'aborder de manière plus évidente, dans le Sud de la France, les dynamiques végétales induites par cette activité et son abandon.

Mots-clés : *Anthracologie, garrigue, charbonnage, action humaine,* Quercus ilex.

Introduction

The presence of holm oak (*Quercus ilex*) in the midst of the Mediterranean deciduous forest, during the Neolithic, was formerly demonstrated by Pons and Vernet (1978), following the analysis of pollen and charcoal remains from sites occupied by populations of shepherds (8000 - 6000 BP).

The occurrence of this species during the Neolithic lead Pons and Vernet (1978) to consider it as an indicator of the earliest changes of the existing natural vegetation under anthropogenic pressures.

Although pubescent oak (*Quercus pubescens*) was the dominant species of the Mediterranean forest during the Neolithic, it appears that human activities might have favoured the spread of *Quercus ilex* from the middle of this period onwards (Vernet, 1973; Vernet *in* Pons *et al.* 1974; Vernet and Thiébault, 1987). From the Iron Age on, holm oak became the dominant species of the Mediterranean forest (Vernet, 1973; Triat-Laval, 1978; Reille *et al.* 1980).

Although the significance of an "anthropisation event" related to a period of clearance ("laying waste") during the Neolithic might have to be reassessed (Blanchemanche and Chabal, 1995; Vernet and Thiébault, 1987), the dominance of *Q. ilex* in the modern garrigue could be a direct consequence of forest clearing and more specialised craft activities since the XIth century, as testified by both historical and palaeobotanical studies (Durand, 1991; Fabre and Gervet, 1992). Amongst human activities related to the development of the Mediterranean forest, charcoal burning is undoubtedly one of the most important (Fabre *et al.* 1992; Fabre, 1996).

The historical remains of numerous charcoal kilns are found today in areas of coppice holm oak, testifying to the existence of woodland management since the XIth century up to the 1960's, when this activity came to an end in southern France.

In this region, charcoal burning sites are generally circular (radii 4 m). Moreover, nearby ruins of the burner's hut ("loge") have made possible an accurate date for the last use of the site on the basis of archaeological remains such as ceramics, recovered inside.

Several features of this economic activity lead us to consider that charcoal burning sites are of great interest to the study of forest dynamics in relation to human activities:
- the periodic use of charcoal kilns provides a record of the vegetation assemblage growing in the adjoining forestry plot through time;
- a great number of charcoal kilns still remain in the Mediterranean areas, which have been abandoned at different periods during the last millennium. Their study will enable us to obtain a history of the development of the woodland and landscape;
- charcoal analysis also allows us to identify the plant species actually employed for this activity.

Few studies based on charcoal analysis (anthracology) have been carried out on this type of archaeological site (Bonhote, 1992; Bonhote and Vernet, 1988; Davasse, 1998; Fabre, 1996; Fabre and Gervet, 1992; Hillebrecht, 1982; Ludemann, 1996; Ludemann and Britsch, 1997; Thommes,

1997). This is mostly due to the lack of empirical evidence which may test the significance of charcoal fragments as representative of the original, exploited vegetation (Bonhote, 1992; Bonhote and Vernet, 1988; Davasse, 1992; Davasse, 1994; Davasse, 2000; Izard, 1992). In this study, we present the results of the first experiment aiming to test the relationship between the botanical composition of the actual forest exploited by charcoal burners and that obtained from charcoal remains. A method is proposed in order to compare the composition of abandoned kiln sites with their surrounding extant vegetation in further analyses. This will allow us to establish potential differences in the forest community between the date of the last use of charcoal kilns and the present. Furthermore, we attempt to answer the following questions:
- Are the charcoal analyses resulting from kiln sites qualitatively and quantitatively representative of the forest area exploited?
- Does the distribution of charcoal fragments within the kiln sites represent any structure in terms of species distribution that might be related to the building structure of the wood rick or any other ethnobotanical aspects?

This experimental approach was rendered possible by the help of two retired charcoal burners who accepted the task of recreating the whole process of charcoal burning.

Material and methods

Experimental stand and area description

During spring 1992, an abandoned charcoal burning kiln was selected by the charcoal burners. According to them, this burning site had not been used since the beginning of the XXth century and the surrouding forest would thus have had the time to recover.

The site chosen is located in the communal forest of Boscorre, 1.5 km east of Puechabon (Hérault, France) (fig. 1). This coppice grows on a relatively flat limestone hill (250-300 m), along a Jurassic thrust-fault. This situation is typical of *Q. ilex* forests in this region (garrigues nord-montpellieraines) where annual rainfall averages 1075 mm and the annual temperature averages 15.6°C (data from the Meteorological Committee of Hérault, recording of Mas de la Serre, J. CH. 75).

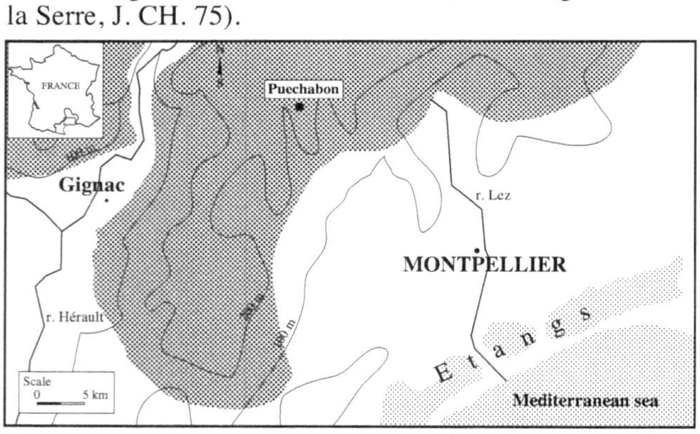

Figure 1: Location of the experimental site of Puechabon (Hérault, France). In grey: limestone plateaux and hills characterised by a "Garrigue" forestry vegetation (essentially holm oak coppices). In white: languedocian agricultural plains (cereal and wine producing districts).

The climate is typically Mediterranean but subjected to the combined effects of altitude and rainfall: cool and sub-humid plains/humid and cold mountains (Dugrand, 1964).

Experimental procedure

Traditionally in this region, charcoal burners used all the trees around the kiln (1100 m^2), with a diameter of more than 5 cm. This diameter may be considered as the only criteria used in the selection carried out in the forestry plot.

During our experiment, wood cutting was controlled and each tree was identified and counted. After cutting, the logs obtained were approximately 90 cm long. These two counts per species (trees and logs), respectively named MT and ML, were to be compared with the charcoal analysis profile obtained at the end of the experiment.

In order to facilitate the assessment of the composition of forests around kiln sites in further analyses, a valuation of this composition was established prior to the experiment. This was done over 533 m^2 corresponding to three equidistant and equal shares of a circular surface of 27 m radius from the centre of the kiln. On this surface, trees more than 15 years old were taken into account (VT) and the potential number of logs of 90 cm per species was estimated (VL). We should note that the surface covered by our estimate does not correspond entirely to the surface actually exploited. In reality, trees found in the close periphery of the charcoal kilns were not usually burnt but they were kept in place as folding screens.

The whole process of charcoal burning lasted one week. Initially, the area where the burning was to take place was cleaned thoroughly, this partly erasing traces of former activities.

The logs were brought in and piled up in order to build the wood rick. The composition of the wood pile was checked (fig. 2). The centre of the wood rick is occupied by a fireplace with a vertical extension similar to a flue. At the beginning of the carbonisation process, water vapour and other gasses are released. Later, the charcoal burner feeds the flue with more wood, only from holm oak. The process of carbonisation depends on the existence of a specific temperature (Kung, 1972). Thermal insulation was provided by a covering of twigs of prickly juniper (*Juniperus oxycedrus*) which was in turn covered by a layer of soil.

The charcoalification lasted for four days. Following this, the pile of charcoal was left to cool for one day. The kiln was then opened and all the charcoal to be sold was removed. The activity of the charcoal burners ended and the site was then abandoned. In former times, the site would be left undisturbed for twenty years. This span of time would enable wood regeneration in the exploited area.

Excavation

Following the removal of the charcoal from the experiment site, the kiln area could be considered analogous to an archaeological site and was divided into square meters, prior to the excavation work. From the centre to the

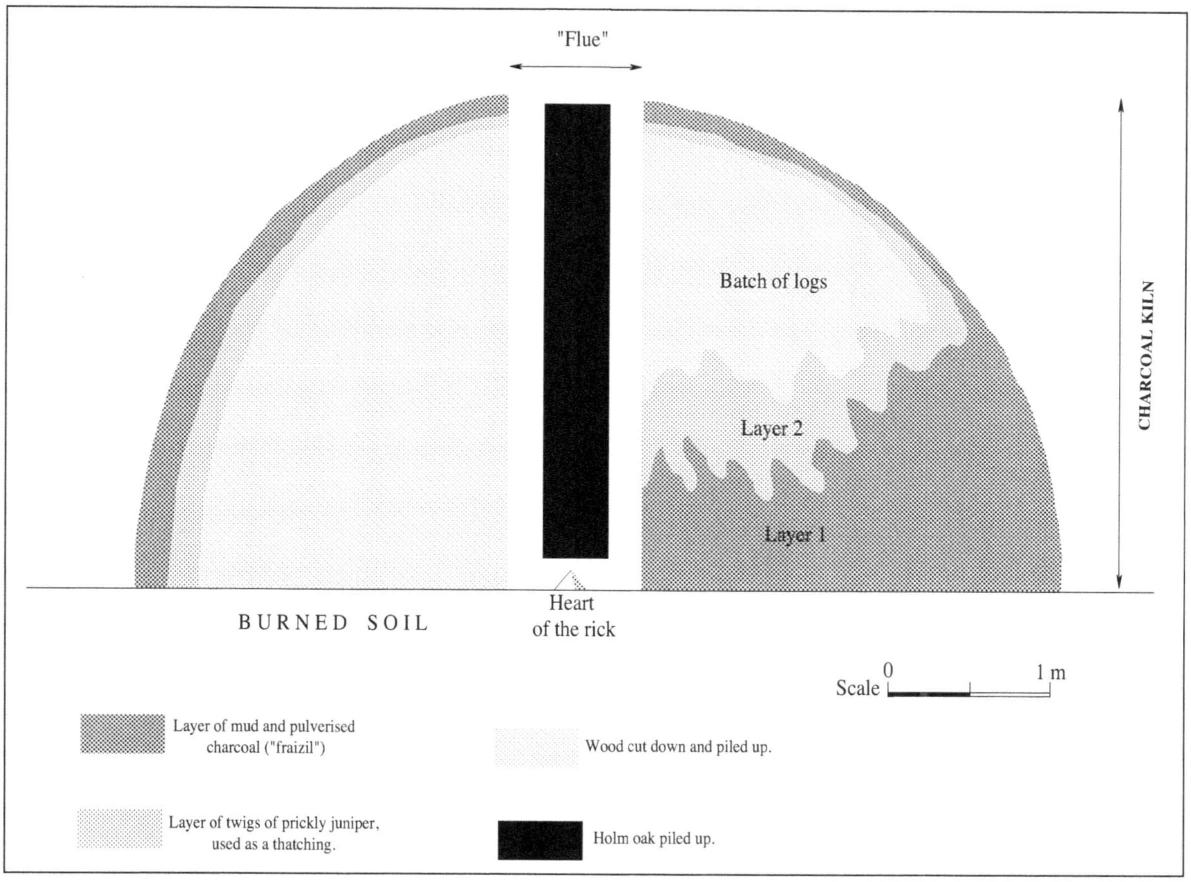

Figure 2: Schematic section of the charcoal kiln.

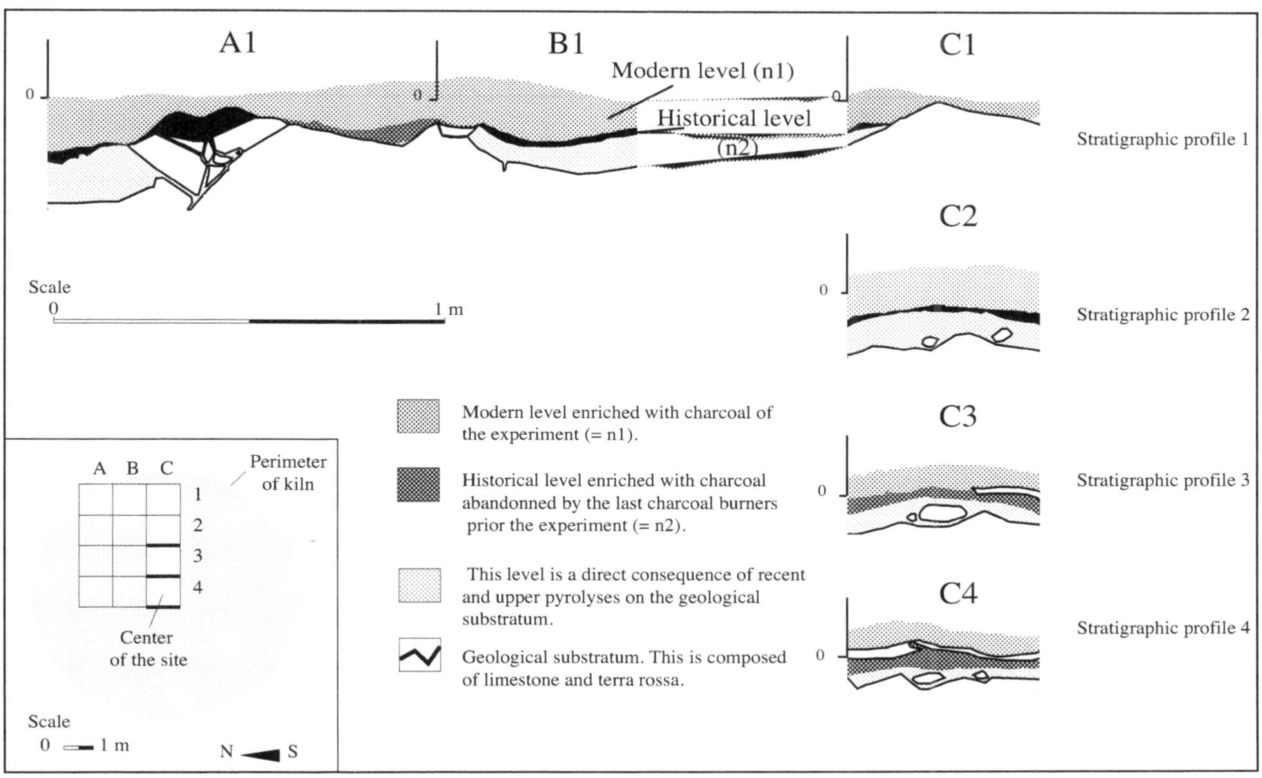

Figure 3: Stratigraphic profiles of the experimental and historical/geological horizons at Puechabon (Hérault, France).

periphery of the site, six archaeological units of $1m^2$ each were established as shown in figure 3. Such a division is justified by the fact that the distribution of remaining charcoal fragments at the site may be heterogeneous and may depend on the structure of the charcoal kiln. Moreover, two stratigraphic levels were differentiated based on the type and colour of the sediments. The top level corresponded to the remaining charcoal of our experiment (modern level), while the lower level represented the remains of previous historical carbonisations. Since this level was partly cleared by the burners previous to the experiment, it is difficult to know if this charcoal represents the last use of the kiln. However, this level was considered as "historical" and will be compared to the top level.

Twelve samples of "sediment" corresponding to the six archaeological units for the two levels were collected and sifted.

Charcoal analysis

Only charcoal fragments larger than $5 mm^2$ were taken into account and identified. Each fragment was fractured by hand according to the three diagnostic planes and the anatomical features were observed under a microscope with reflected light and with a scanning electronic microscope (SEM). The identification of each species (or genus) is achieved by comparison with descriptions from atlases on wood anatomy (Greguss, 1959; Schweingruber, 1990) and with charred extant wood samples from the reference collection of the Laboratory "UPRESA 5059 CNRS". Each archaeological unit of the two stratigraphic levels was characterised by the presence of several species in different proportions. The spectra thus obtained may then be compared to the composition of the forest exploited using correspondence analysis.

Results

Wood-cutting parameters are given in table 1. Standing trunks are refereed to as MT while the designation ML is applied to logs actually chopped down. We observe that: (1) all the species available were used, (2) holm oak is the major element (around 90% of samples). The pile of wood is composed of eight different species: holm oak (*Q. ilex*), box tree (*Buxus sempervirens*), prickly juniper (*J. oxycedrus*), turpentine tree (*Pistachia terebinthus*), strawberry tree (*Arbutus unedo*), mock privet (*Phillyrea latifolia*), mastic (*Pistacia lentiscus*) and mahaleb cherry (*Prunus mahaleb*).

Taxa	(MT)	(MT) %	(ML)	(ML) %
Quercus ilex	279	91,78	2063	95,73
Arbutus unedo	1	0,33	1	0,05
Juniperus oxycedrus	6	1,97	42	1,95
Phillyrea latifolia	2	0,66	3	0,14
Buxus sempervirens	10	3,29	17	0,79
Pistacia terebinthus/lentiscus	3	0,99	21	0,97
Prunus malaheb	3	0,99	8	0,37

Table 1: Results of wood-cutting parameters.

The valuation of the number of trees (VT) and potential logs (VL) established in order to facilitate the assessment of the extant forest to be compared with the charcoal remains (in further analyses) is also given in table 2. Although the surface covered by our estimation (VT) is smaller than that actually exploited by the charcoal burners, the values of VT are usually higher than those of MT. This is due to the fact that the trees in the close proximity of the charcoal site were not cut down and burned (see experimental procedure). However, the number of logs (VL) estimated from VT is smaller than ML. This implies that the mean number of logs obtained from one tree was much higher than for the one onto which VT and VL were estimated (*i.e. Quercus ilex*: ML/MT=7.4, VL/VT=3.2). Such result suggests that the trees chopped down to be burned were more productive than those from the close vicinity.

About 5000 charcoal fragments were recovered and identified. From a qualitative point of view, the results of the charcoal analysis from the upper stratigraphic level (modern level = n1) show that all species used by the charcoal burners were identified. The list obtained from the lower level (historical level = n2) is enriched with *Q. pubescens* and Fabaceae. The frequencies of charcoal fragments belonging to each species per archaeological unit for both levels is represented in table 3. These anthracological results combined with MT, ML, VT and VL were submitted to a multivariate statistical routine. Since these results correspond to a contingency table, correspondence analysis (Benzecri and Benzecri, 1980; Saporta, 1990) appeared to be the method most suited for this statistical analysis and enabled us to compare samples according to their profiles.

The analysis lead to the emergence of three principal axes explaining respectively 80.4, 10.1 and 5.3 % of the total variation (fig. 4). The first axis clearly separates the samples from the modern level from those of the historical level. A MANOVA (multivariate analysis of variance) carried out on the co-ordinates of the modern and historical samples on the three first axes clearly shows the difference between these two levels (Wilk's λ: 0.119; Fs=19.807; df1=3; df2=8; $p<0.001$***). The modern level appears to be enriched with *Q. ilex*, while, as noted above, in the historical level, the presence of *Q. pubescens* and Fabaceae is noted. These two elements are not present in the extant vegetation.

Within each of these levels, it is possible to describe a common structure. In the space defined by the three first axes, the generalised distance matrices between samples within each level were established. The correlation between these two distance matrices clearly suggest a common structure

Taxa	(VT)	(VT) %	(VL)	(VL) %
Quercus ilex	534	93,36	1702	95,08
Arbutus unedo	6	1,05	28	1,56
Juniperus oxycedrus	5	0,87	22	1,23
Phillyrea latifolia	2	0,35	10	0,56
Buxus sempervirens	23	4,02	19	1,06
Pistacia terebinthus	2	0,35	9	0,50

Table 2: Results of the valuation of number of trees ans potential logs.

Taxa	A1n1	%	B1n1	%	C1n1	%	C2n1	%	C3n1	%	C4n1	%
Quercus ilex	261	83,39	286	81,02	278	82,74	253	86,05	356	98,61	387	99,74
Quercus pubescens	0		0		0		0		0	0,00	0	
Arbutus unedo	1	0,32	2	0,57	2	0,60	5	1,70	2	0,55	0	
Juniperus oxycedrus	39	12,46	47	13,31	43	12,80	17	5,78	2	0,55	1	0,26
Phillyrea sp.	2	0,64	4	1,13	3	0,89	10	3,40	1	0,28	0	
Buxus sempervirens	6	1,92	9	2,55	3	0,89	2	0,68	0		0	
Pistacia sp.	4	1,28	3	0,85	5	1,49	7	2,38	0		0	
Prunus sp.	0		2	0,57	2	0,60	0		0		0	
Fabaceae	0		0		0		0		0		0	

Taxa	A1n2	%	B1n2	%	C1n2	%	C2n2	%	C3n2	%	C4n2	%
Quercus ilex	118	40,69	162	52,43	139	46,64	90	31,47	251	73,18	253	70,87
Quercus pubescens	0		0		0		8	2,80	4	1,17	7	1,96
Arbutus unedo	28	9,66	27	8,74	33	11,07	19	6,64	7	2,04	9	2,52
Juniperus oxycedrus	62	21,38	58	18,77	44	14,77	73	25,52	26	7,58	31	8,68
Phillyrea sp.	22	7,59	17	5,50	24	8,05	31	10,84	16	4,66	8	2,24
Buxus sempervirens	35	12,07	19	6,15	23	7,72	42	14,69	34	9,91	43	12,04
Pistacia sp.	14	4,83	22	7,12	27	9,06	12	4,20	3	0,87	6	1,68
Prunus sp.	9	3,10	4	1,29	7	2,35	7	2,45	1	0,29	0	
Fabaceae	2	0,69	0		1	0,34	4	1,40	1	0,29	0	

Table 3: Results of frequencies of charcoal fragments belonging to each taxa per archaeological unit and level (n1 = modern, n2 = historical).

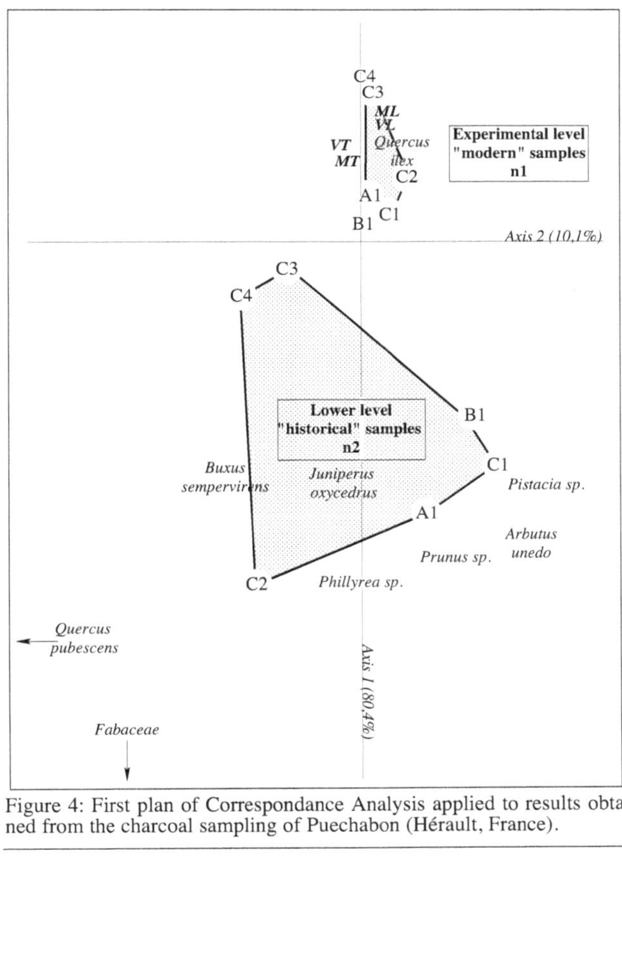

Figure 4: First plan of Correspondance Analysis applied to results obtained from the charcoal sampling of Puechabon (Hérault, France).

of the samples in the two levels (Mantel t-test: t=2.99, p<0.01**). Figure 4 shows that, in both levels the "archaeological units" are distributed in the same way: squares C3 and C4 are on the top corner of each group and this may to be related to the abundance of *Q. ilex*. As a matter of fact, C3 and C4 correspond to the centre of the kiln, *i.e.* to the flue, which is always located in the same place during the successive use of the kiln, when its location is indicated by the presence of a tetrahedral stone. In the experimental section, we have already noted that the flue was exclusively comprised of holm oak and this may explain why these central squares are mostly represented by this species. This also suggests that this species was used by charcoal burners in previous carbonisations. Besides, the third axis (not shown here) representing 5.3% of the total variability is clearly determined by three species, *i.e.* prickly juniper, strawberry tree and box tree. Along this axis and within the modern samples, A1, B1 and C1 are clearly separated from the other samples of the modern level and this has to be related with the higher proportion of prickly juniper. A1, B1 and C1 correspond to the periphery of the kiln suggesting that this zone is enriched in this species. This fact may be easily explained if we take into account the disturbance caused by the loggers during the removal of the charcoal. All the "thatching" (composed of juniper) is removed, stacked aside and, since it has no commercial value, it is abandoned *in situ* increasing the representation of juniper at the periphery of the site.

The quantitative estimates of trees and logs prior to the experiment (MT, ML, VT, VL) were compared with all the modern level squares by χ^2 tests. They were also compared to the estimation of the total number of charcoal recovered on the modern level (by summing the number of charcoal of all squares after having indexed them by the surface they actually represent on the site). All these tests failed to show a clear dependence between the specific numbers of trees and logs counted and estimated before the experiment and those of the charcoal remains (all χ^2***). This could be due to the several causes. Firstly, the branches, essentially *Juniperus*, used for the thatching of the kiln could not be effectively counted previous to the experiment but were recovered as charcoal afterwards. Secondly, the fragmentation of charcoal may depend on the species considered and on the temperature of charcoalification. Noteworthy, from a descriptive, statistical point of view MT, ML, VT, and VL emerge clearly within modern-level samples in the correspondence analysis suggesting the relevance of these parameters to express the global composition of the forestry plot exploited by charcoal burners.

Discussion

Our results suggest that:
- The charcoal remains recovered on the burning site itself represent the plant composition of the forest exploited by the burners.
- The distribution of species within the site is heterogeneous, and depends on several potential ethnobotanical processes, especially in the way species are used to build the kiln and to maintain the burning. Evidence of these biases appears to be similar in both levels.
- The specific composition of the modern and historical levels are qualitatively and quantitatively different.

These results provide the first evidence of the representativity and significance of charcoal remains from kiln sites in relation to the forest exploited by charcoal burners. However, we should note that the structure of the distribution of species inside the kiln depends on some ethnobotanical features. In this respect, all spatial sections of such sites, from the centre to the periphery, should be taken into account in attempts to elucidate the source vegetation of anthropologically exploited areas. Conversely, the structure of species distributions over such sites may be informative in understanding ethnobotanical practices. During the experiment, we observed that the flue was fed exclusively with holm oak and this was clearly reflected in the high proportion of charcoal remains of this species near to the centre of the site for the modern level. The same pattern observed in the historical level suggests that this practice also occurred previously. Additionally, the juniper used for covering the experimental kiln is over-represented at the periphery of the site. This was not observed in the historical level where the peripheral area is enriched with strawberry tree. According to the charcoal burners, either juniper, strawberry tree or box could be preferentially used to cover the kiln depending on their availability around the site, thus corroborating the fact that strawberry tree might have been used to cover one of the last kilns to have been used prior to our experiment.

Remarkably, a similar ethnobotanical kiln structure was observed in the historical level. This suggests that the taphonomic processes influencing the distribution of charcoal remains since the last use of the present charcoal burning site did not apparently disrupt this structure. However, we do not have any evidence that the fragmentation of charcoal through time is homogeneous among species. If heterogeneous, this fragmentation would modify the proportion of each species represented in the charcoal record.

The frequencies of standing trees (MT) and cut logs (ML) that were burned are closely related in the correspondence space to their estimates VT and VL based on a sampling area around the kiln. This suggests that these two estimates may properly represent the composition of the forest exploited by burners. However, in the results section we have noted that the surface used for this estimation is not rigorously representative of the cleared surface. Indeed, in the close periphery of the kiln, the charcoal burners do not cut the trees which are used as protective screens. In the present experiment, the cleared surface overlapped a stone wall encircling the work surface, *i.e.* the burning place and the close periphery. It could then have been possible to estimate the actual cleared surface. However, at most of the numerous historical burning sites that are noted in the French garrigue, containing walls have never been identified, which could have provided evidence of the distance between the burning site and the overall area cleared. Therefore, the estimation has to be done from the site itself and outwards. This methodological shortcoming leads us to note that the mean number of logs per tree is higher in the cleared area than in the close periphery of the kiln. This could be due to the higher regeneration and wood productivity of holm oaks in the cleared areas (Ducrey and Toth, 1992). This result is informative on the management techniques conducted by charcoal burners who improve the productivity of the forest by periodic felling of the same forestry plot.

This method appears to be instructive in establishing the botanical/ecological changes occurring between the date of the last use of the burning site and the extant vegetation around. In our experiment, this could be achieved by comparing the variables VT and VL with samples from the historical level. In this level, two taxa (*Q. pubescens*, Fabaceae) which have been identified in the charcoal were not found in the extant vegetation surrounding the kiln site. The proportion of holm oak has apparently increased in the extant vegetation when compared with the results from the historical levels even though this species seems to have been used to feed the flue at previous uses of the charcoal burning site. This also suggests an impoverishment of the range of species useful for this specific process in the vegetation around the kiln.

This method may now be applied to numerous other burning sites of the French Mediterranean area in order to establish the past impact of man on the origin and the maintenance of holm oak coppices in the garrigue, and more generally on the surrounding forest dynamics. The garrigue, which may be thought as a mosaic landscape, could have been rejuvenated by the work of the charcoal burners and could have resulted from a particular forestry management in space and time: on the one hand, the breakdown of the forest area into individual forestry plots, regular wood production units, and on the other hand, the homogenisation of the forest in a simple *Q. ilex* coppice induced by the steady turnover of exploited plots to charcoal production.

For further analyses, combined archaeological approaches are essential in order to date the sites on the basis of archaeological remains recovered, a procedure that will permit further understanding of changes in the vegetation over time.

The generalised charcoal burning activity in the French garrigue has been abandoned over the last few decades, as well as most of the other traditional activities related to this forest type (pastoral activities, craft glassworking, ironworks, etc.). This abandonment favours natural processes in the evolution of this environment. The consequences of the abandonment of human activities on the biodiversity and on the dynamics of this forest are not yet well established. In this respect, the anthracological analysis of charcoal burning sites could be informative on the potential different effects of "traditional" and "current" human activities and their effects on the vegetation.

ACKNOWLEDGEMENTS
The authors are grateful to I. Figueiral and N. Rowe for their very helpful comments on an early version of the manuscript.

References

BENZECRI J.-P., BENZECRI F. 1980 - *La pratique de l'analyse des données : analyse des correspondances, exposé élémentaire*. Dunod, Paris.

BLANCHEMANCHE P., CHABAL L. 1995 - Potentialités forestières et activités humaines de la fin de la Préhistoire à la période historique dans le midi de la France: dégradation ou socialisation du milieu ? In: *L'Homme et la dégradation de l'Environnement*. APDCA, Sophia-Antipolis, p. 1-22.

BONHOTE J. 1992 - *Forges à la catalane et évolution forestière dans les pyrénées de la Haute-Ariège*. Ph. D. Thesis University Toulouse-Le Mirail, Toulouse.

BONHOTE J., VERNET J.-L. 1988 - La mémoire des charbonnières. Essai de reconstitution des milieux forestiers dans une vallée marquée par la métallurgie (Aston, Haute-Ariège). *Revue Forestière Française* XL (3), p. 197-211.

DAVASSE B. 1992 - Anthracologie et espaces forestiers charbonnés. Quelques exemples dans la moitié orientale des Pyrénées. *Bull. Soc. bot. Fr.* 139, Actual. bot. (2/3/4), p. 597-608.

DAVASSE B. 1994 - Les espaces forestiers de la haute vallée du Vicdessos-Pyrénées ariègeoises : analyse écohistorique et étude des charbonnières. Colloques phytosociologiques, XX, *Phytodynamique et Biogéographie historique des forêts*, Bailleul 1991, p 305-314.

DAVASSE B. 2000 - *Forêts, charbonniers et paysans dans les Pyrénées de l'Est, du Moyen Age à nos jours. Une approche géographique de l'histoire de l'environnement*. GEODE, Toulouse.

DUCREY M., TOTH J. 1992 - Effect of cleaning and thining on height growth and girth increment in holm oak coppices (*Quercus ilex* L.). *Vegetatio*, 99-100, p. 365-376.

DUCREY M., TURREL M. 1992 - Influence of cutting methods and dates on stump sprouting in holm oak (*Quercus ilex* L.) coppice. *Annales des Sciences Forestières*, 49, p. 449-464.

DUGRAND R. 1964 - *La garrigue Montpellieraine. Essai d'explication d'un paysage*. PUF, Paris.

DURAND A. 1991 - *Paysages, terroirs et peuplement dans les campagnes du Bas-Languedoc* (X°-XII° siècle). Ph. D. Thesis, University Paris I, Paris.

FABRE L., GRAU-ALMERO E., LALANE J.-F., VERNET J.-L., DURAND A. 1992 - Charbonnières et forêt méditerranéenne à la Boissière (Hérault). *In*: GDR-ISARD 881-CNRS/CIMA-URA 366 CNRS (eds), Protoindustries et Histoire des forêts. *Les Cahiers de l'Isard*, 3, p. 237-255.

FABRE L., GERVET X. 1992 - Première approche du charbonnage et de l'histoire de la forêt méditerranéenne : l'exemple du canton d'Aniane (Hérault). *Bull. Soc. bot. Fr.*, 139 Actual. bot. (2/3/4), p. 617-625.

FABRE L. 1996 - *Le charbonnage historique de la chênaie à Quercus ilex* L. *(Languedoc, France) : conséquences écologiques*. Ph. D. Thesis, Université Montpellier II, Montpellier.

GREGUSS P. 1959 - *Holzanatomie der Europäischen Laubhölzer und Sträucher*. Akademia Kiado, Budapest.

HILLEBRECHT M-L. 1982 - Die Relikte des Holzkohlewirtschaft als Indikatoren für Waldnutzung und Waldentwicklung. Untersuchungen an Beispielen aus Südniedersachsen. *Göttinger geographische Abhandlungen*, 79, Göttingen.

IZARD V. 1992 - L'art du charbonnier : contributions ethno-botanique et géographico-historique à l'étude des paysages métallurgiques d'après l'anthracoanalyse des charbonnières. *Bull. Soc. bot. Fr.*, 139, Actual. bot. (2/3/4), p. 587-596.

KUNG H.C. 1972 - A mathematical model of wood pyrolysis. *Combustion Flame*, 18, p. 185-195.

LUDEMANN T. 1996 - Die Wälder im Sulzbachtal (Südwest-Schwartzwald) und ihre Nutzung durch Bergbau und Kölherei. Freiburg i. Br., Mitt. Ver. forstl. *Standortskunde u. Forstpflanzenzüchtung* 38, p. 87-118.

LUDEMANN T., BRITSCH T. 1997 - Wald und Köhlerei im nördlichen Feldberggebiet/Südschwarzwald. Mitt. bad. *Landesverein Naturkunde Naturschutz,* Freiburg i. Br., N.F. 16 (3/4), p. 487-526.

PONS A., REILLE M., TRIAT H., COUTEAUX M., JALUT G., ÖNER S., PLANCHAIS N., VERNET J.-L. 1974 - Flore du Bassin Méditerranéen. Essai de systématique synthétique. *Les données historiques et l'étude de la flore méditerranéenne*. CNRS (eds), 235, p. 305-326.

PONS A., VERNET J.-L. 1978 - Le passé de la garrigue : quelques interrogations. Résumé de la communication de A. Pons et J.-L. Vernet au Colloque Garrigue. *Annales de la Société d'Horticulture et d'Histoire Naturelle de l'Hérault*, Montpellier, 118 (1), p. 25-26.

REILLE M., TRIAT-LAVAL H., VERNET J.-L. 1980 - Les témoignages des structures actuelles de végétation méditerranéenne durant le passé contemporain de l'action de l'homme. *Naturalia Monspelliensa*, H.S., p. 79-87.

SAPORTA G. 1990 - *Probabilités. Analyse des données et Statistiques*.Technip, Paris.

SCHWEINGRUBER F.H. (1990) *Anatomie europäischer Hölzer*. W.S.L.F.N.P. Haupt.

THOMMES P. 1997 - Wald und Metall. Eine Methode zur Rekonstruktion des Waldschädigung durch ur- und frühgeschichtliche *Meiler- und Verhüttungstechnologie*. *Diss*. Univ. Freiburg i. Br.

TRIAT-LAVAL H. 1978 - *Contribution pollenanalytique à l'histoire tardi- et post-glaciaire de la végétation de la Basse Vallée du Rhône*. Ph. D. Thesis, University Aix-Marseille, Marseille.

VERNET J.-L. 1973 - Etude sur l'histoire de la végétation du Sud-Est de la France au Quaternaire d'après les charbons de bois principalement. *Paléobiologie continentale*, 4 (1), p. 1-94.

VERNET J.-L., THIÉBAULT S. 1987 - An approach to northwestern Mediterranean recent prehistoric vegetation and ecologic implications. *Journal of Biogeography*, 14, p. 117-127.

Charcoal burning remains and forest stand structure - Examples from the Black Forest (south-west-Germany) and the Bavarian Forest (south-east-Germany)

Oliver Nelle

Institut für Botanik - Universität Regensburg - D-93040 Regensburg, Germany.
E-mail: oliver.nelle@epost.de

Abstract

Charcoal burners of the past have left numerous relics containing information about the composition and history of ancient forests. Examples from projects investigating historic charcoal kiln sites in the Southern Black Forest, south-west-Germany, and the Bavarian Forest, south-east-Germany are discussed. The measurement of the minimum diameter of charcoal pieces is a tool to investigate the diameter distributions of ancient forests. It may be possible to use charcoal analysis to gain information about historic forest stand structure. Charcoal of sites from Schauinsland (southern Black Forest) show that woodland dominated by beech (*Fagus sylvatica*) was present at the hilltop area 250 years ago. The charcoal spectra of some sites are conspicuously dominated by small diameter wood, whereas at other sites large diameter wood dominates. The "small diameter sites" may have been due to short growing, degraded beech forests at high altitudes, whereas the charcoal at sites with large diameter wood may have been from mixed beech-fir forests on steeper slopes. The charcoal spectra of kiln sites near Ringelai (Bavarian Forest) show a mixed fir (*Abies alba*)-beech-forest with some spruce (*Picea abies*). Four sites are dated dendrochronologically to the period between 1822-1834. The species spectra and the generally large wood size suggest that there was no intense woodland degradation prior to the 19th century.

Key-words: *Charcoal burners, History of ancient forest, Southern Black Forest, Bavarian Forest, Diameter wood, Charcoal spectra.*

Résumé

Les charbonnières du passé ont laissé de nombreux témoignages de l'histoire des forêts. Les exemples présentés portent sur l'étude de charbonnières historiques situées dans le sud de la Forêt Noire, Allemagne du sud-ouest et la forêt bavaroise, Allemagne du sud-est. La mesure du diamètre minimum des charbons de bois constitue un outil privilégié pour connaître la taille des fûts des forêts du passé. Il est possible d'utiliser l'analyse anthracologique pour avoir des informations sur la structure des forêts historiques. Les charbonnières de Schauinsland (sud de la Forêt Noire) montrent que le hêtre (*Fagus sylvatica*) dominait la forêt il y a 250 ans. Dans les spectres de plusieurs sites, les diamètres de petite taille sont majoritaires, alors que dans d'autres sites les grands diamètres prédominent. Les sites "à petit diamètre" résultent d'une croissance rapide, de la dégradation des hêtraies d'altitude, alors que les sites présentant des grands diamètres proviennent de l'exploitation de la hêtraie-sapinière des versants escarpés. Le spectre anthracologique de la charbonnière située près de Ringelai (forêt bavaroise) montre une hêtraie sapinière avec de l'épicea (*Picea abies*). Quatre sites sont datés par dendrochronologie entre 1822 et 1834. Le spectre et la grande taille du bois suggèrent qu'il n'existait pas de dégradation importante au XIXème siècle.

Mots-clés : *Charbonniers, Histoire des forêts, Sud de la Forêt-Noire, Forêt bavaroise, diamètre des bois, spectre anthracologique.*

Introduction

Before the use of coal spread in the 19th century in middle Europe, the metal industry and other trades were largely dependent on charcoal, which was a very important energy source for craftsmen and everyday life. For these reasons, a great deal of wood was consumed during the Middle Ages and the Modern Age. Charcoal was produced primarily in the mountainous areas where timber was still abundant. Today, the historic charcoal production sites can still be found. The leftover charcoal pieces can be used to reconstruct the composition and history of the ancient forests.

With a renewed interest in anthracology, there are several studies investigating charcoal kiln sites in Europe (*e.g.* Kauder, 1992; Bonhote and Vernet, 1988; Davasse, 1992). The charcoal analysis shows which tree species were used by charcoal burners. Charcoal may contain other information as well. Apart from the attempting to establish the species composition, we hope to answer the following questions:

- Did the charcoal burners use timber from dense, mature forests or from open, young forests?
- Was the forest degraded because of earlier practices? Is species composition the only information available to understand forest degradation?
- What timber did the charcoal burners select? Or did they use all available timber? For example, they might have only used wood that was left over after tree felling for building timber, or they used only certain species.

The analysis of minimum wood diameter of the charcoal pieces is already in use by Ludemann and co-workers from the anthracology work group at Freiburg University (Ludemann, 1996; Ludemann and Britsch, 1997; Ludemann and Nelle, 2002). Krause (1976) studied charcoal kiln sites in the Hochtaunus (Germany). By counting growth rings he concluded that young trees were used from coppiced woods. Hillebrecht (1982) also investigated wood diameters of charcoal derived from sites in south Lower Saxony (Germany). The aim of current research is to find out what information diameter data can provide, and how accurate this information is. In this paper, two examples from the Black Forest and the Bavarian Forest are given. How this method might improve our knowledge of past forest stand structure is also discussed.

Material and Methods

Charcoal analysis

Charcoal pieces of sizes 0.25 cm^3 and larger were collected at historic charcoal production sites (fig. 1). Usually, charcoal is abundant in a layer up to 50 cm in depth. The average number of growth rings of a charcoal piece is around 20, but pieces can have up to 70 or even 100 rings. The minimum sample size was 100 pieces per site. At some sites, more than 300 fragments were collected.

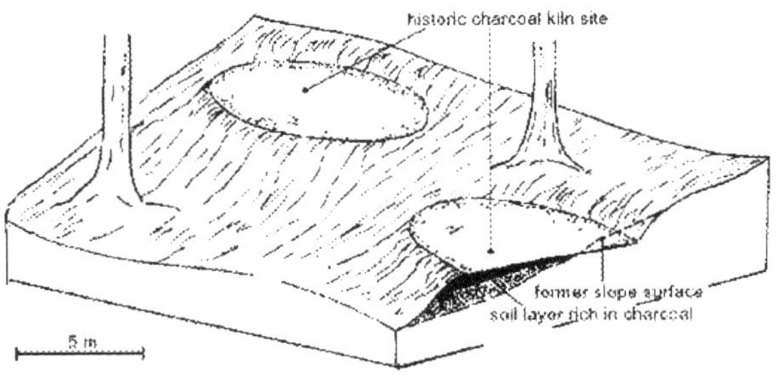

Figure 1: An example of two charcoal kiln sites on a slope.

The pieces were analysed using the key by Schweingruber (1982) and a charcoal sample collection. The minimum diameter of the charcoal pieces was estimated using a diameter stencil. The bending of the growth rings and the angle of the rays make it possible to measure the diameter, if the piece is not too small (fig. 2). The data were put into five arbitrary chosen diameter classes: diameter 0-2 cm, 2-3 cm, 3-5 cm, 5-10 cm and >10 cm. Pieces which were too small or otherwise unsuitable for measuring were not classified by size. If bark is absent (which is the case for most of the material), one cannot tell whether the charcoal piece derived from a branch or from the inner part of a stem. Therefore, it is the minimum diameter that is measured. It is important to consider that during carbonisation wood shrinks about 15-20% radially (Schläpfer and Brown, 1948), but the shrinking is assumed to be proportional. A histogram (fig. 2) shows the frequencies of the five diameter classes for a sample and illustrates the distribution of the charcoal pieces among the diameter classes (size class distribution).

A mean diameter was calculated for each sample using the following formula:

mean diameter [cm] = $[n_{(0\text{-}2cm)}*1 + n_{(2\text{-}3cm)}*2,5 + n_{(3\text{-}5cm)}*4 + n_{(5\text{-}10cm)}*7,5 + n_{(>10cm)}*15] / n_{total}$.

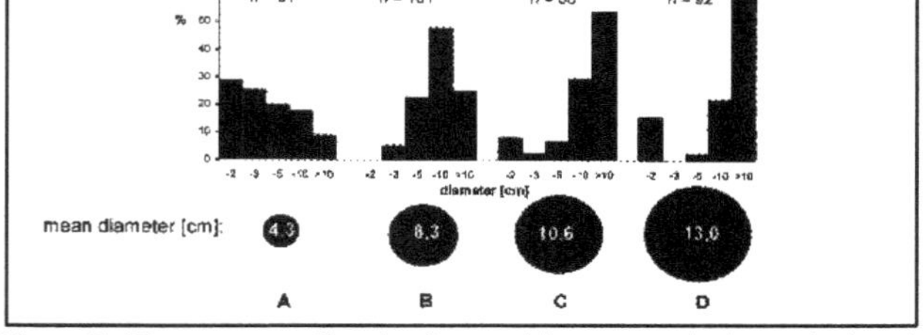

Figure 2: Diameter stencil and reference histograms. **A** charcoal piece is oriented according to growth ring bending and ray angle. The histograms show the distribution of charcoal among diameter classes. A: charcoal from a campfire site (Black Forest); **B**: charcoal from coppice woods (sold by Manufactum Hoof and Partner KG, Waltrop); **C**: charcoal from a contemporary kiln site (Waldmünchen); **D**: charcoal from a historic kiln site (Black Forest); **n**: gives the number of charcoal pieces analysed for each sample.

Mean diameter can theoretically be between 1 and 15 cm. Values vary practically from 4 to 13 cm. The mean diameter is an arbitrary value. It does not tell us about the actual size of the wood that was used, but gives a relative value when comparing different sites (for examples see fig. 2).

Reference histograms

In figure 2, reference histograms of size class distributions are given. They are derived from samples of contemporary charcoal, where the size of the wood prior to charring is roughly known. The campsite fire (fig. 2-A; "campfire type") was made up of branch wood of beech, spruce and fir, and some larger sized timber of beech. In coppiced woods (fig. 2-B; "coppiced wood type"), the trees do not get older than 30 years, so timber (here: oak) for charcoal burning is medium sized. The beech stems that were used at a recently operated site near Waldmünchen (fig. 2-C; "larger wood size type") had diameters between 10 and 30 cm, and the branches were between 3 and 7 cm in diameter. Presumably the timber that was used at the historic site in the Black Forest (fig. 2-D) derived from considerably bigger stems. It represents the highest mean diameter value found so far. These reference histograms are used to compare them to histograms derived from historic charcoal kiln sites, to give a rough idea of the size of the timber used.

Study areas

Southern Black Forest: Schauinsland

Charcoal kiln sites around the Schauinsland mountain (1284 m a.s.l.; *c.* 10 km south-east of Freiburg) were investigated (fig. 3). The study area lies at the western margin of the southern Black Forest. The 19 sites in this study lie around the hilltop of Schauinsland and Hundsrücken, at altitudes between 1000 m and 1200 m. Mean annual temperature is 4.8° C; annual precipitation is 1585 mm (Trenkle and Rudloff, 1982). The current forest composition is highly altered due to spruce (*Picea abies*) plantations. The native species composition is presumably a mixed *Fagus sylvatica-Abies alba* forest which can still be found in some parts of the area. It is unknown whether *Picea abies* was a natural component of the forest, or whether it was limited to higher altitudes further east and spread with the help of humans. The area was used historically for grazing livestock since the medieval settlement. Mining activities (mainly silver) have been verified beginning in the 11th century (Priesner, 1982).

Bavarian Forest: Ringelai

Anthracological studies are currently conducted at the University of Regensburg within the Deutsche Forschungsgemeinschaft Graduate College "Paläoökosystemforschung und Geschichte" (Research on palaeoecosystems and history). One part of the Graduate Colleges' research focuses on vegetation history and environmental change in the Bavarian Forest. This study presents an ensemble of 10 charcoal burning sites near Ringelai (District Freyung-Grafenau; altitude 590-720 m a.s.l., fig. 3). Mean annual temperature is around 5° C; annual precipitation is around 1100 mm. Natural zonal vegetation

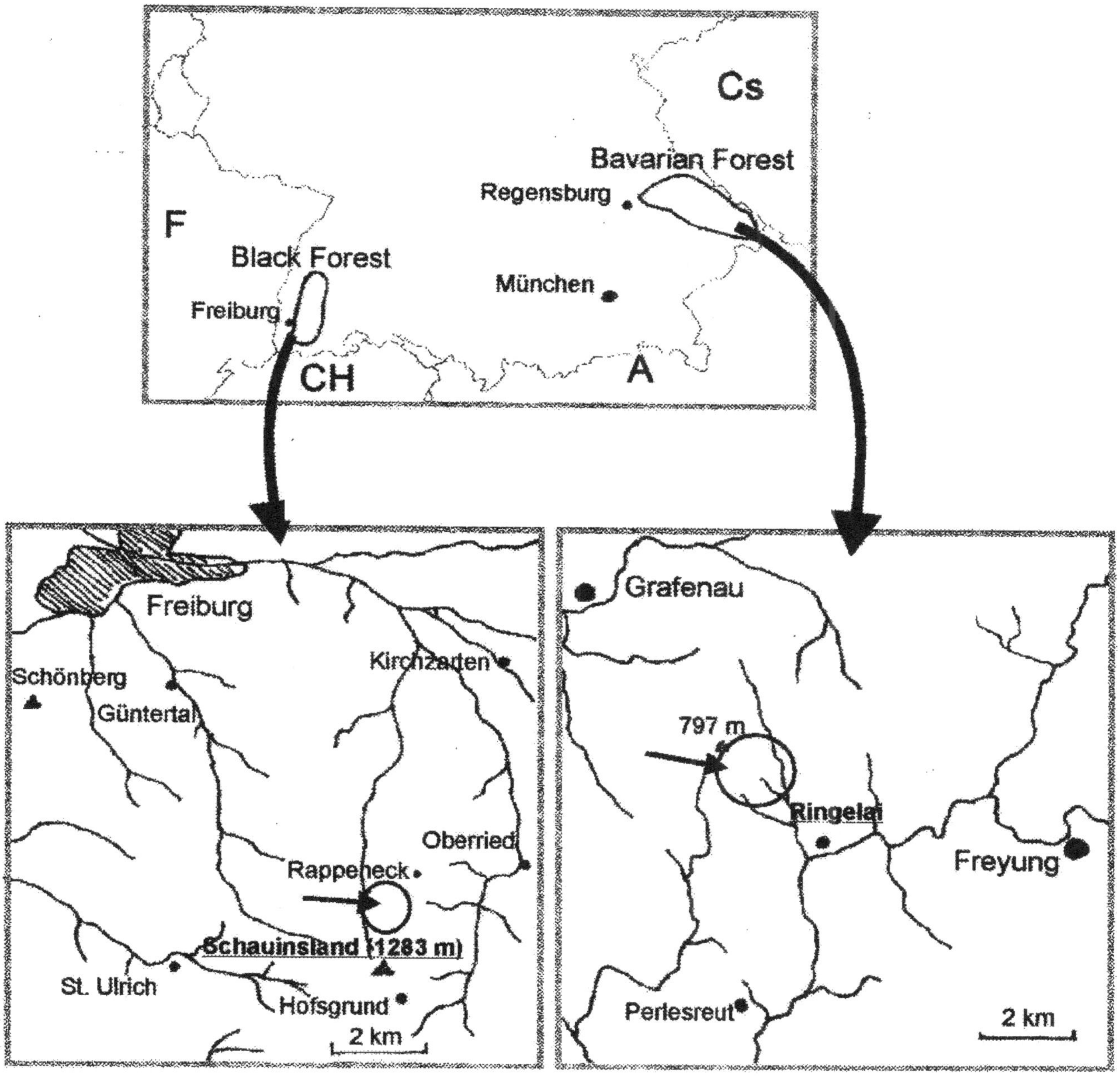

Figure 3: Study areas.

is assumed to be a mixed *Picea abies-Abies alba-Fagus sylvatica*-forest (Stalling, 1987). The area was settled during the high or late Middle Ages.

Results

Southern Black Forest: Schauinsland

The sites at Schauinsland hilltop were probably in use in the 18th century, according to ^{14}C-dating of three sites (all 17th -20th century), and according to the historical record and the mining activities. A total of 2677 charcoal pieces were analysed; 2467 pieces were measured for minimum diameter. Most sites are dominated by *Fagus sylvatica*, with *Abies alba* being the second most abundant tree species. *Picea*-charcoal was found at only three sites around the hilltops (site 679, 681, 675); *Acer* was found at one site (674) (see fig. 4 and tabl. 1 for details).

The charcoal spectra of some sites are conspicuously dominated by small diameter wood, whereas at other sites large diameter wood is prevalent (see fig. 5 for mean diameter values and size class histograms). Mean diameter values range from 4.1 cm to 12.1 cm and cover histogram types from the "campfire type" (fig. 2-A), the "coppiced wood type" (fig. 2-B) and the "larger wood size type" (fig. 2-C and D).

Bavarian Forest: Ringelai

Four sites were dated dendrochronologically during the period 1822-1834 (dating was done by Franz Herzig, Labor für Dendroarchäologie, Bayerisches Landesamt für Denkmalpflege). 1283 pieces were analysed. The minimum diameter of 1145 pieces was determined. The charcoal spectra are dominated by *Fagus sylvatica* (overall 46%) or *Abies alba* (overall 41 %), or show similar portions of the two main tree species (see fig. 6 and tabl. 1 for details). *Picea abies* composes 7% of the overall material and is present at 6 of 10 sites. The other species found are: *Alnus* (3%; probably *Alnus glutinosa*) which was found at sites close to streams or wet areas, *Betula pendula* (2%), and with less than 1% *Corylus avellana*, *Quercus* sp. (*Q. petraea* or *Q. robur*), *Acer* sp. (*A. platanoides* or *A. pseudoplatanus*), *Populus* sp. and *Salix* sp. Mean diameter values are relatively high at almost all sites (10 to 12 cm), except for site 12, which has a slightly smaller value (fig. 7).

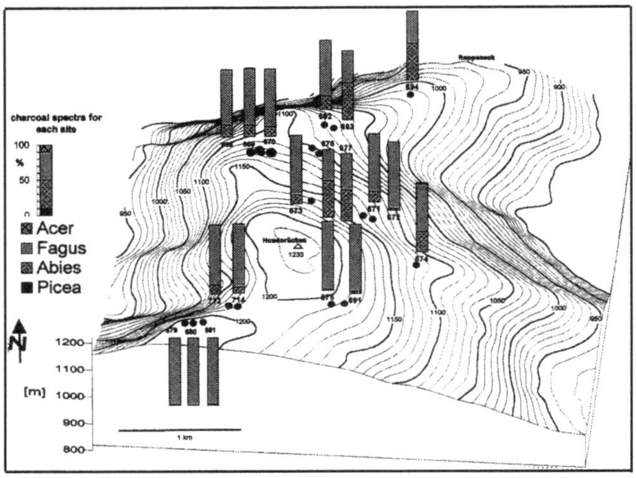

Figure 4: Hilltop area of Schauinsland, charcoal species spectra. One column gives the spectrum for the closest kiln site (black dot). Data basis: 2677 charcoal pieces from 19 sites.

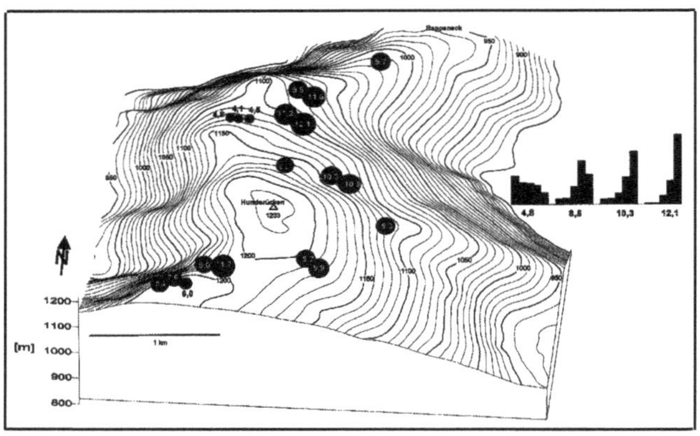

Figure 5: Hilltop area of Schauinsland, spatial distribution of mean diameters. One black circle represents the mean diameter of one kiln site. The circle size is proportional to the value (in cm). Size class histograms of 4 sites with their mean diameter are given. Data basis: 2467 charcoal pieces from 19 sites.

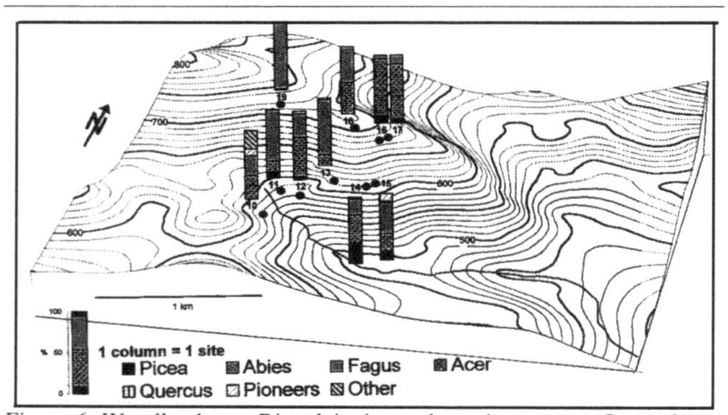

Figure 6: Woodland near Ringelai, charcoal species spectra. One column gives the spectrum for the closest kiln site. Pioneer species: *Betula* cf. *pendula*, *Populus* sp., *Salix* sp. *Corylus avellana*; Other: *Alnus* (*A. glutinosa*). Data basis: 1283 charcoal pieces from 10 sites.

Discussion and conclusions

Schauinsland

A beech (*Fagus sylvatica*)-dominated woodland, with fir (*Abies alba*), maple (*Acer* cf. *pseudoplatanus*) and some spruce (*Picea abies*) grew approximately 250 years ago around the hilltop of Schauinsland. The "small diameter sites" (mean diameters 4 - 9 cm) and "large diameter sites" (mean diameters 10 - 12 cm) show a spatial distribution. "Small diameter sites" are restricted to the ridge, whereas further down the slopes there are "large diameter sites".

This makes the possible explanation of wood selection unlikely. It is more probable that the charcoal burners used the trees they found growing, and therefore the charcoal spectra represent not only species composition, but also the actual size of the trees. The hilltop and the ridge are less steep than the slopes further down. Because of this, the top was grazed for a long time, which possibly resulted in a degraded, short growing beech forest. The fir component was probably reduced at these sites due to grazing. On richer sites further down the slopes there was a mixed beech-fir-forest. Fir was present because growing conditions were better (less grazing, better soil conditions).

Only a few spruce trees grew at the highest point of the hilltop. *Picea abies* was only a minor component of the forest, whereas further east at high altitudes around the hilltop of Feldberg mountain (1493 m) spruce grew in considerable proportions (Ludemann and Britsch, 1997; Ludemann, 1994).

Ringelai

The species composition of the forest about 200 years ago, as derived from the charcoal spectra, is one of a mixed beech-fir forest, with varying portions of beech and fir. The higher portions of *Abies alba* seem to be at sites with more favourable conditions for fir growth, such as moister sites (site 12 and 14 are close to wet sites, and above site 16/17 there is a wet area due to a source). However, the number of charcoal burning sites analysed is not great enough to draw such conclusions. Nonetheless, it seems clear that *Picea abies* was a minor component in the composition of the forest, whereas today it is widely present. The portions of spruce, fir and beech in the charcoal spectra are similar to the values given in old forest management statistics. According to these statistics, the forest composition at comparable localities in the first half of the 19th century was: 60% *Abies alba*, 30% *Fagus sylvatica*, 10% *Picea abies* (Trautmann, 1952).

There is no sign that the charcoal burners used already degraded woodlands. Apart from site 10 and 15, where *Betula* is present with more than 10% in the charcoal spectra, there is no considerable trace of pioneer species that might indicate degradation. Site 10 has also a high portion of *Alnus*-charcoal. At this site, people may have used a stand growing in a wet area along a stream, 50 m from the kiln site.

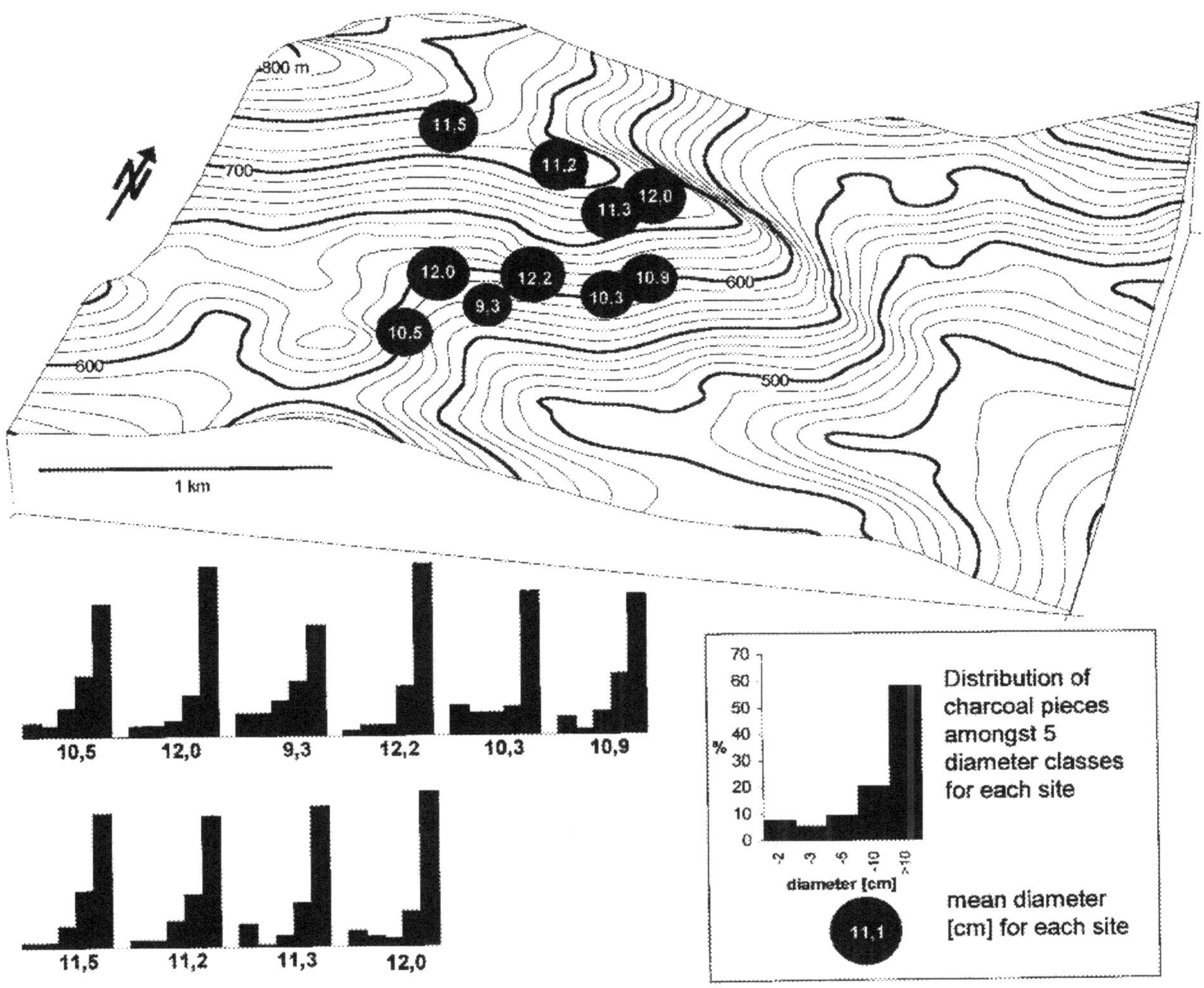

Figure 7: Woodland near Ringelai, spatial distribution of mean diameters. One black circle represents the mean diameter (size is proprotional; value given in cm) of the sample of one site. Data basis: 1145 charcoal pieces from 10 sites.

The mean diameters indicate that the charcoal producers used reasonably big trees. The values and histograms are comparable to the "larger wood size type" as shown in fig. 2 (C and D).

The species spectra and the diameter of charcoal suggest that there was no intense woodland degradation prior to the 19th century. The charcoal burners used timber from a mature, mixed beech-fir-forest without selecting for tree species or wood size.

Reconstructing forest stand structure?

What do the size class distribution (the histograms) and the mean diameter tell us? The charcoal remnants can provide an estimate of the wood size the charcoal burners used, and might represent the size class distribution of the actual stand. However, people might have selected certain size classes for producing charcoal. The "human bias" cannot be ruled out, but it becomes less probable if the spatial distribution shows certain patterns that can be explained by other factors, such as topography.

For the time being, size histograms and mean diameters derived from charcoal analysis of historic kiln sites can be compared to reference histograms, such as those given in figure 2, in order to get an idea of the size of the charred wood.

However, to verify this method, more examples are needed. It is necessary to analyse charcoal samples from charred wood of known timber size.

Forest stand structure includes species composition and size classes, the individual tree size and shape, the spatial arrangement of trees. The stand can be homogenous or heterogeneous, open or dense. So far, we cannot tell forest stand structure of historic woodlands if we do not have an accurately written record.

However, in an attempt to approach some of these aspects with the help of anthracology, we plan to try to compare calculated volume class histograms of contemporary tree specimens with diameter class histograms resulting from charcoal analysis. To do this, a given tree has to be mathematically cut up in wood volume classes (fig. 8).

Further information might be derived from dendroecology. Growth rings on charcoal from charcoal burning sites can be deciphered. It is common to sample charcoal pieces with 20-30 growth rings, and in some case they have 50 to 100

Site	asl	Dating	N	W [g]	n (D)	mD	Picea	Abies	Fagus	Acer	Quercus	Alnus	Betula	Corylus	Populus	Pop/Sal***	Salix
							Charcoal spectra [%]										
Schauinsland, hilltop area (19 kiln sites)																	
668	1125	110 ±30	144	282.3	140	4.8	.	18	82	.	.	.	.	.	.	.	.
669	1125		150	247.3	141	4.1	.	17	83	.	.	.	.	.	.	.	.
670	1125		100	137.0	95	4.5	.	15	85	.	.	.	.	.	.	.	.
671	1130	140 ±30	103	234.7	100	10.3	.	16	84	.	.	.	.	.	.	.	.
672	1130		100	191.1	97	10.9	.	3	97	.	.	.	.	.	.	.	.
673	1160		100	144.1	98	8.5	.	13	87	.	.	.	.	.	.	.	.
674	1120		332	368.9	315	9.3	.	30	69	2	.	.	.	.	.	.	.
675	1185		309	146.9	276	9.9	.	2	98	.	.	.	.	.	.	.	.
676	1100		175	77.8	138	11.2	.	54	46	.	.	.	.	.	.	.	.
677	1100		100	105.6	95	12.1	.	46	54	.	.	.	.	.	.	.	.
679	1200	60 ±30	92	60.7	77	8.4	1	.	99	.	.	.	.	.	.	.	.
680	1200		100	108.2	95	7.5	.	.	100	.	.	.	.	.	.	.	.
681	1200		100	119.5	93	6.0	2	.	98	.	.	.	.	.	.	.	.
691	1180		260	137.1	238	9.2	0.4	3	97	.	.	.	.	.	.	.	.
692	1080		135	135.1	128	9.5	.	19	81	.	.	.	.	.	.	.	.
693	1070		100	86.0	89	11.5	.	54	46	.	.	.	.	.	.	.	.
694	1015		75	29.3	65	9.7	.	53	47	.	.	.	.	.	.	.	.
713	1160		100	91.7	94	9.0	.	13	87	.	.	.	.	.	.	.	.
714	1165		102	80.2	93	11.7	.	9	91	.	.	.	.	.	.	.	.
Sum/average			*2677*	*2783.3*	*2467*	*8.8*	*0.2*	*19*	*81*	*0.1*	.	.	.	.	.	.	.
Ringelai (10 kiln sites)																	
10	580		76	35.7	69	10.5	.	34	29	.	.	26	11	.	.	.	.
11	595		101	78.9	96	12.0	9	16	74	.	.	1	.	.	.	.	.
12	590	1830**	337	567.7	304	9.3	4	57	34	0.6	.	0.6	.	.	0.3	.	3.3
13	615		104	78.2	95	12.2	.	16	81	3	.	.	.	.	.	.	.
14	610	1822**	168	157.5	122	10.3	30	43	24	.	0.6	0.6	.	0.6	.	0.6	.
15	610		102	65.8	97	10.9	13	26	47	2	.	.	11	.	1	.	.
16	680	1825**	128	180.6	118	11.3	12	84	4	.	.	.	.	.	.	.	.
17	680	1825**	118	151.9	116	12.0	6	81	13	.	.	.	.	.	.	.	.
18	715		49	31.4	34	11.2	.	47	53	.	.	.	.	.	.	.	.
19	720		100	104.3	94	11.5	.	2	97	.	.	.	.	1	.	.	.
Sum/average			*1283*	*1451.4*	*1145*	*11.1*	*7*	*41*	*46*	*0.5*	*0.1*	*3*	*2*	*0.2*	*0.1*	*0.1*	*0.3*

asl: Height above sea level [m]
Dating: * 14C absolute BP; ** Dendro Year AD of felling
N: Number of charcoal pieces analyzed
W: Weight of charcoal pieces analyzed
n (D): Number of charcoal pieces measured for minimum diameter
mD: mean diameter [cm]
*** Pop/Sal: possibly *Populus* or *Salix*

Table 1: Data from charcoal kiln sites at Schauinsland hilltop area (Black Forest, SW-Germany) and Ringelai (Bavarian Forest, SE-Germany).

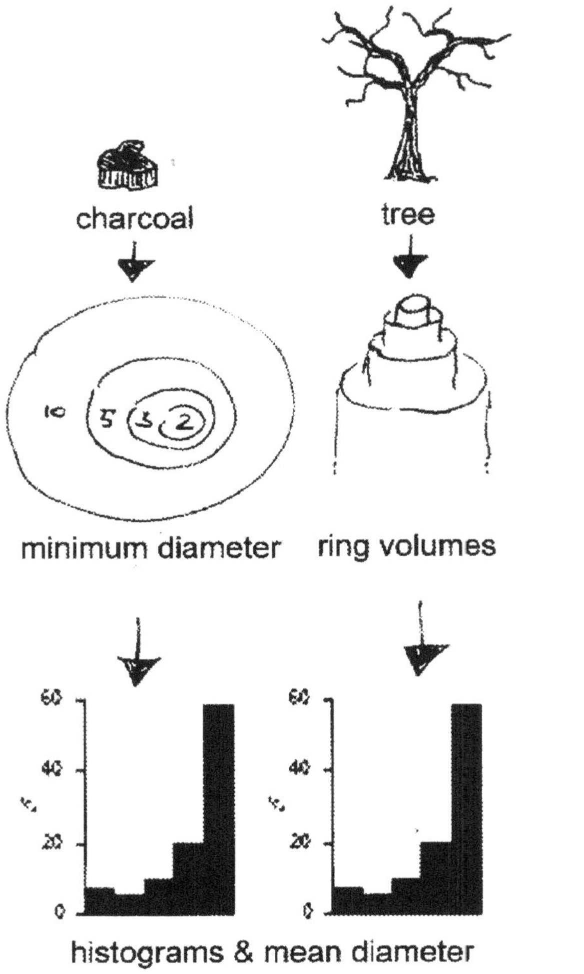

Figure 8: Are histograms and mean diameter derived from charcoal samples and from tree specimen comparable?

growth rings (dated charcoal pieces from Ringelai had 61, 81, 83, 95 growth rings). The width of growth rings contains information on the growing conditions of a tree, which is dependent on numerous factors, one of which is whether the tree grew in an open or dense stand.

The investigation of charcoal production relics contributes to the reconstruction of past environments. Charcoal analysis, in combination with pollen analysis and research on historical documents, enable ecologists to draw a locally and regionally detailed picture of the ancient forest during certain periods. Analysing the information on wood size, in conjunction with dendroecology, might contribute to the understanding of historic forest stand structure.

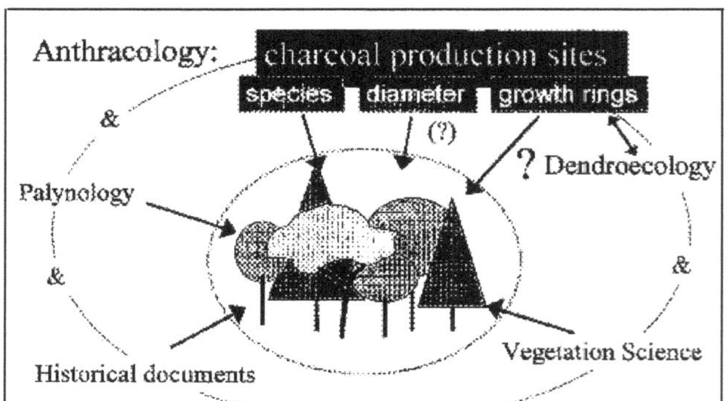

Figure 9: Reconstructing ancient woodlands. The contribution of diameter measurement and growth ring analysis of charcoal from historic kiln sites has yet to be studied.

ACKNOWLEDGEMENTS
The Bavarian part of the work was conducted with a research grant from the Deutsche Forschungsgemeinschaft (DFG). Thanks are due to Thomas Ludemann (University of Freiburg) who directed the Black-Forest work, Paul Freund (Lichtenau), who told me about sites in the forest near Ringelai, Franz Herzig (Labor für Dendroarchäologie, Bayerisches Landesamt für Denkmalpflege, Thierhaupten) for dendrochronologically dating the charcoal from Ringelai sites, Martha Schumann (The Forest Trust, Santa Fe) and Esther Guggenbichler (University of Regensburg) for their advice and comments on the manuscript.

References

BONHOTE J., VERNET J.-L. 1988 - La mémoire des charbonnières. Essai de reconstitution des milieux forestiers dans une vallée marquée par la métallurgie (Aston, Haute-Ariège). *Revue Forestière-Française* 40 (3), p. 197-212.

DAVASSE B. 1992 - Anthracologie et espaces forestiers charbonnés. Quelques exemples dans la moitié orientale des Pyrénées. *Bull. Soc. bot. Fr.* 139, p. 597-608.

HILLEBRECHT M. L. 1982 - *Die Relikte der Holzkohlewirtschaft als Indikatoren für Waldnutzung und Waldentwicklung. Untersuchungen an Beispielen aus Südniedersachsen.* Göttinger Geographische Abhandlungen 79, Verl. Erich Goltze, Göttingen. 157 p.

KAUDER B. 1992 - *Relikte der Waldköhlerei im Winkelhofer Forst bei Ebrach (Steigerwald).* Heimat Bamberger Land Jg. 4 (1), p. 23-28.

KRAUSE A. 1976 - Artenbestimmung an Holzkohlen aus dem Stahlnhainer Grund (Hintertaunus). *Natur u.* Museum 106, p. 45-47.

LUDEMANN T. 1994 - *Vegetations- und Landschaftswandel im Schwarzwald unter anthropogenem Einfluß.* Ber. d. Reinh.-Tüxen-Ges. 6, p. 7-39.

LUDEMANN T. 1996 - Die Wälder im Sulzbachtal (Südwest-Schwarzwald) und ihre Nutzung durch Bergbau und Köhlerei. Mitt. Ver. Forstl. *Standortskunde u. Forstpflanzenzüchtung* 38, p. 87-118.

LUDEMANN T., BRITSCH T. 1997 - Wald und Köhlerei im nördlichen Feldberggebiet/Südschwarzwald. *Mitt. Bad. Landesver. Naturkunde u. Naturschutz* N. F: 16 (3/4), p. 487-526.

LUDEMANN T., NELLE O. 2002 - Die Wälder am Schauinsland (Südwest-Schwarzwald) und ihre Nutzung durch Bergbau und Köhlerei. *Freiburger Forstliche Forschung*, band 15, 139 p.

PRIESNER P. 1982 - *Der Bergbau im Schauinsland von 1340 bis 1954.* Freiburg, Verlag Karl Schillinger, 328 p.

SCHLÄPFER P., BROWN C. 1948 - *Über die Struktur der Holzkohlen.* EMPA-Bericht (Eidgenössische Materialprüfungs- und Versuchsanstalt f. Industrie, Bauwesen und Gewerbe) 153, 121 p.

SCHWEINGRUBER F.H. 1982 - *Mikroskopische Holzanatomie.* Formenspektren mitteleuropäischer Stamm- und Zweighölzer zur Bestimmung von rezentem und subfossilem Material. Birmensdorf/Schweiz, Eidg. Anstalt für das forstl. Versuchswesen, 2nd ed., 226 p.

STALLING H. 1987 - Untersuchungen zur spät- und postglazialen Vegetationsgeschichte im Bayerischen Wald. *Diss. Bot.* 105, p. 1-202.

TRAUTMANN W. 1952 - *Pflanzensoziologische Untersuchungen der Fichtenwälder des Bayerischen Waldes.* Forstwiss. Centralbl. 71 (9/10), p. 289-313.

TRENKLE H., RUDLOFF H.V. 1981 - Das Klima im Schwarzwald. In: LIEHL E., *Der Schwarzwald.* Konkordia-Verlag, Bühl/Baden. 4th Ed. 1989.

Anthracology and forest sites - the contribution of charcoal analysis to our knowledge of natural forest vegetation in south-west Germany

Thomas Ludemann

Forest Research Institute of Baden-Wuerttemberg,
Wonnhaldestr. 4, D-79100 Freiburg, Germany.
E-mail ludemann@uni-freiburg.de

Abstract

Earlier mining and charcoal-burning activities have left charcoal in many locations in the Black Forest. In 1993 we began the systematic analysis of these remains. First, we tried to establish changes in time for selected investigation areas. In addition, spatial differences at a given time were also a key factor. We are attempted to discern differences depending on the different site conditions in the vicinity of the places studied. With the help of charcoal analysis, we hope to obtain information on the natural offering of wood and on the natural frequency of the tree species at different forest sites and in different forest communities. Therefore, we choose sites of charcoal production in upright circular kilns in Modern Times, and we choose study areas, which are characterized by a high density of those sites and at the same time by great differences in the site conditions. The present study deals with the results from 41 kiln sites in one valley of the southern Black Forest. In the charcoal samples analysed, we have found all of the tree species to be expected for the natural conditions. Their frequency also mirrors a natural situation: the tree species of the climax vegetation were mainly used. All other species were quantitatively unimportant. Looking at the results from the individual sites of charcoal burning, we have found many differences in the combinations and frequency of the tree species exploited and regular spatial patterns of the distribution of the species. The spatial patterns can be explained quite well by natural differences in the site conditions and in the forest communities in the vicinity of the places studied. A pronounced dependency of the wood exploitation on the local natural offering is discernible. Indications for selection of distinct tree species and for forest degradation could not be found. Moreover, no indications of shortages of wood or of import of wood or charcoal into the investigation area could be found.

Key-words: *South-west Germany, Black Forest, ancient woodland, charcoal burning, charcoal kiln, forest site, forest exploitation, forest history, forest vegetation, fuel wood, natural vegetation, site conditions.*

Résumé

Dans la Forêt Noire les activités minières et de charbonnage ont laissé des témoins dans de nombreux endroits. Ces restes sont systématiquement analysés depuis 1993 dans le but d'évaluer les changements dans le temps et dans l'espace. Grâce à l'analyse anthracologique nous pouvons obtenir des informations sur la présence naturelle des espèces ligneuses sur les différents sites dans des communautés végétales différentes. L'exemple présenté concerne une vallée du sud de la Forêt Noire. Au regard de chacun des sites de charbonnage nous avons trouvé de grandes différences dans les combinaisons et les fréquences des espèces exploitées et des modèles réguliers de distribution des espèces. Une relation de dépendance entre l'exploitation du bois et la présence des espèces localement est bien discernable. Aucun choix dans l'exploitation des espèces n'a pu être mis en évidence, exploitation qui ne semble pas entraîner de dégradation du milieu forestier.

Mots-clés : *Allemagne du Sud-Ouest, Forêt Noire, charbonnage, charbonnière, exploitation forestière, histoire sylvicole, végétation forestière, combustible, végétation naturelle.*

Zusammenfassung

Früher Bergbau und Köhlerei haben vielenorts im Schwarzwald Holzkohle hinterlassen. Im Jahr 1993 haben wir damit begonnen, diese Rückstände systematisch zu analysieren. Dabei haben wir uns zunächst auf alte Bergbaureviere im südwestlichen Schwarzwald konzentriert und versucht, zeitliche Unterschiede herauszuarbeiten. Andererseits gilt unser besonderes Interesse räumlichen Unterschieden der Holznutzung. Mit Hilfe der Holzkohleanalyse sollen neue Kenntnisse zum natürlichen Holzangebot an den verschiedenen Standorten und in den verschiedenen Waldgesellschaften gewonnen werden. Im Ganzen fanden sich in der analysierten Holzkohle alle Baumarten, die in den untersuchten Gebieten von Natur aus zu erwarten sind. Ebenso spiegelt der Mengenanteil der einzelnen Arten natürliche Gegebenheiten wieder: Hauptsächlich wurden die Baumarten der Klimaxvegetation verwendet. Alle anderen Arten waren mengenmäßig unbedeutend. Betrachten wir die Ergebnisse der einzelnen Kohlplätze, so finden wir viele Unterschiede in der Kombination und Häufigkeit der genutzten Baumarten. Diesen Unterschiede lassen sich anhand von natürlichen Unterschieden der Standortsgegebenheiten und der Waldgesellschaften in der Umgebung der Fundplätze erklären. Eine ausgeprägte Abhängigkeit der Holznutzung vom lokalen natürlichen Angebot ist erkennbar. Hinweise auf Selektion bestimmter Baumarten oder auf Walddegradation konnten nicht gefunden werden. Ebenso wenig lassen sich Belege für den Mangel an Holz oder für den Import von Holz oder Holzkohle in die Untersuchungsgebiete finden.

Schlagwörter: *Südwest-Deutschland, Schwarzwald, Brennholz, Köhlerei, Kohlenmeiler, Kolplatz, potentielle natüriliche Vegetation, Waldstandort, Waldvegetation.*

Introduction

In South-west Germany the forest administration tries to maintain forests as near to natural as possible. Therefore, we need information on the natural situation, on the natural forest vegetation and on the past human influence thereby. Charcoal analysis, beside palynology, vegetation science and studies of written sources, contributes to our knowledge of past forest vegetation (fig. 1). In this way, anthracology, as an applied science, is used by modern forest management today and will continue to be in the future.

In 1993 we started to analyse historical charcoal from the Black Forest systematically. Important criteria (variables) of the anthracological analysis and the analysed samples are age, site and kind of use. First we tried to work out changes in time for selected investigation areas and sample sites. We have concentrated our activities on old mining areas and on the material of mining-archaeological excavations in the south-western part of the Black Forest. The material analysed mainly originates from mining activities in the high and late Middle Ages, including complex settlement activities, and from charcoal production in circular kilns in Modern Times (Ludemann, 1996, 1999a-c; Nelle, 1998; Ludemann and Nelle, 2002).

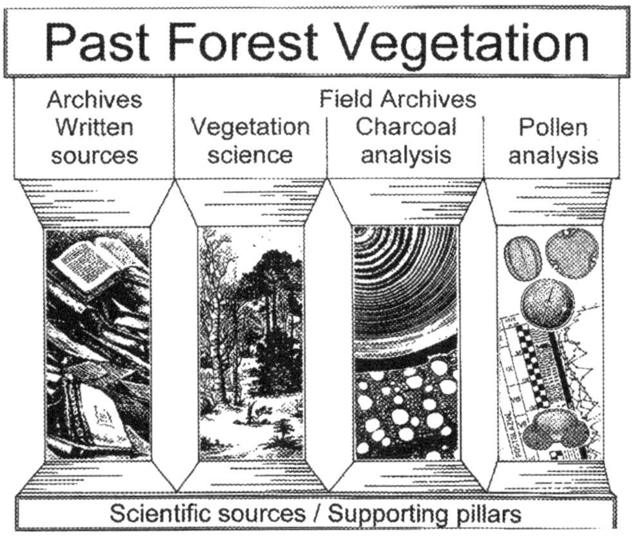

Figure 1: Scientific sources of past and natural forest vegetation.

On the other hand, we are interested in spatial differences at a given time. We are looking for differences depending on the different site conditions in the vicinity of the places studied. With the help of charcoal analysis we are trying to obtain information on the natural offering of wood and on the natural frequency of the tree species at different forest sites and in different forest communities. We want to answer the following questions:

- Which wood was used?
- In what ways did man influence forest vegetation?
- What did the historical forests look like?
- Which tree species grew on the different sites and how frequent were they?

To achieve our aim, we varied the spatial position and thereby the site conditions, while keeping two variables constant, the age and the kind of sample/kind of exploitation. We choose remnants of the charcoal production in upright circular kilns in Modern Times (fig. 2, 3, 4). Figures 2 and 3 show how an upright circular kiln is formed. Figure 4 schematically shows the typical remnants, the typical field structure that can be found even today in numerous instances, especially in mountainous regions.

Figure 2: Les charbonniers (The Charcoal Burners). Charcoal production in upright circular kilns in Modern Times in the Vosges (from the picture-series "La Rouge Myne de Saint Nicolas de la Croix" by H. Gross, about 1550).

Figure 3: Charcoal production in upright circular kilns in Modern Times in the Black Forest. (Historical Metz-photograph from 1905, Gebr. Metz, Tübingen, Germany).

A close correlation among the site conditions can be expected, especially in the analysis of the location of charcoal kilns in the Black Forest, because this kind of field archive has three important advantages for scientific studies:

- 1) Places of charcoal production are involved, not places of charcoal use. An important aim of burning charcoal is to reduce the weight of fuel wood and to simplify the difficulties of transporting fuel wood, especially in forest areas with bad accessibility, far away from settlements and routes. Therefore a close correlation to the site conditions is expected.
- 2) Each can serve to localize historical places of charcoal production nearly everywhere in the Black Forest; they belong to the most frequent remnants of the past fuel wood exploitations we know of. Moreover, a high density of such field archives characterizes many areas. Nearly 2000 localities of old charcoal kilns are already known in the southern part of the Black Forest and we are far from knowing of all such historical places. We are sure there are many thousands in the Black Forest and in other mountainous regions of Central Europe. The maximum density of places of charcoal-burning in our investigation areas is more than 40 places per square kilometre.
-3) Forest exploitation is recorded through the samples of historical charcoal burning in our investigation area, which took place before modern forest management started to work efficiently. Most of the forest stands exploited for charcoal production must have grown up naturally.

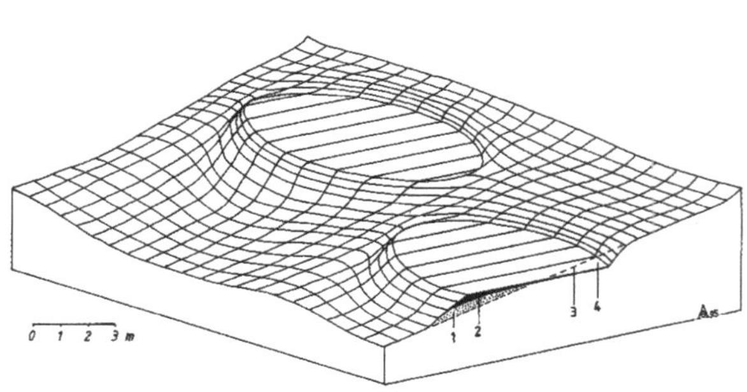

Figure 4: Places of charcoal burning. Typical field structure in slope position of mountainous regions. Circular level grounds with a diameter of 8 to 12 m. 1. soil accumulation downhill. 2. charcoal layer. 3. former slope surface. 4. dug soil uphill.

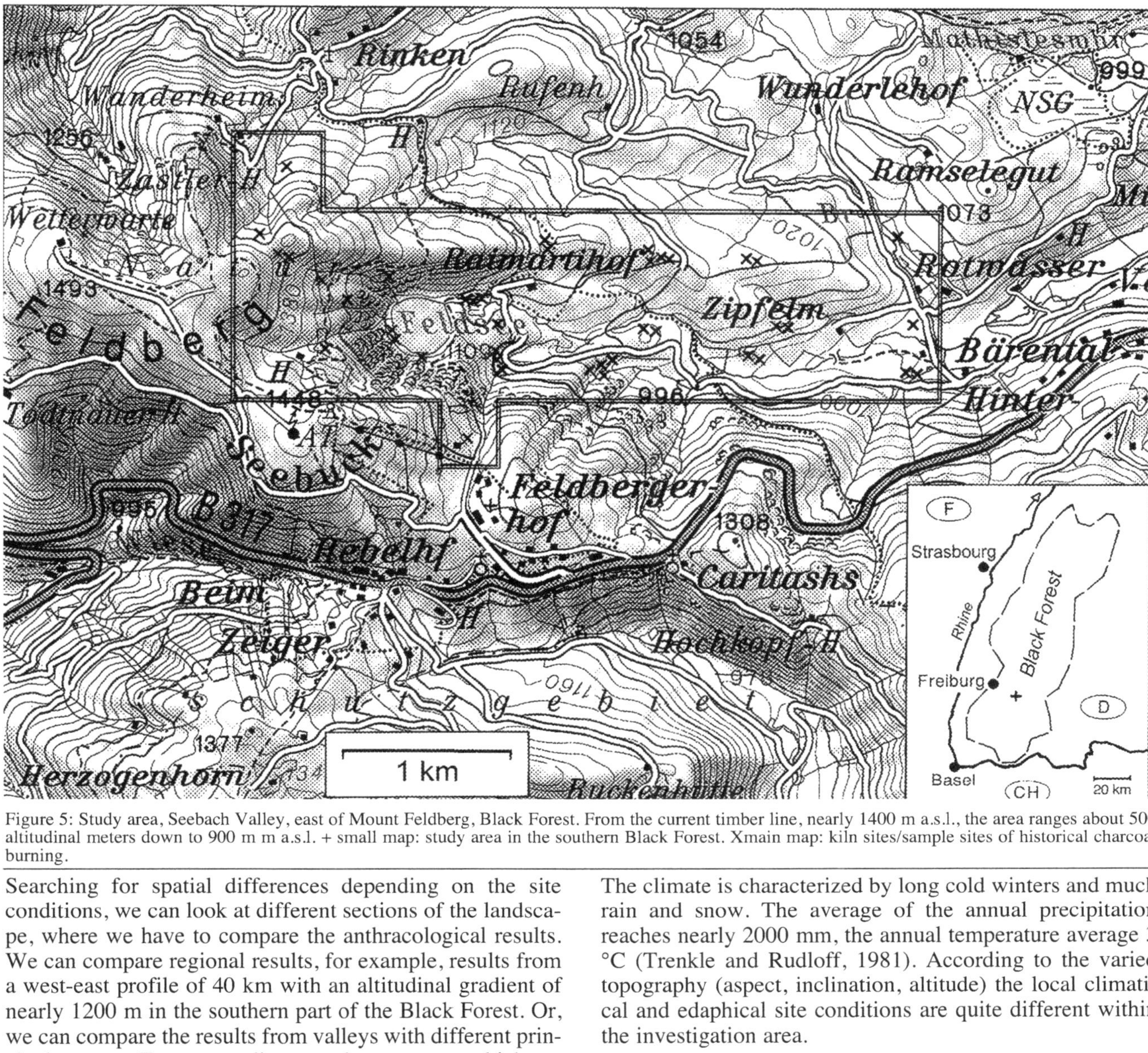

Figure 5: Study area, Seebach Valley, east of Mount Feldberg, Black Forest. From the current timber line, nearly 1400 m a.s.l., the area ranges about 500 altitudinal meters down to 900 m m a.s.l. + small map: study area in the southern Black Forest. Xmain map: kiln sites/sample sites of historical charcoal burning.

Searching for spatial differences depending on the site conditions, we can look at different sections of the landscape, where we have to compare the anthracological results. We can compare regional results, for example, results from a west-east profile of 40 km with an altitudinal gradient of nearly 1200 m in the southern part of the Black Forest. Or, we can compare the results from valleys with different principal aspects. For our studies, we choose areas, which are characterized by a high density of places of charcoal production and at the same time by great differences in site conditions in the vicinity of the places. We will continue to work on these locations for the next two years. The following example deals with a smaller section of the landscape, namely the results from only one valley, the Seebach valley in the southern Black Forest.

Study area

The forest area investigated is located directly east of Mount Feldberg, the highest peak of the Black Forest (1493 m a.s.l.) including a well-prepared glacial end of a valley, the Feldsee-Kar, with steep rocky slopes, which face north, south or east. From the current timberline, nearly 1400 m a.s.l., the study area ranges about 500 altitudinal meters in a horizontal distance of 4 km following the Seebach Valley down up to 900 m a.s.l. (fig. 5).

The climate is characterized by long cold winters and much rain and snow. The average of the annual precipitation reaches nearly 2000 mm, the annual temperature average 3 °C (Trenkle and Rudloff, 1981). According to the varied topography (aspect, inclination, altitude) the local climatical and edaphical site conditions are quite different within the investigation area.

Excavation sites, material, methods

The sites studied are dispersed all over the investigation area. We have analysed samples from 41 locations, where charcoal kilns were burnt in the past. Often two or more of these circular level grounds lay close together, so that summarized there are 24 localities of charcoal burning sites to be studied in total. The samples are dated by the radiocarbon method and by written sources from the 17th to 19th

sample no.	laboratory no.	age a BP –	age (cal 2 Sigma) a AD	century
K 51	Beta 140285	150 50	1650-1955	17.-20.
K 59	Beta 140286	50 40	1690-1955	17.-20.
K 93	Beta 140288	80 50	1655-1955	17.-20.
K 110	Beta 140290	90 60	1665-1955	17.-20.
K 117	Beta 140291	170 50	1645-1955	17.-20.
K 307	Beta 117590	(100 %)	modern	(17.-20.)

Table 1: Results of radiocarbon dating.

century (tabl. 1). From the excavation places and charcoal layers representative samples are taken and subjected to the anthracological analysis. Detailed description of sample taking is given in Ludemann (1996: 94) and Ludemann and Britsch (1997: 493). From the 41 samples, a total of 5380 pieces of charcoal fragments were analysed.

The determination of wood species followed Schweingruber (1978), looking at transversal, tangential and radial fracture surfaces with a stereoscope (Leica MZ 12) and a microscope (Zeiss Universal M III C). In addition to the wood species, the quality of the charcoal as well as the diameter of the wood exploited was determined. Both the number of pieces counted and their weight determined the quantity of a single species in a charcoal sample.

Historical charcoal contains comprehensive dendrological and dendroecological informations. In this study, we only looked at the species used for charcoal production and their quantitative relationship in the samples. In this study are presented only the results of quantitative determinations based on the number of pieces, because the results established by weighing are quite similar to those established by the number of pieces. An example of the analysis of the diameters of exploited wood can be found in the article by Nelle in this volume.

Moreover, in our investigation area the present forest vegetation was recorded by phytosociological methods (Ludemann, 1994a; Schottmüller, 1996, 1997; Ludemann and Schottmüller, 2000). The results of charcoal analysis are compared with the results of these studies of vegetation science and with the current site condition, as well as with results from stand history (Schottmüller, 1996) and palynology (Lang, 1973).

		N pieces	%
Spruce	*Picea*	2478	46
Beech	*Fagus*	2061	38
Silver Fir	*Abies*	490	9
Maple	*Acer*	177	3
Mountain ash	*Sorbus*	119	2
Alder	*Alnus*	20	<1
Pine	*Pinus*	12	<1
Birch	*Betula*	10	<1
Lime	*Tilia*	5	<1
Willow	*Salix*	3	<1
Elm	*Ulmus*	3	<1
Poplar	*Populus*	1	<1
Ash	*Fraxinus*	1	<1
		5380	

Table 2: Anthracological results summarized from 41 places of charcoal burning in Modern Times in the Seebach Valley and its surrounding, east of Mount Feldberg, Black Forest, southwest-Germany. Number of charcoal pieces counted (N) and species percentage (%). Most frequent are *Picea* and *Fagus*. The third to fifth ranks are occupied by *Abies*, *Acer* and *Sorbus*. All other species are quantitatively unimportant, together representing only 1 %.

Results of charcoal analysis

In the material analysed, 5380 pieces of charcoal, we have found 13 wood species: *Picea*, *Fagus*, *Abies*, *Acer*, *Sorbus*, *Alnus*, *Pinus*, *Betula*, *Tilia*, *Salix*, *Ulmus*, *Populus* and *Fraxinus* (tabl. 2). *Picea* and *Fagus* appear most frequently, reaching 46% and 38% respectively. The third to fifth positions are occupied by *Abies* with 9%, *Acer* with 3% and *Sorbus* with 2%. All other species are quantitatively unimportant, together representing only 1%.

Looking at the results from the individual sites of charcoal burning (tabl. 3), we find pronounced differences in the combination and especially in the frequency of the tree species exploited. The single charcoal samples are dominated by *Picea* or *Fagus* with all conceivable combinations of frequency: *Picea*-dominated, *Picea*-dominated with *Fagus*, *Fagus* dominated with *Picea*, *Fagus*-dominated. Moreover we can classify the samples on a lower level (subtypes) using the frequency of *Abies*, *Sorbus*, *Acer* and other species.

Sample	type	*Picea*	*Fagus*	*Abies*	*Sorbus*	*Acer*	*Ulmus*	*Salix*	*Alnus*	*Betula*	*Pinus*	*Tilia*	*Fraxinus*	*Populus*	Pieces n	altitude m
	average %	**44**	**41**	**9**	**2**	**3**	<1	<1	<1	<1	<1	<1	<1	<1	5380	1111
90	**Picea**	**99**		1											138	930
93	**Pc**	**98**		2											57	1040
307		**93**		5	3										189	1380
91		**89**	9	3											70	930
321	**Pc** Ab	**56**	7	29	6	3									258	1270
247	**Pc** So	**86**		5	9										103	1350
308		**85**		7	8										115	1375
61		**83**		5	5	7									175	1330
322		**65**	6	11	11	7									181	1285
246	**Pc** Ac	**65**	2	8	12	12									185	1175
62		**67**	2	10	3	18									103	1325
59	**Pc**Oth	**38**	10	6		9			20	10	1	6			80	920
58		**68**	9	12	1	2		<1	2	1	4		<1	<1	258	950
527	**Pc** Fa	**61**	29	6	<1	4									211	1255
50		**56**	35	9											101	1090
109		**57**	28	11		3									149	1100
110		**61**	20	13	1	5		<1							301	1110
116	**FaPcAb**	**45**	35	19											119	1045
56		25	**41**	34											56	1000
92		27	**47**	25											95	1040
108	**Fa** Ac	34	**43**	11	2	10	<1								250	1105
111		29	**47**	3	2	16	3								58	1110
112		13	**69**	3	3	12									143	1110
55	**Fa** Pc	42	**49**	9											57	1000
57		34	**55**	11	1										141	975
47		26	**63**	10	1			1							145	1080
48		31	**65**	3											86	1080
49		23	**71**	4	1	1									129	1085
114		23	**71**	6											126	1090
115		22	**65**	9	1	3									144	1085
106	**Fagus**	15	**72**	9	2	3									151	1140
362	**Fa**	16	**73**	8	3										118	1105
117		19	**74**	7											129	1045
52		17	**76**	7											159	1045
107		12	**82**	3		4									76	1140
118		12	**84**	4											149	1050
119		12	**84**	3		1									128	1050
113		11	**86**	3											176	1090
51		10	**89**	1											71	1045

Table 3: Anthracological results from 41 places of charcoal burning in Modern Times in the Seebach valley and its surrounding, east of Mount Feldberg, Black Forest, southwest-Germany. Number of charcoal pieces counted (n) and species percentage in the individual sample sites. Classified by the frequency and combination of tree species established. The samples are dominated by *Picea* or *Fagus* with all imaginable combinations of frequency: Pc *Picea*-dominated, PcFa *Picea*-dominated with *Fagus*, FaPc *Fagus* dominated with *Picea*, Fa *Fagus*-dominated. Moreover we can classify the *Picea*- and *Fagus*-dominated samples into subtypes using the frequency of *Abies* (PcAb, FaPcAb), *Sorbus* (PcSo), *Acer* (PcAc, FaAc) and other species (PcOth).

Spatial patterns

The next figures show, where the different sample sites are situated in the field. We have found regular spatial patterns of the frequency distribution of the species. That is explained with the help of two altitudinal profiles (fig. 6), one summarizing map (fig. 7) and five simplified maps of species frequency (fig. 8-12).

Two altitudinal profiles from the top down to 900 m a.s.l. (fig. 6): the results are positioned strictly on the altitudinal gradient. The upper profile includes only samples from sites north of the Seebach, where convex forms of the landscape dominate and which have drier and poorer soils. In the second profile, we find the results from sample sites south of the Seebach, lying mainly in the valley floor and on the lower parts of steep slopes. Here the soils are richer, with a better water balance. Both profiles indicate a clear concentration of spruce charcoal in the highest sites next to the top of the Black Forest on the one hand and in the lowest sites on the floor of the valley on the other hand. The area between 1000 and 1200 m a.s.l. is characterized by *Fagus*-dominated samples. *Sorbus*, the most frequent pioneer species, as well as *Acer*, are concentrated next to the top. *Acer* is found especially on rich, fresh sites, *Sorbus* in the upper profile. *Abies*, *Acer* and *Sorbus* accompany *Picea* farther up, at higher elevations than *Fagus* does.

Figures 7-12 represent six maps of spatial patterns of tree species frequency. The following main features of distribution are to be seen in the maps of tree species frequency:

- *Fagus* and *Picea* appear as antagonists in space and site (fig. 7, 8, 9).
- As can also be seen in the profiles, *Picea* reaches its highest frequency in the highest, subalpine sites next to the top of Mount Feldberg (fig. 8). Its frequency decreases, when we go down east into the montane forest belt. A second area of high frequency encompasses the lowest sites in the east of the investigation area. We find single samples dominated by *Picea* only by exception at an altitude between 1000 and 1200 m o.s. on the floor of the valley near the river bed.
- *Fagus* dominates in the montane-upper montane forest areas north and south of the Seebach valley and of Lake Feldsee (fig. 9).
- *Abies* is of less importance, reaching its highest frequency in climatically protected, moist-air sites between the *Fagus*- and the *Picea*- dominated area (fig. 10).
- *Acer* appears especially in high sites, but not in the highest, where a striking change from *Fagus*-dominated to *Picea*-dominated samples is detected, west and south-west of Lake Feldsee (fig. 11).
- *Sorbus* is found with *Picea* in the highest sites close to the current timberline. In other parts of the investigation area *Sorbus* is missing completely in the historical charcoal (fig. 12).

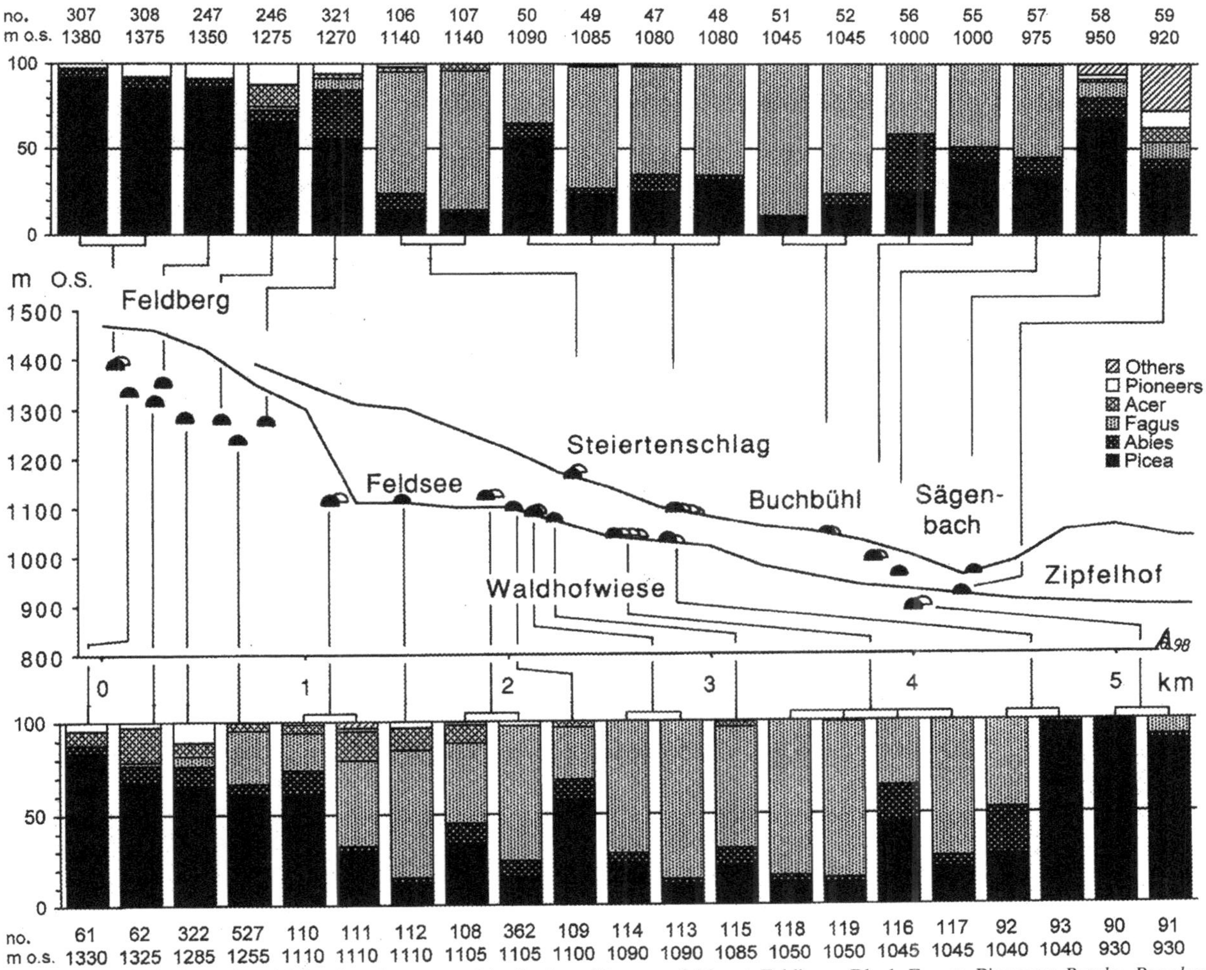

Figure 6: Anthracological results of kiln sites from two altitudinal profile east of Mount Feldberg, Black Forest. Pioneers: *Betula, Populus, Sorbus* (Pomoideae), *Salix*. Others: *Alnus, Fraxinus, Pinus, Tilia, Ulmus*. N°: sample number, m o.s.: altitude of kiln site (meter a.s.l.).

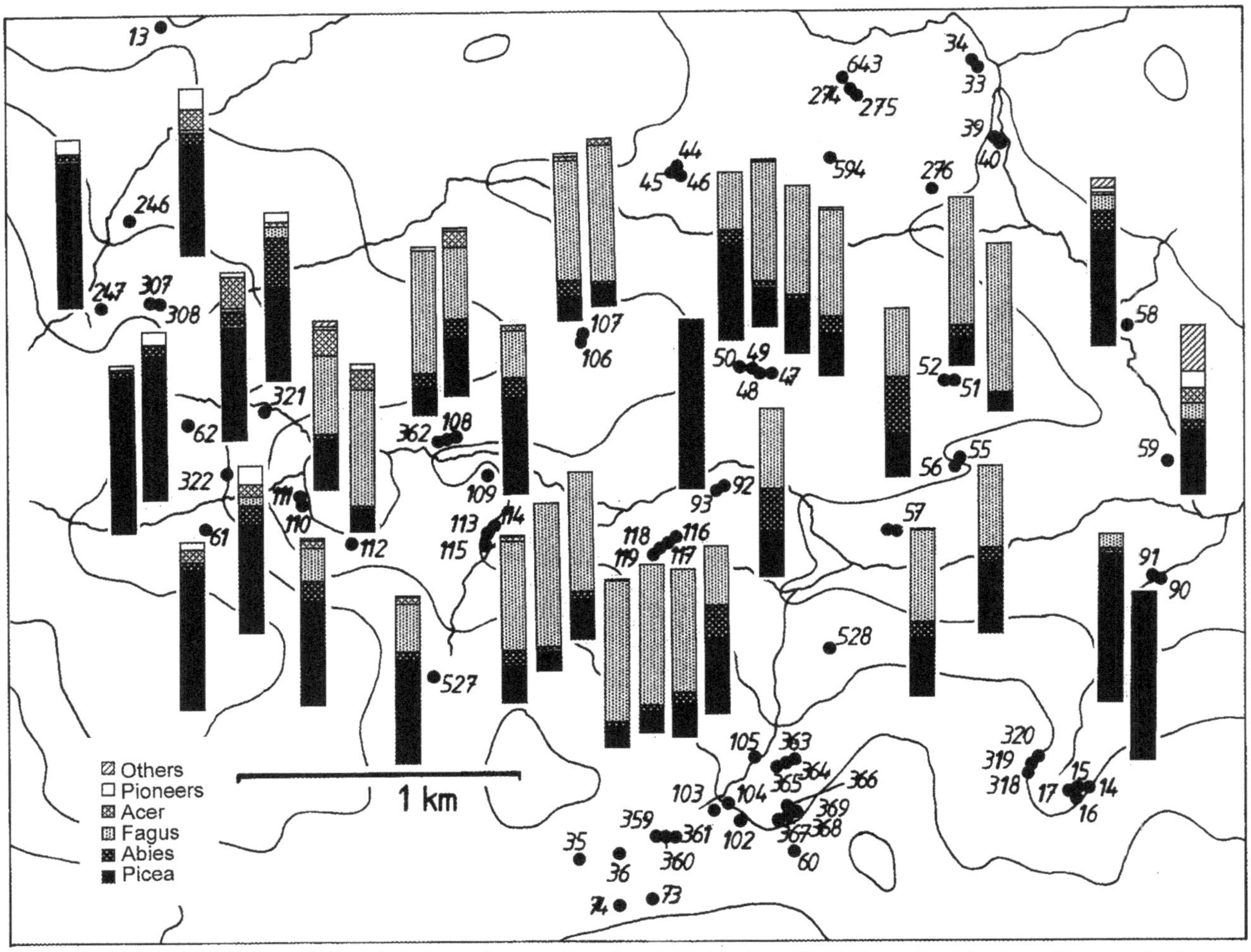

Figure 7: Anthracological results from 41 historical places of charcoal burning. Seebach valley/, Black Forest. Spatial pattern of tree species frequence. Pioneers: *Betula, Populus, Sorbus* (Pomoideae), *Salix*. Others: *Alnus, Fraxinus, Pinus, Tilia, Ulmus*.

Discussion

From the point of view of a vegetation scientist, specializing in the forest communities of the Black Forest, the distribution of wood species in the charcoal samples and the spatial patterns of past distribution in the field, deduced from the anthracological results, can be explained quite well by natural differences in the site conditions (altitude, aspect, inclination, edaphical conditions, etc.) and in the forest communities in the vicinity of the places studied. On the other hand, vegetation science acquires new indications on the natural distribution and frequency of the tree species and forest communities from the anthracological results. This can be seen both in the general features and in many details:
- *Picea*: both at the highest, subalpine sites and at cold-moist sites in depressions and valleys in the montane belt coniferous forest communities, or at least a relatively high percentage of conifers have to be expected in natural forest. The charcoal samples of such sites are characterized in this way, as are also the corresponding forest communities (*Bazzanio-Piceetum, Luzulo-Abietetum*) to be found there today (Schottmüller, 1996, 1997; Ludemann and Schottmüller, 2000).

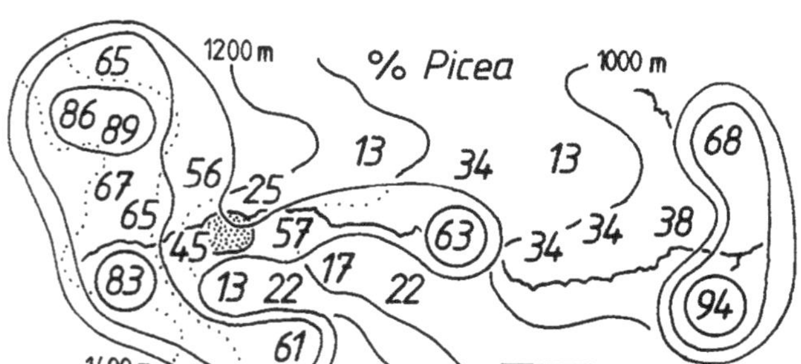

Figure 8: Percentage of spruce in charcoal samples from 41 kiln sites in the Seebach valley, Black Forest, (results of neighbouring sites summarized).

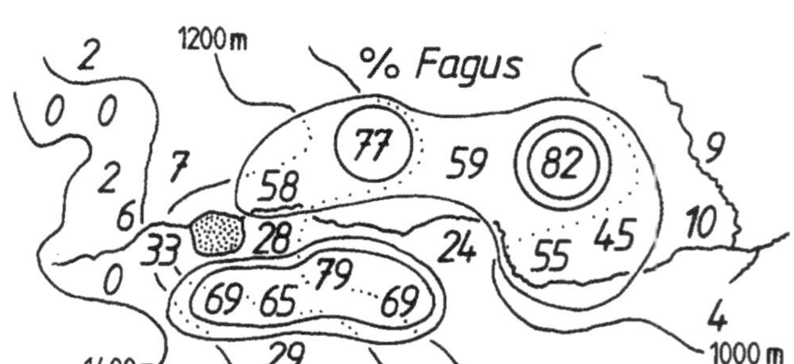

Figure 9: Percentage of beech in charcoal samples from 41 kiln sites in the Seebach valley, Black Forest (results of neighbouring sites summarized).

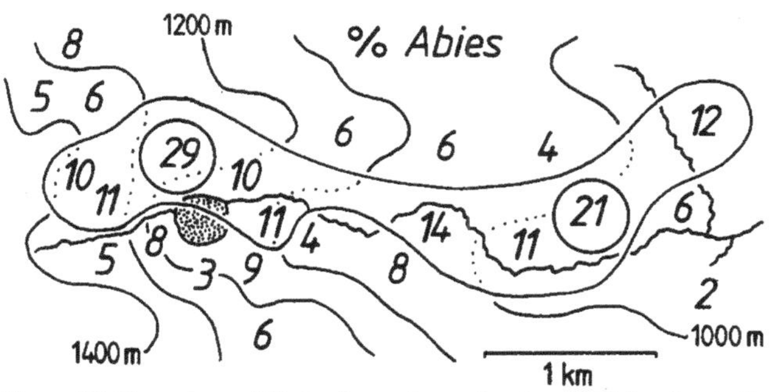

Figure 10: Percentage of fir in charcoal samples from 41 kiln sites in the Seebach valley, Black Forest (results of neighbouring sites summarized).

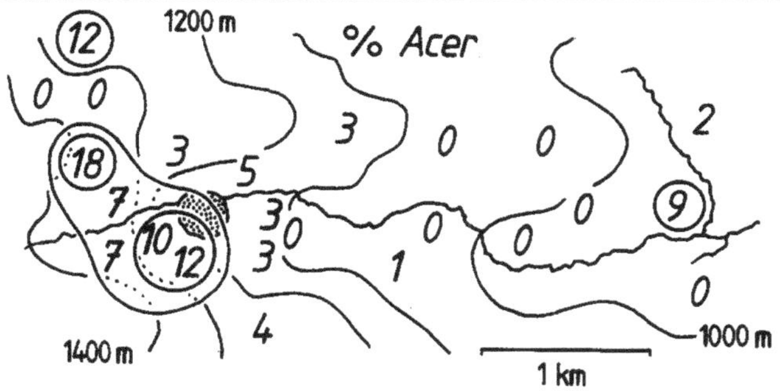

Figure 11: Percentage of maple in charcoal samples from 41 kiln sites in the Seebach valley, Black Forest (results of neighbouring sites summarized).

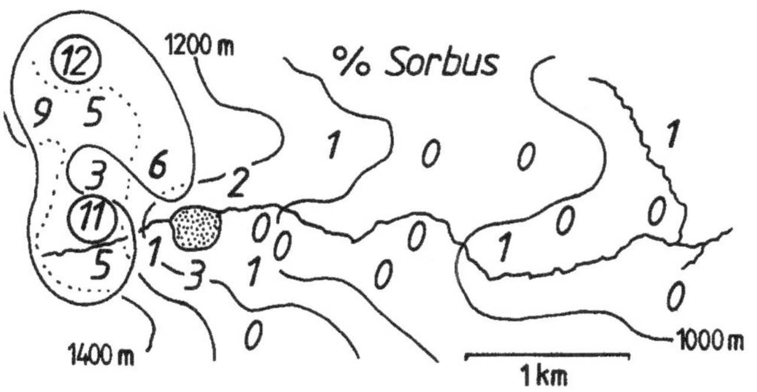

Figure 12: Percentage of mountain ash in charcoal samples from 41 kiln sites in the Seebach valley, Black Forest (results of neighbouring sites summarized).

Moreover, from the presented results from the east side of Mount Feldberg, as well as from anthracological results from its north, west and south side (Ludemann, 1994b; Britsch, 1995; Ludemann and Britsch, 1997; Jung, 1999) we can deduce the existence of a natural *Picea* forest belt in the highest areas of the southern Black Forest. This interpretation is also substantiated by palynological results (Lang, 1973). Along with these findings, charcoal analysis provides a new contribution to answer an old scientific question, an old disputed debate in the field of vegetation science: the question of the natural distribution and frequency of *Picea* in the Black Forest. Without a doubt, man could have indirectely influenced the distribution of *Picea* in the past, well before modern forest management started to work efficiently. Certainly, the natural site conditions, the natural potential of the species and the natural predisposition of site conditions and species character seem to be the most important reasons for the high percentage of *Picea* in some of the historical charcoal samples. Of course there are areas and sites to be found in the investigation areas, that are influenced by man in the same way as those rich in *Picea* charcoal, but still characterized by samples rich in *Fagus* or *Abies*. Descending from the top in every direction, *Picea*-dominated samples disappear and *Fagus*-dominated samples are increasingly prevalent.

- *Sorbus* and *Picea*: in the highest sites *Sorbus* is the most frequent deciduous wood species, not *Fagus*. These sites are not far away from the natural, local-climatically caused timberline, which is formed by *Picea*. Vegetation scientists are familiar with the pre-forest association *Piceo-Sorbetum*. It is a natural pioneer forest community in regards to time and space (Oberdorfer, 1978). Charcoal analysis gives evidence that there have been existing light and low growing stands within the *Picea* forest belt and that *Piceo-Sorbetum* stands were an important part of this vegetation belt in the past.
- *Fagus*: looking at the distribution of *Fagus,* we can deduce a widespread area east of Mount Feldberg with mixed coniferous-deciduous forest rich in this species. *Fagus* ranges not only over the steep slopes at the end of the valley, where *Luzulo-*, *Galio-* and *Aceri-Fagetum* are growing today. It is also found east of this area, it in the forest areas south and north of the valley.
- *Abies*: the contribution of *Abies* in our investigation area expresses its synecological position between *Fagus* and *Picea* and its synsystematical position between *Querco-Fagetea* and *Vaccinio-Picetea* (Oberdorfer, 1992). Certainly, the frequency of *Abies* is lower than expected. Because of its high frequency in other investigation areas and in the corresponding samples of historical charcoal (Ludemann, 1996, 1999a-c; Ludemann and Nelle, 2002), we think that this is caused naturally and not provoked by human intervention. Therefore, as some vegetation scientists have previously asserted, *Abies* is of less importance in the highest sites of the Black Forest under natural conditions (Oberdorfer, 1982).
- *Acer*: the distribution of *Acer* and with it that of *Aceri-Fagetum* stands rich in Acer is narrow-limited to rich moist sites between Lake Feldsee and the top of Feldberg.

Anthracological studies north of Mount Feldberg (Ludemann and Britsch, 1997) point out, that the main distribution of *Aceri-Fagetum* stands rich in *Acer* was in north exposures, where the subalpine site conditions are expressed best; it is precisely there that we find the highest frequency of *Acer* in the charcoal.

This is also confirmed by phytosociological studies of current forest vegetation. North of Mount Feldberg, there are numerous typical *Aceri-Fagetum* stands (Ludemann and Britsch, 1997), whereas east of Mount Feldberg many deciduous forest stands of rich soils make floristic-phytosociological tendencies into the montane *Galio-Fagetum* noticeable. In fact, they would be more appropiately mapped as *Galio-Fagetum* (Schottmüller, 1996, 1997; Ludemann and Schottmüller, 2000). Therefore, the different frequency of *Acer* north and east of Mount Feldberg is quite understandable.

Selection, wood transport, degradation

The fact that the charcoal samples mirror the site conditions in the general features and in many details, at the same time speaks both against selection, far distance transport and forest degradation. Of course, the regular spatial patterns of (climax) tree species established depending on the site conditions would have been outwiped, if such activities or processes had taken place. Further arguments speak against such activities and processes:

- Selection: many studies that have previously taken place in our investigation area and many other regions in Central Europe assert that the charcoal burners only used hard deciduous wood, mainly or exceptionally from beech (*Fagus sylvatica*). Our study shows that this is not a correct assertion. A selection of distinct tree species for charcoal burning is not to be found; selection criteria of the charcoal burners could not be of great importance. In the historical charcoal, we have found all of the tree species to be expected for the natural conditions in the area investigated, including pioneer species and high percentages of coniferous wood from *Abies* and *Picea*. Moreover, the frequency of the species agrees with our expectations for the natural conditions, with the exception of *Abies* (see above).
- Far distance transport: we could not find any species or high frequencies of species in the samples, which could not grow in the vicinity of the places studied, like *Quercus* or *Carpinus* in high altitude or *Fraxinus*, *Tilia* or *Ulmus* on nutrient poor sites.
- Degradation: pioneer species are not frequent and climax species are always dominant. Only one sample was characterized by a high percentage of pioneer species and non-climax species. This sample originated from a locality farthest down in the valley, where the distance to the next village was the shortest and where the most intensive and the longest anthropogenic influences were to be expected.

Conclusion

In summary: a pronounced dependency of the wood exploitation on the local natural offering is discernible. Indications for selection of distinct tree species and for forest degradation could not be found. The natural offering, the complete exploitation of remote forest areas, and the restrained possibilities of wood transport obviously seem to be the important criteria for wood exploitation for charcoal production. Moreover, no indications of shortages of wood or of import of wood or charcoal into the investigation area could be found.

On the contrary, the results and their interpretation are much more complicated in the old mining areas and in the old settled landscapes. Besides the natural site conditions in those areas, selection by man and anthropogenic changes of vegetation are more important than in the presented results of charcoal production in Modern Times in remote, areas that were settled later.

In conclusion, the dependence of wood exploitation on the natural offering and the large numbers and wide distribution of historical sites with analysable charcoal provide together a unique chance to obtain new exact information for larger regions on the former vegetation and the changes therein. The contribution of anthracology to our knowledge of natural forest vegetation and forest history could and should be much more comprehensive.

References

BRITSCH T. 1995 - *Historische Meilerplaetze im noerdlichen Feldberggebiet/Suedschwarzwald. Ergebnisse der Holzkohlenanalyse - Vergleich mit der Waldgeschichte. - Staatsex.* Thesis. Univ. Freiburg, Biol. Faculty, Geobotany, 60 p.

JUNG A. 1999 - *Waldvegetation und Kohlholznutzung auf ausgesuchten Hoehenprofilen im Suedschwarzwald.* - Dipl. Thesis. Univ. Freiburg, Biol. Faculty, Geobotany, 81 p.

LANG G. 1973 - Neue Untersuchungen über die spaet- und nacheiszeitliche Vegetationsgeschichte des Schwarzwaldes. IV. Das Baldenwegermoor und das einstige Waldbild am Feldberg. - *Beitr. naturkundl. Forschung* SW-Dtl. Karlsruhe, 32, p. 31-51.

LUDEMANN T. 1994a - Die Waelder im Feldberggebiet heute. Zur pflanzensoziologischen Typisierung der aktuellen Vegetation. - *Mitt. Verein forstl. Standortskunde u. Forstpflanzenzuechtung* Freiburg i. Br. 37, p. 23-47.

LUDEMANN T. 1994b - Vegetations - und Landschaftswandel im Schwarzwald unter anthropogenem Einfluss. - *Ber. d. Reinh.-Tüxen-Ges.* Hannover. 6, p. 7-39.

LUDEMANN T. 1996 - Die Waelder im Sulzbachtal (Suedwest-Schwarzwald) und ihre Nutzung durch Bergbau und Koehlerei. - *Mitt. Verein forstl. Standortskunde u. Forstpflanzenzuechtung* Freiburg i. Br. 38, p. 87-118.

LUDEMANN T. 1999(a) - Zur Brennstoffversorgung im Bergbaurevier Sulzburg. - *Archaeolog. Nachrichten aus Baden* Freiburg i. Br. 61/62, p. 131-138.

LUDEMANN T. 1999(b) - Holzkohle - Energiequelle für den Bergbau, Informationsquelle für die Wissenschaft. - *Archaeolog. Info. Baden-Wuertt.* 41 (Frueher Bergbau im suedl. Schwarzwald. - Begleitheft zur Bergbauausstellung. Museum Ur- u. Fruehgeschichte Stadt Freiburg i. Br.),Stuttgart. p. 123-129.

LUDEMANN T. 1999(c) - Zur Brennstoffversorgung einer roemischen Siedlung im Schwarzwald. - In: BRATHER S., BUECKER CH. and HOEPER M. (eds.): Archaeologie als Sozialgeschichte. Studien zu Siedlung, Wirtschaft und Gesellschaft im fruehgeschichtlichen Mitteleuropa (Festschrift Steuer). Internationale Archaeologie 9: *Studia honoraria.* Rahden/Westf. (Leidorf). p. 165-172.

LUDEMANN T., BRITSCH T. 1997 - Wald und Koehlerei im noerdlichen Feldberggebiet/Suedschwarzwald. - *Mitt. bad. Landesverein Naturkunde Naturschutz* N.F. Freiburg i. Br. 16 (3/4), p. 487-526.

LUDEMANN T., NELLE, O. 2002 - Die Wälder am Schauinsland und ihre Nutzung durch Bergbau und Köhlerei. - *Freiburger Forstliche Forschung*, Freiburg i. Br.15, 139 p.

LUDEMANN T., SCHOTTMUELLER E. 2000 - Zur Vegetation und Geschichte des Feldseewaldes. - *Ber. Freiburger Forstliche Forschung,* Freiburg i. Br. 24, p. 1-24.

NELLE O. 1998 - *Waldstandorte und Koehlerei am Schauinsland (Suedschwarzwald).*- Dipl. Thesis. Univ. Freiburg, Biol. Faculty, Geobotany: 96 p.

OBERDORFER E. 1978 - *Sueddeutsche Pflanzengesellschaften.* - 2nd ed., Jena, Stuttgart, New York (Gustav Fischer). Part II: 311 p.

OBERDORFER E. 1982a - Die hochmontanen Waelder und subalpinen Gebuesche. - In: *Landesanstalt für Umweltschutz Baden-Wuerttemberg: Der Feldberg im Schwarzwald*, Karlsruhe. p. 317-364.

OBERDORFER E. 1982b - Erlaeuterungen zur vegetationskundlichen Karte Feldberg 1:25000. - *Beih. Veroeff. Naturschutz Landschaftspflege Bad.*-Wuertt. Karlsruhe. 27, 83 p.

OBERDORFER E. 1992 - *Sueddeutsche Pflanzengesellschaften. Teil IV: Waelder und Gebuesche.* - Jena, Stuttgart, New York (Gustav Fischer). 2nd ed. Part IV: 862 p.

SCHOTTMUELLER E. 1996 - *Das Mosaik der Waldgesellschaften im Feldberggebiet und seine Ursachen. - Vegetationskundliche, historische und bodenkundliche Bearbeitung ausgesuchter Flaechen im St. Wilhelmer Napf, auf dem Stuebenwasen, im Feldseekar und im Seebachtal.* - Dipl. Thesis. Univ. Freiburg, Biol. Faculty, Geobotany: 91 p.

SCHOTTMUELLER E. 1997 - Vegetationskartierung Bannwald Feldseewald - oestlicher Abschnitt. - Forstl. Versuchs- u. Forschungsanstalt Bad.-Wuertt., *Abteilung Botanik und Standortskunde*, 16 p. & vegetation map.

SCHWEINGRUBER F. H. 1978 - *Mikroskopische Holzanatomie. Formenspektren mitteleuropäischer Stamm- und Zweighölzer zur Bestimmung von rezentem und subfossilem Material.* - Eidg. Anstalt für das forstl. Birmensdorf/Schweiz. Versuchswesen. 226 p.

TRENKLE H., RUDLOFF H.V. 1981- Das Klima im Schwarzwald. - In: LIEHL and SICK (eds.): *Der Schwarzwald.* - Veroeff. Alemannisches Inst. Buehl/Baden. Freiburg 47, p. 59-100.

Charcoal kilns and environmental history in the eastern Pyrenees (France). A methodological approach

Jérôme Bonhote, Bernard Davasse, Claude Dubois, Véronique Izard, Jean-Paul Métailié
CNRS-UMR 5602, GEODE,
Maison de la recherche, Université Toulouse-le Mirail, 31058 Toulouse, France.

Abstract
Since 15 years, an interdisciplinary research program on long-term forest history and the impact of metallurgy has been undertaken in the eastern Pyrenees. The objective is to describe the spatial evolution of forests and pastoral landscapes and the principal historical events and their causes. Our methodology is based on large-scale regressive studies of various characteristic sites (in valleys or forests) and includes the application of history, phytogeography, palynology, anthracololgy and archaeology. Remains of charcoal kilns are numerous in the mountain forest areas, whether these are actual or potential, all over the eastern Pyrenees and in particular in Ariege and in the Catalan mountains. Charcoal kilns can thus be used as a tool for reconstructing environmental history and ancient woodlands. Two different methods are described here: an anthraco-biogeographic method, based on a geosystemic approach, comparative studies and spatialisation of results and an archeo-environmental method, that allows a precise diachronic study of limited forest units and charcoal kiln sites.

Key-words: *Eastern Pyrenees, Forest history, Charcoal kiln, Phytogeography, Palynology, Anthracology, Archaeology.*

Résumé
Un programme interdisciplinaire de recherche est conduit depuis 15 ans sur l'histoire des forêts et l'impact de la métallurgie dans les Pyrénées de l'est. L'objectif est de décrire l'évolution spatiale des paysages forestiers et pastoraux, les principaux seuils historiques et leurs causes. La méthodologie est basée sur l'étude régressive à grande échelle de sites caractéristiques (vallée ou unités forestières), en utilisant l'histoire la phytogéographie, la palynologie, l'anthracologie et l'archéologie. Dans toutes les Pyrénées de l'est, en particulier en Ariège et dans les montagnes catalanes, les restes d'anciennes charbonnières sont innombrables dans l'ensemble de l'aire de la forêt, actuelle ou potentielle. Les charbonnières ont donc été utilisées en tant qu'outil pour l'histoire de l'environnement et la reconstitution des forêts du passé. Deux méthodologies sont décrites : la méthode anthraco-biogéographique, qui est basée sur une démarche géosystémique et permet une approche comparative et spatialisée ; la méthode archéo-environnementale, qui permet une étude diachronique précise d'unités forestières limitées et des sites de charbonnage.

Mots-clés : *Pyrénées de l'est, histoire des forêts, charbonnière, phytogéographie, palynologie, anthracologie, archéologie.*

Charcoal kiln records: a tool for reconstructing environmental history

Research on charcoal kiln sites and their role as a palaeoecological source for the reconstruction of the history of Pyrenean forests was initiated in the upper Ariege in 1986-87. This area is one of the main centres of metallurgy in the Pyrenees and the initial research revealed the scientific potential of this type of sites. The very numerous charcoal-burning sites are distributed in all the forests, and even in mountain pastures, where they represent the evidence of disappeared forests (Bonhôte and Vernet, 1988). Research on forests was developed since the beginning of the 1990s' in the whole eastern part of the range (fig. 1), in the framework of several Ph.D. theses and interdisciplinary programs, which have now yielded significant results on the long-term history of Pyrenean environments. The results concern at least four different fields of research:

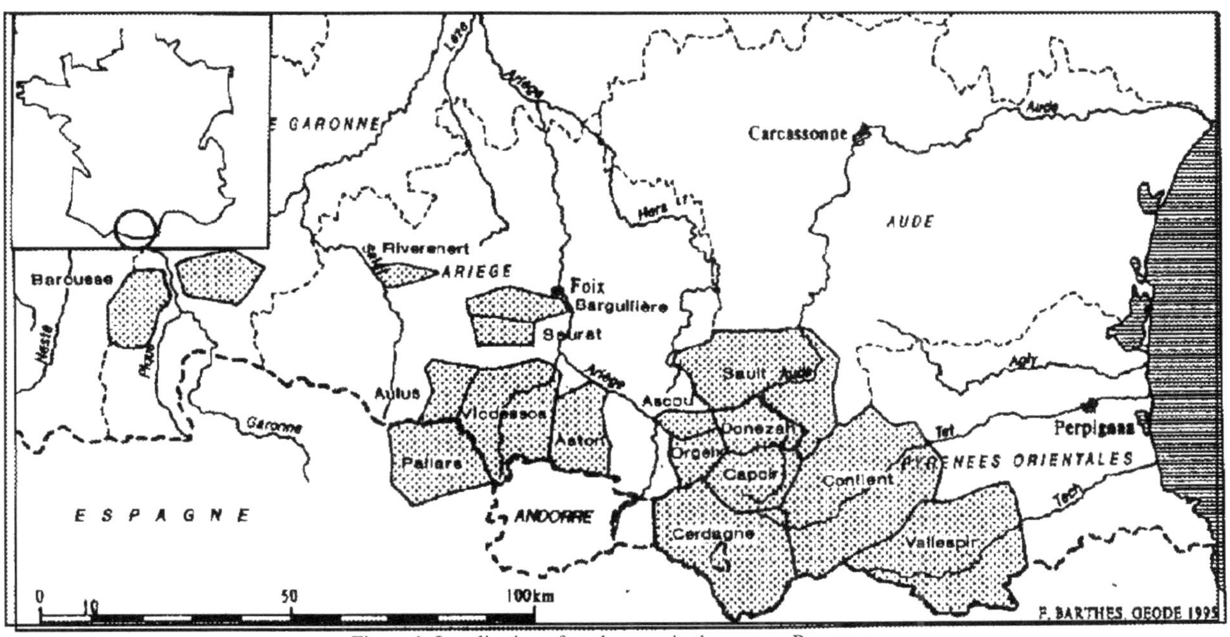

Figure 1: Localisation of study areas in the eastern Pyrenees.

Historical cartography of the environment

In the valleys of the Catalan Pyrenees and in the upper Ariege region, the development of metallurgy, most often starting in the Middle Ages but sometimes as early as in the Classical period, entailed an extensive exploitation of forests for charcoal production. No forest has escaped from this, whatever its altitude or topographic situation. Charcoal burning sites thus constitute a reliable indication of the past exploitation of forests and their period of activity can be determined by radiocarbon dating. The study of numerous charcoal kilns, located in current pasture areas, allows us to map the extension of former woodlands and to establish their specific composition. The extension of past deforestation can be quantified and its impact on the upper forest limit can be determined. This limit is now situated at an average altitude of 1400-1600 m a.s.l. while its potential altitude would range between 1800 and 2400 m a.s.l. Ancient charcoal kilns are found at altitudes of 2300-2450 m a.s.l. (Bonhôte and Vernet, 1988; Bonhôte and Métailié, 1993).

The chronology of the formation of landscapes and deforestations

The data obtained from archaeology, biogeography, geomorphology, archive sources and palaeoenvironnemental studies have allowed the elaboration of a chronological model for the eastern part of the Pyrenees. This model reconstructs various phases of landscape formation and highlights the major turning points in the relationship between human societies and the environment (Bonhôte *et al.* 2000). Pastoral colonisation starts during the Neolithic (*c.* 4500 BC) and the generalisation of pastoral practices takes place during the Bronze Age (*c.* 2200 BC). Metallurgy and charcoal production have an impact on the landscapes of the Catalan Pyrenees as early as in the Classical period, even though the intensification of these activities does not occur in the valleys of Ariege until the medieval period. In the beginning of the XIVth century, the introduction of hydraulic forges (mouline, iron mill) represents a major event, resulting in a first forest crisis and the promulgation of regulations. A second limit is reached in the XVIIth century, due to population growth, metallurgy (invention of the Catalan forge) and agro-pastoral practices. During the XVIIIth century the deforestation at higher altitudes (*i.e.* above 1700 m) comes to an end.

Evolution of the specific composition of the forests

Early historical studies have already pointed to the role played by human activities in the expansion of *Fagus* and the rarefaction of *Abies* (Fruhauf, 1980; Bonhôte and Fruhauf, 1990). The time and place of similar processes could be determined in the various valleys, where the expansion of beech always marks an intensification of the human impact. However, the evidence is not unequivocal: there are, for example, cases of protected forests "bedats" from the end of the Middle Ages, in which beech was replaced by fir. In the last centuries, oak woodlands have been favoured by an intensified exploitation of the lower slopes. Anthracological research has also revealed the probable existence of vast forests of *Pinus sylvestris* in the upper Ariege until the XVIIth century. Until now, this phenomena was unknown to phytogeographers who considered the climate too humid for such forests to grow there.

The role of metallurgical activities for the promulgation of forest regulations

In the Catalan mountains, the impact of metallurgical activities on the forest cover led the lords to announce protective promulgations in the XIIth century, and overall regulations from the XIVth century (Izard, 1999). In Ariege, various local regulations date from the XIVth to the XVIIIth century (Verna, 1994; Davasse, 2000). The exploitation of woodlands was organised well before the intervention of the central royal power, which starts during the Reformation of 1669 (and hardly touches the province of Roussillon before the XIXth century). The archaeological analysis of charcoal kilns is part of an interdisciplinary research program, based on the analysis of sites at different scales, spanning from the archeological site to the valley, forest unit or moutain slope. The preferred scale for anthracological-biogeographical studies was until now the "charcoal site" that constitutes a coherent unit of exploitation. The chosen unit corresponded, in the majority of cases, to a slope, a secondary small valley or a forest, while historical and phytogeographical studies were done at the scale of the valley. Before presenting a critical synthesis of our research methods, we will first discuss, in a more general manner, various aspects of charcoal burning in the Pyrenees.

Charcoal production in the Pyrenean forests: data from historical sources and fieldwork

Different types of carbonisation processes and their traces

The two types of charcoal burning structures used elsewhere in Europe (kilns or pits) are also known from the Pyrenees, although they are not always easy to distinguish from each other in the written sources. Moreover, they were probably used simultaneously, from the beginning of iron metallurgy until the XIXth century. The main difficulty in studying these processes is the lack of descriptions of ancient practices, as the charcoal production did not start to interest engineers until the end of XVIIIth century.

Charcoal pits (charbonnières en fosse)

The production of charcoal in pits seems to have been most widespread during the times of the low furnace metallurgy and was described by Biringuccio in his book "Della pyrothecnica" (1540). This method, known from many European countries until the XIXth century and from Africa (Bielenin, 1993) consisted of the filling of a cylindrical hole with wood. In the Pyrenees, we found mentions of "pits". In Catalonia, in the XIVth century, they were named "sitges cubertes". In order to avoid forest fires, a royal decree from 1345 prohibited the use of charcoal kilns. The pit process is used in Catalonia until the XVIIth century (Izard, 1999). In Pays de Sault (Aude), charcoal sites named fosses (pits) are mentioned in 1730. These generally consist of small structures, corresponding to 1/4-1/2 pile (1-2 m^3 of wood). Some of them are somewhat larger, attaining 3-4 m^3, but still remain relatively small (Bonhôte and Fruhauf, 1990). We know nothing about their form or depth. In Ariege, pits are mentioned until the XIXth century. For example in An 13, a 6-7 m long pit is described from the forest of Gourbit (Bonhôte, 1998). The description suggests that the pits were not very deep but elongated, probably in order to facilitate ventilation.

The charcoal production in pits is still not very well known, especially as no site has yet been found and excavated in the Pyrenees. An explanation to this may be their small size, mentioned in the texts, which would have left less traces in a mountainous context where colluvial processes quickly obliterate superficial features. Some sites presenting considerable deposits of charcoal (up to 80 cm) and which have been cored might correspond to small pits, but they have not yet been excavated.

Charcoal kilns (charbonnières en meule)

Since Theophrastus (IIIrd century BC), there are numerous mentions of charcoal kilns (meule) in the texts. There are also numerous remains of this type of structures along the moutain slopes. Nevertheless, we still do not know very much about the practices and their evolution through time. Indeed, descriptions of charcoal kilns and their functionning from the XIXth century are not directly transposable to medieval or more ancient practices. For example, the use of the term meule was introduced by engineers from les Mines in the XIXth century. Before that, a charcoal kiln was called "furnace" (fourneau) or "faulde" in Ariege and "farregans" in Catalonia.

In the XVIIIth century, De Gensanne (1775) described charcoal kilns of 15-20 steres (*i.e.* 4-5 piles). This corresponds to kilns with an average diametre of 4-5 m and with platforms measuring 5-7 m. The most recent and best preserved sites noted during fieldwork belong to this size class. François (1843) also quotes kilns of 10-20 steres in the mountains and 20-40 steres in the plain. The kilns built illicitly by charcoal burners or inhabitants of the countryside, and described in the texts, were generally much smaller and sometimes produced 2-5 bags of charcoal from no more than 1-2 m^3 of wood. The resulting remains are in these cases scarce and therefore more difficult to find and interpret.

During surveys and excavations, an inventory of all sites was made, independently of their size, altitudinal position and topographical context. Both bigger (6x7 m or more, multiple sites with a surface of several hundred m^2 can be found) and smaller (2x4 m or less) sites were included in this inventory. The results are in accordance with descriptions of charcoal kilns in written sources. The presumably oldest sites (badly preserved, colluvial depositions) seem smaller, but this could be an effect of the colluvial deposition. In fact, after excavation, medieval sites that were badly preserved seemed to be of "normal" size, *i.e.* in average 5x5 m.

Logging and transport of wood; localisation of charcoal kilns in mountain slopes

We do not know much about the practices of forest exploitation by charcoal burners in the Pyrenees, except for the XVIIIth and XIXth centuries, when controlled exploitations in state forests or seigniorial domains appeared. In the Pays de Sault, for example, beech wood was preferentially chosen for the production of charcoal in order to make place for fir trees. In the Aston valley, the exploitation of forests was done without precise rules (irregular coppices or clearing of high stands) until the XVIIIth century. From the XIXth century on, coppices were managed in a regular manner (Bonhôte, 1998). Summary descriptions exist in the official reports on forest offences, in particular from the XVIIIth and XIXth centuries. In the sediments of the kilns, we often found big charcoal pieces that indicate the use of whole trunks, especially when the charcoal kiln was large. Photographic documents or investigations concerning recent practices also confirm this. It is likely that roots and branches of fir were charred even in areas where the exploitation of fir was prohibited. The archives confirm this practice and, for example in the crown forest of Lercoul, charcoal from fir is found in kilns that date to the XVIIIth century, *i.e.* to a period when this species was particularly protected (Dubois *et al.* 1996). In XIVth century Catalonia, regulations protect the fir tree, but nevertheless authorise the use of its branches. The same regulations protect pioneer species such as *Betula* and *Fraxinus* (Izard, 1999). For the Middle Ages, forestry practices can sometimes be known through palaeoenvironmental research. For example, in the Aston valley, palynological and anthracological analyses have revealed long-term exploitation cycles, certainly involving forest clearing (Davasse, 2000; Galop, 1998).

Topographic constraints seem to have limited the transport of wood in the Pyrenees. Practices of launching wood down steep slopes or transporting it over short distances, when a site was not favourable for the construction of charcoal burning kilns, are known. The transporting techniques were probably not very different from those still used in the XXth century in various mountainous areas where a simple, forked branch allowed the carrying of a faggot on the shoulder. Long-distance transport would have been possible when a charcoal kiln was constructed illicitly and required lesser quantities of wood. Some doubts about the precise origin of the wood could arise concerning charcoal kilns situated near mountain ridges, as the raw material could have come from either side of the ridge. Wood could also have been stolen in neighbouring forests.

The distribution of the charcoal kilns is a function of the topography. If it was soft and regular and if the substrate was easy to work, the charcoal burners built many kilns in the same area in order to transport the wood as little as possible. For example, in the valley of Barguiliere, there are up to 15 charcoal kilns per ha (Davasse, 2000). In the Picaussel forest (Pays de Sault), a land surveyor noted in 1737 the presence of 95 charcoal kilns on 14 acres, *i.e.* an average of 13 kilns per ha (Bonhôte and Fruhauf, 1990). If the topography was abrupt, sites were concentrated at the base of slopes where wood was launched through natural "corridors". In the field, we noted that slopes exploited in this way could rise to a height of 200 m, with a potential area in the shape of a triangle. Even though the re-use of charcoal burning localities is frequent over the centuries, the kilns in the Pyrenees are not established with the same intentions of durability as those of the Languedoc garrigues during modern times (Fabre, 1996).

Preparation and construction of the kiln

The lack of precise data on techniques used in the Pyrenees, leaves us with much incertitude about the way charcoal kilns were built at various periods. We therefore have to investigate the remains of ancient kilns and the few historical sources that are available.

Construction of a charcoal kiln in a steep mountain slope required careful preparation of the site. The site consisted of platforms and embankments resulting from earthwork. The slope was often consolidated with stones or even by actual stone walls, attaining 1,20 m in particularly steep slopes. Regular stone bases are often noted but their precise function remains unknown: were they used for ventilation of the wet ground or for insulation? Theoretically, charcoal kilns are not built on wet ground as humidity can provoke a bad combustion or even explosions. In spite of this, charcoal kilns on peat bogs are not rare. The exact construction techniques for kilns are badly known. In France, kilns mostly comprise a central chimney for firing and aeration. Kilns without chimney still exist in Liguria while, in Portugal, the lighting of the fire is done through a tunnel at the base of the kiln. In the XXth century Aragon, the production of small-size charcoal for braziers "picòn" was still made by putting fire to a heap of faggots, that was covered with earth once it had started to burn. It is thus unlikely that only one technique of charcoal production was employed from the Classical period until the XIXth century.

After the construction of the kiln, the wood was normally covered by a layer of plant material (leaves, grass, moss, branches...) and a layer of fine soil. In the Pyrenees, charcoal from *Sarothamnus*, *Rhododendron* and *Calluna* seems to come from this insulating layer. Another technique consisted of the posing of pieces of turf directly on the wood, the grassy side towards the interior. Contemporary charcoal burners prefer to re-use the soil that covered ancient charcoal kilns "fraisil" and a transfer of soil from one site to another is thus possible. Such a practice, leading to the mixing of charcoal pieces of different ages and origins, can of course trouble the anthracological and chronological interpretation of a deposit. However, this was probably done only when two sites were close to each other and more or less contemporaneous, which would then diminish the risk of error. The heavy colluvial deposition, which generally occurs in mountain slopes, prevents the recovery of charcoal from sites that were abandoned since a long time. Excavations also show that ancient sites were re-used without the removal of colluvial layers and the earlier charcoal deposits thus most often remain intact. Even if the degree of mixing of the layers is difficult to evaluate, it probably remains limited.

The distribution of charcoal on a burning site depends on the methods used for the construction and dismantling of the kiln. In the centre, we will find charcoal from the chimney (when this type of burning was used). Sometimes, this charcoal consists of dry branches that can come from far away. *Juniperus* wood has for example been found here. At the periphery of the kiln, the remains result from the spreading and cleaning out of charcoal, but they can also correspond to shrubs and branches used to cover of the kiln (*Genista, Rhododendron, Calluna, Buxus*). It is thus preferable to study charcoal from a transect crossing the whole site in order to obtain a complete image of the species used.

The representativeness of charcoal

Even though we do not really know how the charcoal kilns were constructed or operated, we observe an important and rapid accumulation of charcoal debris on Pyrenean sites. In general, the deposits are 20 cm thick and cover the entire surface of the site. Often, these deposits can also consist of successive layers, 40-50 cm thick. Layers of 80 cm have also been noted. This accumulation of refuse is not common everywhere. For example in Liguria, the meticulous cleaning of the site after each burning event does not leave the charcoal *in situ*. Instead, it is found at the foot of the site in refuse dumps resulting from the cleaning out of the kiln. In the Pyrenees it is on the contrary possible to undertake a stratigraphic analysis of the deposits, extremely rich in charcoal fragments.

In order to better control the combustion, charcoal burners normally used only one species for the construction of a kiln. Anthracological analysis shows that the choice of this species depended rather on availability than on combustion properties. Thus, any species available in the surroundings was exploited, especially during the last centuries when wood resources had become more rare: pine (*P. uncinata, P. sylvestris*), *Abies alba, Fagus sylvatica*, oak (*Q. petraea, Q. pubescens, Q. ilex*), *Corylus, Tilia, Prunus, Crataegus, Alnus, Castanea, Ilex, Acer, Fraxinus, Sorbus, Juniperus, Betula* and *Salix*. This "eclectism" is attested since the Classical period by the study of charcoal from slag deposits from low furnaces dating to the 1st century (Izard, 1989).

We consider that the Pyrenean charcoal kilns reflect the surrounding forest environment. However, the problem of wood transportation is likely, in certain cases, to disturb the representativeness and interpretation of the samples and should not be underestimated (see the example of Tignalbu, charcoal kiln 1, fig. 3).

Consequences of the charcoal representativeness on the field surveys

The active slope dynamics of the Pyrenees, resulting in important colluvial deposits in the Mediterranean area, constitute a problem for surveys. Remains of charcoal kilns, buried under colluvial deposits of 50-1,50 m and sometimes stones, can frequently be seen along roads and tracks in the Catalan valleys and in particular in the lower and steep mountains of Vallespir, Aspres and Conflent. The discovery of such charcoal kilns during a survey is practically impossible and one may presume that most charcoal sites older than the XVIIth century belong to this category or have simply been eroded (Izard, 1999). In Ariege, on the Atlantic side, the erosion is less important and the platforms are generally visible, even those from the Middle Ages. However, colluvial deposits exist, even in forests, and require the coring or systematic excavation of layers located uphill from the visible site: frequently, the oldest charcoal layers are covered by colluviums and the more recent site has being shifted downhill (Dubois *et al*. 1998). In areas used for pasture, the trampling of herds often results in charcoal layers without any apparent topography. In the upper mountain grasslands with *Festuca eskia*, the accumulation of organic matter can be significant and hide charcoal layers under a 20-40 cm thick layer of roots.

The survey of the sites includes direct observations of the topography, a reconstruction of different factors involved in the establishment of the kiln (possibilities of launching, proximity to water etc.) and a systematic or random research on the soil, carried out with the help of an earth auger or a shovel. Generally, the soils of charcoal kilns are black except when the structure is old or little used. However, the presence of charcoal, even small-sized, reveals the existence of a charcoal burning site. It is true that most soils in the Pyrenean mountain slopes contain charcoal fragments when observed from a micro-morphological point of view. However, the presence of charcoal pieces measuring a few millimetres and observable in the soils with the naked eye always reveals the presence of a charcoal kiln situated uphill. There can be some doubts, however, when few and small-sized charcoal fragments are noted at unlevelled sites and in soils that otherwise are not very dark. This configuration may rather correspond to forest fires or slash-and-burn agricultural practices. Repeated pastoral fires can also be responsible for the presence of charcoal in the soils, but characteristic taxa such as *Genista* and Ericaceae should then be present in the anthracological assemblages.

Analysis of charcoal kiln sites: interest and limits of the methods used in the Pyrenean context

The methodology used during our work in the Pyrenees is based on a precise spatial unit - the charcoal burning site. Our research aims at understanding environmental dynamics at the scale of a geographical unit, defined through preliminary studies in topography, biogeography (current observations), history (archives, forest maps etc.) and archaeology (survey of surface remains whether pastoral or metallurgical) as well as through a reconstruction of forest exploitation practices.

The spatial unit defined in a mountainous zone corresponds to a slope or part of a slope, pastoral or covered by forest, with an average size of 10-50 ha and presenting an altitude gradient of 200-300 m. A site can also be delimited according to administrative boundaries of a forest or a forest parcel.

Once this delimitation has been made, a complete inventory of charcoal burning kilns is carried out within the perimeter of the site, by boring with an earth auger or by using a shovel. A certain number of charcoal kilns are selected, described from a stratigraphic point of view and then sampled for charcoal.

At this point, we can choose to apply one of two investigation methods, according to the objectives and the scale of the research: an "anthraco-biogeographic" method or an "archaeo-environmental" method.

The "anthraco-biogeographic" method

The aim of this method is to conduct a spatial study at the scale of the valley or more. Comparisons are made between different slopes, altitudes or sites (Davasse, 2000). Preferentially, the research should be carried out by one person only and given the amount of work involved both in the field and in the laboratory, too many surveys cannot be included.

Therefore, a limited number of charcoal kilns are selected (5 kilns on average, but this number can vary according to study areas) and studied in detail. The sounding, a square with 50 cm sides, is located in the tallest part of the kiln, previously identified by the help of an auger. The sediment is then excavated by 5 cm layers until the base of the kiln. It has to be noted that intermediate soils are numerous and sometimes thick. It is also necessary to verify that no charcoal layers are present deeper down.

The sediment from each 5 cm layer is sieved through a mesh of 2 mm, a procedure that provides sufficiently large samples (about 500 charcoal pieces) for the anthracological study. Various analyses show that the optimal sample size in order to study the vegetation history of the Pyrenees is 60-120 fragments. In the eastern Pyrenees, we studied around 25 charcoal sites of this type during the last decade.

The study of Tignalbu, in the high valley of Vicdessos, is a representative example of the "anthraco-biogegraphic" approach (fig. 2). It is located at 1400-1600 m a.s.l., in an area of summer pastures, like two other charcoal sites stu-

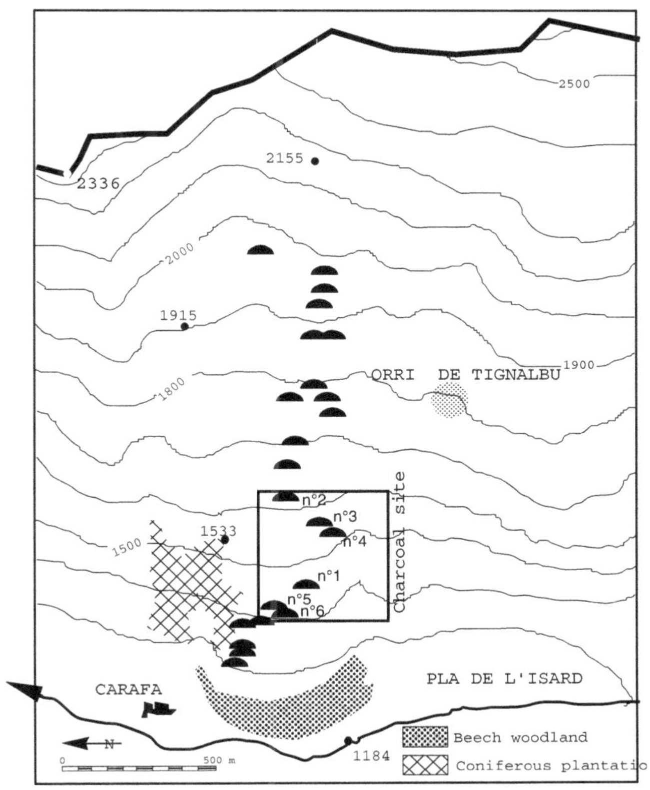

Figure 2: Site of Tignalbu (Ariège, Vicdessos) and localisation of charcoal kilns.

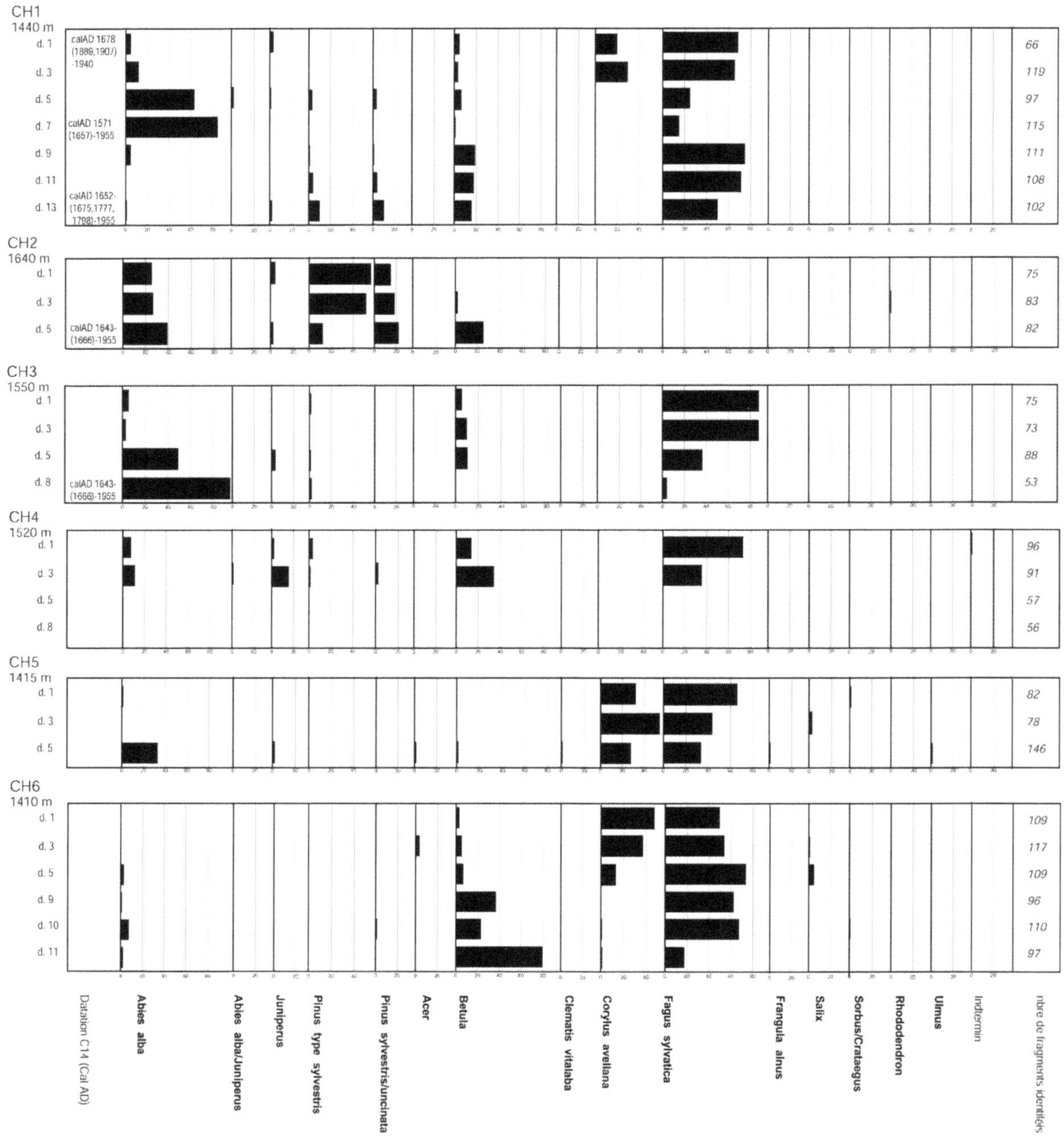

Figure 3: Anthraco-analysis of charcoal kilns at Tignalbu

died in this high glacier valley (Davasse 2000, p. 173 and following). The upper timber line, situated at *c.* 1200-1400 m a.s.l., is particularly low at this locality. The radical deforestation is not only due to the persistence of important pastoral activities since 1800 BC (Galop and Jalut, 1994), but also to charcoal production (Davasse 1992, 1993a). In Tignalbu, the charcoal from six kilns was studied, allowing the identification of sixteen taxa. A general evolution of the vegetation cover can be noted (fig. 3): beech is favoured by the forest exploitation to the detriment of fir. This evolution is observed in most charcoal assemblages in the sector, with the example of charcoal kiln 3 being particularly speaking. At the beginning of charcoal exploitation, fir was represented by 94,3% and beech by 3,8% of the identified charcoal fragments from this kiln. In the end of the sequence, the relative frequencies of these species were instead 5,3% and 88% respectively. The considerable proportions of pine should also be noted: in layer 1 of charcoal kiln 2, located at *c.* 1600 m a.s.l., 51,8% of the fragments belong to *Pinus* type *sylvestris*; in layer 13 of charcoal kiln 1, located at 1440 m a.s.l., this taxon is represented by 21,7%. Do we deal here with *Pinus sylvestris* or *Pinus uncinata*? Even though the distinction between these two pine species, both currently present in the Pyrenees, is difficult from an anatomical point of view it is likely, due to the high altitude of the kiln, that the exploited conifer was rather *Pinus sylvestris* (Davasse, 1992).

Comparing the frequencies of *Betula* and *Corylus* highlights another type of dynamics: birch is generally present in the beginning of the sequences while hazel dominates towards the end. This is particularly obvious in charcoal kiln 6 where the frequencies of birch and hazel pass from 79,4% and 2,8% respectively to 1% and 46,8%. The ecological requirements of these two species are, however, similar: they constitute secondary woodlands or are part of recolonizing taxa. Birch (*Betula pendula*) is a heliophilous species and a pioneer of moorlands and cleared forests. Consequently, it disappears when the forest cover becomes denser. It was the principal element of the forest recolonization of deforested slopes in the valley of Vicdessos throughout the XXth century. Hazel occupies abandoned lands where it constitutes relatively stable thickets. This species is also introduced into degraded beech coppices. At present-day, hazel is abundant below 1400-1500 m while birch thrives at higher altitudes. It is difficult to explain their dynamics. Perhaps it is necessary to take into account that one species is more heliophilous than the other? In that case, different exploitation patterns could be the explanation, as clear cutting would favour the growth of birch and coppicing that of hazel. In any case, the presence of these two species in the beginning of the sequence indicates that the surrounding woodlands were already exploited at this point.

When looking into details, the interpretation becomes complicated. The case of charcoal kiln 1 (fig. 3) is revealing. Three different phases could be distinguished in the charcoal layer. The first phase is characterised by the preponderance of beech (50-74,8 % of the fragments) and the presence of birch and pine. In the second phase, fir dominates with up to 84,3% of the fragments. During the third phase, beech and hazel are present. These phases had initially been interpreted as exploitation cycles because of the presence of sterile layers in the profile (Davasse, 1993a). However, radiocarbon dates seem to indicate only two periods of exploitation, the first occurring during the XVIIth century and the second more or less a century later. Colluvial deposits separate these two periods, represented by layers 4 and 5.

The contemporanity of beech and fir in the charcoal assemblages can only be explained by the exploitation of separate forest parcels, which would suggest a management of woodlands. However, as seen above, the protection of fir stands for building purposes was the principal form of forest management conducted by local communities. The results obtained thus allow two different interpretations: either high fir trees were cut down for construction wood and their branches used for charcoal, or whole stands of fir were exploited indifferently for charcoal burning. This happened at a time when offences to the forest regulations became more and more frequent, due to an increased demand for wood resources.

The method described here allows us to study vast spaces and to make comparisons between several sites. It also includes a spatial dimension, taking into account the results on the scale of a slope, a great forest, a large valley etc., all while following the principles of geosystemic classification. By adding the radiocarbon dates obtained at various sites, this procedure allows us to validate or invalidate historical and biogeographic hypotheses.

The small size of a charcoal kiln excavation constitutes the main limiting factor to the understanding of the forest history in a given place. All structures cannot be revealed and

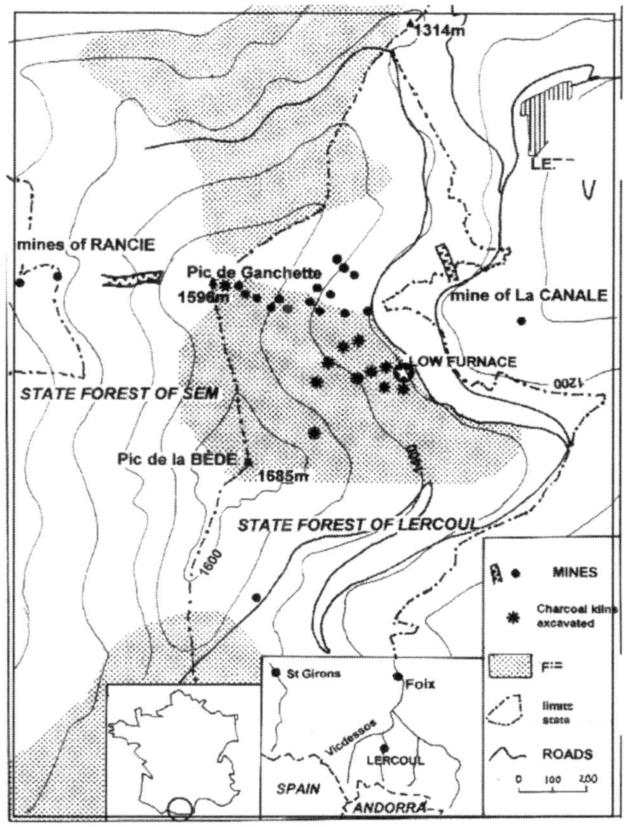

Figure 4: Localisation of the forest of Lercoul.

stratigraphies are difficult to correlate from one kiln to another. Inevitably, the sampling does not take into account topographic shifts occurring through time. The charcoal site can be analysed, not without difficulty, as a diachronic feature, reflecting successive practices.

The "archeo-environmental" method

The aim of this method is to carry out a complete analysis of the vegetation history and wood use at a single charcoal site, considered as "global" at the scale of a geo-complex. After a first survey, similar to that described above, a limited number of charcoal kilns are selected. These are excavated by the removal of successive strata, aiming at determining the various charcoal layers, their extension, intermediate soils and colluvial deposits in order to understand the diachronic evolution of the site. It should be recalled that charcoal kilns in mountain areas are frequently covered by colluvial deposits that destroy the upper part of the kiln. The excavation of such a site can therefore be particularly constraining due to the considerable volume of sediments to be evacuated. However, it is seldom useful to excavate the whole charcoal kiln site, especially if the aim of the study is to reconstruct the exploited vegetation through charcoal analysis. Instead, a system with two perpendicular trenches, as broad as a shovel and meeting at the supposed summit of the kiln, can be used. The limits of the charcoal layers are much clearer when seen in a stratigraphic section than at the surface. The two trenches, forming a cross, allow the observation of each stratigraphic layer and the estimation of its relative significance. From these observations, an excavation strategy can be established where 1/4, 1/2 or 3/4 of the site will be excavated. 1/4, at least, of the site will always be kept untouched.

Moreover, the distinct stratigraphic sections are easy to sample. Sediment samples are taken from arbitrary 5 cm layers, as in the "anthraco-biogeographic" method, and sieved through a 2 mm mesh.

This method was applied to the site of the Royal Forest of Lercoul (Ariege) (fig. 4) (Dubois *et al.* 1996; Dubois and Métailié, to be published). Today, this forest consists of high fir stands, attested by the archives since the XVIIth century. Research was initiated at this site after the discovery of a low-furnace from the Classical period (IIIrd century). The slag heap of this furnace contained only beech charcoal and a quick anthracological survey of 25 charcoal kilns in the area confirmed the predominance of beech in the past. These first results brought us to excavate, in an exhaustive manner, 10 charcoal kilns situated within a perimeter of 7 ha. 17 radiocarbon dates were obtained, spanning between the Vth/VIth centuries, for the oldest charcoal kiln, and the XVII/XVIIIth centuries, for the most recent one. This represents the longest available sequence for forest history in the Pyrenees and probably for French mountains in general. The anthracological results made it possible to describe the evolution of a woodland that was already extensively exploited in Antiquity. While this woodland was mainly characterised by the presence of beech in the earliest periods, fir stands took over from the end of the Middle Ages (XVth-XVIth centuries), when protective measures were taken by the communities and the Earl of Foix (Verna, 1994) at the time of expansion of hydraulic metallurgy.

Charcoal kiln 30 summarizes the forest history of Lercoul (fig. 5). It is located at 1540 m a.s.l., in a stand of high fir trees. A few beech trees are growing on the rocky slope above. Even if these rocks collapsed several times on the charcoal burning site, it was re-used several times by charcoal burners. The oldest charcoal layers (9, 34) date to the Vth century and contain 85%-98% of *Fagus*, as did the IIIrd century slag heap (*Fagus*, 90-97%). In layer 8, dated to the XI-XIIIth centuries, *Fagus* is represented by 45% of the identified fragments and *Abies* by 43%. In other medieval layers at this complex site, the average frequency *Fagus* attains 70%. The presence of other taxa, such as *Corylus*, *Betula* and *Juniperus*, indicates a cleared forest. The most recent layer (1), from the XVII-XVIIIth centuries, contains mainly *Abies* (53%). The history of this charcoal kiln, as reconstructed from charcoal analysis, is in perfect accordance with the data obtained at other sites in the same forest.

This method allows a diachronic reconstruction of charcoal production and woodland evolution in a homogeneous context.

The disadvantage is the considerable excavation work involved as each charcoal kiln may require several days of digging, especially if the stratigraphy is complex (for example in the case of charcoal kiln 30). This method requires team work and does not allow an in-depth study of a great number of charcoal kilns at several sites. It should thus be applied to particularly interesting sites that have been adequately inventoried at an archaeological as well as at a historical level. These sites should also be of limited size.

Conclusion

In conclusion, we suggest the following methodological procedures, adapted to the objectives and the geographical context of the research:

- A fast and systematic survey of charcoal burning sites in a

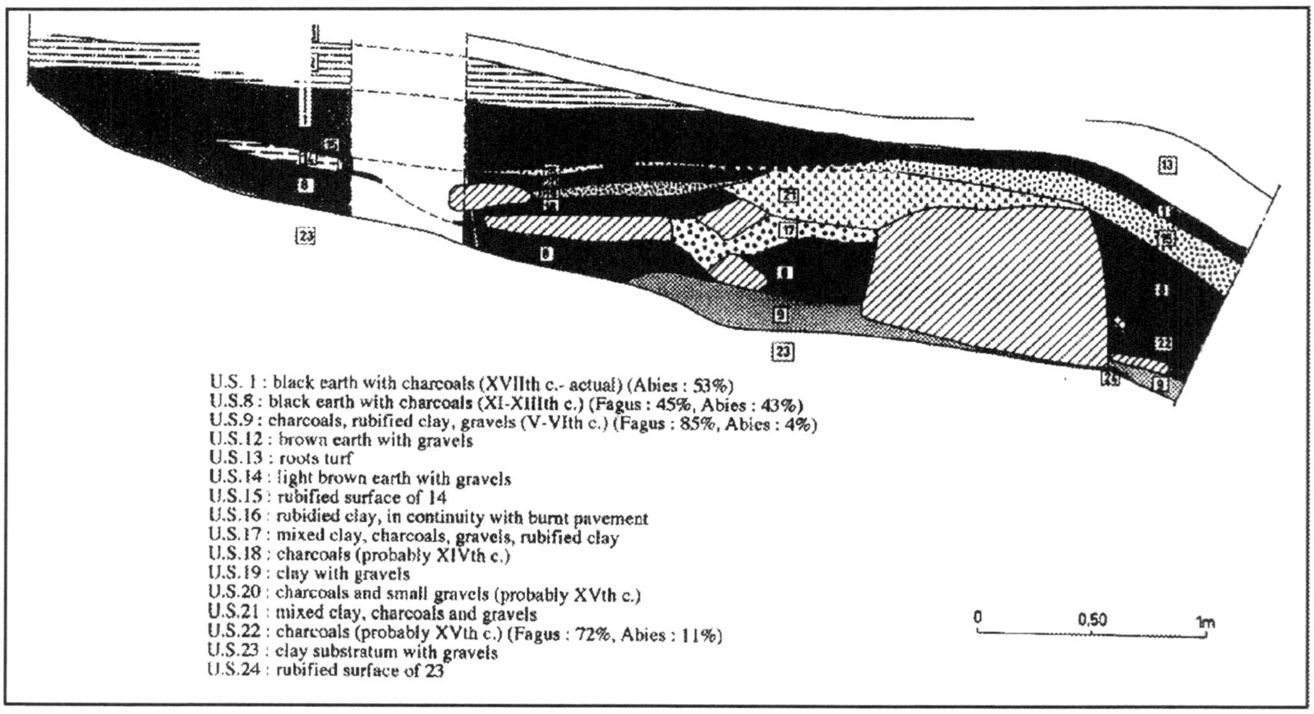

Figure 5: Charcoal kiln 30 at Lercoul, North-South section.

still provide an overall image of the sites. This survey will either precede a more thorough analysis of a chosen site or provide a general inventory of a sector. Practically, the survey is made with the help an auger. Five points, forming a cross, are sampled by coring at all of the charcoal burning sites. Anthracological analysis is limited to the study of 10 charcoal pieces from the lower and superior levels of the stratigraphy. In this way, the stratigraphy of a site can be established rather quickly and we can get a first idea of the specific composition of charcoal assemblages as well as of the vegetation history. The use of a GPS allows a rapid localisation of the charcoal kilns on a map.

- "Anthraco-biogeographic" sites: After the first phase, described above, a selection of charcoal kilns considered as representative is made. Samples are taken from a square sounding (50x50 cm) in the centre of the charcoal kiln, *i.e.* where the thickest and most complex layers are located. The approach is based upon a geosystemic analysis of the area and allows us to inscribe the data in a precise spatial context.

- "Archeo-environmental sites": Charcoal kiln sites, selected according to their particular interest, revealed by archaeology, historical sources or surveys (implying sampling and radiocarbon dating), are excavated in stratigraphic cross-sections by the removal of successive 5 cm layers. The anthracological analysis is based on samples of 50-120 charcoal fragments per layer. This approach focuses on the reconstruction of the diachronic evolution of woodlands.

References

BIELENIN K. 1993 - Le combustible de la sidérurgie antique des Monts de Sainte-Croix en Pologne. Actes du Colloque "Proto-industries et Histoire des forêts", 11-12 octobre 1990, Foix, *Cahiers de l'Isard*, 3, p. 317-329.

BONHOTE J. 1998 - *Forges et forêts dans les Pyrénées ariégeoises*. Pyrégraph ed., 337 p.

BONHOTE J., VERNET J.-L. 1988 - La mémoire des charbonnières. Essai de reconstitution des milieux forestiers dans une vallée marquée par la métallurgie. *Revue Forestière Française*, p. 197-212.

BONHOTE J., FRUHAUF Ch. 1990 - La métallurgie au bois et les espaces forestiers dans les Pyrénées de l'Aude et de l'Ariège. In : *Forges et forêts. Recherche sur la consommation proto-industrielle de bois* (sous la direction de D. Woronoff), EHESS éd., Paris, p. 151-212.

BONHOTE J., METAILIE J.-P. 1993 - La limite supérieure de la forêt dans une vallée métallurgique (vallée d'Aston, Ariège, France). Actes du Colloque "Proto-industries et Histoire des forêts", 11-12 octobre 1990, Foix, *Cahiers de l'Isard*, 3, p. 271-285.

BONHOTE J., DAVASSE B., DUBOIS C., GALOP D., IZARD V., METAILIE J.-P. 2000 - Histoire de l'Environnement et cartographie du temps dans la moitié est des Pyrénées : pour une "chrono-chorologie". In : Barrue-Pastor M., Bertrand G. (éd.), *Les temps de l'Environnement*. Toulouse : Presses Universitaires du Mirail; p. 501-515.

DAVASSE B. 1992 - Anthracologie et espaces forestiers charbonnés. Quelques exemples dans la moitié orientale des Pyrénées, *Bulletin de la Société Botanique Française*, vol. 139, n°2-3-4, p. 597-608.

DAVASSE B. 1993(a) - Les espaces forestiers de la haute vallée du Vicdessos (Pyrénées ariégeoises) : analyse écohistorique et étude des charbonnières, in Dubois J.-J., Gehu, J.-M., Rameau J.-P. (dir), *Phytodynamique et biogéographie historique des forêts*, Berlin-Stuttgart, J. Cramer, p. 305-314.

DAVASSE B. 1993(b) - Aspects méthodologiques de l'anthracoanalyse des charbonnières. Histoire des forêts de la vallée de Soulcem (Vicdessos, Pyrénées ariégeoises). Actes du Colloque "Proto-industries et Histoire des forêts", 11-12 octobre 1990, Foix, *Cahiers de l'Isard*, 3, p. 207-222.

DAVASSE B. 2000 - *Forêts, charbonniers et paysans dans les Pyrénées de l'Est, du Moyen Age à nos jours. Une approche géographique de l'histoire de l'environnement*. Toulouse, GEODE, 287 p.

DAVASSE B., GALOP. D. 1994 - Impact des activités pastorale et métallurgique sur les forêts d'altitude dans les Pyrénées ariégeoises (France), Actes du Colloque International Ecologie et Biogéographie Alpines, La Thuile (Italie), 1990, 1-6 septembre, *Revue Valdotaine d'Histoire Naturelle*, supplément au n°8, p. 151-160.

DAVASSE B., GALOP D., RENDU C. 1997 - Paysages du Néolithique à nos jours dans les Pyrénées de l'est d'après l'écologie historique et l'archéologie pastorale, in Burnouf J., Bravard J.-P., Chouquer G. (éds), *La dynamique des paysages protohistoriques, antiques, médiévaux et modernes*, Sophia Antipolis, APDCA, p. 577-599.

DUBOIS C., METAILIE J.-P. (à paraître 2001) - Anthropisation et dynamique forestière dans les Pyrénées ariégeoises à l'époque gallo-romaine. L'exemple de la forêt de Lercoul. in : *Les ressources des Pyrénées*, Sablayrolles (ed), p. 7-19.

DUBOIS C., IZARD V., METAILIE J.-P. 1993 - Forêts charbonnées et archéologie métallurgique en Ariège (Pyrénées). Une méthodologie interdisciplinaire pour l'histoire de l'environnement. *Archéométrie et achéologie en paléométallurgie. La sidérurgie ancienne dans l'Est de la France dans son contexte européen*. Besançon, 11, p. 93-106.

DUBOIS C., IZARD V., METAILIE J.-P. 1997 - Archéologie de la forêt charbonnée : questions et méthodes, illustrées par l'exemple de Lercoul (Ariège). Colloque *La dynamique des paysages protohistoriques, antiques, médiévaux et modernes*, XVIIème rencontres internationales d'archéologie d'Antibes, 1996. APDCA, p. 525-540.

FABRE L. 1996 - *Le charbonnage historique de la chênaie à* Quercus ilex *L. (Languedoc, France) : conséquences écologiques*. Thèse, Montpellier II, 2 t. 164 p + annexes.

FRUHAUF C. 1980 - *Forêt et société : de la forêt paysanne à la forêt capitaliste en pays de Sault sous l'Ancien Régime (vers 1670 —1791)*. CNRS éd., Toulouse, 302 p., 3 cartes h.t.

GALOP D. 1998 - *La forêt, l'homme et le troupeau. 6000 ans d'anthropisation des Pyrénées de la Garonne à la Méditerranée. Contribution palynologique à l'histoire de l'environnement*. Toulouse, GEODE, FRAMESPA, LET, 285 p.

GALOP D., JALUT G. 1994 - Differential human impact and vegetation history in two adjacent Pyrenean valleys in the Ariege basin, southern France, from 3000 BP to the present. *Vegetation history and Archeobotany*, 3, p. 225-244.

IZARD V. 1993 - La typologie des charbonnières : méthode d'étude diachronique du charbonnage pour une contribution à l'histoire des forêts. Actes du Colloque Proto-industries et Histoire des forêts, 11-12 octobre 1990, Foix, *Cahiers de l'Isard*, 3, p. 223-236.

IZARD V. 1995 - Le charbonnage des forêts dans la vallée de la Lentilla. Paléo-métallurgie, anthracologie et histoire de l'environnement (Conflent, Pyrénées-Orientales). In : *La farga catalana en el marc de l'arqueologia siderùrgica*. Simposi internacional sobre la farga catalana, Ripoll, 1993. Andorra, p. 177-190.

IZARD V. 1999 - *Les montagnes du fer. Eco-histoire de la métallurgie et des forêts dans les Pyrénées méditerranéennes (de l'Antiquité à nos jours)*. Thèse, Université Toulouse-le Mirail, 2 t. 560, 250 p.

KENLA J.V., JALUT G. 1979 - Déterminisme anthropique du développement du hêtre dans la sapinière du Couserans (Pyrénées ariégeoises, France), durant le subatlantique. *Geobios*, 12, p. 735-738.

METAILIE J.-P. 1984 - La forêt paysanne dans les Pyrénées centrales. *Revue Géog Pyrénées Sud-Ouest*, 55 (2), p. 231-238.

METAILIE J.-P. 1996 - La forêt du village et la forêt charbonnée. La mise en place des paysages forestiers dans la chaîne pyrénéenne. In : *L'uomo e la foresta*, seccXIII-XVIII. Istituto Datini, Atti della settimane di studi, 27. Le Monnier, p. 397-421.

METAILIE J.-P., JALUT G. (Dir.). 1991 - *La forêt charbonnée. Histoire des forêts et impact de la métallurgie dans les Pyrénées ariégeoises au cours des deux derniers millénaires*. Rapport, URA 366 CNRS, Toulouse, 240 p.

METAILIE J.P., BONHOTE J., FRUHAUF C.,1988 - A thousand years of forest history in the french Pyrenees mountains: the Ariege example. *Human influence on forest ecosystems development in Europe*, Bologne, ESF FERN-CNR, p. 159-167.

VERNA C. 1994 - *Le temps des moulines. Le fer et son exploitation du comté de Foix à la vicomté de Béarn (fin XIIème-fin XVème siècles)*. Thèse, Paris I, 2 vol, 501 et 196 p.

Anthracology and landscape planning

Loïc Gaudin
CNRS-UMR 6566, Civilisation Atlantiques et Archéosciences,
Université de Rennes 1, Bât. 25, 35042 Rennes Cedex, France.
E-mail: Loic.gaudin@univ-rennes1.fr

Abstract
Over the last decades, a good deal of effort has gone into the development of leisure activities and into the promotion of the historical and natural heritage. The latter has largely benefited from available archaeological data. The integration of monuments in their immediate environment can now be enhanced by archeobotanical research which supplies essential contextual data. For example, the anthracological research carried out on the Lannec-er-Gadouer site in the Morbihan (western France) has supplied information about the paleoenvironnement of this area during the Neolithic era. Information on the quality of the landscape canopy and on the structure and evolution of the vegetation is now available. The identification of the causes and processes of site degradation can in fact be partly derived from current ecological findings. The diachronic nature of archaeobotany provides a long-term perspective on the interactions between Man and Nature. Site restorations have an environmental impact. Landscape analyses can supply new and essential information in the choice of restoration techniques and in the decision-making process relative to long-term planning.

Key-words: *Archaeology, Charcoal analysis, Ecology, Environment, Landscape, Landscape planning, Open-plan offices, Restoration, Cultural heritage.*

Résumé
Durant les dernières décennies, parallèlement au développement des loisirs, les efforts d'aménagement du patrimoine historique et naturel se sont accrus et ont bénéficié, dans la mesure du possible, des données archéologiques. Désormais, afin d'insérer tant soit peu le monument dans un environnement cohérent, l'archéobotanique peut apporter des repères indispensables. Les travaux anthracologiques tels que ceux effectués sur le site de Lannec-er-Gadouer dans le Morbihan (France) nous ont ainsi apporté un certain nombre d'informations sur le paléoenvironnement de ce secteur au Néolithique. Des renseignements sur la qualité du couvert paysager mais aussi sur la structure et l'évolution de la végétation sont maintenant disponibles. En effet, l'identification précise des causes et des processus de dégradation d'un site peut en partie relever d'études environnementales actuelles. L'archéobotanique, par sa composante diachronique, permet d'autre part une mise en perspective des interactions entre l'Homme et la Nature sur le long terme. La restauration d'un site entraîne un impact sur l'environnement. Or ces analyses paysagères peuvent constituer de nouveaux arguments essentiels dans le choix des techniques de restauration et dans l'orientation des actes décisionnels concernant l'aménagement durable.

Mots-clefs : *Aménagement, anthracologie, archéologie, écologie, environnement, patrimoine, paysage, restauration.*

Introduction

At first glance, the link between anthracology and environmental management may seem rather tenuous. The current trend of research, in spite of talk of a "pluridisciplinary approach", leans generally towards increasing specialization. In order to avoid such difficulties, it was important to define a broad framework that would allow connections to be made between apparently remote fields, which nevertheless both refer to the same context, *i.e.* the landscape. Landscape ecology supplies such a framework.

This study has been organized around two main themes:
- Integrating the anthracological data into landscape ecology;
- Improving the characterization of anthracological data in order to enhance site-planning proposals.

The landscape of the ecologist

Two approaches are commonly used to define a landscape. A first definition generally results from the perception of an environment. This is the concept that inspired painters such as Poussin, Monet, Millet, or Courbet. Of course, in these cases, their pictures include the structural realities of the landscape in the form of vegetation, rivers, relief features, or anthropic elements (such as buildings, paths, etc.). However, such descriptions remain subjective, as the search for aestheticism, the experience and the social environment of an artist strongly determine the way he or she describes and qualifies a landscape.

A second definition derived from the emergence of landscape ecology. The term "ecology", which was first suggested by Haeckel in 1866, refers to the science that studies the relationship between living creatures and their environment. The areas studied by ecologists became more and more complex, ranging from the individual to the landscape which integrated humans and their activities (Barnaud and Lefeuvre, 1992).

The term "landscape ecology" was first introduced by Troll in 1939, three years after Tansley introduced the "ecosystem" in 1935. This German biogeographer attempted to combine geography and ecology, or in other words, to connect spatial landscape structures with ecological processes (Burel and Baudry, 2000).

Over the last decades, industrial countries have become much more environmentally aware: in France, the Department of the Environment was created in 1971. An environmental protection act was passed in 1976 in response to major changes in the economy but also in landscape planning (agricultural land reorganization).

The dual nature of landscape ecology makes it possible both to analyze contemporary landscape changes, such as those caused by excessive urbanization, excessive fertilizations, or climatic fluctuations, and to provide solutions to these changes.

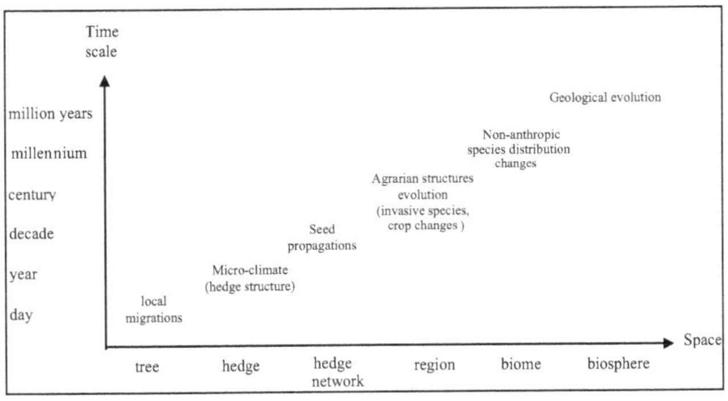

Figure 1: Control factors for hedge flora according to different space and time scales (taken from Burel and Baudry, 2000).

Before analyzing a landscape, the spaces to be studied must first be defined. Landscapes are complex systems in which different functional units subject to their own space-time scale changes can be identified. There is a correlation between time and space scales; phenomena evolving over wide areas also evolve over long periods of time (fig. 1).

A landscape comprises a whole, and a variety of process-determining scales exists. These scales can be classified according to their duration and range of action. Once these spatial and functional units have been established, three main axes are generally used to describe a landscape:
- **Structure**: involving the three following parameters:
- Biotic factors: fauna, flora;
- Abiotic factors: climate, geology, pedology, hydrography, geography.These parameters form the underlying landscape features;
- Human factors: these have obviously become significant over the last few thousand years, particularly due to cultural adaptation.
- **Function**: This refers to the intrinsic functional parameters but also to the interactions between the biotic, abiotic and anthropic parameters. Together, these parameters make up the driving force behind landscape dynamics. (fig. 2);
- **Dynamics**: This derives from the intrinsic variations in each parameter but also from the interaction between each one of those parameters. A landscape observed at a time "T" is only the transitory result of these dynamic interactions. It makes us aware of the ephemeral nature of landscapes, including present-day landscapes.

Of course, the use of models based on archaeological data to anticipate the negative effects of landscape transformations is likely to interest many local authorities. However, such modeling is often very complex for two reasons: first, because each landscape parameter will have its own specific effect and changes. For example, geomorphological parameters change over the very long term (millions of years), whereas a landslide seems almost instantaneous on a millennial time scale. Second, because multiple interactions existing between each of the landscape parameters (fig. 2).

It will always be preferable to combine the different fields of activity. However, outlining vegetation changes very often means outlining landscape changes (at least its broad features). Unlike fauna associations or human activities, plant associations are determined by their immediate environment (fig. 3). As a result, defining vegetable dynamics means taking into account climatic, geological, pedological and human changes.

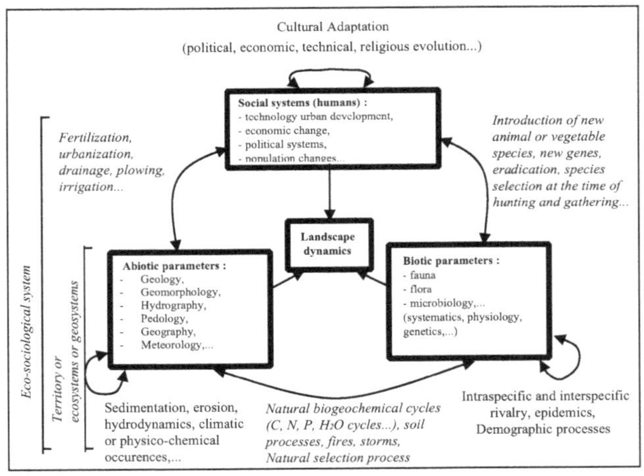

Figure 2: Main structural divisions and interactions involved in global landscape dynamics.

While the identification of landscape dynamics can be very complex due to the various interactions between the different parameters, the study of vegetable dynamics is more appropriate, as it is determined by all the other factors. However, it is very often necessary to resort to other fields of study to identify the factors responsible for vegetation changes.

Using anthracological data: case study of the hearths preserved in the Lannec-er-Gadouer Neolithic barrow (Erdeven, western France)

The use of anthracological data was deemed useful in gaining a better understanding of landscape changes. The anthracological study of the hearths preserved in the Lannec-er-Gadouer Neolithic barrow was a good example of the use of this type of data (Gaudin, 1999; Gaudin and Marguerie, 2000).

The excavation of the cairn between 1994 and 1997 by S. Cassen (Boujot and Cassen, 1997a and 1997b) revealed two hearths:
- The first dates to the final Mesolithic, (5770±80 BP, or [4803-4457] cal BC (2σ) (N°labo AA#9240);
- the second, more recent, dates to the Neolithic, (5640±80 BP, or [4687-4339] cal BC (2σ) (N°labo AA#20130).

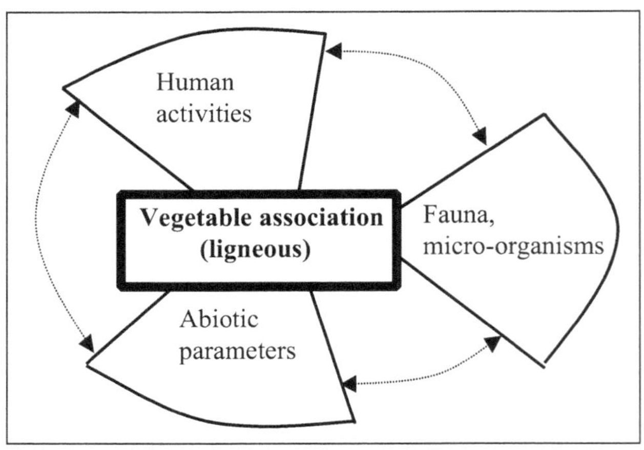

Figure 3: Vegetable associations potentially involve human, biotic and abiotic factors.

Charcoal determination was carried out by observation with a binocular lens (x 7,5 to 75) and by observation with a reflected light microscope with episcopic lenses (x 500). The binocular lens was used for the dendrology measurements and the transversal section of charcoals, whereas the other sections were observed through the light microscope. A charcoal collection and reference books (Schweingruber, 1982 and 1990; Jacquiot *et al.* 1973; Wilson and White, 1986) were regularly consulted for determination purposes. The palaeoecological interpretation was carried out using a number of published ecological studies (Ramade, 1994, Rameau *et al.* 1996).

Sampling and sieving were carried out by the archaeologists. 400 charcoal pieces were studied from the first hearth and 550 charcoal pieces from the second.

Archaeological interpretation gave prominence to domestic uses over craft-based uses, leading to a paleoecological interpretation.

Anthracological histograms were established for each hearth, using charcoal analysis (fig. 4).

Typically, quantitative data (Y coordinate) and qualitative data (X coordinate) were obtained. However, within the framework of this study, while the qualitative data truly reflect the paleoecological conditions in the collection areas (as shown by the taxa collected), the quantitative data are explained both by paleoecological conditions but and by the use of kindling techniques as well as the duration of fires. Therefore, in order to obtain landscape simulations of the site, the study focused only on the qualitative data, *i.e.* anthracological associations: Three flora associations were identified and characterized (Rameau *et al.* 1996):
- a deciduous oak forest, typical of the Atlantic period;
- a shrubby heliophilous taxa formation, probably an intermediate or younger stage of the *Quercus robur* or *Quercus petraea* series;
- a hygrophilous and heliophilous taxa formation (fig. 5).

The two anthracological associations were mainly differentiated by the detection of three heliophilous taxa in hearth 2 (*Salix* sp., *Ulmus* sp. and *Prunus* sp.). A dendrological analysis recommended by D. Marguerie was conducted in parallel with these anthracological studies. It consisted in establishing the average widths of the tree rings of oaks with both low curvature and regular growth rates (Marguerie, 1992). Average widths of 1.16 mm for hearth 1 and 1.53 mm for hearth 2 were thus obtained. These results are coherent with an improvement in growth conditions, probably followed by a reduction in vegetation densities.

Once the vegetable formations are identified, a better characterization of the ecosystems to which these vegetable formations belong can be obtained thanks to a "reductive process", commonly used in phytoecology. General information on fauna, edaphic or climatic conditions can thus be estimated.

Finally, the reframing of anthracological data according to the three descriptive axes of landscape ecology is indeed quite useful for the following three reasons:
- Data on the vegetable structure of the collecting area is available: three vegetable formations (based on the ligneous family...) have thus been identified including the dense structure climacic oak forest.
- Indications regarding the changing functions of the past landscape can also be derived from the data: the fact that a climacic forest lay only a few kilometers away from the seashore is pertinent because it shows the interactions between the marine environment and the coastal vegetation. The study also shows signs of at least sporadic human influence on the three types of vegetable formations.
- Finally, data on the dynamics of the past landscape can be estimated since the opening of the forest can be seen in the vegetable remains found in hearth 1 and 2. Three interpretations can be given for this phenomenon:
- a climatic interpretation, which can easily be dismissed as the palynological profiles of the sector (Visset *et al.* 1995, 1996) shows no signs of climatic changes in this period;
- a shift in the collecting area, which can also be discounted as the three types of vegetation detected in both hearths seems to show that prehistoric inhabitants collected their materials from the same areas;
- an anthropic impact, which remains the most likely hypothesis. While the increase in the number of heliophilous

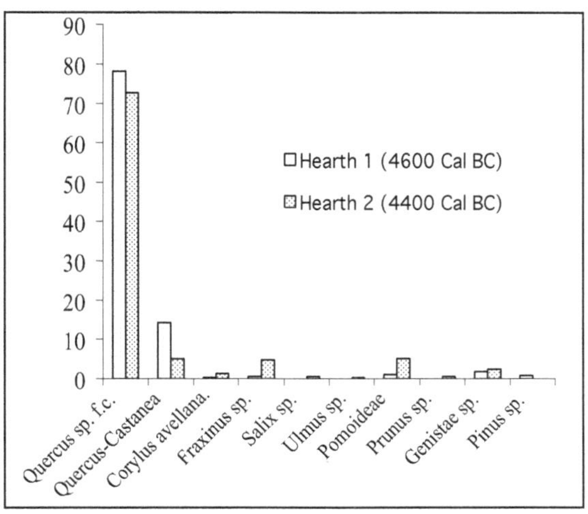

Figure 4: Anthracological histogram based data derived from hearths 1 and 2.

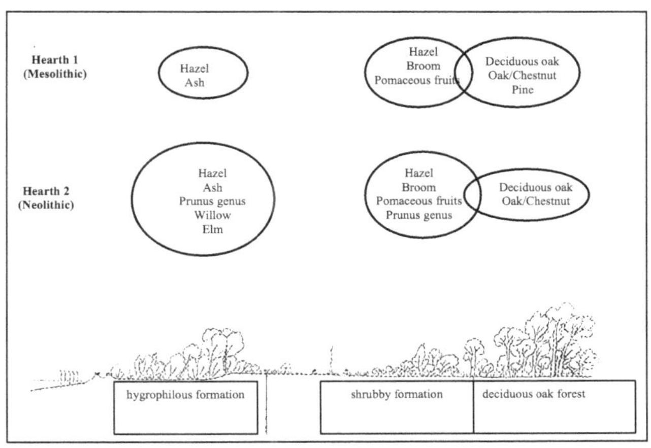

Figure 5: Major types of vegetation derived from anthracological studies of fire structures preserved under the Neolithic barrow of Lannec er Gadouer in Erdeven (Morbihan).

taxa between hearths 1 and 2 suggests an opening-up of the timbered environment, the increase in average ring widths seems to suggest clearly that this opening took the shape of a thinning out of the vegetation.

Several human activities can explain this phenomenon. First of all, the thinning of the vegetation may have been caused by regular trampling of the collecting area due to hunting, gathering or pasturage activities. In addition, maintenance of the funerary monument may have had an indirect impact on the vegetation in the area. Finally, successive periods of clearing and neglect of the vegetation may have once again left a detectable impact, seen in the ring growth of the collected wood.

Whatever the hypothesis, this analysis of the limited anthracological data of the Lannec-er-Gadouer site, has given an insight into the nature of human impact on the wooded environment. Without doubt, this shows a particular usefulness of anthracology in demonstrating the existence of traditional coherent geographical area management systems.

Paleoecological research, new ideas regarding landscape planning, weak and strong points

The use of archeobotanical research data in landscape planning has a number of advantages but also implies a number of drawbacks which must be taken into account. The following points summarize this paradox (tabl. 1).

In the Lannec er Gadouer site example, the modern vegetation consists of pine plantations and "regressive" associations (heaths dominated by purple moor-grass (*Molinia caerulea*), common broom (*Cytisus scoparius*), gorse (*Ulex europaeus*) and shrub formations (Rosacea)). Pine plantations have been common since the 19th century as a way of exploiting poor heathland soils. The deciduous oak forest (probably also the potential natural vegetation) described by anthracology is no longer immediately present.

This landscape degradation is probably the result of intensive past and contemporary "landuse" beginning in the Neolithic period.

Economic factors are currently essential in the decision-making process and in overall project management. Yet, these factors are not generally suited to long term planning (Montgolfier and Natali, 1987). The question is how to update the values and costs of a physical planning project over a ten, fifty or two hundred year period. Ecology, archaeology and the sciences can not provide perfect answers to the questions raised by the future implementation of sustainable planning projects (Archives des actes du colloque de Melrand, 1991). However, reconstituting the rationale of past landscape dynamics (*i.e.* the interaction between humans and biotic/abiotic factors) based on a few past unchanging structural aspects of our territorial environment, should provide examples of coherent management of geographical areas and even allow us to avoid some of the previously mentioned pitfalls.

Disadvantages	Advantages
- an additional investment of a few thousand francs in the short term (heritage analysis, i.e archaeological, archeometrical and ecological assets...)	- an identification of anthracology and "archeo- sciences" by both the general public and other scientific fields
- a shift between the restoration aim and the "aesthetic" expectations of the general public, which limits the approach to sites with a heritage potential (*i.e* historical sites or natural areas)	- studying the history of landscape changes is related to cultural and identity building values, which are not marketable but play a significant role in the promotion of local or regional projects
- technical and ethical difficulties related to the restoration aims (i.e reintroduction of species, new impacts and ecological imbalances).	- restoring on the basis of paleoenvironmental (or anthracological) images allows decisions not to be purely determined by: - planning methods - contemporary ecological conditions (which are sometimes wrongly anthropized...)
	- paleoecological reconstitution is a very coherent objective for sites with a heritage potential (*i.e* historical and natural sites)
	- archaeological wood study can provide archeo-ethnobotanical information relating to past craft industries, architecture and also environmental management (*i.e* voluntary fires, forest management, pruning, etc.)
	- with other paleoenvironmental sciences, anthracology helps to determine to what extent, when and how, deliberate human intervention has altered environment potential. Seeing the landscape in a long-term perspective is a good approach to organising and planning the landscape on a sustainable basis.

Table 1: Paleoecological research and landscape planning, advantages and disadvantages.

Indeed, geomorphological, pedological, hydrographical parameters as well as changing vegetation cover and animal territories have traditionally determined territorial planning. On the one hand, agricultural technology and current environmental management allow planning to disregard the abiotic landscape constraints (through irrigation, drainage, fertilization or intensive farming). On the other hand, many environmental problems probably linked to these excesses can now be seen in the shape of uncontrollable floods, soil erosion, climatic disorders (*i.e* the greenhouse effect or exceptional storms), various forms of pollution caused by fertilizers, nitrates, phosphates, energy consumption problems and hazardous development sites.

In other words, paleoenvironmental sciences, in particular anthracology, should also help to provide examples of coherent or incoherent geographical space management, and thus influence current and future landscape planning.

References

Archives des actes du colloque de Melrand, 1991 – *Archéologie du paysage*, Penn ar Bed, 148/149, p. 18-83.

BARNAUD G., LEFEUVRE J.-C. 1992 – L'écologie avec ou sans l'homme ? In : collectif. Entre nature et société. *Les passeurs de frontières*. CNRS. Paris, p. 69-112.

BOUJOT C., CASSEN S. 1997a – *Le tertre néolithique de Lannec er Gadouer Erdeven (Morbihan)*. Document de synthèse d'une fouille programmée trisannuelle, Service Régional de l'Archéologie Bretagne, Rennes UMR 6566 du CNRS, Laboratoire de Préhistoire, Nantes Commune d'Erdeven Département du Morbihan, 35 p.

BOUJOT C., CASSEN S. 1997b – Néolithisation et monumentalité funéraire : exploration du tertre de Lannec er Gadouer à Erdeven (Morbihan, France). In : RODRIGUEZ CASAL, O *Neolithico atlantico e as orixes do Megalithismo*, Université de Santiago, Actas do coloquio internacional Santiago de Compostela, p. 211-232.

BUREL F., BAUDRY J. 2000 – *Ecologie du paysage, concepts, méthodes et applications*, Editions TEC & DOC.

GAUDIN L. 1999 – *Paléoenvironnement de la région du Golfe du Morbihan : étude archéobotanique des foyers et de la nappe charbonneuse conservés sous le tertre néolithique de Lannec er Gadouer (Erdeven)*. Mémoire de DEA, Université de Paris X, 122 p.

GAUDIN L., MARGUERIE D. 2000 – In : Cassen S. – *Eléments d'architecture, Exploration d'un tertre funéraire à Lannec-Er-Gadouer (Erdeven, Morbihan), Construction et reconstruction dans le néolithique morbihannais, propositions pour une lecture symbolique*. Chauvigny, France, éd. Chauvinoises, 400 p., 160 fig.

JACQUIOT C., TRENARD Y., DIROL D. 1973 – *Atlas d'anatomie des bois des Angiospermes*. Centre technique du bois, Paris, 2 tomes, 175 p., 72 pl.

MARGUERIE D. 1992 - *Evolution de la végétation sous l'impact humain en Armorique du Néolithique aux périodes historiques*. Travaux du Laboratoire d'Anthropologie Rennes, 40, 262 p.

MONTGOLFIER J., NATALI J.-M. 1987 – *Le Patrimoine du futur, approches pour une gestion patrimoniale des ressources naturelles*, Collection économie agricole & agro-alimentaire.

RAMADE F. 1994 – *Elément d'écologie : écologie fondamentale*. Ediscience international, 2e édition, 517 p.

RAMEAU J.-C., MANSION D., DUME G. 1996 – *Flore forestière française : guide écologique illustré*. T.1 Plaines et collines. Paris, Institut pour le développement forestier, 1785 p.

SCHWEINGRUBER F.H. 1982 – *Anatomie microscopique du bois*. Ed. Zürcher, Birmensdorf, 2ème édition, 226 p.

SCHWEINGRUBER F.H. 1990 – *Anatomy of European Woods*. Ed. Haupt, Stuttgart, 800 p.

TANSLEY A.G. 1935 – The use and abuse of vegetational concept and terms. *Ecology*, 16, p. 284-307.

TROLL C. 1939 – *Luftbildplan und okologische Bodenforschung*. Zeistschraft der gesellschaft fur erdkunde zu Berlin, p. 241-298.

VISSET L., SELLIER D., L'HELGOUAC'H J. 1995 – Le paléoenvironnement de la région de Carnac, sondage dans le marais de Kerdual, La Trinité-sur-Mer (Morbihan). *Revue Archéologique de l'Ouest*, 12, p. 57-71.

VISSET L., L'HELGOUAC'H J., BERNARD J. 1996 – La tourbière submergée de la pointe de Kerpenhir à Locmariaquer (Morbihan). Etude environnementale et mise en évidence de déforestations et de pratiques agricoles néolithiques. *Revue Archéologique de l'Ouest*, 13, p. 79-97.

WILSON K., WHITE D.J.B. 1986 – *The anatomy of wood: its diversity and variability*. Ed. Stobart &son LTD, London, 309 p.

Linking anthracology and historical ecology: suggestions from a post-medieval site in the Ligurian Apennines (north-west Italy)

Carlo Montanari, Sara Scipioni, Gilberto Calderoni, Giovanni Leonardi and Diego Moreno
c/o L.A.S.A. - Corso Dogali 1 M - 16126 Genova, Italy.
E-mail: geobotge@unige.it / envhist@lettere.unige.it

Abstract
The anthracological studies performed during a five year project devoted to the environmental history and archaeology of the Scrivia and Trebbia valleys in the Ligurian Apennines allow us to assess how anthracological evidence can act as "historical evidence" when studied in a chronologically precise historical and archaeological context (*i.e.* post-medieval, 1650-1950 AD). This indeed appears to be the case for the remains recovered in our site and by extension, in the Ligurian mountains, since the Bronze Age. This "historical evidence" (or better, "evidence of evidence") helps us establish a detailed history of the local systems and practices of production and exploitation of the vegetation cover (providing information on local land use systems and changes, practices of coppicing, pruning, domestic chestnut productions, etc.). In contrast with previous historical and archaeological assessment of charcoal evidence, we cannot prove the disappearance (or grubbing out) of the woodland resources in the studied area during the last 350 years of continuously documented charcoal production. On the contrary, we can suggest some recent, small variations in composition and shape of the tree cover of the same area. Finally, the expertise of the most recent local charcoal burners was used as a reference to trace and record the charcoal production by means of traditional techniques.

Key-Words: *Scrivia valley, Trebbia valley, Apennines, Liguria, Post-Medieval History, charcoal, charcoal kiln.*

Résumé
La recension des études anthracologiques, effectuées dans le cadre d'un projet de cinq ans dédié à l'archéologie et à l'histoire de l'environnement des vallées du Scrivia et du Trebbia dans les Apennins ligures, nous permet d'interpréter les témoins anthracologiques comme des témoins historiques en les étudiant dans un contexte historique et archéologique chronologique précis (c'est-à-dire post médiéval, 1650-1950 AD). Ces restes sont situés dans notre aire d'étude, mais également dans toutes les montagnes Ligure, essentiellement depuis l'Age du Bronze – et constituent une trace historique ou mieux la "trace de trace" d'une histoire détaillée des systèmes et des pratiques locaux de production et de l'exploitation du couvert végétal (informations sur les méthodes et les changements de l'utilisation du terroir local, pratiques du taillis, taille des arbres, production des châtaigneraies domestiques). Contrairement à la précédente évaluation historique et archéologique il n'est pas possible de démontrer la disparition des ressources forestières dans l'aire étudiée pendant les 350 dernières années de production du charbon de bois ; à l'inverse, nous pouvons proposer des variations récentes dans la composition et dans la structure de la couverture forestière de cette aire. Enfin, l'analyse de la pratique du dernier charbonnier local a été utilisée pour suivre et enregistrer la production de charbon de bois avec des techniques traditionnelles.

Mots-clés : *vallée du Scrivia, vallée du Trebbia, Apennins, Ligurie, histoire post-médiévale, gestion forestière, charbons de bois, charbonnière.*

Introduction

Some results and perspectives, which came out studying evidence of charcoal production in different periods and examined in a precise environmental and cultural framework, will be illustrated. The data were collected over the past five years, during field courses of environmental history and archaeology that were promoted by the Laboratory of Environmental Archaeology and History (LASA: degree course in cultural heritage conservation at the University of Genova), with the intervention of colleagues of the GEODE (University of Toulouse). Many students attended the digs, and two of them (Scipioni, 2000; Prono, 2000) carried out the most of the anthracological analysis. In this paper, we give an assessment of the anthracological evidence collected in several fieldwork programs (1996-2000) devoted to a project of history and archaeology of mountain environmental resources (woodland, pasture, meadows, etc.) in the area of the Antola Regional Park, in the hinterland of Genoa, NW-Italy (Montanari *et al.* 2000). During the fieldwork, the archaeology of charcoal burning sites and the present practices for charcoal production were studied. The historical ecology at each site was studied at a topographical scale and with a historical regressive approach. However, considering the present position and aims of historical ecology studies, the observational side of this research lies squarely between palaeoecological and neoecological field studies and techniques. We have assumed that the development of the historical approach (Rackham, 1976) to environmental studies brings into question:

- conventional palaeoecologists (such as anthracologists, palynologists, and sedimentologists) and their need for higher resolution analysis;
- vegetation neoecologists and their need for precise studies in time and space;
- historians and archaeologists in a sort of "experimental approach" in the evaluation of the conventional historical sources for the study of the past society (peculiarly those concerned in rural history, cf. Moreno and Raggio, 1999).

Pollen based studies in the Mediterranean mountains, especially in the Ligurian Apennines, have stated that confined practices (such as charcoal production) and unconfined fire practices have shaped woodland and pasture resources since the Bronze Age (Lowe *et al.* 1994, 1995). Local agro-silvo-pastoral systems have undergone a long and complex evolution with different stages, evolving until through medieval and post-medieval ages and lasting to the present desertion of the mountain areas.

The question that the present study addresses is: how can "anthracological evidence" act as a "historical evidence" when studied in such a precise historical and archaeological, post-medieval context? In the sites dating from the Bronze Age in the Ligurian mountains, anthracological evidence indeed appears to be historical evidence, or at least provides a better historical picture of the local systems and practices of production (giving information on local land use systems and changes, practices of coppicing, pruning, domestic chestnut productions, etc.). Contrasting with previous assessments of the interest of charcoal pits for archaeologists (*e.g.* Lugli and Pracchia, 1995), the historical ecology approach allows us to suggest a specific connection between the exploitation of the local woodland resources (*e.g.* composition, structure, etc.) and a 350 year-long period of charcoal production.

Charcoal burning sites for the environmental history and archaeology of the Ligurian Apennines

Until today, only economic historians have been concerned with the production of charcoal in the Ligurian Apennines, studying the role it played for the local economy and its similarities and differences from other areas (Di Stefano, 1986; Baraldi *et al.* 1992; Viola, 2000). An article by Lugli and Pracchia (1995) provides an ethnoarchaeological point of view of the subject, on a global and national scale, and gives an example of a site with stratigraphic evidence of several stages of charcoal burning.

In the upper Scrivia and Trebbia valleys, peculiar circumstances and the contribution of researchers with a variety of backgrounds, students, and local institutions and inhabitants have made a multidisciplinary approach to the subject possible. Thanks to this collaboration, evidence of recent charcoal burning sites (possibly used over the last 50 years) has been studied and the anthracological spectra were matched with the present vegetation at each site. Moreover, a soil profile containing evidence of ancient charcoal production in different times was described, sampled and dated with archaeological methods; afterwards, a cross-section, subdivided into three parts, was pulled out from the slope with archaeological techniques and adventurously transported by cable up to the Castello della Pietra, where it will be integrated in a museum of local history. We have tried to answer several questions, which arose in connection with the interpretation of these traces, by making a traditional charcoal burning and recording each stage, even by means of a professional film.

The archaeology of a charcoal burning site

In the Vobbia Valley, at 420 m a.s.l. in the northern Apennines (fig. 1), traces of charcoal burning were found superimposed one upon the other, preserved under colluvial sediment and more or less recent soil layers. The stratigraphic excavation (named UT 1) was directed by one of the authors (G. L.), producing a detailed analysis and interpretation of the sequence (fig. 2). A series of radiocarbon

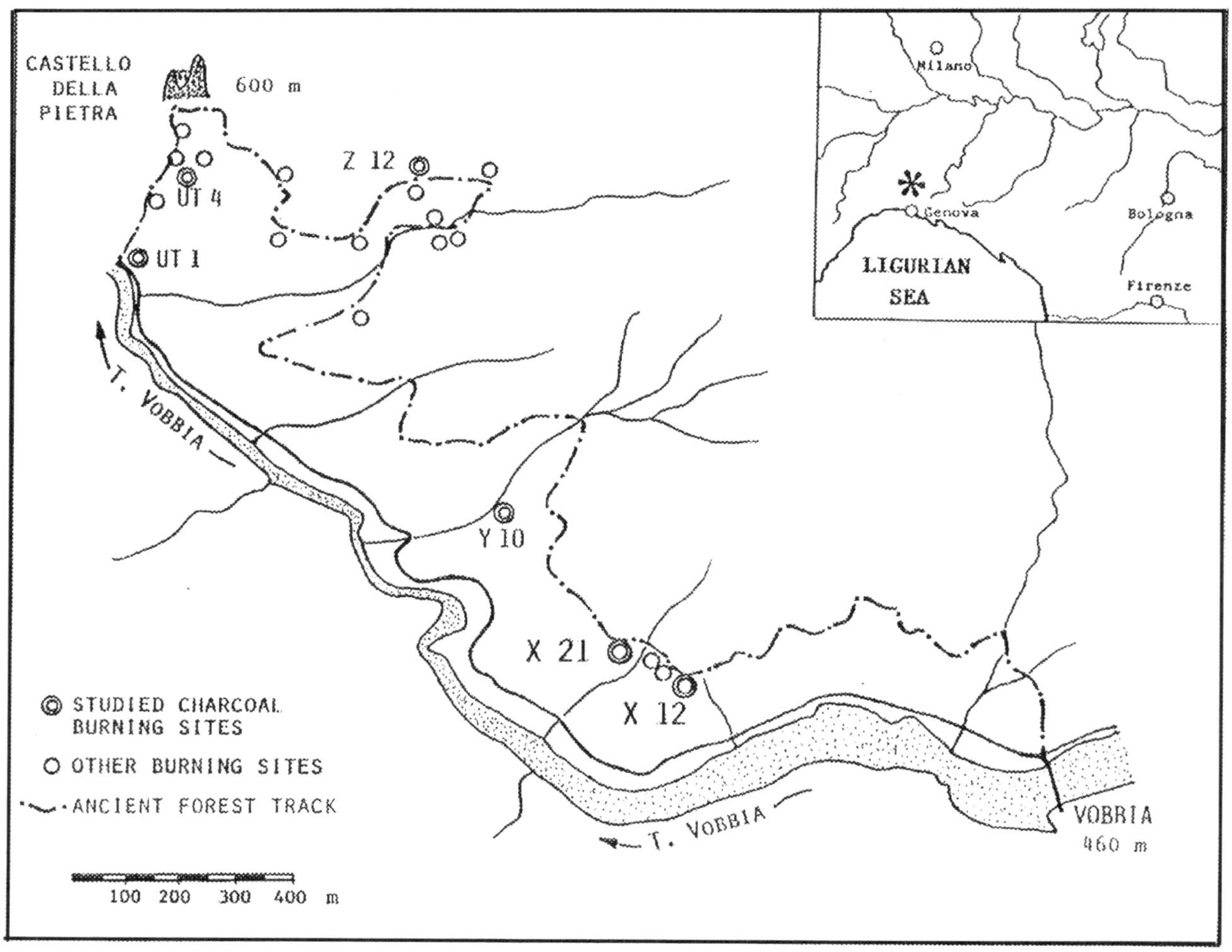

Figure 1: Location and map of the study area in which several charcoal-burning sites are marked.

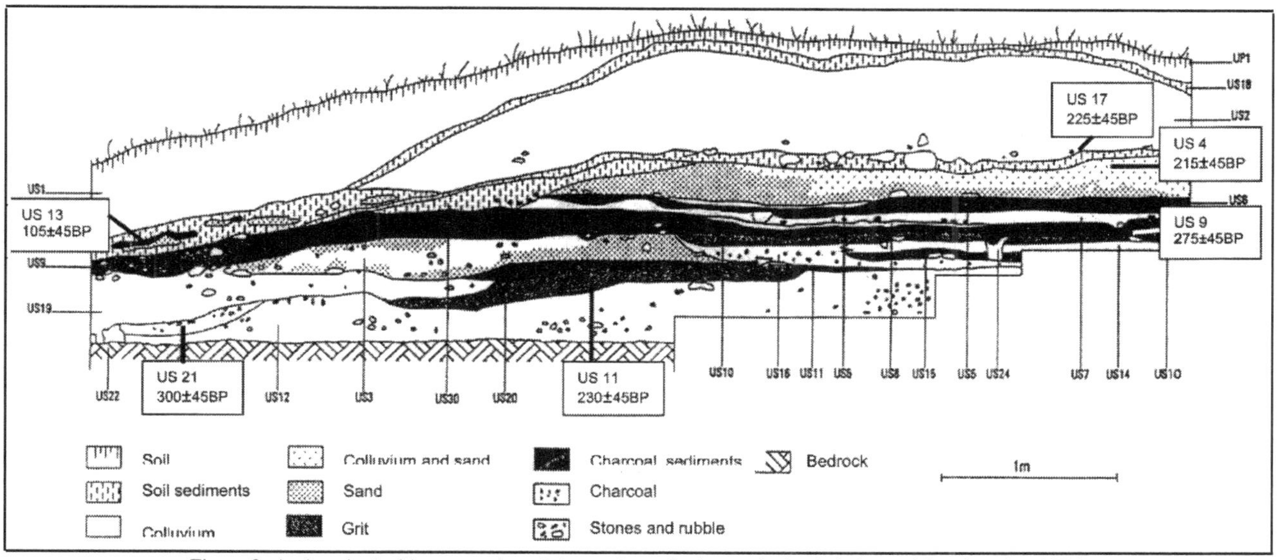

Figure 2: Archaeological interpretation of the section referring to the buried charcoal-burning site (UT 1).

datings, (carried out by G. C.), allowed us to place this series between the 17th and the 19th centuries. At least three different charcoal burning stages were identified, which, briefly, have the following features (fig. 2):

- US 11 (230 ± 45 BP: Rome 1076-1079): This level contains almost exclusively sweet chestnut charcoal (fig. 3), it is therefore probably due to cultivation practices (*e.g.* pruning of branches) of a fruit-chestnut grove or to its felling. Two different types of charcoal were observed, with wide and narrow annual rings respectively; the latter were prevalent, having mainly 3-5 annual rings (perhaps pruned branches?).
- US 9 (275 ± 45 BP: Rome 1077): The anthracological analysis of a few hundreds of fragments indicated the presence of 11-14 taxa, with *Ostrya*, *Castanea* and *Carpinus* prevailing (fig. 4).
- US 6: In this case as well, the anthracological analysis of several hundred charcoal fragments revealed about 10 taxa, among which *Ostrya*, *Castanea*, *Carpinus* prevail (fig. 5). Moreover, the fragments observed in the US 9 and US 6 layers have more annual rings than those from the US 11, which could suggest longer cutting cycles, about 10-12 years. Although the datings suggest a more recent age for the US 11 with respect to the US 9, we can regard it as an older layer, bearing in mind the margin of error and the stratigraphy (fig. 2).

Anthracological analysis was carried out also on other stratigraphic units in which charcoal was found (*e.g.* US 21 and US 13):

- US 21 (300 ± 45 BP: Rome 975): Although only few fragments were found, the results of the anthracological analysis are similar to those from US 11, *i.e. Castanea*, prevails.
- US 13 (105 ± 45 BP: Rome 1074): The anthracological spectrum of this layer is similar to that of US 9 and US 6 (fig. 5). It provides the most recent evidence of a burning site in the profile, linking an older layer (US 9) with the recent ones (20th century), which, as we will see later, have rather similar anthracological spectra. These charcoal remains could be interpreted as coming from one of the last charcoal burnings, of which nothing else was found.

When small branches with the bark were found, it was concluded that the felling season was springtime (this corresponds to the present local practices). Afterwards, the site underwent repeated landslides and was no longer used for charcoal burning, though the superimposed soil contains some charcoal fragments, coming from upper sites.

The present vegetation of the site is qualitatively similar to that of the traces found in the profile, starting from US 9 (that is, mixed coppice). It is a deciduous coppiced wood with *Carpinus, Ostrya, Corylus, Tilia,* and *Fraxinus*, in

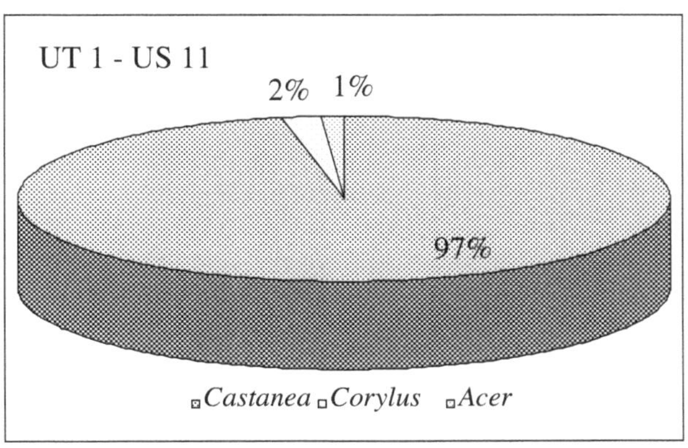

Figure 3: Anthracological spectrum of the layer US 11 (UT 1) in which chestnut charcoal prevails.

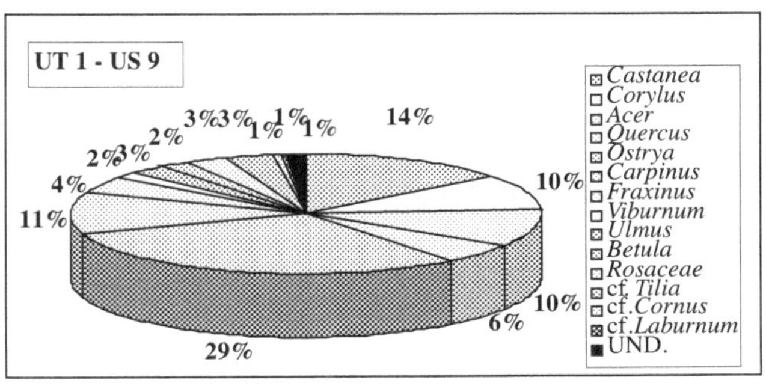

Figure 4: Anthracological spectrum of the layer US 9 (UT 1) showing a large number of different taxa.

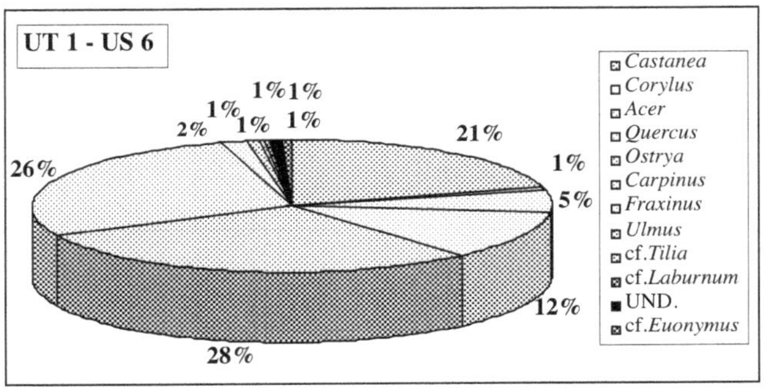

Figure 5: Anthracological spectrum of the layer US 6 (UT 1), qualitatively very similar to that of the layer US 9.

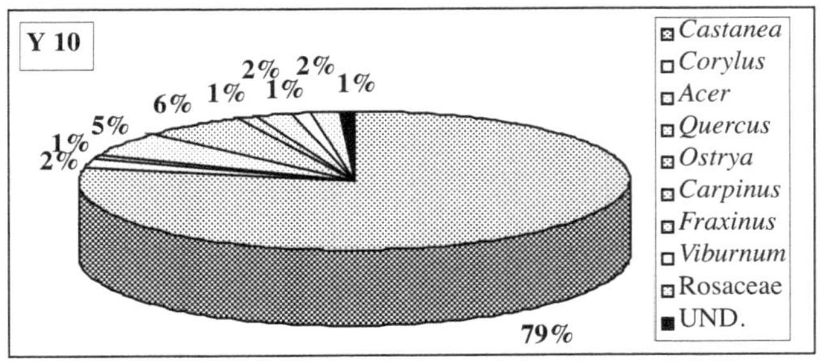

Figure 7: Anthracological diagram of the sub recent charcoal site Y 10; the abundance of chestnut charcoal can be linked to the management or to the abandonment of a sweet chestnut grove.

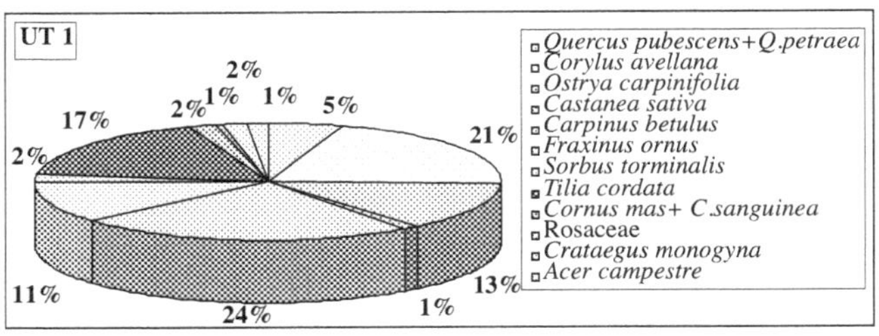

Figure 6: Diagram of the present vegetation cover near the site UT 1: many taxa occur which are found in the most recent anthracological spectra (US 9, US 6).

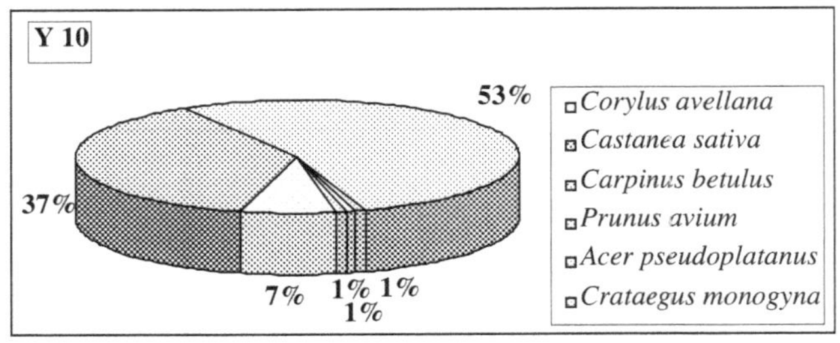

Figure 8: Diagram of the present vegetation cover near the site Y 10: a recent invasion by hornbeam is clearly shown within an old sweet chestnut grove.

which hornbeam prevails locally, as compared with hop-hornbeam (fig. 6). Therefore, a gradual transition becomes evident, possibly starting with a fruit-chestnut wood, which was exploited for charcoal production and then substituted in the following centuries with mixed coppiced wood, expressly exploited for the charcoal production. This kind of woodland is locally named "scabbia", a common toponym that means "a hardwood woodland exploited for charcoal production, coppiced every 7-15 years" (Cevasco and Poggi, 2000).

Davasse (1991) published an article in which a series of charcoal burning sites were also studied stratigraphically; nevertheless, in that case the charcoal sediment was sampled every 5 cm in depth, allowing vegetation changes to be pointed out. On the contrary, in this case the charcoal layers - archaeologically detected - are separated by mineral sediment, so that each burning stage is not in contact with the subsequent stages (the charcoal composition showed no appreciable difference within the single layers).

Sub-recent charcoal sites and present wood composition

The anthracological analysis of 6 charcoal burning sites used in the first half of the 20th century, laying within a radius of 1 km from the UT 1 site, allows us to get an idea of the features of the anthracological evidence at the slope level and of its correspondence with present vegetation. The anthracological spectra show a good qualitative homogeneity, with a constant presence of at least 5-6 (up to 12) different taxa; quantitatively, there is an almost equal distribution in three typologies:

1)- *Castanea* prevailing (70-78%), with *Ostrya, Quercus, Corylus*, etc. (*e.g.* Y 10, fig. 7);
2)- *Quercus* cf. *cerris* prevailing (66-70%), with *Castanea, Ostrya, Fraxinus*, etc. ;
3)- *Ostrya* prevailing (40%), with *Castanea* and *Fraxinus, Quercus, Carpinus*, etc.

Sub-recent spectra compared with present vegetation show a variable quantitative correspondence. This could be due both to the local development of the forest communities (at present, little-exploited and much more aged than in the past) and to the difficulty of circumscribing a reliable catchment area for the charred wood. Also, the oral tradition attests to a wide variability concerning the wood supply area, owing to various circumstances (*e.g.* the use of common woodland; availability of means of transport, for example, mules; species selection; ground features, etc.).

With the exception of one case (site Z 3 = UT 4), these charcoal sites were not studied stratigraphically.

One charring site in particular (Y 10) seems to be similar to the US 11 of the UT1 site, from an anthracological point of view: The charcoal of *Castanea* (68-76 %) clearly prevails, and the modern vegetation covering shows old chestnut groves abandoned for some time, in which *Carpinus* prevails, among other broadleaved species (fig. 8). Therefore, a common picture emerges from the majority of the studied sites, showing a transition stage as a consequence of the abandonment of the sweet chestnut growing, and the expansion of a recently developed woodland with hornbeam, hop-hornbeam, hazel and wild chestnut.

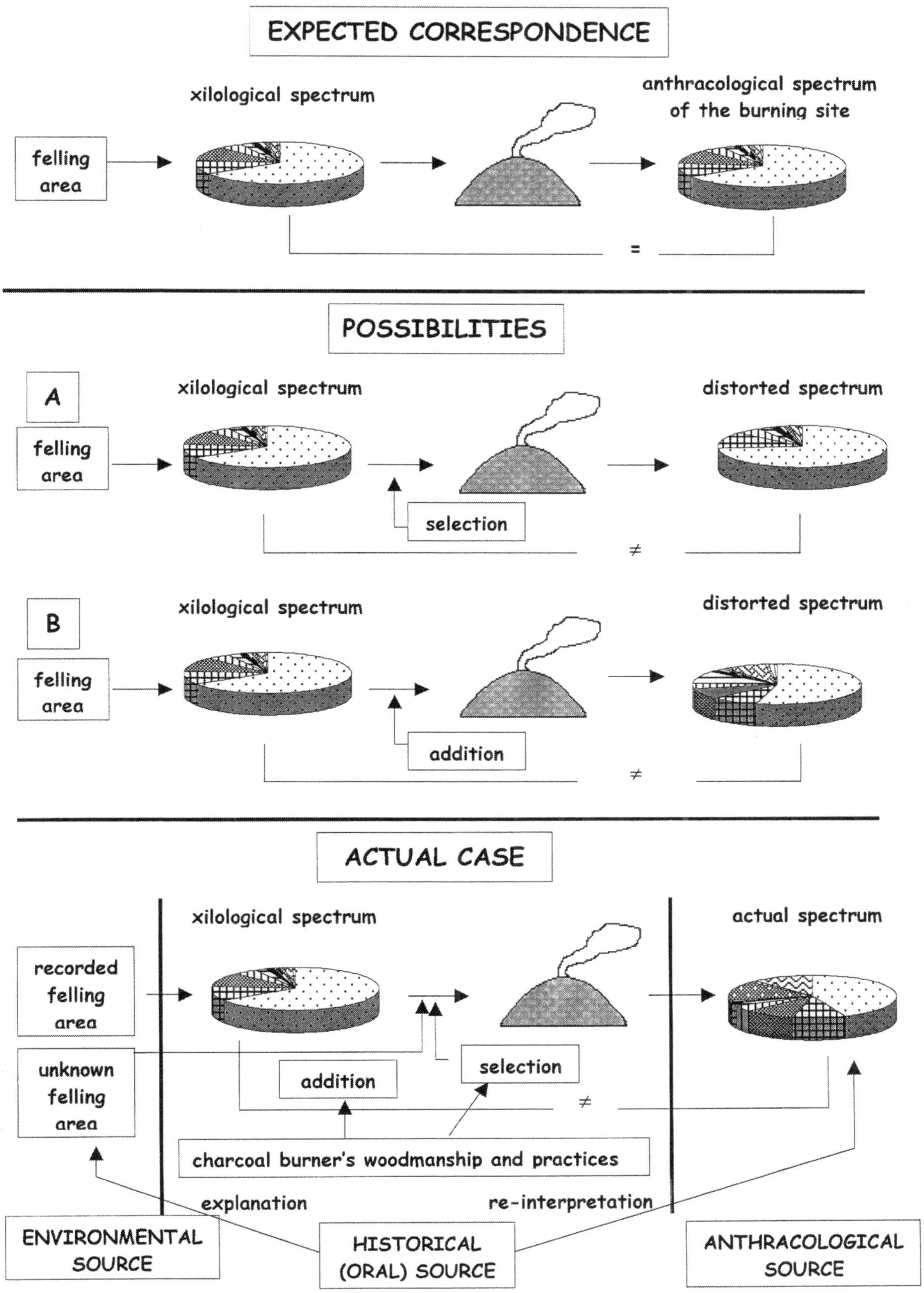

Figure 9: Suggestions and evaluation of historical sources coming from an experiment of forest archaeology that is the reconstruction of a traditional charcoal kiln.

A charcoal kiln as experimental archaeology

The skill and expertise of the last local charcoal burners was exploited to follow and record charcoal production by means of traditional techniques. For many peasants, this activity was only a seasonal complement to their annual income; nevertheless, there were also groups of professional charcoal burners, recruited by a contractor, and they were often from Veneto and Lombardia (northern Italy).

All of the stages of charcoal production were recorded, even by means of a professional film (distributed in French and Italian versions) by J.-P. Métailié and B. Bastard (Poggi and Métailié, 1999).

The main stages of the experimental charcoal production were:
1) planning the work and the film;
2) felling of the wood; recording quality and quantity, season, weathering;
3) choice of the burning site; use of an existing bank, with subsequent troubles with draught;
4) building of the stack; typology, disposition of the wood, covering;
5) ignition, burning; management, ordinary and exceptional interventions, temperature recording;
6) charcoal recovery; distribution on the site;
7) cooling and packing into sacks;
8) study of the site after burning: surface sampling, stratigraphy, plant settlement;
9) filming; projection of the film, recording of reactions and comments, confrontation of methods; renewal of interest (to further charcoal burning with an ethnographic aim).

Besides the qualitative and quantitative features of the charcoal remains on the burning site, a quantitative and clear dimensional distribution of the fragments was pointed out, which is visible even to the naked eye and was confirmed by a line sampling; this is invaluable for assessing the sampling methods.

At this point, with the help of a synthetic illustration, we will suggest only a few reflections as a result of this project. In figure 9, the relationships between the anthracological evidence and the ecological and historical local features are stressed. The first example shows the most simple and theoretical case of a close correlation between the felling area and its anthracological evidence. More frequently, modifications occur due to the burner's unpredictable intervention, by adding or selecting the wood (second example).

In our experimental charcoal burning, an inconsistency was found between the xylological spectrum and the anthracological one (lacking or unexpected presence of some species). Historical (oral or textual) sources have been useful to explain the inconsistencies caused by the charcoal burner's practices (third example).

Conclusions

The parallel and integrated use of information sources and of study techniques gives a wider view than the picture provided by single traditional disciplines (archaeology, palaethnology, history, archaeobotany, archaeometry, etc.). This type of multidisciplinary approach provides useful contributions for the understanding and the interpretation of the evidence of the various activities carried out by local people exploiting environmental resources. Moreover, the interaction of data obtained by means of various specialised research methods provokes questions and yields assumptions which can lead to new conclusions or can be tested through parallel and complementary research lines. For instance, in the case presented, the oldest burning stage contains almost exclusively chestnut charcoal with very close annual rings, which suggested the pruning of a sweet chestnut grove. This point is rather important because it could indicate a change in the local resource management, *i.e.* a shift from the production of chestnuts to charcoal, which was mostly exported. Archival sources (Baraldi *et al.* 1992) testify the demand of chestnut charcoal (which is poor in sulphur) for the metallurgy, with an increase in the consumption of the "domestic" one during the 17th century. On the contrary, the chestnut tree is locally regarded as hardly suitable for charcoal production. Nevertheless, in spite of this local belief, this species is constantly found and often prevails in the examined burning sites: the archaeobotanical evidence seems to contradict local traditional knowledge. In any case, the recording of charcoal production practices is basic for the study and interpretation of the burning sites. For instance, it gives information about the building of chimneys and burning management, temperature, the arrangement of the charcoal in the collection and cooling stage, the effect of the heating on the ground and the charcoal traces left on the site.

In summary, we can conclude that:
- anthracology has once more proven itself a sensitive tool even in detecting small modifications of the surrounding vegetation at a local and precise temporal scale (1650- 1950 AD);
- the topographical use of textual and oral historical sources focused on local wood use practices is essential to understand the actual meaning of traces of historical land use (*i.e.* forest management);
- charcoal burning over a period of at least 300 years at the same site did not cause an exhaustion of the resources (in an Apennine valley with mesophilous vegetation).

AKNOWLEDGMENTS
The authors are grateful to the Ente Parco Antola and GAL Antola and Penna Leader for founding the project, Presidi Universitari delle Valli dell'Antola, to the teaching staff of the L.A.S.A., to the students who joined in the environmental history campaigns, to the charcoal burners, and to the local institutions and folks who allowed this study to be carried out.

References

BARALDI E., CALEGARI M., MORENO D. 1992 - Ironworks economy and woodmanship practices. Chestnut woodland culture in Ligurian Apennines (16-19th C.). Protoindustries et histoire des forêts. *Les Cahiers de l'ISARD*, t. 3, p. 135-149.

CEVASCO R., POGGI G. 2000 - L'alpe, l'arbre et le lait. Sud-Ouest Européen, *L'environnement entre nature et paysage*, t. 7, p. 35-48.

DAVASSE B. 1991 - Les espaces forestiers de la haute vallée du Vicdessos - *Pyrénées ariégeoises : analyse écohistorique et étude des charbonnières*. Colloques phytosociologiques, XX, p. 305-314.

DI STEFANO P. 1986 - Castagneti aggregati a massarie: trasformazioni nella castagnicoltura a Voltaggio nella seconda metà del '700. AA.VV. 1986. Studi in memoria di T.O. De Negri; *Bollettino Ligustico*, n° speciale, (SAGEP Genova), p. 124-137.

LOWE J.J., BRANCH N., WATSON C. 1994 - The chronology of human disturbance of the vegetation of the northern Apennines during the Holocene. In Biagi P. and NANDRIS J. (Eds.) Highland Zone Exploitation in Southern Europe, *Monografie di Natura Bresciana*, t. 20, p. 169-187.

LOWE J.J., DAVITE C., MORENO D., MAGGI R. 1995 - Stratigrafia pollinica olocenica e storia delle risorse boschive dell'Appennino settentrionale. *Rivista Geografica Italiana*, t. 102, p. 267 - 310.

LUGLI F., PRACCHIA S. 1995 - Modelli e finalità nello studio della produzione di carbone di legna in archeologia. ORIGINI. *Preistoria e protostoria delle civiltà antiche*, t. 18, p. 425-479.

MORENO D., RAGGIO O. 1999 - Dalla storia del paesaggio agrario alla storia rurale. L'irrinunciabile eredità scientifica di Emilio Sereni. *Quaderni storici*, t. 100, p. 89-104.

MONTANARI C., PRONO P., SCIPIONI S. 2000 - The study of charcoal-burning sites in the Apennine Mountains of Liguria (NW Italy) as a tool for forest history. In Agnoletti and Anderson (Eds.) *Methods and approaches in forest history*. Cambridge, CABI Publishing, 281 p.

POGGI G., METAILIE J.-P. 1999 - *Carbonai dell'Antola, videocassetta 40'*, co-produzione CAV-UTM/Ente Parco Antola, realizzazione B. Bastard.

PRONO P. 2000 - *Studi archeobotanici su piazzole da carbone dell'entroterra ligure*. Tesi di laurea, Facoltà di Lettere e Filosofia dell'Università di Genova, aa. 1998 - 1999, 223 p., 58 fig.

RACKHAM O. 1975 – *Trees and Woodland in the British Landscape*. London

SCIPIONI S. 2000 - *Ricerche di archeologia ambientale su resti di carbonaie nel Parco dell'Antola*. Tesi di Laurea, Facoltà di Scienze Matematiche, Fisiche e Naturali, Università di Genova, aa. 1999 - 2000, 245 p., 81 fig.

VIOLA M., 2000 - *Storia della produzione di carbone vegetale nell'Appennino Ligure. L'Alta Valle Scrivia*. Tesi di laurea, Facoltà di Lettere e Filosofia dell'Università di Genova, aa. 1998 - 1999, 172 p., 2 fig.

Gathering of firewood during the Palaeolithic

Isabelle Théry-Parisot

CNRS-UMR 6130, CEPAM, 250 rue A. Einstein, 06560 Valbonne, France.
E-mail: thery@cra.cnrs.fr

Abstract

The anthraco-analysis rarely allows us to know the practices of firewood gathering during the Palaeolithic. Here we present some reflections on this subject, taking into account the teachings of ethnology, the combustion properties of wood and the particular life conditions prevailing during this period. Ethnological studies reveal the diversity of habits from one community to another. The fuel properties do not always determine the choice of wood. Moreover, the type of wood used is not alone responsible for the quality of the combustion. The technological context (stone tools) was not a limiting factor to the wood gathering during the Palaeolithic. It seems that general life conditions and site functions were more important in determining, for example, the controlled management of green wood, brought to and stored in habitation sites or the random collection of dead wood by inhabitants of specialized settlements occupied seasonally. The hypothesis of a controlled management of wood resources is reinforced by the intentional use of other types of fuel (lignite, animal bones) at certain sites. This hypothesis does not interfere with the palaeoecological significance of charcoal analysis.

Key-words: *Palaeolithic, Firewood, Gathering, Charcoal, Economy.*

Résumé

L'analyse anthracologique nous permet rarement de remonter directement aux pratiques de la collecte du bois de feu au Paléolithique. Nous présentons ici une réflexion sur ces pratiques qui prend compte tant les enseignements de l'ethnologie, les propriétés combustibles du bois, que la spécificité de cette période en termes de moyens, de modes de vie et de milieu. L'ethnologie nous renseigne essentiellement sur la diversité des pratiques rencontrées d'une communauté à l'autre. Les propriétés combustibles du bois, qui ne dépendent pas uniquement du facteur espèce, ne sont pas toujours le critère de choix décisif. Le contexte technique des groupes paléolithiques ne représente pas une limite pour l'acquisition du bois. En revanche il semble que le mode de vie des groupes, comme la fonction des sites soient autant de facteurs déterminants : gestion contrôlée de stocks de bois dans les sites d'habitat, ramassage aléatoire du bois mort dans les sites spécialisés où la durée d'occupation est plus courte. L'hypothèse d'une gestion réfléchie du bois de feu, qui prend en compte un nombre important de variables, se justifie d'autant plus qu'il a été montré que le recours à d'autres combustibles était plus intentionnel que déterminé par une pénurie de bois. Cette hypothèse ne remet pas en cause la représentativité paléoécologique des charbons de bois. Entre choix des espèces et ramassage du bois mort, la palette des essences prélevées, conformément à la diversité des fonctions d'un foyer, doit recouvrir la quasi-totalité des taxons présents autour d'un site.

Mots-clés : *Paléolithique, bois de feu, collecte, charbons, économie.*

Charcoal fragments are often well preserved in archaeological sites. They are essentially studied from a palaeoecological perspective, aiming at characterising the plant formations and their transformations during various climatic cycles. By this approach, various aspects of the gathering of firewood are rarely taken into account. In this paper, we focus on this particular issue of the economy of prehistoric societies. How and according to which criteria was wood chosen? Could we expect that a selection of species was operated even when the anthraco-analysis cannot prove this? We try to reconstruct the habits of firewood collection in the Palaeolithic, by taking into account all the factors that could have influenced the management and acquisition of firewood. We then consider the impact of firewood gathering on the palaeoecological record as reflected by charcoal analysis.

The teaching of ethnology

Diversity of practices

The selection criteria used in the gathering of firewood are well documented by ethnological studies. In various communities, the selection is done according to traditional parameters, fixed by the group and not always related to the concept of botanical species. According to Nicholson, the aboriginals of Australia carefully selected their firewood according to its morphology without ever taking into account the particular species (Nicholson, 1981, *in* Smart and Hoffman, 1987). Athapaskan Indians (Britannic Colombia, Canada), traditionally hunters and gatherers but today sedentary, selected their firewood as much according to the species as to the state of the wood: diameter, physical and phonological state (Beyries, personal communication). Inuit of the river Kobuk in Alaska lived in a forest environment but used primarily driftwood, without selecting for a particular species (Giddings quoted by Alix, 1998).

The variability of traditional practices also shows the subjectivity of selection. According to Osgood (1940, 1958, *in* Smart and Hoffman, 1987) Ingaliks preferentially burnt spruce (*Picea*) in winter because of its thermal qualities while they chose poplar (*Populus*) and willow (*Salix*) in spring, when they needed less heating. However, when referring to the calorific values, it becomes clear that the actual difference between different wood types is not that significant. In the archipelago of Kiribati in Micronesia, the inhabitants of two nearby islands, Nikunau and Tabiteuea, have at their disposal and in similar proportions seeds of *Pandanus* and endocarps of coconut. For making fire, members of the first group use exclusively coconut endocarps while those of the second group only burn *Pandanus* seeds. No one in these two communities could explain their particular choice (Di Piazza, pers. communication). The examples could be multiplied but, apparently, the criteria used for the selection of firewood vary almost as much as the communities, described by ethnologists. Moreover, the choice of a species does not always rely on objective parameters concerning the combustion properties. It would therefore be risky to establish a simple correlation between the choice of a wood species by prehistoric populations and its combustion properties.

On the notion of species

It is doubtless that the prehistoric populations had a perfect knowledge of their plant environment. Studies of traditional societies teach us, however, that the modern concept of species should be considered with caution (Levi-Strauss, 1962). Our taxonomical classification, due to Linné, is based on botanical characteristics (fruits, flowers...) that are probably very remote from the thinking of Palaeolithic communities. Their "classification" was certainly based on other criteria. For example, Inuit populations from Greenland gave different names to floated wood according to its quality and its use, without considering the botanical species (Alix, 1998). Also, on the north coast of Alaska, the local population use a specific term, "kun.na.ta.kin", designating indiscriminately all species used for firewood (Alix, according to Petersen, 1986). It could be expected that the prehistoric "classification" of wood was based on the specific use of a species, for example by distinguishing manufactured wood from firewood. For firewood, a second level of classification could have taken into account the functioning of various hearths. Two botanically different species could thus, surprisingly enough, have been designated by the same name as long as they presented similar firing properties.

The function of hearths and the choice of wood

What is a good fuel?

The selection of wood by prehistoric societies probably took into account the notion of "good fuel". But what is a good fuel? In our societies, the fire is essentially an element of pleasure and its limited use has restricted our idea of good fuel to a few criteria such as an easy ignition or the continuity of flames, to the detriment of all other combustion properties. Often one hears that "this type of wood is a good firewood, while this other type is bad". During the Palaeolithic, hearths were used for very varied and essential purposes: lighting, heating, cooking, transformation of raw materials, smoking of food and protection (Olive and Taborin, 1989). Each of these purposes requires the use of an appropriate fuel. A wood type that produces high flames is adapted to lighting up a space but not to produce large quantities of incandescent charcoal. In the same way, a wood that produces a lot of smoke can be considered "uncomfortable" except when wild animals have to be kept away. So, the notion of good fuel is intimately related to particular needs. In the Palaeolithic, there was no "good fuel" but a range of wood types whose properties were useful for different purposes. We should take into account the capacity of the prehistoric populations to adapt their choice of wood to an intended use.

The function of hearths and heat transfer

We often believe that the temperature of combustion is the main factor determining the choice of a specific wood species. However, when thinking in terms of actual needs of energy, we observe that maximal temperatures of 250°C are sufficient for most activities. All species provide such temperatures when burning and they produce quantities of energy that are much superior to basic needs. The adaptation of the wood to the various uses of the hearths does not depend on the combustion temperature but on the heat transfer. The transfer of heat is a process by which energy is exchanged between bodies or spaces having different initial temperatures. Three simultaneous, fundamental physical processes are involved in these exchanges: conduction, convection and radiation (fig. 1). Each of these, alone or combined with one of the others, responds ideally to one or several activities; for example, the radiation provides light, the convection and the radiation allow drying etc. Even if these processes always occur simultaneously, one of them can be more effective than the other during the combustion. Intrinsic and extrinsic factors, such as the shape of the hearth or the presence of wind can influence on the heat transfer. The combustion properties of each species also determine the kind of heat transfer obtained. The choice of the wood should then take into account the type of heat transfer preferred for the functioning of the hearth. Thus, the knowledge of the combustion properties of species identified from prehistoric hearths could inform us about their function. But, establishing strict correlations would be excessive. Cultural choices certainly also influenced the gathering, without any consideration of fuel properties. Moreover, the wood properties depend on many different parameters.

Fuel wood properties

It is usually admitted that the burning qualities of the wood depend on the species. However, the particular properties of each species are largely influenced by the level of humidity, the morphology of the log (size and diameter) and its physiological state (fresh or altered) (Amy, 1961; Briane and Doat, 1985; CEMAGREF ed, 1990; CTBA ed., 1973; Escudie, 1989; Trabaud, 1976, 1989) (fig. 2). For example, the combustion properties of a green twig from beech (*Fagus*) are closer to those of fresh oak (*Quercus*) than to those of beech wood that has dried for several months. Similarly, the combustion properties of a twig of dead wood that has been altered by microorganisms are very different from those of a healthy twig from the same species. Therefore, it is only possible to compare the properties between species that present the same rate of humidity, morphology and state of alteration. Consequently, the use of

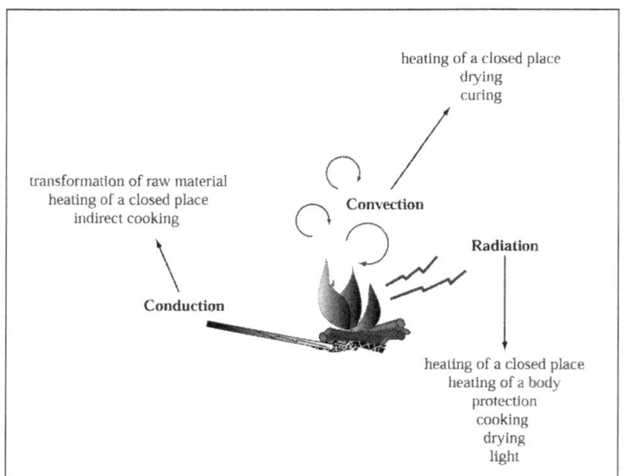

Figure 1: Heat-transfer and hearth functions.

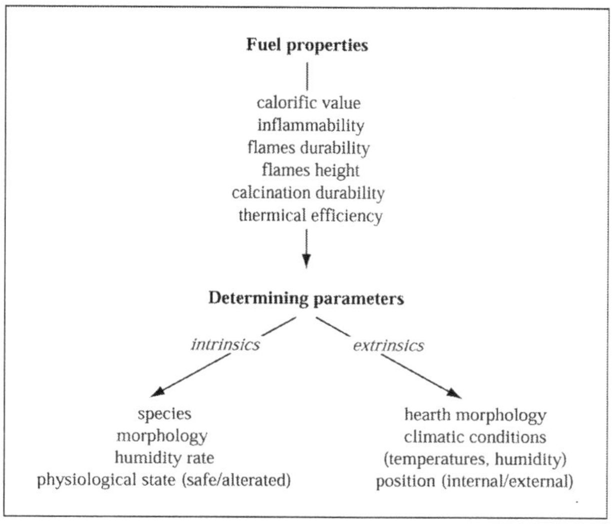

Figure 2: Factors determining the combustion properties.

wood species for different purposes implies the management of stocks of healthy and dry wood, preferentially dried under shelters during minimum eight months after cutting (Briane and Doat, 1985). There is an alternative to this type of management. It is indeed possible to fulfill all the specific requirements of the hearth by taking into account only the shape and state of the wood and not considering the species, for example by the selection of altered or small branches for the ignition, by the use of healthy and dry wood for the maintenance of the fire and by the choice of green wood in order to obtain smoke. We may ask ourselves about the necessity, for Palaeolithic groups, to manage stocks of healthy and dry wood when dead wood, presenting a wide range of states and perfectly convenient for the requirements of every kind of combustion, was available in the environment. Thus, the choice of certain species for firewood appears more to be a possibility than a necessity. However, a wide range of possibilities, between the collection of dead wood and the management of green wood, certainly existed depending on the specificity of every group in terms of technological level, ways of life and environmental context.

Choice of wood and the specificity of the Palaeolithic

The cutting of wood

The question if the quality of the stone tools limited the gathering practices normally arises. The use of stone tools "en percussion lancée" requires the presence of a haft that is solid enough to support important and repeated mechanical constraints. One may suppose that the haft is a very ancient invention, although direct proofs are rare because of the bad preservation of wood in traditional archaeological contexts. The discovery of hafted tools in the peat bog site of Kärlich (Middle Palaeolithic, Germany) is exceptional (Beyries and Hayden, 1993). Still, use-wear analysis yields evidence of hafting of stone tools as well as characteristic use-wear due to the cutting of green wood (Anderson-Gerfaud and Helmer, 1987; Beyries, 1987, 1990, 1993; Handled and Topfer, 1973). Moreover, experimental studies show that a well-handled tool, presenting an attack angle and a carefully prepared sharp edge allows an experienced user to cut down trees (Anderson-Gerfaud, 1987; Beyries, 1987; Beyries and Boëda, 1983). Finally, several ethnographical examples confirm the efficiency of stone tools for the gathering of wood (Hayden, 1979). According to Beyries and Plisson (personal communication), Lower and Middle Palaeolithic tool kits include stone tools whose morphology would have allowed the felling of trees of any shape or size. In the Palaeolithic, the technological level did thus not constitute a limiting factor to the acquisition of wood. The whole biomass, without limitation of size or shape, was potentially accessible.

The management of firewood by nomads or semi-nomads

One of the main limits to the felling of trees for nomads or semi-nomads can be the necessity of drying the wood. The wood of living trees always presents a humidity rate of 40 to 85 % (Briane and Doat, 1985; CTBA ed., 1977; Kauman, 1983). The ideal humidity rate for most combustions is situated between 12 % and 25 % (Trabaud, 1976). To reduce the humidity rate, freshly cut wood has to be dried for 8 to 18 months under shelter or for 8 to 30 months outdoors. This period of drying can be perceived as a constraint for non-sedentary populations, although the mobility pattern of Palaeolithic groups does not totally exclude such a procedure. One can indeed advance the hypothesis of a real anticipation and an intended stocking of wood between two occupational periods. However, the main question is not to know if this was possible but rather if such a procedure was useful as dead wood is easy to collect and naturally presents a very low rate of humidity. The variety of alteration states found in dead wood allowed a quick gathering of wood adapted for every type of use. Driftwood, easily dried in a hearth within some hours, can be added to this category. Could the collection of dead wood be combined with a second sorting based on the properties of a particular species? The determination of a species from an isolated dead twig can be difficult, but for an experienced observer for whom the identification moreover presented an interest, this effort might have been done with the help of precise criteria. Did this supplementary effort present an interest when it was possible to adapt the fuel to the intended use of the hearth by skillfully managing the morphology and the phonological and physiological state of the wood? Besides, the variability of combustion properties of a same species, due to caliber, humidity and alteration of the wood, is such that the collection would essentially have concentrated on the state of the stubs. The gathering of dead or floated wood is thus a practice that almost excludes the selection of species.

The management of fire wood and site functions

The various economic approaches in prehistoric research now allow us to distinguish different types of sites, for example hunting camps, seasonal base camp etc. Was the management of firewood in a given territory related to the function of the sites? In seasonal camps, where a variety of activities were represented, it is possible to imagine the felling and storing of carefully selected wood. On the contrary, in specialized settlements, occasionally used by small groups, the gathering of dead wood, *i.e.* without felling and drying, appears to be more adapted. This hypothesis was verified in la Combette (Mousterian, France). All the studies carried out at this site converge towards its definition as a hunting camp, occupied only sporadically (Texier *et al.* 1998). In this shelter, all the charcoal pieces from level E present anatomical features characteristic of alterations occurring in dead wood. This reflects the preferential gathering of dead wood (Théry-Parisot, 1998, 2001). The same phenomenon is observed in level 14 (early Aurignacian) of the Abri Pataud shelter (Lower Palaeolithic, Dordogne, France). On the contrary, in Les Canalettes (Mousterian, France) that seems to be a seasonal base camp (Meignen, 1993), the proportion of dead wood is low (Théry-Parisot, 1998; Théry-Parisot and Meignen, 2000). If a correlation between the management of firewood and the function of sites really exists, the modes of acquisition and management of wood resources could be diagnostic of a site's function.

Choice of wood and plant environment

Availability of wood resources

One can also wonder about the influence of the available wood resources on the management of firewood. Shackelton and Prins think that in an environment presenting less floristic diversity, the collection was more selective, due to increased constraints (Shackelton and Prins, 1992). The systematic selection of one or two species, either by felling trees or by gathering dead wood, leads to an extension of the catchment area and consequently to an increase in the effort and time invested. This effort will be even greater if the biomass is restricted and if only dead wood is collected. From a rational point of view, one is tempted to apply "the law of the least effort" to the gathering of firewood. However, the research on current traditional societies show that what we consider logical and rational might not have been seen in the same way by our ancestors (Pétrequin, 2000). If the selection of a particular species presented some interest to a prehistoric population, it is likely that the scarcity of this species and the distance to be crossed in order to find it did not constitute a real obstacle. For the moment, the extensive anthraco-analysis carried out at Palaeolithic sites does not allow us to demonstrate a correlation between the selection of a species and its availability in the surrounding vegetation cover.

Wooden scarcity and the use of other types of fuel

There is a recurrent idea according to which the rarefaction of wood resources during the coldest periods of the Palaeolithic led to the use of other types of fuel. It seems today that this type of binary interpretation is partially erroneous. First, the notion of abundance or scarcity of wood should be used with caution as it is related to the duration of an occupation and a group's needs of energy. Even a very poor environment can be a land of abundance for a small group of hunters. We have been able to show that the use of brown coal (lignite) and animal bones as fuel in two Palaeolithic rock-shelters did not ensue from a scarcity of wood resources. In the Mousterian shelter of the Canalettes

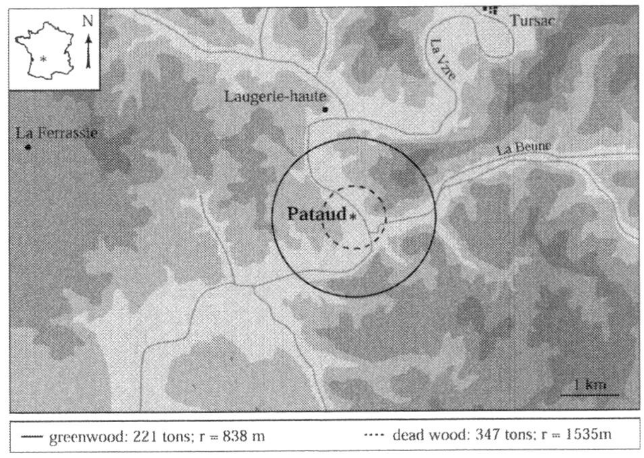

Figure 3: Optimal radius of the catchment area.

(Larzac, France), the use of brown coal occurs during a temperate period of the Pleistocene when wood resources were plentiful (Théry *et al.* 1995, 1996; Vernet, 1994; Théry-Parisot, 1998; Théry-Parisot and Meignen, 2000). In the ancient Aurignacian levels of Pataud (Dordogne, France) we observe a massive use of animal bones as fuel at the same time as a very weak exploitation of wood. The estimation of the available biomass, based on the anthraco-analysis and the simulation of a group's energy needs, show that even for an intense occupation in an environment poor in woody species, the gathering of wood was practiced within a radius of 1500 m (fig. 3). Even if the area is not circular, the distances are not very important and anyhow lower than the distances covered for the gathering of other raw materials (Théry-Parisot, 1998, 2001, in press). It is therefore necessary to be cautious when establishing a correlation between the scarcity of wood and the use of other types of fuel.

Firewood and palaeoecology

The problem of wood gathering leads us to examine the palaeoecological significance of carbonized wood remains. It seems evident that the modes of gathering during the Palaeolithic have an influence on the palaeoecological significance of charcoal. Nevertheless, we have to admit that this influence does not always distort the image that we obtain of the vegetation.

We know that the prehistoric populations had the capacity to select species according to their properties. As fires were used for different purposes, the choice of wood did not focus on one species but on a range of species suited for various activities. Therefore, in a seasonal camp where a number of different activities took place, almost every type of wood would have been collected. Residual charcoal is therefore likely to represent a quite accurate picture of the surrounding woody vegetation. On the contrary, in a specialized site, where the use of hearths was less diversified, we can expect an over-representation of some preferred species. However, in such a seasonal camps we believe that dry wood was preferentially gathered, no matter the species. The choice of wood for a specific use was made on the basis of its morphology and state (phonological and physiological). As any species is likely to present the looked-for form and alteration state (Gerfaud, 1991) no conscious selection of species would have taken place at temporary campsites where the use of dry wood prevailed. Moreover, a combination of different strategies of firewood gathering - the selection of particular species on the one hand and the selection of wood presenting specific morphological or physiological characteristics on the other - would yield a good qualitative synthesis of the environment.

However, in some cases the modes of gathering can cause a distortion to the vegetation picture. For example, allochthonous species can be introduced into assemblages of indigenous species when driftwood from far away is gathered (Fain and Rigolio, 1996). A very short occupational period of a site is the main cause of distorsion to our knowledge of the plant environment. When the occupation is short, the forest surface exploited for wood is restricted and the chances for the charcoal analysis to reflect the vegetation cover are weak. By using rigorous sampling methods and by the careful examination of the results, it is possible to estimate the palaeoecological significance of charcoal.

The ecological significance of properly sampled charcoal assemblages is reinforced by three arguments (Chabal, 1991, 1992, 1994, 1997):
- the importance of the species spectrum: when the number of species is high, the chances that a specific choice was made are low (not forgetting, of course, the climatic context knowing that few woody species were present during the coldest periods of the Palaeolithic);
- the recurrence of the data: it is important to compare, when possible, the number of taxa found in several, similar archaeological contexts (for example at two contemporary sites);
- the coherence of the results (qualitative and quantitative) compared to present-day ecological situations.

Finally, it is recommended to confront, if possible, the results of the anthraco-analysis with those of other types of palaeoenvironmental analysis.

By following these methodological lines, it has been shown that the anthraco-analyses carried out at Paleolithic sites do supply reliable paleoecological information (Badal Garcia, 1990, 1992; Badal Garcia *et al.* 1994; Bazile-Robert, 1979, 1980a, 1980b, 1980c, 1981, 1983; Figueiral, 1990, 1995a, 1995b, 1996; Heinz 1995; Heinz and Thiébault, 1998; Heinz and Barbaza, 1998; Krauss-Marguet, 1981; Machado-Yanes *et al.* 1997; Rodriguez Ariza, 1992; Théry-Parisot, 1998, 2000; Thiébault, 1995, 1997, 1999; Thiébault and Bintz, 1997; Thiébault *et al.* 1995; Thiébault and Vernet, 1992; Uzquiano, 1992; Vernet, 1994; Vernet and Thiébault, 1987).

Conclusion

For a long time, the gathering of firewood was thought to have been based on the choice of particular species. It now seems clear that this idea is too simplistic, especially as the species alone does not determine the combustion properties. Moreover, the choice of a specific type of wood depends on a number of variables such as availability, location, facility of collection, size and caliber, species, humidity rate and,

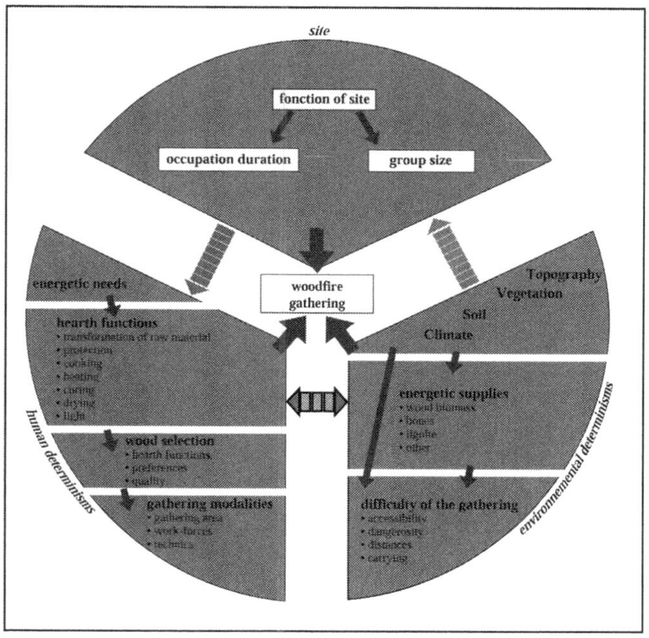

Figure 4: The system of firewood gathering.

finally, cultural preferences of various groups. The modes of wood gathering in the Palaeolithic reflect the specificity of this period in terms of means, life-styles and needs. The gathering of wood was certainly dependent on the specific needs of every group. The purposes and the modes vary according to numerous parameters (fig. 4): the "settlement" factor, a level of cultural "determinism", a level of environmental "determinism". It thus seems vain to establish a general model that could be applied to all sites. Moreover the Palaeolithic does not constitute a technical continuum but the environment underwent numerous structural modifications, in terms of vegetation changes. Every site presents its own specificity and every group its own identity. It is therefore necessary to admit that during the Palaeolithic, there was not one strategy of firewood gathering but a variety of practices, maybe as numerous as the studied sites.

References

ALIX C. 1998 - Provenances et circulations des bois en milieu arctique : quels choix pour les Thuléens? *Revue d'Archéométrie*, n°22, p. 11-22.

AMY L. 1991 - Les bases physico-chimiques de la combustion de la cellulose et des matériaux ligneux. *Cahiers du Centre Technique du Bois*, n°45, série 1.

ANDERSON-GERFAUD P., HELMER D. 1987 - L'emmanchement au moustérien. In : Stordeur D. (dir.), La main et l'outil, manches et emmanchements préhistoriques. Table ronde CNRS, Lyon 26-29 nov. 1984. *Travaux de la Maison de l'Orient* n°15, p. 37-46.

BADAL GARCIA E. 1990 - Méthode de prélèvement et paléoécologie du Néolithique d'après les charbons de bois de la Cova de les Cendres (Alicante, Espagne). In: Hackens T., Munaut A.V., TILL C. (éd.), *Wood and Archaeology: Bois et archéologie: First European Conference*, Louvain-la-Neuve, October 2nd-3rd 1987, PACT, 22. Louvain-la-Neuve.

BADAL-GARCIA E. 1992 - L'anthracologie préhistorique : à propos de certains problèmes méthodologiques. *Bulletin de la Société Botanique Française*, n°139, p. 167-189.

BADAL GARCIA E., BERNABEU AUBAN J., VERNET J.-L. 1994 - Vegetation changes and human action from Neolithic to Bronze age in Alicant, Spain (7000-4000 BP) based on charcoal analyses. *Vegetation History and Archaeobotany*, n°3, p. 155-166.

BAZILE-ROBERT E. 1979 - *Flore et végétation du sud de la France pendant la dernière glaciation d'après l'analyse anthracologique*. Thèse de IIIème cycle, Université des Sciences et Techniques du Languedoc, Montpellier II, 154 p.

BAZILE-ROBERT E. 1980 (a) - Flore et végétation des gorges du Gardon à la moyenne vallée de l'hérault, de 40000 à 9500 BP, d'après l'anthracologie. *Paléobiologie continentale*, t.XII, fasc.1, p. 79-90.

BAZILE-ROBERT E. 1980 (b) - La Baume de Vallorgues Analyse anthracologique. *Études Quaternaires Languedociennes*, Cahier n°1, p. 15-18.

BAZILE-ROBERT E. 1981 - Flore et végétation en Bas-Languedoc à la fin du Pléistocène et au début de l'Holocène d'après l'anthracoanalyse. *Études Quaternaires Languedociennes*, Numéro spécial, Excursion de l'AFEQ, 28-30 mai 1981.

BAZILE-ROBERT E. 1983 - La baume de Montclus (Gard), étude anthracologique, *Études Quaternaires Languedociennes*, n°3, p. 25-32.

BEYRIES S. 1990 - Problems of interpreting the fonctional results for ancient periods. Aun 14. *Proceeding of the International Conference on Lithic Use-wear Analysis*, 15th-17th February 1989 in Upsala, Sweden, p. 71-76.

BEYRIES S. 1993 - Analyse fonctionnelle de l'industrie lithique du niveau CA : rapport préliminaire et directions de recherches. In : Tuffreau A. (dir.), Riencourt-lès-Bapaume (Pas-de-Calais). Dossier de *Documentation Archéologique Française*, n°37, p. 53-61.

BEYRIES S., BOËDA E. 1983 - Étude technologique et fonctionnelle sur des "éclats débordants" de Corbeheim (Pas-de-Calais) *Bulletin de la Société Préhistorique Française*, t. 80, fasc. 9, p. 275-279.

BEYRIES S., HAYDEN B. 1993 - L'importance du travail du bois en Préhistoire. In : *Traces et fonction : les gestes retrouvés*. Colloque international de Liège. Éditions ERAUL, vol. 50, p. 284-315.

BRIANE D., DOAT. J. 1985 - *Guide technique de la carbonisation*. La fabrication du charbon de bois. Édisud, Aix-en-Provence, 180 p.

C.T.B.A. (ed.) 1973 - *Le chauffage au bois*. Cahier n°74, Paris, 13 p.

C.T.B.A. (ed.) 1977 - *Conseils pratiques pour le séchage des bois*. Cahier n°56, Paris, 13 p.

CHABAL L. 1991 - *L' homme et l' évolution de la végétation méditerranéenne, des âges des métaux à la période romaine: recherches anthracologiques théoriques, appliquées principalement à des sites du Bas Languedoc*. Thèse de IIIe cycle, Université de Montpellier II.

CHABAL L. 1992 - La représentativité paléo-écologique des charbons de bois archéologiques issus du bois de feu. *Bulletin de la Société Botanique Française*, n°139, p. 213-236.

CHABAL L. 1994 - Apports récents de l'Anthracologie à la connaissance des paysages passés : performances et limites. *Histoire et Mesure*, n°11, fasc. 3/4, p. 317-338.

CHABAL L. 1997 - Forêts et sociétés en Languedoc (Néolithique final, Antiquité tardive). L'anthracologie, méthode et paléoécologie. *Documents d'Archéologie Française*, n°63, 189 p.

CHABAL L., FABRE L., TERRAL J.-F., THERY-PARISOT I. 1999 - L'Anthracologie. In : Alain Ferdière (dir.) *La Botanique*, collection "Archéologiques", Editions Errance, 208 p.

ESCUDIE J.-C. 1989 - *Analyse théorique et expérimentale du comportement thermique de générateurs de chaleur à bois*. Thèse de L'École Normale Supérieure des Mines de Paris, 235 p.

FAIN I., RIGOLIO F. 1996 - *Identification anatomique et provenance des bois rejetés par la mer entre Maguelone et les Aresquier (Hérault) dans la semaine du 9 octobre 1995*. Mémoire d'initiation à la recherche, Université de Montpellier II, 25 p.

FIGUEIRAL I. 1995 - Charcoal analysis and the History of *Pinus pinaster* (cluster pine) in Portugal. *Review of Paleobotany and Palynology*, n°89, p. 441-454.

FIGUEIRAL I. 1995 - Charcoal analysis for environmental change during the interval late Bronze Age to Roman, at the archaeological site of Castro de Penice, N.W. Portugal. *Vegetation History and Archaeobotany*, n°4, p. 93-100.

FIGUEIRAL I. 1996 - Wood ressources in north-west Portugal: their availability and use from the late Bronze Age to the Roman period. *Vegetation History and Archaeobotany*, n°5, p. 121-129.

FIGUEIRAL I. 1990 - *Le Nord-Ouest du Portugal et les modifications de l'écosystème, du Bronze final à l'époque romaine, d'après l'anthracoanalyse de sites archéologiques*, Thèse de doctorat, Montpellier II, 203 p.

GARFANT J. (ed.) 1991 - *Atlas des forêts de France*. Paris, J.-P. Monza ed., 240 p.

HAYDEN B. 1979 - *Palaeolithic reflections: lithic technology and ethnographic excavation among aborigines studies*. Camberra, Humanities Press, 181 p.

HEINZ C. 1995 - Quelques exemples Pyrénéens illustrant les transformations du paysage végétal des quinze derniers millénaires, p. 79-85. In : *Cultures i Medi de la Prehistoria a l'Edat Mitjana. 20 anys d'Archeologia Pirinenca*, Homenatge al Professor J. Guilaine. X Colloqui Internacional d'Archeologia de Puigcerda. Del 10 al 12 novembre de 1994, Puigcerda i Osseja.

HEINZ C., THIEBAULT S. 1998 - Characterisation and palaeoecological significance of archaeological charcoal assemblages during Late and Postglacial phases in southern France. *Quaternary Research*, n°50, p. 56-68.

HEINZ C., BARBAZA M. 1998 - Environmental changes during the Late Glacial and Post-Glacial in the central Pyrenees (France): new charcoal analysis and archaeological data, *Palaeobotany and Palynology*, n°104, fasc. 1, p. 1-17.

KRAUSS-MARGUET I. 1981 - Analyse anthracologique du gisement post-glaciaire de La Poujade (Millau, Aveyron). *Paleobiologie Continentale*, t. XII, fasc. 1, p. 93-110.

KAUMAN W.G. 1983 - Quelques aspects pratiques du séchage du bois. In: *Courrier de l'Exploitant et du Scieur*. Études du CTBA, p. 1-11.

LEVI-STRAUSS C. 1962 - *La pensée sauvage*. Paris, Plon, 389 p.

MACHADO YANES M. C., ARCO AGUILAR M. C., VERNET J.-L., OURCIVAL J.-M.. 1997 - Man and vegetation in northern Tenerife (Canary Islands, Spain), during the prehispanic period, based on charcoal analysis. *Vegetation History and Archaeobotany*, n°6, p. 187-195.

MEIGNEN L. 1993 - L'abri des Canalettes. Un habitat moustérien sur les Grands Causses (Nant, Aveyron). *Monographie du CRA, n°10*, Paris, CNRS Éditions, p. 63-67.

NICHOLSON P.-H. 1981 - Fire and the Australian Aborigine - an enigma. In: *Fire and the Australian biota*, ed. Gill A.M., Groves R.,H., Noble I.,R., p.,55-76. Camberra: Australian Academy of Science.

OLIVE M., TABORIN Y. 1989 - *Nature et fonction des foyers préhistoriques*, Actes du coll. International de Nemours, 12-13-14 mai 1987. Mémoires du Musée de Préhistoire d'Ile de France, n°2, APRAIF (éd.), 334 p.

PETREQUIN P., PETREQUIN A.-M., WELLER O. 2000 - Cuire la pierre et cuire le sel en Nouvelle Guinée : des techniques actuelles de régulation sociale. In: P. Pétrequin, P. Fluzin, J. Thiriot, P. Benoit (ed.), *Arts du feu et productions artisanales*. Actes des XXe Rencontres Internationales d'Archéologie et d'Histoire d'Antibes, Octobre 1999.

SHACKELTON C.-M., PRINS F. 1992 - Charcoal analysis and the "Principle of Least Effort". A conceptuel model. *Journal of Archaeological Science*, n°19, p. 661-637.

SMART T.-L., HOFFMAN E.-S. 1987 - Environnemental Interpretation of Archaeological Charcoal. In: C.A. Hastorf and V.S. Popper (eds), *Current Paleoethnobotany: Analytical methods and cultural interpretations of archaeological plant remains*. Univ. Chicago Press, Chicago and London, p. 167-205.

TEXIER P.-J., BRUGAL J.-P., LEMORINI C., WILSON L. 1998 - Fonction d'un site du Paléolithique moyen en marge d'un territoire : l'abri de La Combette (Bonnieux, Vaucluse). In : *Economie préhistorique: les comportements de subsistance au Paléolithique*, XVIIIe Rencontres Internationales d'Archéologie et d'Histoire d'Antibes, Éditions APDCA, Juan-les-Pins, p. 325-348.

THERY-PARISOT I. 1998 - *Économie du combustible et Paléoécologie en contexte glaciaire et périglaciaire, Paléolithique moyen et supérieur du sud de la France (Anthracologie, Expérimentation, Taphonomie)*. Thèse de doctorat, Université de Paris I, 500 p.

THERY-PARISOT I. 2001 - Économie des combustibles au Paléolithique. *Dossier de Documentation Archéologique*, n°20, CNRS Éditions, 195 p.

THERY-PARISOT I. sous presse - Fuel management (bone and wood) during the lower Aurignacian in the Pataud rock shelter (Lower Palaeolithic, Les Eyzies de Tayac, Dordogne, France). Contribution of experimentation. J*ournal of Archaeological Science*.

THERY I., GRIL J., VERNET J.L., MEIGNEN L., MAURY J. 1995 - First use of coal. *Nature*, n°373, p. 480-481.

THERY I., GRIL J., VERNET J.-L., MEIGNEN L., MAURY J. 1996 - Coal used at two prehistoric sites in southern France: Les Canalettes (mousterian) and Les Usclades (mesolithic). *Journal of Archaeological Science*, n°23, p. 509-512.

THERY-PARISOT I., MEIGNEN L. 2000 - Économie des combustibles dans l'abri moustérien des Canalettes, de l'expérimentation a la simulation des besoins énergétiques. *Gallia Préhistoire*, n°42, p. 45-55.

THIEBAULT S. 1988 - L'homme et le milieu végétal. Analyse anthracologique de six gisements des Préalpes au tardi et postglaciaire. *Documents d'Archéologie Française*, 15, 110 p.

THIEBAULT S. 1995 - Dégradation et/ou substitution du milieu végétal au Néolithique en Provence. In : *L'homme et la dégradation de l'environnement* : XVe Rencontres Internationnales d'archéologie et d'histoire d'Antibes, p. 185-194.

THIEBAULT S. 1997 - Early-Holocene vegetation and the human impact in central Provence (Var, France) Charcoal analysis of the Baume of Fonbregoua. *The Holocene*, 7, 4 p. 343-349.

THIEBAULT S. 1999 - L'Aulp du Seuil (Saint-Bernard-du-Touvet, Isère, France), un site d'altitude du Mésolithique et du Néolithique ancien, études préliminaires. In : Bintz P. et Thevenin, A. (ed.), *L'Europe des derniers chasseurs*. Paris, Comité des travaux historiques et scientifiques, p. 611-616.

THIEBAULT S., BINTZ, P. 1997 - Les grottes du Vercors et des massifs subalpins : relations homme et milieu végétal. Documents d'archéologie en Rhône-Alpes, 15, p. 23-38. *Dynamique du paysage*. Table Ronde, Lyon, 1995.

THIEBAULT S., VERNET J.-L. 1992 - Végétations méditerranéennes et civilisations préhistoriques : le cas de Font - Juvénal. *Bulletin de la Société Botanique Française*, n°139, p. 441-450.

THIEBAULT S., GIRARD M., BUI THI MAI 1995 - Le milieu végétal du Tardiglaciaire à l'holocène d'après les analyses de charbons de bois et de pollens. In : *Épipaléolithique et Mésolithique en Europe : préhistoire et quaternaire en Vercors* , livret guide de l'excursion. Congrès international UISPP XIIe commission. Grenobre 1995. p. 32-37.

TRABAUD L. 1976 - Inflammabilité et combustibilité des principales espèces des garrigues de la région méditerranéenne. *Eocologia Plantarum*, t. 11, n°2, p. 117-136.

TRABAUD L. 1989 - *Les feux de forêts. Mécanismes, comportements et environnement*. France- Sélection (ed.), 278 p.

UZQUIANO P. 1992 - *Recherches anthracologiques dans le secteur Pyrénéo-Cantabrique*. Thèse de 3ème cycle, Université de Montpellier II.

VERNET J.-L. 1973 - Étude sur l'histoire de la végétation du sud-est de la France au Quaternaire, d'après les charbons de bois principalement. *Paléobiologie Continentale*, Montpellier, n°4, fasc. 1, 90 p.

VERNET J.-L. 1993 - Analyse anthracologique de l'abri des Canalettes. In : L'abri des Canalettes - *Un habitat moustérien sur les grands Causses (Nant, Aveyron)*, Meignen L. (dir.). Monographie du CRA, n°10, Paris, CNRS Éditions, p. 63-67.

VERNET, J.-L., THIEBAULT S. 1987 - L'analyse anthracologique : une approche du paléoenvironnement végétal, In : *Géologie de la Préhistoire, méthodes, techniques, applications*, p. 619-630.

"They plant what they will burn" – the impact of traditional burning on the landscapes of Wallis and Futuna (western Polynesia) – proposals for an ethnoarchaeological and palaeoenvironmental study

Hélène Guiot

15, rue Pierre Moulie - 94200 Ivry-sur-Seine, France.
E-mail: hguiot@hotmail.com

Abstract

The purpose of this article is to establish the basis of a research project that will associate paleo-environmental studies with the ethnoarcheologic approach. In Wallis and Futuna, as anywhere else in Polynesia, inconsiderate deforestation induced by slash-and-burn farming and integrated with farming practices is thought to have acompanied human settlement. Horticultural overfarming due to population growth is considered to be the cause of the irreversible formation of a damaged fern land "toafa". This theory is based especially on the inappropriate analysis of contemporary observations. Indeed, recent behaviours that appeared trough contact with the Western world have not been distinguished from pre-European practices. The ancient religious concept of the functioning of the insular ecosystem still exist. Insulars view the functioning of the ecosystem as a whole, and forests play an essential role in it. Practices aiming at regenerating the forest cover went along with these traditional representations. This management of space thus makes large-scale damaging of original botanical landscapes unlikely and goes against the idea of the fern land being linked to human activity. Religious activities aiming at ensuring insular vitality have left archaeological traces (for example, a wall enclosing the taboo forest of Wallis, a ritual site called Lalolalo located in Futuna). Further research will enable us to determine the history of these structures and of the forest cover.

Key-words: *Ethnoarchaeology, palaeoenvironment, slash and burn, forest, ecosystem, Wallis, Futuna, Polynesia.*

Résumé

Cet article a pour but de poser les bases d'un projet de recherche qui intégrera à l'approche ethnoarchéologique des études paléoenvironnementales. A Wallis et à Futuna, comme dans toute la Polynésie, le peuplement se serait accompagné d'une déforestation inconsidérée, induite par les brûlis intégrés aux pratiques culturales. Une surexploitation horticole, due à l'augmentation de la population, aurait provoqué la formation irréversible d'une lande à fougères dégradée "toafa". Cette théorie est notamment basée sur l'analyse inadéquate d'observations actualistes. En effet, les comportements récents, apparus en raison du contact avec l'Occident, n'ont pas été distingués des conduites pré-européennes. En dépit de l'implantation du catholicisme, les conceptions religieuses anciennes sur le fonctionnement de l'écosystème insulaire ont perduré. Les insulaires envisagent le fonctionnement de cet écosystème dans sa globalité, la forêt y occupant un rôle fondamental. Ces représentations traditionnelles s'accompagnaient de pratiques visant à la protection et la régénération du couvert forestier. Cette gestion de l'espace rend donc improbable une dégradation de grande ampleur des formations végétales originelles, et va à l'encontre d'une origine anthropique du "toafa". Les activités religieuses visant à assurer la vitalité insulaire ont laissé des traces archéologiques dans le paysage (mur enserrant la forêt taboue de Wallis, site cérémoniel Lalolalo à Futuna). De plus amples recherches permettront de déterminer l'histoire de ces structures, ainsi que celle du couvert forestier.

Mots-clés : *Ethnoarchéologie, paléoenvironnement, brûlis, forêt, écosystème, Wallis, Futuna, Polynésie.*

Inconsiderate deforestation induced by slash-and-burn farming and integrated in traditional farming practices is thought to have accompanied human settlement in Wallis and Futuna, as anywhere else in Polynesia. Moreover, on the islands –which seem to us to be isolated areas–, progressive and inevitable growth of the initial population is said to have led, among other things, to horticultural overfarming. This brought the irreversible formation of a highly damaged moor, characterized by fern (*Dycranopteris linearis*) and called "toafa" by Wallisians and Futunians.

This theory is based especially on the inappropriate analysis of contemporary observations. Indeed, recent behaviours adopted by some islanders –because of the contact with western models during the 19th century– have not been distinguished from pre-European ones.

Wallisians and Futunians' representations of their environment

In contrast to the assertion made above, however, the distinction between current and older behaviours can be established by considering the degree of coherence between observed horticultural and silvicultural practices and traditional representations of the environment.

For instance, despite the spread of Catholicism beginning in 1837, the ancient religious concepts of the functioning of the ecosystem still exist. Wallisians and Futunians view the functioning of this ecosystem as a whole, and forests play an essential role in it: the tall trees of the forest attract atmospheric humidity and change it into "ma'uli", the vital principle which allows all animate beings to grow and multiply. Animate beings "'ulufenua" include plants, animals and minerals.

A basin –the top of the mountain range in Futuna and a crater lake in Wallis– enables inhabitants to stock rain water, thus ensuring a regular and equal distribution of "mau'li" (vitality and fertility) resources, hence the necessity for Wallisians and Futunians to protect the forest.

On the whole, the ancient polytheist religion aimed at maintaining the fertility of insular ecosystems. Chieftaincy representatives were, and still are, guarantors of fertility thanks to their religious power, and used to be responsible for fertility through a system of taboos. In Wallis, the "ma'uli" basin, Lake Lalolalo, stands at the heart of a taboo forest called "vao tapu", from which all human intervention is banned, especially horticulture. Strong religious connotations are associated with this forest: it shelters numerous gods from the traditional religion who still arouse awe. In Futuna, all forests considered as ancient are subjected to taboos.

On both islands, the rules are imposed by the king and controlled by the chieftaincy.

Silvicultural and horticultural practices aiming at the regeneration of the forest cover

In regard to horticultural activities, these traditional representations are paired with practices aiming at protecting and regenerate cultivated areas of the forest cover, such as control of burn-over areas and preservation of certain ligneous species.

Horticulture in Wallis and Futuna, as elsewhere in the Pacific, is set in irrigated plantations such as taro fields, or pluvial ones, such as yam fields. The processing work chain of planting is slightly different from one island to another.

In Wallis, grubbing is conducted in the following manner: in a corner of the plantation, a fire of "fau" (*Hibiscus tiliaceus*) branches is stoked continuously. Fire-brands are taken from it and put around stocks which thus calcinate superficially; trees and shrubs lose their leaves, which enrich the soil, and they die in a few days without being cut down. Then yams are planted around the stumps. Ashes enrich the area, and the yam roots are supposed to attract soil humidity and enrich the soil by decaying; the trunks and the remaining branches are used later on as supports for yam props made of hard, healthy wood.

Horticulturists do not blindly destroy all plants during the grubbing. They save ligneous plants which can be eaten, like "mei" (*Artocarpus altilis*), "viteletele" (*Spondias dulcis*) or "kafika kai" (*Syzygium malaccense*); those which can be used for naval or domestic (like "pou muli", *Securinega flexulosa*) carpentry and carving; and those whose leaves, bark, flowers and fruits are used for pharmacopeia, or to make necklaces once intended for gods, like "mapa" (*Diospyros major*), "koli" (*Syzygium neurocalyx*) or "pua'ueva" (*Fagraea berteriana*).

At the end of a culture cycle, horticulturists often stimulate the new growth of the forest cover by planting "fau" (*Hibiscus tiliaceus*) or "lepo" (*Macaranga harveyana*) cuttings, famous for their easy regeneration and their quick growth. These plantings are conducted at new moon; according to local horticulturists, cuttings are full of sap and therefore grow better at this time. Even if fallowing is not long enough for a real forest to regrow back, new shrubs will allow a new grubbing (usually every 10 or 15 years).

In other words, Wallisian horticulturists think ahead: they plant what they will burn, as one of our interviewees said.

In Futuna, the course of grubbing varies a little. On the chosen plot, horticulturists cut down trees and shrubs and let them dry for two or three weeks. Then they gather the branches and the leaves in a pile and set them on fire. During the grubbing, the felling of trees is always directed towards the plot so that the edge of the forest will not be damaged and the dead burning trunks will not spread the fire towards the forest. Moreover, the trunks are laid on the ground rather than being maintained upright between other trees of the forest's edge so that they burn out with so little flame that the risks of fire spreading is limited. Combustion in piles rather than on the whole surface of the plot allows stumps to be preserved, which ensures a quick regeneration of plants due to the possibility of new shoots. Moreover, it is important for Futunian horticulturists to preserve the stumps, the more so because their sap, an important element of "ma'uli" spreads towards food-supplying plants set in between the trunks.

Futunian and Wallisian horticulturists have thus created a balanced horticultural ecosystem, as they are constantly careful not to damage it or the neighbouring forests. In both cases, it should be noted that slash-and-burn cultivation is controlled according to specific work chains, which reduce the global impact on the environment as a whole as much as possible. This technique also preserves utilitarian species (food plants, material plants...).

Spatial distribution of burn-over areas

In Futuna, as in Wallis, even though slash-and-burn cultivation is practiced on forest areas, it does not concern just any plot. A specific vocabulary classifies the ligneous groups according to age, thus indicating their presumed degree of fertility:
- "vao lelei": any plot likely to be cultivated;
- "vao mui" in Wallis and "lusa" in Futuna: retaking forestation of fallowing land (10 to 12 years); this type of plot is always included in the culture cycle;
- "vao matu'a" in Wallis and in Futuna: ancient forest, which is cultivated only if absolutely necessary.

In other words, only certain forests –retaking forests– are traditionally cultivated. Finally, some areas of the territory are never meant to be cultivated, like the "vao tapu" (taboo forest) in Wallis, as well as the forests neighbouring the "toafa", even luxuriant ones, in order to avoid spreading of the moor. Those different practices thus limit the hold of slash-and-burn farming over vegetation.

Formation of the "toafa"

In Wallis and Futuna there are large surfaces of deep ferrallitic soils which are located in interior plateaus (fig. 1) and which reveal a high degree of damage, with scattered and low vegetation, sometimes consisting only of fern cover (*Dicranopteris linearis*), the "toafa".

This fern land might have been formed because of inappropriate farming practices denounced by the usual model, *i.e.* ancient horticultural over-farming of forests.

The word "toafa" exists in the language of all the islands neighbouring Wallis and Futuna (Tonga, Samoa, Niue...) and has the same meaning. Moreover, the word "toafa" exists in the reconstituted proto-Polynesian language, which may have been used around 3000 BC ("upland area of leached, infertile soil, desert" -personal communication of the linguist Cl. Moyse-Faurie). This shows how old this type of landscape is and the extent to which it is integrated into representations and everyday life of the islanders.

In everyday life, Wallisians and Futunians cannot allow such a sterile environment to expand. That is why Futunian horticulturists do not plant next to the "toafa"; on the contrary, they put a crown of "vao matu'a" around it –a buffer zone called "talavao". The ban of all cutting down in this zone avoids the weakening, even temporary, of its dammingy functions. Futunians, when speaking French, say they have to protect ("protéger" in French) the "toafa" so that it does not expand. The chieftaincy forbids setting the "toafa" and the "talavao" on fire under this pretext. If such an event occurs, the individual who set the fire is punished. Of course this coercive method does not entirely preclude the possibility of accidental or criminal fires, but those are also fined.

Further, recent phytosociological and pedological studies confirm the natural, rather than human, origin of the "toafa". Its constitution is the result of the "damaging by fire [the damaging being not necessarily voluntary] of a more open and low vegetation than the dense, humid and sempervirent forest, which might have developed during drier periods in the Quaternary era" (Brisse and Hoff, 1990: 36) on the more damaged lateritical soils. Several arguments bear out this hypothesis:
- "toafa" has an ancient origin, since it is a rich floristic botanical landscape with numerous species such as *Dicranopteris linearis*, *Decaspermum fruticosum*, and *Mussaenda raiatensis*. On the whole, one counts, in Wallis and Futuna, 56 species characteristic of this floristic grouping.
- "toafa" is an original, more or less natural, landscape. It is essentially made up of indigenous plants –not ruderals, which grow naturally in environments damaged by man– and because there are practically no intermediate landscapes and no interpenetrations, between plots and "toafa".

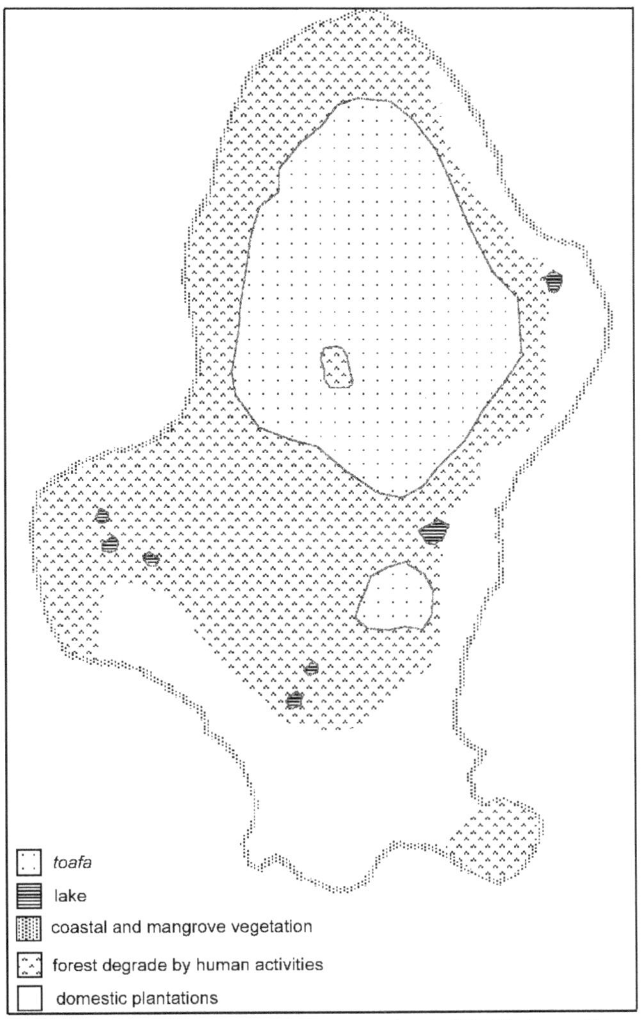

Figure 1: Schematic map vegetation of the island of Wallis (according to J. Barrau, 1963, 159)

On the geological level, Wallis is an emerged portion of an underwater volcano of the Tertiary era, very close to the beginning of the Quaternary era. In the North, the lava flows formed on the generally faded plane surfaces. Red deep ferrallitic soils, washed by tropical rainwater and acid humus are found there. It is in this zone of degraded soils that the "toafa" is located. Volcanic activity of an explosive nature recently resumed and has set up created a certain number of cones scories starting from a score of eruptive mouths (in the South and on part of the western coast of the island). It is mainly in the southern area of the island, around the craters, that one finds brown, youthful and rich soils.

Pedological data tell us that "toafa" is located on the oldest volcanic grounds, the ones which have suffered damage from tropical heat and humidity for the longest period of time. Human intervention, even if it started with the establishment of man, that is to say around 1000 BC in Wallis, cannot cause such consequences as dozens of meters of damaged soils to develop (M. Bourgeon).

"It is then clear that the "toafa" in Wallis is not the result of an agricultural over-farming," even if we can most probably proclaim that "man is a maintaining factor of "toafa"." (Brisse and Hoff, 1990: 36, 37).

Perspectives

Religious activities that aimed at ensuring insular vitality have left archaeological traces on the landscape. It is especially the case for the ceremonial site of Lalolalo on the Asoa plateau in Futuna and for the "kaupalua" (the wall enclosing the taboo forest) in Wallis. Moreover, research in this forest has enabled me to locate a group of structures, in which a series of clues tends to indicate that they might be the vestiges of a place of cult associated with rituals of the preservation of fertility.

For the reason cited above and, within the framework of the comprehension of modalities of human establishment and its consequences on insular territory, we wish to plan a project which will associate ethnoarchaeology with anthracological and pedological analyses, one which will derive its inspiration from the work already carried out in tropical areas (as, for example, the work of Ch. Tardy, 1998). These studies will attempt to obtain informations about the constitution of the past forest cover and its evolution, the history of the forest soils (posing such question as: did they undergone episodes of fire, slash-an-burn farming? When did the "vao tapu" forest begin?) and of the soils of the "toafa". Excavations will jointly be carried out on the archaeological structures implanted in the forest belts in order to try to determine the age and the history of these structures, as well as their functional relations.

AKNOWLEDGEMENT
The author would like to thank M. Benguigui for the french to english translation.

References

BARRAU J. 1963 - L'agriculture des îles Wallis et Futuna. *Journal de la Société des Océanistes*, n°19, p. 159-171.

BOURGEON G. 2000 - Sols : des questions à creuser. *Te Fenua Fo'ou*, n°221, Wallis, p. 7.

BRISSE H., HOFF M. 1990 - Contribution à l'étude des groupements végétaux des îles Wallis et Futuna. *Documents phytosociologiques*, N. S., Vol. XII, Camerino.

GUIOT H. 1998 - Forêt taboue et représentations de l'environnement à 'Uvea (Wallis). Approche ethno-archéologique. *Journal de la Société des Océanistes*, n°107, p. 179-198.

GUIOT H. 1999 - Rapport de terrain 1999, Wallis et Futuna, Octobre 1999, in *Sauvegarde du patrimoine sur les îles Wallis et Futuna*, Année 1999, n°7, IRD, Nouméa.

GUIOT H. 2000 - Gestion traditionnelle des espaces forestiers à Futuna (Polynésie occidentale) : Contenu idéel et pratiques associés, in *Journal de la Société des Océanistes*, n°110, p. 19-33.

KIRCH P.V. 1989 - *Cultural Adaptation and Ecology in Western Polynesia: an ethno-archaeological study*. Yale University, Ph. D. Thesis, U.M.I, Michigan.

KIRCH P.V. 1994 - *The Wet and the Dry: Irrigation and Agricultural Intensification in Polynesia*. Chicago, The University of Chicago Press.

TARDY C. 1998 - *Paléoincendies naturels, Feux anthropiques et Environnements forestiers de Guyane française du Tardiglaciaire à l'Holocène récent. Approches chronologique et anthracologique*. Thèse de Doctorat, Montpellier II, 400 p.

Dendroarchaeology and charred wood from the late Neolithic site of Pléchâtel (Brittany, Western France). A chronological, palaeoenvironmental and architectural contribution

Vincent Bernard and Cristel Thibaudeau
CNRS-UMR 6566, Laboratoire d'Anthropologie/Archéométrie,
Université de Rennes 1, Campus de Beaulieu, Bât. 24-25, F-35042 Rennes cedex, France.
E-mail: vincent.bernard@univ-rennes1.fr

Abstract
A 3.5 hectare excavation, undertaken since 1992 at the late Neolithic site of "La Hersonnais" at Pléchâtel (Ille-et-Vilaine), has revealed four distinct, large houses. Two of them are particularly monumental, 100 m long with fortified fences and posts with a section of 80 cm. Building D, which seems to be the most recent, is also the least well preserved. The other buildings, A, B and C, had been destroyed by fire and the buried wood was thus more well preserved. Based on the analysis of approximately 1700 oak charcoal fragments, sampled exclusively from post holes, this archaeodendrological study exploits in an optimal way the information contained in the charred wood. By comparing data from archaeology (shape, depth and diameter of the post holes...), dendrology (physiological and anatomical characteristics of the wood, relative correlations, absolute dendrochronological dates...) and dendrochronology, we are able to follow, for the first time in a dry archaeological site, the simultaneous evolution of woodland potential and architecture between the XXVIIth and the XXVIth centuries BC in the Armorican massif.

Key-words: *Archaeology, Dendrology, Dendrochronology, Neolithic, Armorican massif.*

Résumé
La fouille extensive menée depuis 1992 à l'emplacement du site d'habitat néolithique final de "La Hersonnais" à Pléchâtel (Ille-et-Vilaine) couvre environ 3,5 ha. Elle a révélé quatre ensembles architecturaux distincts, parmi lesquels deux présentent, avec leur centaine de mètres de long, leur enceinte fortifiée et des poteaux pouvant atteindre 80 cm de section, un caractère monumental très marqué. L'ensemble qui apparaît comme le plus récent (ensemble D) est le moins bien conservé ; en revanche les bois des ensembles A, B et C incendiés ont été bien préservés dans leur partie enterrée. Sur la base d'environ 2000 échantillons de chêne issus exclusivement des trous de poteau, nous avons engagé une approche archéo-dendrologique visant une exploitation maximale de l'information contenue dans ces vestiges de bois carbonisés. Ainsi, en croisant les données archéologiques (forme, profondeur et diamètre des trous de poteau...) à celles issues de la dendrologie (caractères physiologiques et anatomiques des bois, calages relatifs et datations dendrochronologiques absolues...), on est en mesure de suivre, pour la première fois sur un site en milieu sec, l'évolution du potentiel forestier et de l'architecture à la charnière des 27e et 26e s. av. J.-C. dans le Massif armoricain.

Mots-clés : *Archéologie, dendrologie, dendrochronologie, Néolithique, Massif armoricain.*

Introduction

A first aim of this dendrochronological and dendrological study is to carry out tree-ring dating on archaeological structures by the use of charcoal remains. A second aim is, as indicated by the use of terms such as "dendrotypology" or "archaeodendrology" (Billamboz, 1987), to characterise the selection criteria for wood used for architectural elements. Finally, this paper provides information on the environmental context and the structure of surrounding woodlands.

We have deliberately limited this first study to oak timber (*Quercus* sp.) for the following reasons:
- the high proportion of oak (*c*. 90%) on the site compared to other taxa;
- the good preservation of the oak samples;
- the availability of local tree-ring chronologies and dendrological references for the species in Brittany (Bernard, 1998; Marguerie, 1992).

The other species will be studied later and have been grouped under the generic term of "whitewoods".

Presentation of the site

The site

The site of Pléchâtel is located approximately 30 km south of Rennes (Ille-et-Vilaine, Brittany) and consists of a large 3.5 ha late Neolithic settlement, excavated mainly between 1991 and 1999 by J-Y Tinevez (1995). The four large, rectangular buildings (the largest is 104 m long by 11 m large) are all surrounded by fences. Building A and B have monumental entrances with posts measuring up to 60-80 cm in diameter. A particularly defensive and ostentatious aspect is rendered to the site by the size and the organisation of its buildings. 128 post holes have provided remains of charred wood corresponding to the remnants of burnt architectural elements (fig. 1). All this emphasises the scientific importance of the Neolithic settlement, which is the only large building site from this period known and excavated in western France so far.

Relative chronology and radiocarbon dating

A stratigraphic study shows that building A is older than building C and that building B is older than building D. However, the lack of common features for buildings B and C (at the time of excavation) did not allow the elaboration of a general chronology of the site. The apparent similarity between A and B might suggest that a rather short period of time separates the two buildings. The technology and the design used in the construction of B seem however to be more elaborate than in building A, and thus probably later in date.

In order to establish the chronological framework of the site, several charcoal fragments from buildings A, B, C and D were selected for radiocarbon dating. The six ^{14}C dates obtained cover a period from 3000 to 2500 BC (tabl. 1) and therefore confirm that the archaeological structures of Pléchâtel belong to the late Neolithic period. However, given the wide interval of the ^{14}C dates it was not possible to situate the different buildings chronologically (fig. 2).

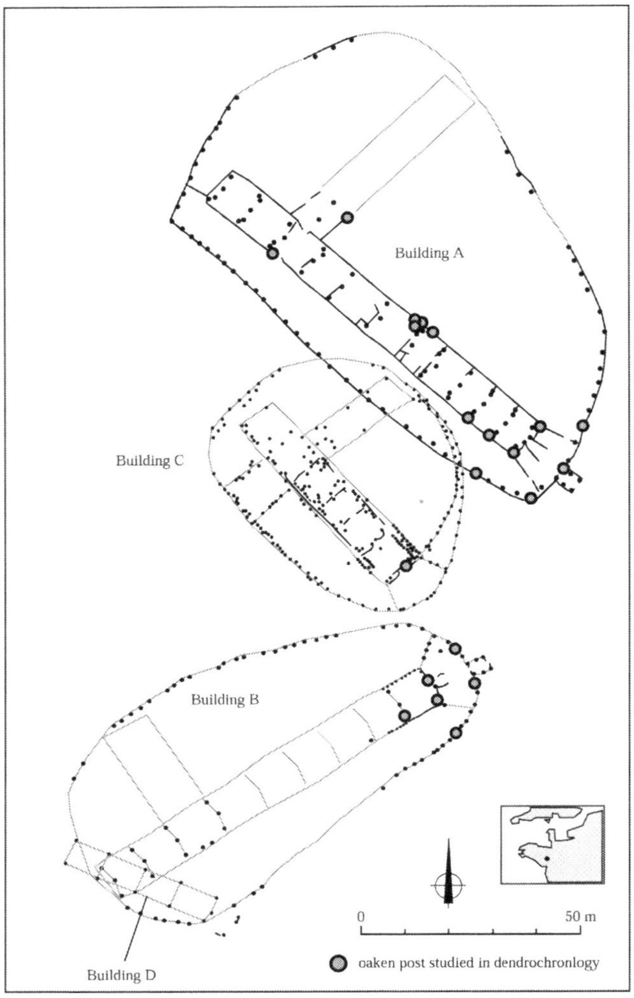

Figure 1: Pléchâtel, "La Hersonnais". The main structures of the late Neolithic settlement.

A relative dating of the structures has instead been suggested by the archaeologist on the basis of typological criteria (Tinevez, 1995). Accordingly, the oldest building seems to be structure A, followed in order by B, C and D.

Building	*lab. Code*	*conventional date*	*age interval* *
A	Gif.9119	4070±50 BP	[2859-2503 BC]
	Gif.9647	4290±50 BP	[3035-2698 BC]
	Gif.10880	4255±50 BP	[3015-2625 BC]
B	Gif.9905	4070±50 BP	[2865-2466 BC]
	Gif.9904	4090±50 BP	[2869-2484 BC]
	Gif.10881	4005±45 BP	[2560-2469 BC]
C	Gif.10882	4030±25 BP	[2585-2469 BC]
	Gif.10884	4150±50 BP	[2878-2585 BC]
	* confidence level : 95%		

Table 1: ^{14}C datings on charcoal fragments.

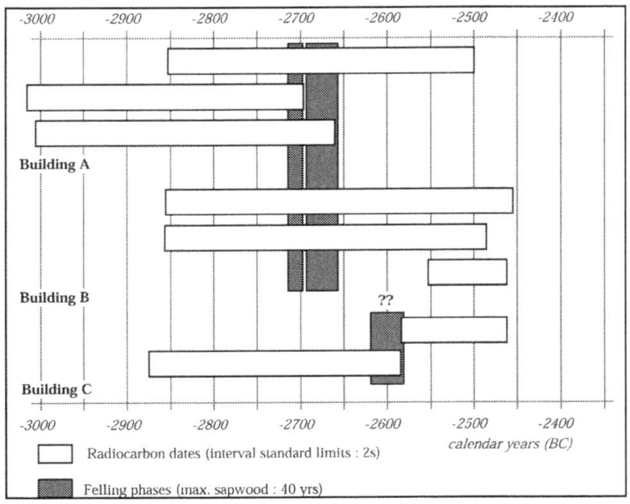

Figure 2: Relative positions of main radiocarbon dates. The tree-ring related felling phases are positioned in relation to the absolute chronology (BC).

Material and methods

Almost 1700 charcoal fragments, retrieved from the homogenous fill of post holes, have been studied. Additional charcoal remains were considered to be out of the scope of this study and were thus excluded.

First, the following dendrological observations were made (Marguerie, 1991):
- identification of the wood species
- growth type (fast, medium or slow)
- ring curves (weak or strong)
- ring numbers.

The section, cut and depth of the 128 charred posts (or their traces) were also recorded.

Secondly, samples displaying enough rings to be studied (*i.e.* more than 15-20) were put apart. The dendrochronological and dendrotypological method was then applied to the charred material. Each tree-ring measurement was carried out on a 1/100 mm basic scale after having obtained a transversal section by breaking or cutting the charcoal piece. The first stage of the dendrochronological study consisted in reconstructing the average radius per sample, *i.e.* restoring the growth of each tree used as a post ring by ring, from the pith to the bark. However, as the largest of the charcoal pieces measured less than 3 cm in diameter, a lot of measurements were required in order to reconstitute the increment of a 40 to 60 cm section tree. Some posts required up to 30 measurements before a partial radius could be reproduced. In this respect, a 1 mm threshold was very useful for the identifying of charcoals from the same trunk.

Only 36 posts out of 128 could be used for measurements and for a partial reconstruction of the annual growth pattern. Two samples presented a pith and two others had partial sapwood. The sapwood is important for dendrochronologists as it helps to determine felling dates and thus the different phases of building and restoration of Neolithic structures.

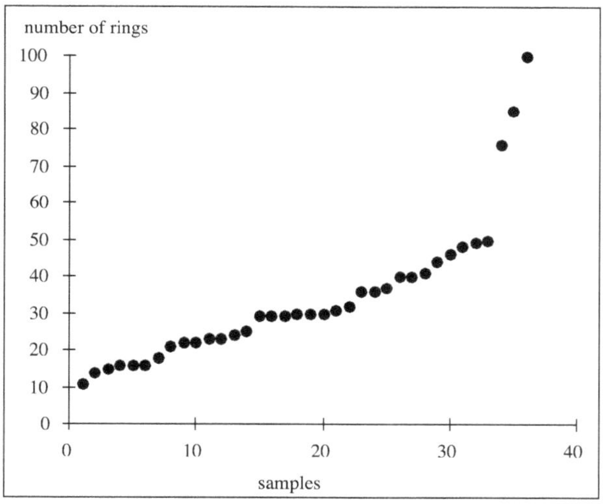

Figure 3: Total number of rings per sample.

European dendrochronologists use a standard test called the "Eckstein test" or "W test" (Eckstein, 1969) for statistical correlations. This coincidence test (from "Gleichläufigkeit"; Schweingrüber, 1988) measures the percentage of coincidences between the interval trends of two chronologies. Further, it identifies event years and provides evidence for similar growth trends. When the sequences are longer the cross-dating obtained is more reliable. This method is nevertheless limited by an 80-year statistical threshold, implying that results are not significant for sequences shorter than 80 years (Gassmann, 1996). The study of 36 tree-ring sequences from Pléchâtel produced an average 34 annual rings. Moreover, only 3 of the sequences presented more than 60 rings (fig. 3) and most synchronisations of individual series had therefore to be operated visually.

Results

Tree-ring dating

Only 13 of the 36 reconstructed tree-ring sequences have been synchronised so far. Some of the 23 remaining sequences might also be synchronised in accordance with our local tree-ring mean chronology. However, it has to be stressed once again that this work is limited by the small number of rings per sample.

By successive correlations of 13 individual series, a 123 year long mean chronology has been elaborated. This chronology represents the mean growth of the oak trees used for construction during the late Neolithic at Pléchâtel and can now be compared with other local or standard chronologies for this period.

In 1998 the excavation of a fossilised channel in the centre of Rennes (30 km north of Pléchâtel) brought to light about 90 subfossilised oak trunks. This find has provided us with one of the most important standard chronologies for the late Neolithic in western France (Aubry *et al.* 1999). The comparison of the tree-ring dating of the mean chronology at "La Hersonnais" with more relevant European standard chronologies is very instructive: the curves from Pléchâtel, Rennes and the Isle of Wight (Hillam, 1994) all show a characteristic peak of more than 1 mm and particular good years between 2727 and 2725 BC (fig. 4). While the correlation of the tree-ring dates between Brittany and Great-Britain seems evident, the links with the Paris basin or the Jura are more tenuous. The discrepancy in dating between these two neighbouring regions, western and northern France, undoubtedly reflects the prevailing geoclimatic differences; with an atlantic influence in the Armorican massif and a continental influence in the Paris basin.

Architectural and palaeoenvironmental data

According to the general chronology of the site, the first occupation is synonymous with the construction of building A and its fence, dated to between 2719 and 2699 BC. Indeed, the dendrochronology, the radiocarbon dating and the ceramic analysis all give similar results for this first phase. However, during this period (*c.* 2700 BC) the construction of building B equally takes place, a fact that completely changes the global understanding of the site based on archaeological evidence (Tinevez). According to the latter, building B succeeded A, at least by judging from the technical innovations and the lightening of B's superstructure. In fact, it has now been established that two social groups, with family links and similar cultures, might have lived and undertaken construction works at the site simultaneously. This changes the initial view of the site, as not one but two large collective houses seem to have coexisted around 2700 BC. This fact of the matter might have had both demographic and environmental consequences. If a population twice as big as the one we first thought lived in the settlement, the human influence on the landscape and in particular on the woodlands must have been more important.

During a less precise period than the previous (2695-2660 BC), restoration work was undertaken on building A and its fence. This first occupation of the site seems to have ended with a large conflagration (accidental or not?) sometime during the second part of the XXVIIth century.

During the above mentioned period, the average growth rate of the oak samples is similar for building A and B, *c.* 1.25 mm a year, which reinforces the idea of two contemporaneous structures. The growth rate further suggests that relatively dense forest stands, where the regeneration cycle had not yet been disturbed by humans, were exploited during the second half of the XXVIIth century (fig. 5).

Before the rebuilding of a new village, the old settlement must have disappeared completely. The soil recovered and the coppices, previously exploited for one or two generations, regenerated and could provide new building material. It is estimated that this would have taken maximum 50-80 years, which means that the site could have been reoccupied at the latest between 2620 and 2595 BC.

A wider range of wood types was used during the first construction phase of building C:
- trees with slow growth rates (1-1.5 mm per year),
- large trees ensuring forest regeneration (probably from the previous phase),
- trees with a much faster growth rate (nearly 4 mm per year) originate from coppice shoots.

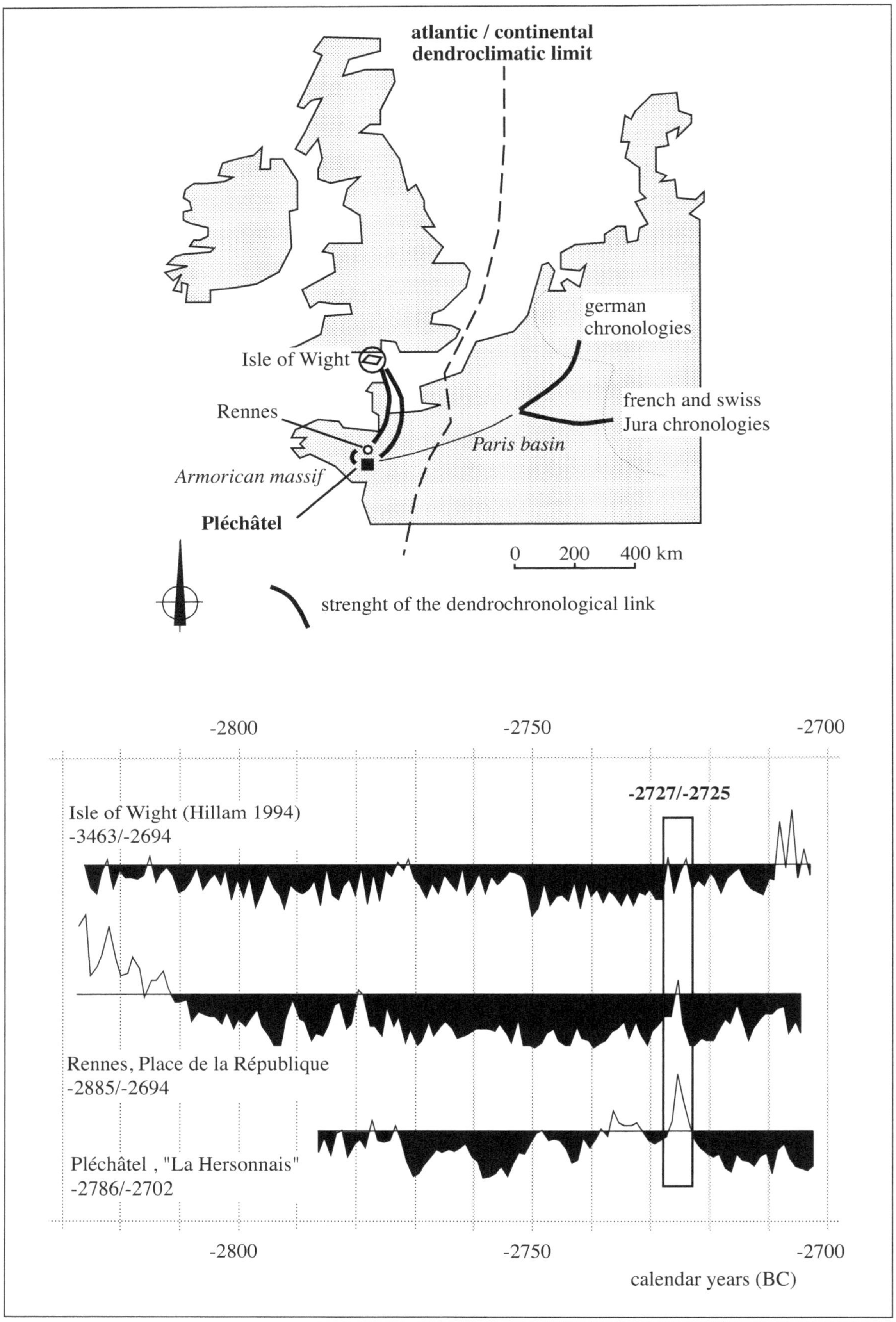

Figure 4: Proceedings for tree-ring dating. (a): location of the main chronologies used in this study; (b): crossdating between the chronologies from Pléchâtel, Rennes and the Isle of Wight.

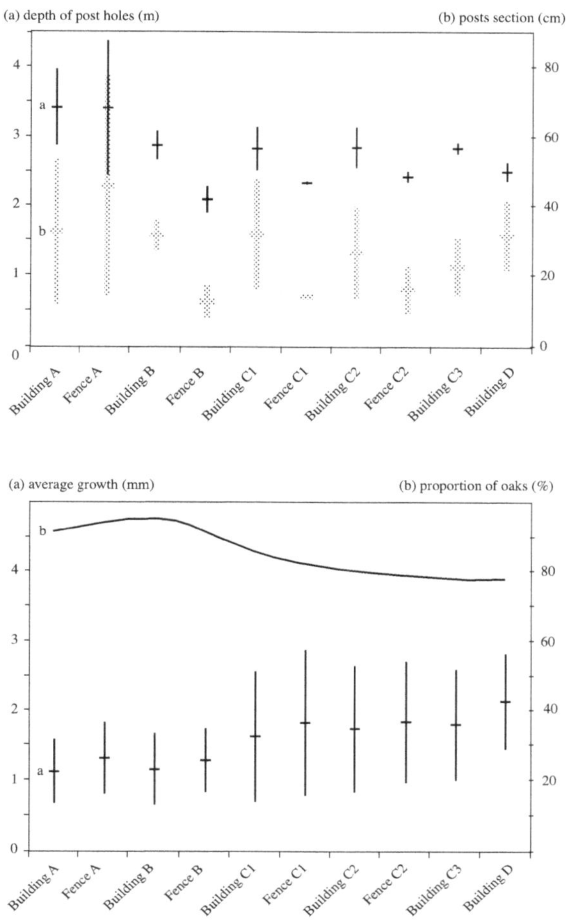

Figure 5: Archaeodendrologic characteristics of wood used at Pléchâtel during the Neolithic.

Indications from different parts of the site suggest that the diameter of the trunk was one of the main selection criteria for wood. From the early stage of building C, we can observe the use of coppice shoots, presenting a faster radial growth as a result of deforestation, for the construction of lighter and smaller buildings. This architectural evolution reflects an attempt to maintain traditional identity (through the shape and internal arrangement of the house) all while adapting to the physiological and mechanical changes in the wood available in the natural environment. During that period, forest exploitation cycles seem to have become shorter and shorter and the Neolithic settlers at Pléchâtel had to adapt to shorter, harder and knottier wood with smaller sections. Consequently, buildings became smaller and the walls previously made of joined oak posts were replaced by thinner wattle walls. Moreover, a decrease in the proportion of oak in comparison with whitewood (hazel and willow) can be observed, even though oak still predominates. The more frequent use of heliophilous species for buildings and wattle equally supports the view that techniques changed due to transformations of the woodland.

The evolution of the dendrological characteristics of oak wood used during the various building stages of building C is in accordance with the general evolution of houses A and B. It seems that a single forest was exploited during the whole occupation of the site. It was probably located at the periphery of the settlement and exploited within a radius that varied according to demographic changes.

References

AUBRY L., BERNARD V., COZIC N., GEBHARDT A., MARGUERIE D., POUILLE., VERTONGEN S. 1999 - *Rennes " Place de la République", Réalisation d'une station de métro souterraine,* Rennes (DFS de sauvetage urgent, SRA de Bretagne/AFAN), 55 p., 34 fig.

BERNARD V. 1998 - L'homme, le bois et la forêt dans la France du Nord entre le Mésolithique et le Haut Moyen Age, Oxford (*BAR International Series* 733), 190 p., 146 fig.

BILLAMBOZ A. 1987 - Archéodendrologie des palafittes du sud-ouest de l'Allemagne - Le bois raconte l'histoire des hommes et de la forêt, *Archeologia*, n° 222, p. 30-38.

ECKSTEIN D. 1969 - *Entwicklung und Anwendung der Dendrochronologie zur Alterbestimmung der Siedlung Haithabu* (Thèse de doctorat, Université de Hambourg), 113 p.

GASSMANN P., LAMBERT G.-N., LAVIER C., BERNARD V., GIRARDCLOS O. 1996 - Pirogues et analyses dendrochronologiques, in : ARNOLD B. dir., Pirogues monoxyles d'Europe centrale : construction, typologie, évolution, T. 2, *Archéologie neuchâteloise*, 21 (Musée cantonal d'archéologie, Neuchâtel, Suisse), p. 89-127.

HILLAM J. 1994 - *The dating of oak timbers from the Wootton Quarr Survey, Isle of Wight*, London (Ancient Monuments Laboratory Report, English Heritage,10/94), 20 p.

MARGUERIE D. 1992 - *Evolution de la végétation sous l'impact humain en Armorique du Néolithique aux périodes historiques, Rennes* (Travaux du laboratoire d'Anthropologie de Rennes, 40), 313 p., 181 fig.

SCHWEINGRUBER F.H. 1988 - *Tree rings. Basic and applications of dendrochronology,* Dordrecht / Boston / London (Kluwer Academic Publishers), 276 p.

TINEVEZ J.-Y. 1995 - La Hersonnais à Pléchâtel (Ille-et-Vilaine) : un vaste ensemble du Néolithique final – Résultats préliminaires, *Revue Archéologique de l'Ouest*, suppl. 7, Rennes, p. 293-317.

Elements for a cultural history of wood in Southern France (X^{th}-XVIth centuries)

Aline Durand

Department of History, University of Provence, Aix-Marseille I
29 avenue R. Schuman F.-13621 Aix-en-Provence cedex, France.
E-mail :adurand@mmsh.univ-aix.fr

Abstract

Until now, the study of the role of wood in rural architecture in southern France, during the Middle Ages, has remained a marginal subject of investigation since stone has always been considered omnipresent. Charcoal analysis of building remains allows us to prepare the ground for an history of wood during the Middle Ages. The spectrum of wood used in southern rural buildings is quite large as illustrated by recognized taxa: *Castanea sativa*, deciduous Mediterranean *Quercus*, *Fraxinus* sp., *Fagus sylvatica*, *Ulmus minor*, *Salix-Populus* sp., *Pinus* type *sylvestris*, *Abies alba*. However, and according to our study, from the year 1000 until the beginning of modern times, only one species was mainly used, for building purposes, as a result of technical, economical and biogeographical reasons: deciduous oak. The other species were used very occasionally, to answer specific needs, or as a result of their abundance and proximity. In fact, all species used were growing in the vicinity of the settlements: the use of timber from more distant regions has not been documented. Villagers, builders or other specialised workers adapted their needs to the local possibilities: they chose the most suitable woods for construction but also the cheapest ones. Certain observations of wood transverse sections and measurements of wood diameters prove that specimens were also carefully chosen. On occasion, wood management strategies seem to have been developed. Further charcoal analyses will hopefully support and enrich these first data, and a better understanding of wood uses during the medieval period will be achieved.

Key-Words: *Southern France, Middle Age, Charcoal analysis, Building remains, Deciduous oak.*

Résumé

Longtemps la place du bois dans l'architecture rurale méridionale a été jugée marginale au Moyen Age. L'analyse anthracologique de vestiges ligneux carbonisés offre les premiers jalons pour une histoire de la culture matérielle médiévale du bois. L'éventail des bois de construction méridionaux s'avère assez ouvert puisque près d'une douzaine de taxons différents ont été reconnus : châtaignier, chêne pubescent, frêne, hêtre, orme, peuplier, pin sylvestre et sapin. Cependant, depuis l'an Mil jusqu'au seuil de l'époque moderne, malgré cette diversité relative, une seule essence est massivement et universellement employée pour des motifs à la fois techniques, économiques et biogéographiques : le Chêne à feuillage caduc méditerranéen. Les autres ne le sont que ponctuellement pour répondre à des utilisations spécifiques ou pour leur abondance et proximité du chantier. En effet, toutes les essences utilisées proviennent de l'environnement immédiat avec les nuances biogéographiques propres à chaque gisement : l'origine exogène du bois n'a pas été mise en évidence. Les villageois, bâtisseurs ou autres travailleurs spécialisés ont adapté leurs besoins aux possibilités locales et tirent partie au mieux des ressources en choisissant les bois les plus aptes à la construction au plus faible coût économique. Un certain nombre d'observations anatomiques et de mesures de calibre prouvent qu'ils ont soigneusement examiné les arbres à abattre et parfois même développé comme à Cabrières, une véritable politique forestière avec des traitements sylvicoles adaptés. Il faudra évidemment que des analyses ultérieures viennent enrichir ces premières données.

Mots-clés : *Sud de la France, Moyen Age, anthracologie, construction, chêne à feuillage caduc.*

In memory of Jean-Marie Pesez

In the seventies, an extensive European archeological survey of deserted medieval and modern villages was carried out throughout Europe (Villages, 1965; Pesez, 1965; Démians d'Archimbaud, 1980). From this perspective, it appears natural to the supporters of material culture, to attract interest to the building materials of peasant's houses. In France, this subject has been gradually integrated into archeological medieval investigations. Important changes have taken place between the first meeting of Besançon in 1972 (Besançon, 1973) and the recent synthesis "One hundred medieval houses" (Esquieu and Pesez dir., 1998), as the scientific community lost interest in the monumental construction and recognized the importance of simple soil remains. The carbonized woody remains of rural architecture are often found charcoalified during excavation work: laths, beams, battens, frames, fences, even agricultural tools and so on. Charcoal analysis of such artifacts contributes to a better understanding of how plant species were used, of forest management and, occasionally, of conditions of transport.

At first sight, the role of wood in rural architecture of the southern regions seems very marginal, as stone has been considered as an omnipresent building material. In fact, this is not true. First of all, written documents from the XIIth century mention wooden houses, *i.e.* completely built with wood, both in Camargue and in the hills of the "Biterrois" (the area around Béziers) (Durand, 1998). Also, during the last decade, excavation work carried out in southern France uncovered hut floors dating back to the VIIIth-XIIth centuries entirely built with perishable materials, both in rural and urban contexts (Ginouvez, 1995; Garnier *et al.* 1995; Mercier and Raynaud, 1995; Schneider *et al.* 1995). Indeed, the use of wood as building material seems to have been more important than previously thought by historians.

Southern chrono-typology (X^{th} - XVIth centuries) (tabl. 1, fig. 1, fig. 2)

The first feudal fortified sites (X^{th}-XIIth centuries)

Until today, fortified sites from the first feudal age provided very few samples of this kind. As a consequence, the study of these sites proves to be particularly interesting.

A study of this kind has been carried out at Le Mourrel, a feudal moat and bailey castle at Molleville, in the "Lauraguais" region (Aude, 11). On the summital platform, 19 pieces of charcoal were found inside a post hole. All of them were identified as deciduous oak.

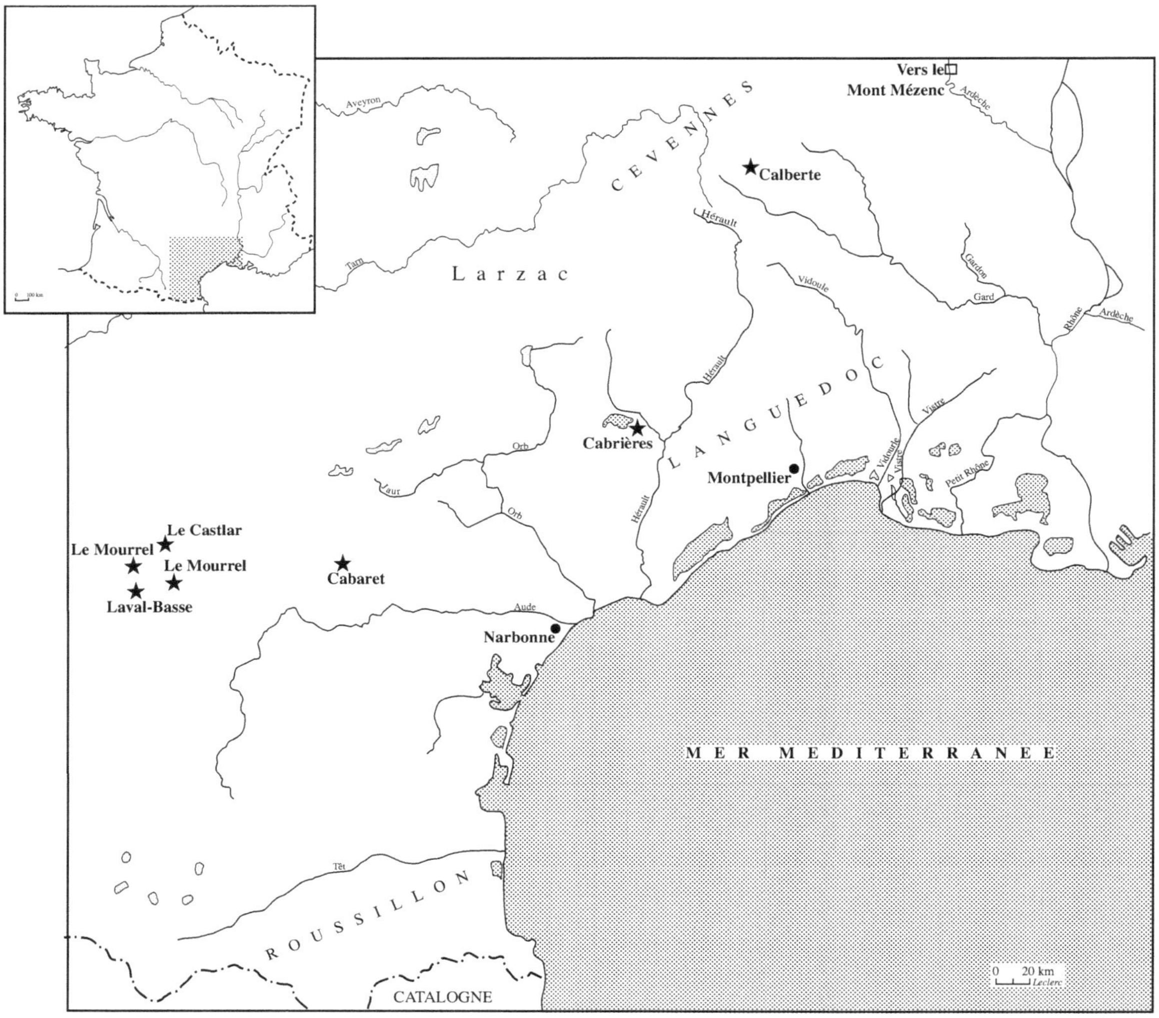

Figure 1: Location of sites in Languedoc.

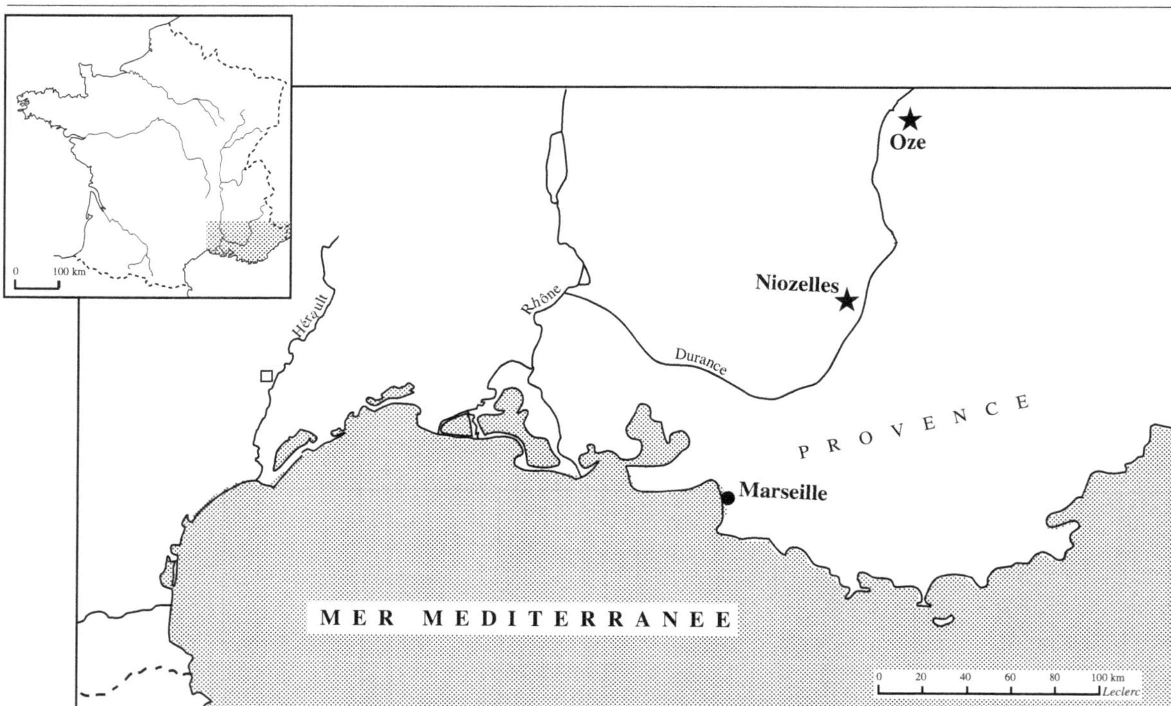

Figure 2: Location of sites in Provence.

TAXA	**Xth-Xth c.** **Niozelles** Remains of construction	**XIth-XIIth c.** **Oze** Remains of construction	**XIth-XIIth c.** **Le Mourrel** Post hole	**XIIth c.** **Cabaret** Ligneous artefact	**XIIIth-XIVth c.** **Laval-Basse** Levels of destruction	**XIIIth-XIVth c.** **Laval-Basse** Remains of structure	**XIVth c.** **Le Castlar** Timbering of store house	**XVIth c.** **Cabrières** Remains of construction
Abies alba		11						
Acer sp.				4				
Acer monspessulanum		4						
Angiospermae	1							
Buxus sempervirens	1							
Castanea sativa-Quercus pubescens								9
Castanea sativa								52
Corylus avellana				2			24	
Crataegus sp.-Sorbus sp.	3							
Erica sp.						1		
Fagus sylvatica		130			1		19	
Ficus carica					1			
Fraxinus sp.		2						
Fraxinus oxyphylla.								5
cf. Juniperus sp.		1						
Juniperus sp.		2						
Leguminosae		6		1				
Ligustrum vulgare							5	
Pinus type *sylvestris*		53						
cf. Populus sp.					1			
Populus sp.					2			
Prunus sp.					1			
Prunoïdeae type *Prunus avium*		18						
Prunus avium					1			
Prunus amygdalus					1			
Prunus cf. amygdalus					1			
Prunus spinosa					1			
Quercus ilex-coccifera				14				16
Quercus deciduous			19		15		1497	
Quercus sp.-Castanea sativa		1						
Quercus type *pubescens*	297	174		534		15		395
Quercus type *sessiliflora-pedunculata*					2			
Salix sp.-Populus sp.		3						
Salix sp		72					9	
Sorbus sp.					1		4	
Tilia sp.							1	
Ulmus minor					12			4
cf. Viburnum sp.				3				
Vitis vinifera sp.						1	4	
Bark or knot	1	3		11	1			
Undeterminate		1						
Undeterminable		1		13	3			
TOTAL	303	482	19	582	45	17	1563	481

Table 1: Absolute frequencies of taxa (remains of construction).

East of Forcalquier (Basses-Alpes, 04) in the French Prealps, the moated site of Niozelles was also recently excavated. It may be considered as contemporary to the generation of the first feudal fortifications since one of the occupation phases has been dated to the Xth-XIth century. Several phases of occupation have been distinguished by the excavation work. The charred wood comes from layers that correpond to the transformation of the site: it belonged to the wooden outhouse near one of the habitation structures situated on the top of the moated site. Charcoal analysis results are based on the identification of 303 charcoal specimens from this outhouse. 297 of them were identified as deciduous Mediterranean oak (*Quercus* type *pubescens*). The specific discrimination of deciduous oaks is very difficult, often quite impossible, based on wood anatomy alone. However, and based on site ecology and biogeography, the samples recognized here as deciduous oak probably refer to a Mediterranean deciduous oak (*Quercus* type *pubescens*).

In the Buech valley in the "Gapençais" area (Hautes-Alpes, 05), the rural dwelling in Oze provided charred elements of a wooden construction. They are considered to be remains from beams or battens found on the occupation soil dated to Xth-XIIth. Charcoal fragments have been photographied and drawn as they were found in the site. The study of 482 fragments allowed us to identify 10 different taxa. Most of the fragments have been assigned to four species: *Pinus* type *sylvestris*, *Salix* sp., *Fagus sylvatica* and *Quercus* type *pubescens*. Two major results can be pointed out and corroborate archeological observations. First, charcoal analysis has identified only very few taxa (10). This relative specific poverty and the dominance of only four taxa (beech, oak, pine and willow) is characteristic of an ethnobotanical spectrum, not of a palaeoenvironmental one. Anthracological data agree with the archeological data in considering these remains as vestiges of a building and not of domestic fires. Secondly, clear traces of wood work have been noticed: cut in edge, quadrangular cuts with characteristic decentring of wood medulla. The gymnosperms (pine and fir) present anatomical deformations of ligneous tissues. Ligneous rays are close to the growth ring boundary and are deviated obliquely in earlywood and tracheids, which are compressed but without rupture. Latewood appears often unchanged; however, in some cases cellular bands of latewood are also deviated obliquely. According to I. Théry (Théry 1983), this kind of deformation is classified as "type 1b" and is not a result of carbonization nor specific growing conditions. At Oze, these features are very rare and observed on soft wood only: it could be suggested that the wood had supported a heavy load. These data may provide evidence that we are dealing with artifacts, not with scattered charcoal of domestic fires. In summary, deciduous oak is the major species used as timber in these three southern sites dating back to *c.* 1000 AD.

The "castral" villages (XIIIth-XIVth)

During this period, villages were well established and developing; no changes in the choice of building material can be observed. Previous anthracological spectra concerning timbers of sites occupied during the middle feudal period are not different from spectra based on early feudal sites.

At the moated village of Laval-Basse (Aude, 11), destruction levels, composed of yellow clayey soil, stones, tiles, charcoal and broken artifacts, have been identified. The study of 45 charcoal specimens allowed us to identify *Quercus* type *pubescens*, *Quercus* type *sessiliflora-pedunculata*, *Salix* sp., *Sorbus* sp., *Ulmus minor* and *Vitis vinifera*. Large pieces of deciduous oak found in one structure (n°371) may be considered as remains of carpentry work. In this site, two different types of deciduous oak may be distinguished based on wood anatomy features: one, Mediterranean, with early wood vessels arranged in only one layer and having a rounded cross section of 200-300 μm, the other, Atlantic, with early wood vessels arranged in two or three layers and having an elliptic cross section of 300-400 μm.

In Cabaret, at one of the four castles of Lastours (Aude, 11), excavations have taken place in the lower part of the village, near the river. One of the buildings, dated to the XIIth century, provided a concentration of large pieces of charcoal. At first, these remains were believed to come from a beam. However, further research showed that they belong to a single ligneous artifact. This object had a round or oblong shape with a concave side which provide evidence of a carpenter's or joiner's work. The study of 582 charcoal fragments testifies to the great importance of the deciduous Mediterranean oak.

On the most westerly point of the Black Mountain, at Castlar of Durfort (Tarn, 81), archeologists have uncovered a store-house, dated to the end of the XIIIth century/beginning of the XIVth century. All the harvested grain was in place and stored in heaps on the site (Ruas, in press). The charcoal fragments analysed result from the use of the structure as timbering which after had fallen down due to a fire. This charcoal concentration was covered by a layer of tiles. One of the beams had a diameter close to centimeters. Again, deciduous oak (*Quercus* type *pubescens* or *Quercus* type *sessiliflora-pedunculata*) is the major taxon among the 1497 charred fragments identified. *Fagus sylvatica*, *Corylus avellana* and *Salix* sp. are also present. The discrete, but clear presence of these taxa is believed to result from repairs or minor additions. They are considered marginal in relation to the whole timber structure.

Modern Times

Data from modern times confirm the predominance of deciduous oak in the southern peasant architecture. At Cabrières (Hérault, 34), an ancient castral site, new dwellings were built at the bottom of the hill. In the sixteenth century, several houses were aligned along a small street. Remains of wooden beams or battens have been discovered in one of them, under a layer of tiles. Some of these beams present very clear traces of nailing. Nails have actually been found in place. Among the 481 charcoal pieces analyzed, 395 have been identified as deciduous Mediterranean oak. We must underline, however, that batten were also made out of *Quercus ilex-coccifera*, *Ulmus campestris*, *Fraxinus ornus-oxyphylla* and *Castanea* sp. This last species had never been found before; it is the first sure attestation of medieval chestnut used as timber in France. *Castanea sativa* has not been identified as timber wood, neither in the "Bassin Parisien" (Dietrich, 1989) nor in Roman times in southern France (Chabal, 1997).

Conclusion

Anthracological studies concerning wood and its uses in medieval rural constructions have provided some significant results, which must however be enriched and refined by further work. It is clear that between the beginning of the Xth century and the end of the XVIth century, deciduous oak, and not chestnut, as usually refered to in the literature, was chosen as building material, in southern medieval villages. Our results clearly show that, in spite of its mechanical and physical properties, *Castanea sativa* is not used as timber during the Middle Ages. Timbers in medieval buildings, believed to belong to chestnut wood are, in fact, an illusion. O. Rakham (Rakham, 1980) provides a very simple explanation: the two genera have very similar wood anatomical features (only medullary rays distinguishs *Castanea sativa* from deciduous *Quercus*) and may be easily mistaken. The following question must now to be considered: why is deciduous oak dominant?

Medieval timber in southern areas: from felling to use (Xth - XVIth centuries)

Some anatomical characters observed during the identification work allow us to understand the uses of medieval timber and the "forest culture" of the peasant population.

The dominance of deciduous oak

The constant use of deciduous oak in southern rural buildings during medieval times may be easily explained. Its technical and physical qualities are well known: the very hard wood of deciduous oak does not warp, can bear pressures, bending and crushing and offers very long monoxylem bearings; it is also quite resistant to bad weather, humidity and time; last, but not least, its duramen, very rich in tannins, protects the wood against fungi and insect attacks. Because of these properties, this tree provides an excellent wood for carpentry and joinery work. However, these qualities are not sufficient to explain the systematic use of oak. The possibilities offered by the environment must also be considered. In fact, deciduous oak is the key-taxon in different vegetation levels (mesomediterranean, supramediterranean) of all the sites studied. Peasants and craftsmen selected the most suitable species for construction in the vicinity of their villages. This was also the cheapest choice: gathering of wood stocks from neighbouring areas allowed them to avoid costly and lenghty transport. For isolated and perched castral sites, this was an indispensable condition. In the plain also, where villages were situated near markets and integrated in commercial and economical relationships, the same choices for wood gathering were made (even in the XIVth-XVIth centuries).

Anthracological diagnosis also reveals that builders did not use any kind of oak tree. The meticulous observation of charcoal transverse section proves this. At Le Mourrel, the average width of growth rings was 3 mm, which was the result of a normal growing rate. In this case, wood may have little value due to lower technological qualities (such as cutting up and function), but, on the other hand, it has good mechanic properties. Growth ring widths did not present significant variations, which is an important quality standard (Venet, 1974).

At Oze, traces of squaring had been clearly observed. However, the tool used to carry out this work left no characteristic traces on remains. The same was observed at Niozelles (tabl. 2), but further information was obtained. Usually, almost all vessels of the deciduous oak were obstructed with tyloses which, when not traumatic, are characteristic of old wood. Growth ring arches are not very marked, which may indicate the presence of trunks rather than branches. It is possible that wood was collected in forest stands and mature formations.

Some charred fragments were very well preserved and allowed further conclusions. At Niozelles, the orientation of ligneous fibers and growth rings boundaries towards the orientation axis of artifacts are characteristic of a specific way of carving wood known as "cutting up on slabbing offcut". This kind of wood work consists of cutting the wood along the length of the trunk instead of following the rays. For J.-Y. Hunot (Hunot, 1992), this procedure is characteristic of sawing with parallel traces of a saw. This corroborates the interpretation put forward by the archeologist, D. Mouton, according to whom these are remains of timber (beams, butten...). At Cabrières, in the XVIth century, the preservation of carbonized pieces is such that an estimation of wood sections and sizes has been made. The observation and measure of growth ring curvatures makes it possible to calculate minimal diameters (fig. 3). This is obviously an estimation because the charcoal pieces may come from a bigger trunk or be warped while burning. Whenever diameter estimations were possible, the size of oak pieces was always the most important. In the same way, construction remains made of deciduous oak have greater dimensions. In the Mediterranean area, this tree can easily grow up to twenty meters high providing that the formation is not treated as a coppice or a coppice under forest. In the latter case, the shoots are small (under 10 m high) and have a middling section. At Cabrières, bark came either from "spring from coppiced stools" with 15 or 20 years between each cutting or from a protected vegetal formation treated as fully-grown trees.

All these observations emphasize the idea that timber had been carefully chosen and seriously examined. Therefore, peasants adapted their needs to their environmental possibilities: either they had preserved the forest or they developed

	Datation	*Angiospermae* undeterminable	*Buxus sempervirens*	*Quercus* type *pubescens*	*Crataegus sp. Sorbus sp.*	Knot	TOTAL
SE1 layer 138	beginning of XIth c.			11	1	1	13
layer 139	late Xth c.			32			32
E1 layer 224	late Xth c.	1	1	16	2		20
SE2 layer 141	begining of XIth c.			216			216
SO1 layer 215	beginning of XIth c.			22			22

Table 2: Observation of charcoal transverse sections at Niozelles (X-XIth).

strategies of forest management to obtain long bearings.

The other timbers

Anthracological analysis of timber remains show that the remaining species used for building appear only occasionally. They are not used in the majority of artifacts; they are not recognized in all archeological sites analysed. In sites close to the biogeographic mountainous level, conifers are present. At Oze, *Abies alba* and *Pinus* sp. are identified, while at Mézenc (Ardèche, 07), only *Abies alba* has been found so far. Compression wood has been observed in nearly all the samples: the alpine conditions of the region may explain the formation of these tissues. In southern sites, the presence of *Fagus sylvatica*, *Ulmus minor*, *Salix* sp. and *Castanea sativa* is attested. Important frequencies of riverside taxa are noted. These species might have been carefully selected as they provide good wood and very long monospecific bearings. *Ulmus minor* and *Fagus sylvatica* are suitable for the heavy structure of buildings, but not for roof pieces. At Cabrières, trees assigned to these species seem to have reached average dimensions that were smaller than

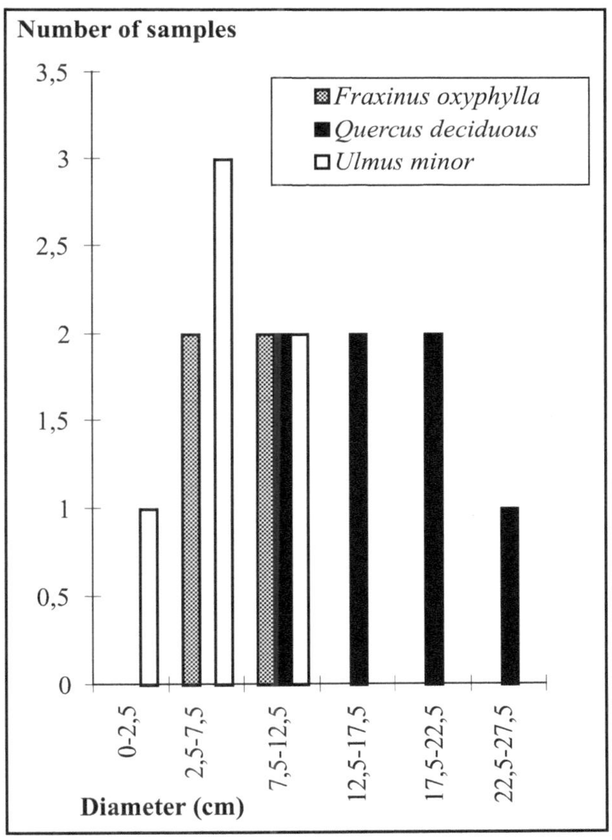

Figure 3: Minimal diameter of timbers at Cabrières (XVIth).

deciduous oak. Under Mediterranean climatic conditions, beech wood becomes heavy and hard (classes 7, 8, 9), and can be treated as a timber (Venet, 1974). On the other hand, willow and poplar are very soft woods, inadequate to the mechanical needs of heavy building. Their selection might have answered specific and punctual needs, such as minor armature, pegs, repairs, penury, mortise-and-tenon. The flexible wood of ash, with fine texture, was usually chosen for this last coupling. The wood of Mediterranean ash is dense (classes 7 and 8), and therefore hard and may be used in carpentry (Venet, 1974).

At Niozelles, three other species were identified: *Buxus sempervirens* and *Sorbus* sp./*Crataegus* sp. These wood types have a short durability but are hard enough to be used in joinery or cabinet work. However, hawthorn is marginal in this use. The three samples identified as Rosaceae are well cut: the section (2 cm) is quadrangular, 2-3 cm long, in the shape of a peg. *Buxus sempervirens* has a good durability and is a very hard wood. These remains are smaller than those of deciduous oak. The use seems specific and unusual, probably to join together other different pieces or to complete a structure.

Conclusion

The spectrum of wood used in southern rural buildings is quite large, as shown by the taxa recognized: *Castanea sativa*, *Quercus* deciduous, *Fraxinus* sp., *Fagus sylvatica*, *Ulmus minor*, *Salix* sp./*Populus* sp., *Pinus* type *sylvestris*, *Abies alba*. However, despite this relative diversity, one species is mainly used as a result of technical, economical, biogeographical reasons: deciduous oak. The others were used occasionally to answer specific needs or because of their abundance and proximity. In fact, all species used grew in the vicinity of the sites: exterior origin of timber has not been documented. Villagers, builders or other specialist workers adapted their needs to the local possibilities; they choose the most suitable woods but also the cheapest ones. They have also chose carefully and examined timber seriously. They seem to have developed forestry strategies, with management forest practices. Further charcoal analyses will confirm and enrich these first data and a better understanding of wood uses during the medieval period will be achieved

References

BESANÇON 1973 - *La construction au Moyen Age*. Histoire et archéologie, Actes du 2ème congrès de la Société des Historiens médiévistes de l'Enseignement supérieur public tenu à Besançon les 4-6 juin 1972, Besançon, Annales littéraires de l'Université de Besançon, Paris, Les Belles Lettres.

CHABAL L. 1997 - L'anthracologie, de l'échantillonnage des charbons de bois à l'interprétation du paysage du Néolithique final à la période romaine en Bas-Languedoc, Paris, Maison des Sciences de l'Homme, *Documents d'Archéologie Française*, 63, 192 p., 64 fig., 14 tabl.

COLIN M.-G., DARNA S I., POUSTHOMIS N., SCHNEIDER L. (coord.), 1996 - La maison du castrum de la bordure méridionale du Massif central, *Archéologie du Midi médiéval*, supplément n° 1, Carcassonne, Centre d'Archéologie médiévale du Languedoc, 221 p.

DÉMIANS D'ARCHIMBAUD G. 1980 - *Les fouilles de Rougiers* (Var), Paris, C.N.R.S., 724 p.

DIETRICH A. 1989 - *Problèmes d'archéologie et d'ethnohistoire liés à la conservation des bois (étude des bois archéologiques médiévaux du Bassin parisien)*, Thèse de l'Université de Paris I Panthéon-Sorbonne, 2 vol., 477 p.

DURAND A. 1998 - *Les paysages médiévaux du Languedoc (Xe-XIIe siècles)*, Toulouse, Presses universitaires du Mirail, 491 p., 61 fig., 34 tabl.

ESQUIEU Y., PESEZ J.-M. (dir.) 1998 - *Cent maisons médiévales en France du (XIIe au milieu du XVIe siècle)*. Un corpus et une esquisse. Paris, C.N.R.S. Editions, Monographie du C.R.A., 450 p., 406 fig.

GARNIER B., GARNOTEL A., MERCIER C., RAYNAUD C. 1995 - De la ferme au village : Dassargues du Ve au XIIe siècle (Lunel, Hérault). *Archéologie du Midi médiéval*, t.XIII, p. 1-73.

GINOUVEZ O. 1993 - Des maisons excavées à Narbonne autour de l'an Mil. *Archéologie du Midi médiéval*, t. XI, p. 53-68.

HUNOT J.-Y. 1992 - Charbons de bois recueillis à l'intérieur de vases funéraires. Approche ethnobotanique. Les charbons de bois, les anciens écosystèmes et le rôle de l'homme (Montpellier 10-13 septembre 1991), *Bulletin de la Société botanique de France*, t.CXXXIX, 2-3-4, p. 651-658.

MERCIER C., RAYNAUD C. 1995 - L'habitat rural en Gaule méridionale méditerranéenne aux VIe-VIIe s. Approche régionale et étude de cas. Lorren C. et Perin P. (ed). *L'habitat rural du Haut Moyen Age* (France, Pays-Bas, Danemark et Grande-Bretagne), Actes des Journées internationales d'Archéologie mérovingienne Guiry-en-Vexin et Paris, 4-8 février 1993, Rouen, Association française d'Archéologie mérovingienne, p. 193-206.

PESEZ J.-M. 1965 - Les villages désertés en France : vue d'ensemble. *Annales Economies, Sociétés, Civilisations*, t. XX, n° 2, mars-avril, p. 267-290.

SCHNEIDER L., PAYA D., FABRE V. 1995 - Saint-Sébastien-de-Maroiol (34) et l'histoire de la proche campagne du monastère d'Aniane (V-XIIIe s.). *Archéologie médiévale*, t. XXV, p. 133-181.

RACKHAM O. 1980 - *Ancient Woodland, its history, vegetation and uses in England*, London, E. Arnold, 402 p.

RUAS M.-P, in press - Productions agricoles, stockage et finage en montagne noire médiévale. Les récoltes incendiées au XIVe siècle du grenier castral de Durfort (Tarn), Paris, Maison des Sciences de l'homme, collection *Documents d'Archéologie française*.

THERY I. 1983 - *Les déformations anatomiques des charbons de bois : implications archéologiques (sites des Canalettes et des Usclades, Aveyron)*, D.E.A., Université Paris1., 80 p.

VENET J. 1974 - *Identification et classement des bois français*, Paris, Engref, 300 p.

VILLAGES 1965 - *Villages désertés et histoire économique XIe-XVIIIe siècles*, Paris, SEVPEN (Les hommes et la Terre, 11).

Analysis of burnt building structures of the Ambato valley (Catamarca, Argentina)

Maria Bernarda Marconetto

Museo de Antropología - UNC
Hipólito Yrigoyen 174, 5000 Córdoba - Argentina.
E-mail: marconet@ffyh.unc.edu.ar

Abstract

This work is part of an archaeological project being developed in the Ambato Valley, in the southern Andean area of north-west Argentina. The present research aims at reconstructing, from a multidisciplinary point of view, the social, economic and symbolic processes that led to the formation of a complex society, known in Argentine archaeology as the Aguada culture (600-1000 AD). Against this background, the present study intends to bring together analytical elements concerning the management of forest resources, departing from the idea that as social structures became more complex, they influenced on the way resources were exploited. In this article, we will deal with the problem of species selection according to use, and how the anatomical properties of different wood types were taken advantage of when building the settlements of Piedras Blancas and Iglesia de los Indios.

Key-words: *Northwest Argentina, Southern Andean area, Ambato valley, Aguada culture, Forest resources, Charcoal analysis.*

Résumé

Ce travail s'inscrit dans le programme archéologique de la vallée d'Ambato, dans les Andes méridionales, au nord ouest de l'Argentine. L'objectif principal de ce programme est l'étude des processus de différentiation sociale qui débouchèrent sur une configuration inégale du pouvoir au sein de la culture de La Aguada (600-1000 AD). Dans ce cadre, notre projet de recherche a pour objectif d'apporter des éléments d'analyse concernant la gestion des ressources forestières, en partant de l'idée que la complexification des structures sociales interfère dans le mode d'exploitation de ces ressources. Au cours du travail présenté ici, je me concentrerais tout particulièrement sur le problème de la sélection des espèces en fonction des usages et l'utilisation des propriétés liées aux structures anatomiques des différents bois utilisés lors de la construction des sites Piedras Blancas et Iglesia de los Indios.

Mots-clés : *Argentine du nord-ouest, Andes du Sud, vallée d'Ambato, Culture d'Aguada, Ressources forestières, Anthracologie.*

Introduction

From research done on processes leading to more differentiated and complex social structures, it has been acknowledged that one way to better know these processes is by the study of how resources were managed. With this in mind, my work intends to add elements of analysis to the issue, focusing on the exploitation of forest resources. It also tries to understand how this type of resources were managed in a particular society.

Considering that timber and wood are key factors for survival, settling and daily life of human groups, it is logical to suppose that the cultural processes involving these products were, like other aspects of the society, subject to social, economic an political changes. We agree with the hypothesis, put forward by several anthracologists (Castelletti and Zimerman, 1985; Pearsall, 1989; February, 1992; Kreuz, 1992; Shackleton and Prins, 1992; Pique i Huerta, 1999), considering that charcoal assemblages reflect the wood that was selected and transported to a site by man and that the way wood resources were managed was determined by cultural choices.

Therefore, charcoal is seen here as part of the archaeological record just in the same way as any other archaeological items, and thus reflecting the organisation and the strategies of a given society. Even though the taxonomic composition of an analysed sample is mainly the result of anthropogenic factors, it does to some degree also reflect the environment.

This work focuses on the remains of burnt wood belonging to building structures. It is interesting to realise that in several sites related to the Aguada culture, remains of roofs were preserved as they burnt and collapsed. As carbonized wood does not decay, these fires have proved to be very useful for our understanding of how the forest resources were managed and for the identification of the taxa used for building purposes. Wood remains from a fire reveal, more than fuel wood, how wood was selected. Therefore, we can study the technology used for a building, the properties of a certain wood type (endurance, resistance, flexibility, hardness, etc.) and if these were taken into account when choosing wood for a specific purpose. It is also possible to determine the catchment area for the exploitation of wood resources.

Biogeographical presentation

The Ambato valley, where the studied sites are found, is situated in the centre of the Catamarca Province in northwestern Argentina. This valley is located in the southern part of the Andean region, between the eastern Andes and the lowlands of Chaco. Due to its location, the valley belongs to the phytogeographical province of Chaco serrano (Cabrera, 1976). To the north-east we find the biogeographical region of Yungas, belonging to the Amazonian complex (rainforest at 1000-2000 m a.s.l), and to the northwest, the provinces of Prepuna and Monte. Situated in a mountainous region, the vegetation of the Ambato valley shows several altitudinal successions (fig. 1).

The archaeology of the Ambato Valley

Of the 160 sites in the Ambato area currently surveyed, eight have been excavated. They date from the beginning of the Christian era to 1000 AD (Pérez Gollán *et al.* 1996-1997). Several studies indicate that different aspects of the society were subject to specialization and standardisation: pottery production, metalworking, large-scale agriculture (practised in a system of terraces), public building and cere-

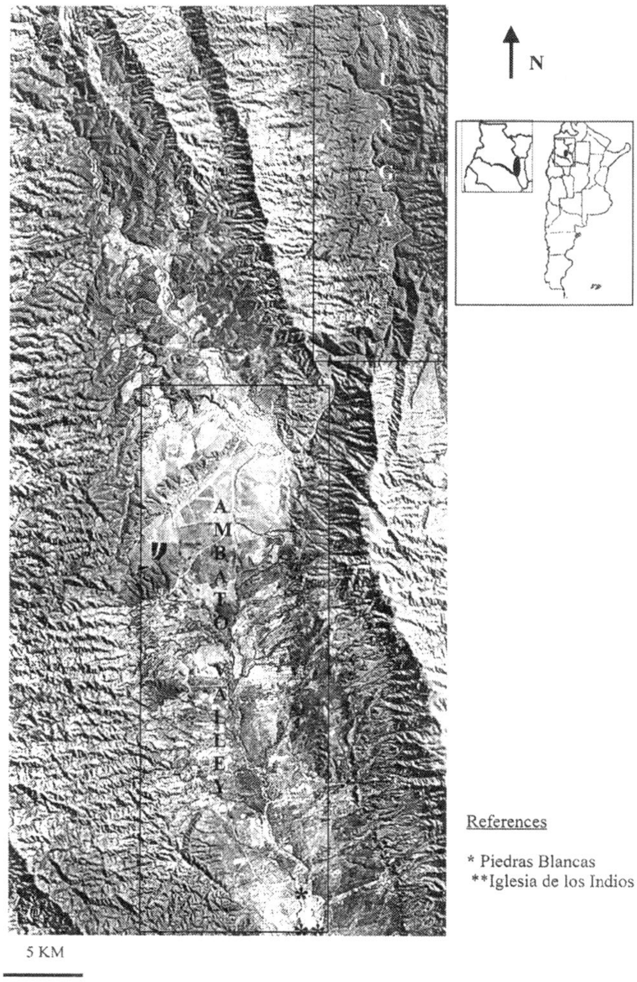

Figure 1: Localisation of the Amabato valley, Catamarca, Argentina.

monial centres. Studies of burying rites and religious practices show that the society was changing into a more complex socio-economic organisation.

Provenance of analysed samples and procedures

Most of the wood on the studied sites has disappeared. However, fire, whether intentional or not, affected the majority of the sites associated with the Aguada culture. Therefore we have a record, not only of the wood used as fuel, but also of the timber used for building purposes.

The analysed material was recovered during the excavation of Piedras Blancas, a site of the "elite", and also from a ceremonial site called Iglesia de los Indios (Gordillo, 1994). Both sites were affected by fire, and we could therefore recover part of the building material. At the site of Piedras Blancas, a structure of carbonized logs and branches that had fallen on burnt pottery vessels on an occupational floor was excavated, at a depth of 30 to 60 cm, in an area called the south wall, belonging to a perimetral wall. In another part of the site, called the south-eastern enclosure, logs and other elements associated with the roofing were located at same depth as the above mentioned remains from the south wall area. However, these were sealed by a later 10-20 cm thick occupation layer.

A sample from the fourth structure of the eastern area at Iglesia de los Indios was analysed. An enclosure of 7x6 m was totally excavated. The occupational level was covered by the burnt remains of a roof constructed of logs, intertwined branches, twigs, mud and straw. The original intertwining could still be observed in some places. As in Piedras Blancas, burnt pottery was associated with the occupational level.

The charcoal from the above-mentioned structures was analysed by the use of a binocular microscope and a scanning electronic microscope (SEM). Ten samples from large logs, nine from middle-size logs and 60 from twigs were analysed. As one of the objectives of our work was to study construction techniques, we divided the analysed material into three categories according to their size: big logs (diameter >10 cm), small logs (diameter <2-10 cm), and twigs (diameter <2 cm). Thereby, we assume that the size of the wooden piece was chosen according to its use in the building structure: big logs, used as posts and beams, would support roof structures and small logs would be part of the cover together with smaller branches. The way the combustion affected these structures and some finds of straw, suggest that the roofs were covered by mud and straw.

The obtained data was presented by presence/absence, regardless of the number of analysed fragments, and this information was compared to size variations. It should be mentioned, however, that only charcoal pieces belonging to measurable logs were taken into account. We discarded other fragments, even when those belonged to the same taxa as the logs, as we cannot be positive that they actually came from the roof structure.

Results

The wood taxa used for the roof structures (as their diameters were large enough) and identified by the charcoal analysis are the following:
Acacia sp. (fig. 2, 3)
Several species of *Acacia* are found in the valley, not very far from the studied sites. *Acacia visco* (locally called viscote) has the straightest stem of these. The other species present more or less crooked stems and numerous branches. All of them are characterised by a hard and heavy wood that is difficult to cut due to a high content of crystals. Pieces up to 4 m long and 0.30 m in diameter can be obtained.

Prosopis spp. (fig. 4, 5)
Prosopis alba (algarrobo blanco), *P. nigra* (algarrobo negro) and *P. torcuata* (tintitaco) have been identified. All of these *Prosopis* species grow in the valley today, although in a reduced number. The tintitaco has a straight stem, even when short, and is therefore much appreciated for short posts. The wood of the algarrobos is heavy and can be worked, even though the high content of crystals in the cells uses the tools. The *Prosopis* species provide hard wood that dries well *i.e.* without changing its structure. Short logs (2-3 m) can be obtained.

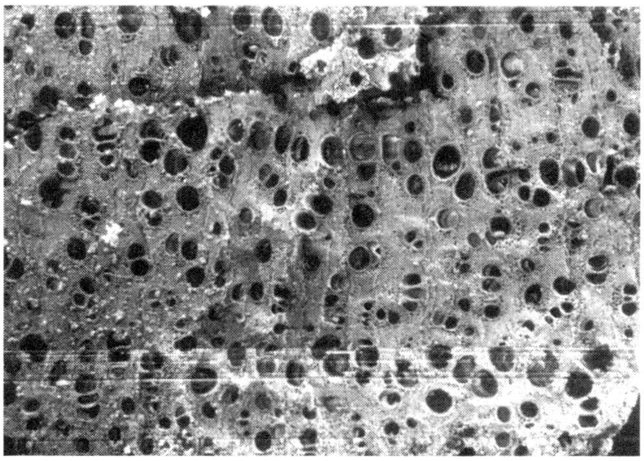

Figure 2: *Acacia visco*, transverse section x40.

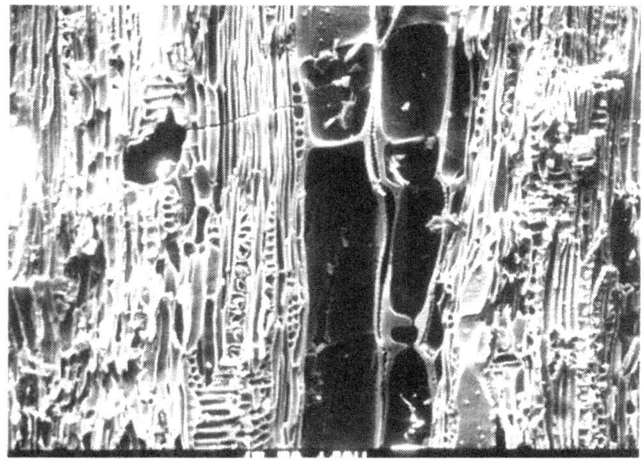

Figure 3: *Acacia visco*, longitudinal tangential section x160.

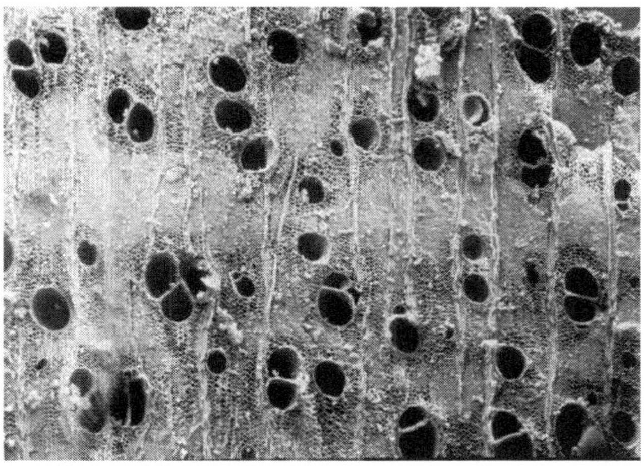

Figure 4: *Prosospis nigra*, transverse section x40.

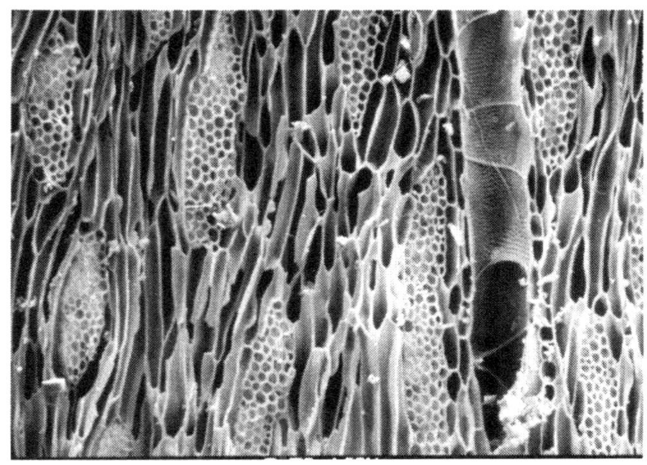

Figure 5: *Prosopis nigra*, longitudinal tangential section x160.

Phoebe porphyria (fig. 6, 7)
This taxa, called laurel de la falda or San Antonio is not indigenous to the Ambato valley. It grows in the Yunga zone and is found in the rain forest of the humid hillsides, situated at 1200-1500 m a.s.l. to the north-west of the valley. Its anatomical structure shows characters that are typical for species growing in areas rich in nutrients and water (scarce parenchyma, indistinct growth-ring boundaries). The wood of *Phoebe porphyria* can be worked without difficulty. It is semi hard and not very heavy. It also presents a fine and homogeneous texture and straight fibres. The wood of *Phoebe porphyria* does not resist well to the attack of fungi and insects and is therefore not adapted to being in direct contact with the ground. It requires careful drying to avoid cracks and warping. Good pieces of up to 6 m long and 0.30 m in diameter can be obtained from this species.

Alnus sp. (fig. 8, 9)
Alnus or aliso del cerro does not grow in the valley, just like *Phoebe porphyria*. It provides a soft and light wood with a fine and homogeneous texture and straight fibres, easy to work. The wood of the stem does not crack but is not very durable. The pieces can be up to 5 m long and 0.20 m in diameter.

Interpretation of the results

The results of our analyses show that a selection of the wood used for constructions took place at the studied sites. Species presenting hard wood were used for supporting elements such as beams and posts (fig. 10). Among the species growing in Ambato valley and identified from the archaeological record, *Prosopis* and *Acacia* stand out as providing the most durable wood. Although these taxa would not be the most suitable for roofs, beams or posts because of their weight, hardness, short stems and resistance to stone or bronze tools, they are nevertheless the most appropriate trees present in the area and this thanks to their strength, durability and resistance against attacks of arthropods and fungi.

Taxa such as *Phoebe* and *Alnus* would have provided longer wooden elements, susceptible of covering wider rooms. This could not have been obtained from the shorter stems of the *Acacia* and *Prosopis* species. *Phoebe* or *Alnus* wood is certainly less durable but easy to work and lightweight. As mentioned above, the latter taxa grow in the Yungas forest zone which is located at 40 km distance, through a mountainous landscape, from the sites of Piedras Blancas and

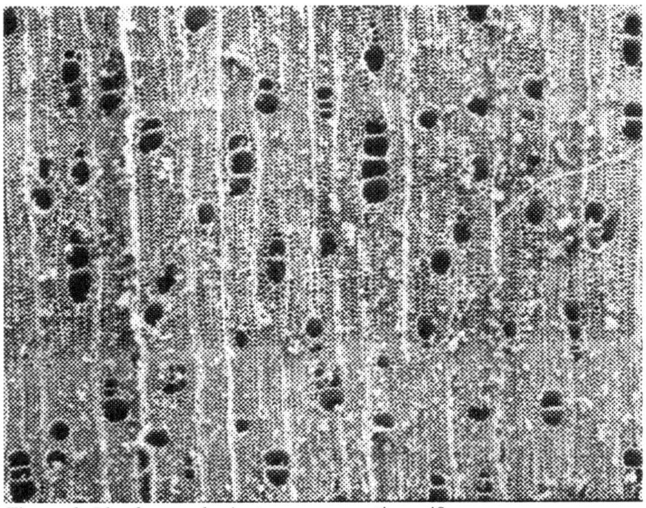
Figure 6: *Phoebe porphyria*, transverse section x40.

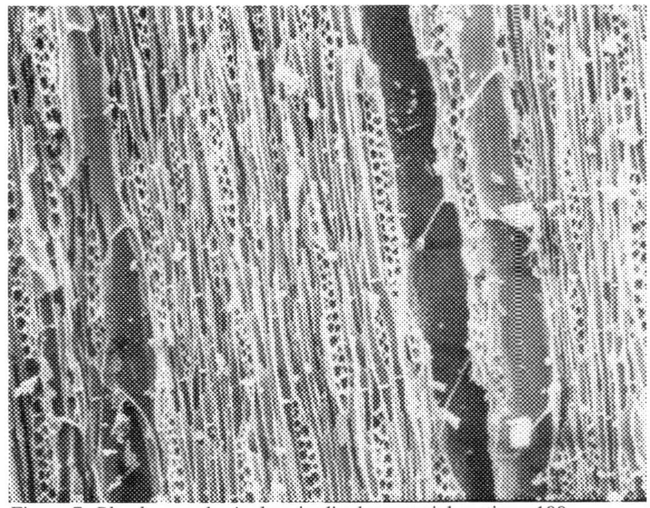
Figure 7: *Phoebe porphyria*, longitudinal tangential section x100.

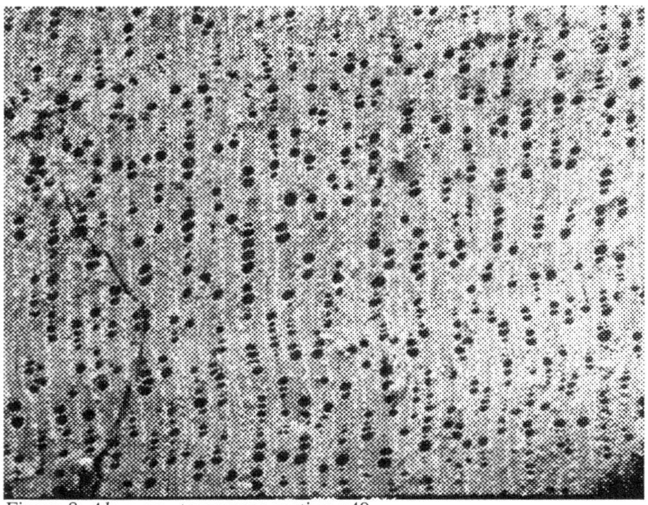
Figure 8: *Alnus* sp., transverse section x40.

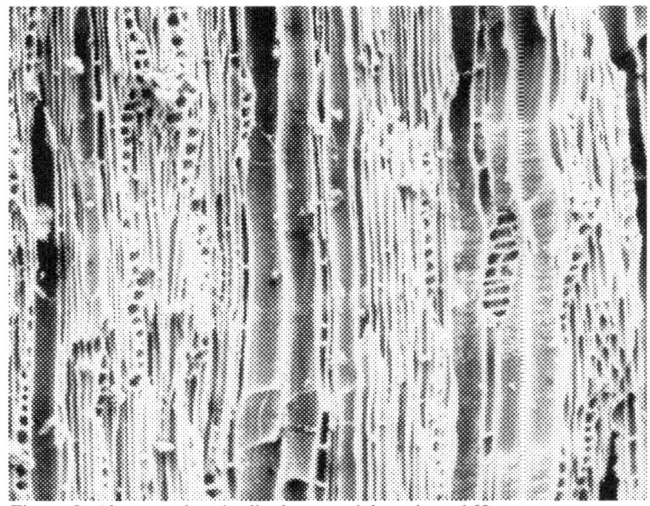
Figure 9: *Alnus* sp., longitudinal tangential section x160.

Iglesia de los Indios. Palaeoenvironmental studies, carried out in the southern Andean region, have established that the current distribution of the vegetation appeared around 2000 BC (Markgraff, 1985). The charcoal samples from domestic hearths abound in *Prosopis* and others species that are indigenous to the valley (Marconetto, 1999). These do not grow in the same type of habitat as *Phoebe* and we therefore discard environmental changes as an explanation to the presence of *Phoebe* and *Alnus* in the archaeological record of the Ambato valley.

We consider that socially based variables influenced on the taxonomic composition of the samples. In our study area, significant changes leading to a new social and economic organisation and a new relationship between people, things and nature, took place at the beginning of the Christian era. This new way of life resulted in a more complex economy, political inequalities and a diversification of social roles (Pérez Gollán, 1991). In this general context, the presence of non-indigenous wood, brought to the Ambato valley from the forests of the Yungas area, seems relevant. It should be stressed that the cebil-tree (*Anadenhantera peregrina*), whose pods yield a hallucinogenic substance used by pre-Hispanic populations of the Andean region (Pérez Gollán and Gordillo, 1993), also grows in this forest. The controlling of this resource by groups living in the Ambato valley could be one possible explanation of the process of increased social complexity in the area. As the Ambato valley is located on the western border of the Yungas area it has been suggested that the area may have been under the control of the Ambato people or an allied group (Pérez Gollán and Gordillo, *op.cit.*). The control over a sacred and highly prized resource could have contributed to the political development of the valley. The presence of *Phoebe* and *Alnus* in the archaeological record reinforces this hypothesis as the timber of these two taxa probably was exploited in the forest area of the Yungas.

A considerable effort had to be made to transport wooden logs from the rainforest to the sites, situated at a distance of 40 km. This indicates a certain level of organisation, equally visible in the agricultural practices and the construction of public and religious structures. This also implies the diversification of tasks and the use of labour.

By the identification of wood used for constructions at Piedras Blancas and Iglesia de los Indios, we have been able to study the use of various species and to integrate the anthracological data into a general archaeological picture. Anthracological analysis can thus contribute to the solving

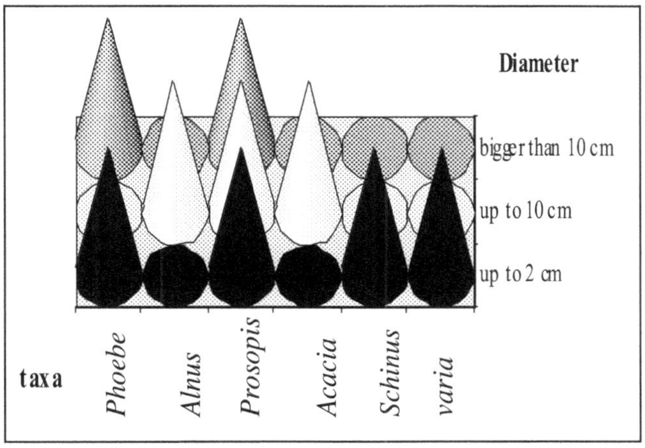

Figure 10: Average diameters of identified taxa.

of archaeological problems and provide references for taxonomical identification.

References

CABRERA A. 1976 - *Regiones fitogeográficas argentinas. Enciclopedia Argentina de Agricultura y Jardinería*. ACME. 2da ed. Buenos Aires, vol 2, p. 1- 85.

CASTELLETTI L., ZIMMERMAN A. 1985 - Seriation for a spatial factor. An interpretation of the distribution of wood species in a settlement as a hypotesis of the use of the settlements of neighbourhood. *To pattern the past: Proceedings of the Symposium of Mathematical Methods in Archaeology*. Worrips A. Loving S. (eds). Amsterdam, p. 111-118.

FEBRUARY E. 1992 - Archaeological charcoals as indicators of vegetation change and human fuel choice in the late Holocene at Elands bay, western cape province, South Africa. *Journal of Archaeological Science* 19, p. 347- 354.

GORDILLO I. 1994 - *Arquitectura y religión en Ambato, organización socio espacial del ceremonialismo*. CIFFyH, Universidad Nacional de Córdoba. Publicaciones 47, p. 27-39.

KREUZ A. 1992 - Charcoal from the early Neolithic settlements in Central Europe and its interpretation in terms of woodland management and wildwood resources. Les charbons de bois, les anciens ecosystèmes et le rôle de l'homme. *Bulletin de la Société Botanique de France*, 139, Actualités Botaniques, p. 383-394.

MARCONETTO M. B. 1999 - *Las leñas del Jaguar. En los tres reinos: Prácticas de recolección en el cono sur de América*. Eds. Aschero, Korstanje y Vuoto. Ediciones Magna publicaciones, p. 179-186.

MARKGRAF V. 1985 - Paleoenvironmental history of the last 10000 years in Norwestern Argentina. *Zbl. Geol. Paleont*, Stuttgart, p. 1739-1749.

PEARSALL D. M. 1989 - *Paleoetnobotany. A handbook of procedures*. Academic press inc. San Diego. 700 p.

PEREZ GOLLAN J. 1991 - *La Cultura de La Aguada vista desde el Valle de Ambato*. CIFFyH, UNC. Publicaciones 46, p. 158-173.

PEREZ GOLLAN J.A., GORDILLO I. 1993 - Religión y alucinógenos en el antiguo Noroeste Argentino. *Ciencia Hoy*. Buenos Aires, vol 4, 22, p. 23-33.

PEREZ GOLLAN J., BONNIN M., LAGUENS A., ASSANDRI S., FEDRICI L., GUDEMOS M., HIERLING J., JUEZ S. 1996-97 - Proyecto Arqueológico Ambato: Un estado de la cuestión. Universidad Nacional de Catamarca. *Shincal 6*, p. 141-152.

PIQUE I HUERTA R. 1999 - Producción y uso de combustible vegetal arqueológico: Una evaluación arqueológica. *Treballs d'Etnoarqueología*, 3. Universidad Autonoma de Barcelona. Consejo superior de investigaciones científicas Madrid. 307 p.

SHACKELTON C. M., PRINS F. 1992 - Charcoal analysis and the principle of least effort. A conceptual model. *Journal of Archaeological Science*, 19, p. 631-637.

TINTO J. 1978 - *Aporte del sector forestal a la construcción de viviendas*. Folleto técnico forestal, 44. 2da Ed. Instituto Forestal Nacional. 145 p.

Wood from ancient Egypt: the Ramesseum and the Valley of the Queens (18th Dynasty-Roman period). A preliminary report

María Victoria Asensi Amorós

Universidad de Alicante, Spain.
Laboratoire de Paléoécologie et Paléobotanique, Université Pierre et Marie Curie-Paris VI, France.
4, Place Violet, 75015, Paris France
E-mail: Victoria.Asensi@snv.jussieu.fr

Abstract

Sixteen samples of wood discovered in the Ramesseum ("Mansion of Millions of Years" of Rameses II) and in the Valley of the Queens in Egypt have been analysed. This is the first study of wood used for manufactured objects from this sector on the western side of the Nile river. The objects are dated from the 18th Dynasty (XVth century BC) until the Roman period (I^{st} century BC-IVth century AD). Most of them belong to the Third Intermediate Period, *i.e.* to the 22nd-25th Dynasties (945-656 BC), a period during which a necropolis was built in the Ramesseum complex. The objects for which a botanical identification has been made are the following: a fragment of a box made of cypress wood (*Cupressus* sp.), a piece of a cane made of acacia (*Acacia* sp.), an arrow point made of oak (*Quercus* cf. *ilex/coccifera*) and small bags containing cedar sawdust (*Cedrus* sp.) dated to the 18th Dynasty; two fragments of coffins made of cedar and eitght coffins made of pine (*Pinus* cf. *halepensis*, *P.* cf. *sylvestris* and *P.* cf. *sylvestris/nigra*) dated to the Third Intermediate Period and two fragmentary objects made of sycamore wood (*Ficus sycomorus* L.) dated to the Roman period. The analysis shows that most of the studied objects were manufactured from wood that was not indigenous to Egypt. In fact, only three objects out of sixteen were made of local wood species (*Acacia* sp. and *Ficus sycomorus* L.). This shows the importance of the trade of wood during Antiquity.

Key-words: *Wood, Egypt, Ramesseum, Trade, Antiquity.*

Résumé

Seize échantillons de bois découverts au Ramesseum (Temple de "Millions d'années" de Ramsès II) et dans la Vallée des Reines en Egypte ont été analysés. Il s'agit de la première étude de bois d'œuvre provenant de ce secteur situé sur la rive occidentale du Nil. La datation des échantillons s'étend de la 18ème dynastie (XVème siècle av. J.C.) jusqu'à l'époque romaine (Ier siècle av. J.C-IVème siècle ap. J.C.). La plupart des échantillons appartiennent à la Troisième Période Intermédiaire, c'est-à-dire de la 22ème à la 25ème dynastie (945-656 av. J.C.), époque où fut installée une nécropole sacerdotale sur le temple de Ramsès II. Les pièces ayant fait l'objet d'une identification botanique sont les suivantes: un fragment de boîte en cyprès (*Cupressus* sp.), un fragment de bâton en acacia (*Acacia* sp.), une pointe de flèche en chêne (*Quercus* cf. *ilex/coccifera*) et des sachets contenant de la sciure de cèdre (*Cedrus* sp.) datée de la 18ème dynastie; les fragments de deux sarcophages en cèdre et de huit sarcophages en pin (*Pinus* cf. *halepensis*, *P.* cf. *sylvestris* et *P.* cf. *sylvestris/nigra*) datés de la Troisième Période Intermédiaire et deux fragments d'objets en figuier sycomore (*Ficus sycomorus* L.) de l'époque romaine. L'analyse met en évidence la prééminence d'essences n'appartenant pas à la flore égyptienne par rapport aux espèces indigènes. En effet, seuls trois des 16 fragments étudiés se rapportent à des espèces locales (*Acacia* sp. et *Ficus sycomorus* L.), ce qui démontre l'importance du commerce du bois dans l'Antiquité.

Mots-clés : *Bois, Egypte, Ramesseum, Commerce, Antiquité.*

Introduction

This preliminary report presents the results of the wood identification made on the request of a Franco-Egyptian team bringing together the National Centre for Scientific Research (CNRS, URA 1064) and the Centre for Studies and Documentation of Ancient Egypt (CEDAE). The wood samples come from two Egyptian localities: the Ramesseum (memorial temple of Rameses II, at the edge of the Theban flood plain) and the Valley of the Queens (southernmost necropolis of western Thebes), both located on the western side of the Nile river facing the modern town of Luxor. The excavations of these two localities is carried out by the Franco-Egyptian team since 1980. The results presented here are the first to have been obtained from these sites.

The Ramesseum

The so-called Ramesseum is in fact the "Mansion of Millions of Years", the memorial temple of Rameses II (1279-1212 BC). This monument for eternity was constructed not far from Rameses II's tomb in the Valley of the Kings. The Rameses II cult in this temple forms a link between the actual temple, the economic complex temple, and the tomb. The construction of the memorial temple began in the first year of Rameses II's reign and was completed in the twenty-first year of the same reign. It covers approximately five hectares and it was the largest royal mortuary complex of its time, both in religious and economic terms. Today it is the only example of a mortuary temple conserving a part of the economic complex. The sanctuary is located in the centre of the complex, surrounded on three sides by annexes and storehouses (fig. 1).

The history of the Ramesseum does not end with the reign of Rameses II and the monument seems to have remained in activity throughout the Ramesside period (19th-20th Dynasties, XIIIth and XIIth centuries BC). The decay of the complex probably started during the 21st Dynasty (X^{th} century BC, beginning of the Third Intermediate Period). Throughout the Third Intermediate Period (22th-25th Dynasties, 945-656 BC), the Ramesseum, and in particular the western and northern annexes, were occupied by a sacerdotal necropolis. During the Late Period (664-332 BC), the site was quarried for stone, and during the Coptic era a church was built in the area of the second hypostyle hall.

The excavation of the "Mansion of Millions of Years" of Rameses II was first undertaken by Champollion and Lepsius in 1843 and 1845 (Lepsius, 1900), and then by Quibell (1898). In 1900-1906, Carter and Baraize worked on the site (Carter, 1901) followed by the Oriental Institute of Chicago in 1931-1932 (Hölscher, 1932). In 1969 the archaeological exploration of the Ramesseum was reinitiated by the Franco-Egyptian team (Hassanein *et al.* 1986; Babled and Leblanc, 1991; Lecuyot, 1999). The Association for Safeguarding the Ramesseum was created in 1989, in order to continue the research at the site.

It should be taken into consideration that the site has been disturbed by different events that have changed the stratigraphy in some of the sectors: first by the installation of the Third Intermediate Period necropolis and then by robberies, ancient excavations etc.

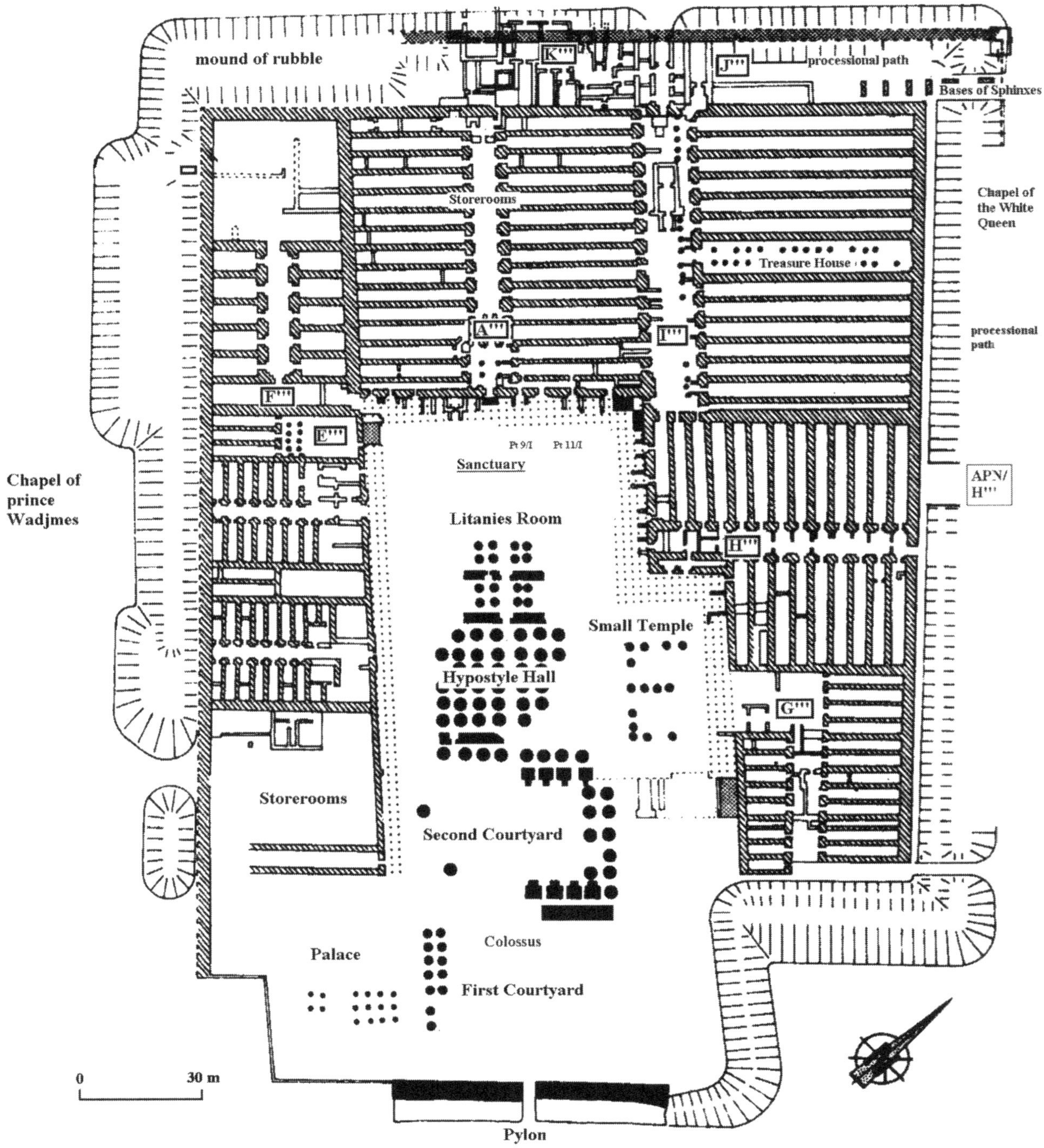

Figure 1: General Plan of the Ramesseum Complex (G. Lecuyot).

The Valley of the Queens

This necropolis is situated in an arid wadi of the Libyan mountain chain, next to the Valley of the Kings and not far from the Ramesseum. The Valley of the Queens contains numerous tombs of queens, princesses and princes from the New Kingdom, especially from the 19th and 20th Dynasties (1295-1069 BC). As for the Ramesseum, the locality continued to be used until the beginning of the Arab period (Leblanc, 1989). The Roman period has provided the greatest amount of archaeological material.

The exploration of the valley started with Champollion in 1828 and 1829. The formal excavations began in 1903 with the Italian Archaeological Mission (Schiaparelli, 1924) and the work done by the Franco-Egyptian team began in 1970 with the exploration of numerous tombs from all of the periods represented in the necropolis (Lecuyot, 2000).

The wood identification

Material and methods

We prepared microscope slides from sixteen samples of archaeological wood. The wood samples were cut manually along three sections (a transversal section, a tangential section and a radial section) using a microtome blade. The sections were mounted in euparal on microscope slides. The sawdust was mounted directly on slides without cutting.

Identifications were made with the help of the wood reference collection of the Laboratory of Palaeobotany and Palaeoecology (Laboratoire de Paléobotanique et Paléoécologie) at the University of Pierre and Marie Curie-Paris VI, the author's reference collection and several bibliographical references (Fahn *et al.* 1986; Greguss, 1955; IAWA Committee, 1989; Neumann *et al.* 2001; Schweingruber, 1990).

In this preliminary report we have often restricted the identification to the genus level and we only occasionally suggest a species name as a more detailed publication on wood from the Third Intermediate Period Ramesseum is in preparation. Nevertheless, concerning the identification of *Pinus* sp., we have used the diagnostic features provided by Hudson (1960), Greguss (1955), Grosser (1977), Schweingruber (1990) and Fahn *et al.* (1986).

Samples from the Ramesseum

The samples from the Ramesseum come from three different areas of the Third Intermediate Period necropolis (fig. 1): area J''' (between the storerooms and the processional path, north-west to the temple), area APN/H''' (northern part of the processional avenue surrounding the temple) and the sanctuary (in the rear of the Litanies Room, tombs Pt 9/I and Pt 11/I). The samples received for analysis were badly preserved as the tombs had been robbed in antiquity and the mummies and their sarcophagus had been burned in order to obtain valuable objets.

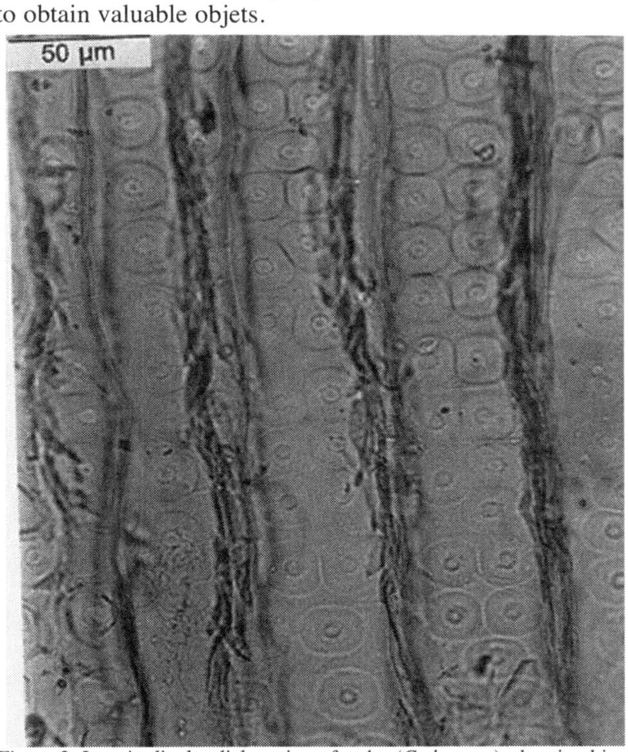

Figure 2: Longitudinal radial section of cedar (*Cedrus* sp.), showing biseriate tracheids pits with scalloped tori (fringed torus margins): sample J'''II.

The tombs of the Third Intermediate Period generally consist of a vaulted chapel with one or more shafts. These vertical shafts were not more than four or five meters deep and connected with one or two chambers. To construct these tombs, building material recovered from the Rameses II's temple as well as from other surrouding temples was used. The material discovered in the tombs gives an idea about the type of objects they may have contained originally. Written documents (papyrus, stelae), canopic jars, ushabtis (faience, stone or wooden statuettes representing funeral servants), amulets and fragments of wooden coffins were found. The wooden coffins were of three types: painted in black with yellow lines, undecorated (or occasionally with painted blue hieroglyphs) or painted in white colour upon gesso with blue or black hieroglyphs.

From the Meritamon tomb (J'''II; Leblanc and Nelson, 1997, p. 67; Hassanein *et al.* 1986, p. 17) two fragments of wood from coffins were identified: one of them had a mention of a "priest of Maât" and the other was decorated with a yellow line. Both of them were made of cedar wood (*Cedrus* sp., fig. 2).

A fragment of a box of cypress wood (*Cupressus* sp.) and a piece of cane made of acacia wood (*Acacia* sp.) were found in area J''' (IV/A and B ; Leblanc and Nelson, 1997, p. 68-69). These two items seem to date to the end of the 18th Dynasty (XIII century BC), *i.e.* prior to the installation of the Third Intermediate Period necropolis (Hassanein *et al.* 1986, p. 18).

Two pieces of a sarcophagus from the Third Intermediate Period tombs (Pt 9/I and Pt 11/I; Lecuyot, 2001) were also found in the sanctuary. These were identified as pine wood

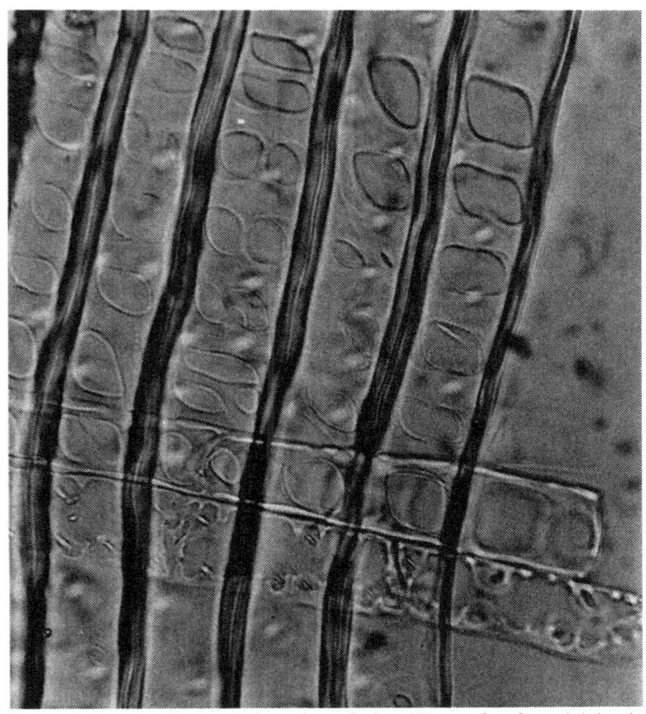

Figure 3: Longitudinal radial section of pine (*Pinus* cf. *sylvestris/nigra*), showing 1-2 very large pinoid, window-like pits, per cross-field. Ray tracheids arranged in numerous rows, either marginal or interspersed among the parenchyma cell rows, and with thick and dentate walls inlaid with numerous small, bordered pits: sample Pt 11.

(*Pinus* cf. *halepensis* and *Pinus* cf. *sylvestris/nigra* respectively, fig. 3).

In area APN/H''' (north of the temple) a funeral complex with six tombs from the Third Intermediate Period was discovered (Kalos *et al.* 1996). Six fragments of coffins were analysed and identified as pine wood (*Pinus* cf. *halepensis* and *Pinus* cf. *sylvestris*).

Samples from the Valley of the Queens

From this necropolis, samples from two tombs, nos. 72 and 33 (next to Nefertari's tomb), were studied.

Tomb 33 (QV 33), excavated by Ch. Leblanc (CNRS), revealed the name of an unknown queen, Tanedjemy who lived in the beginning of 19th Dynasty. Nevertheless, our material (a piece of manufactured wood and a piece of sarcophagus) has been dated to the Roman period. Both of these pieces were made of sycamore wood (*Ficus sycomorus*, fig. 4).

In tomb 72 (QVs 72), a vase containing fruits of *Ficus sycomorus* or *Ficus carica* and inscribed with the name of prince Baki of the 18th Dynasty (1550-1295 BC) was found. The studied material (apparently from the 18th Dynasty) was different from that of tomb 33. An arrowpoint made of evergreen oak (*Quercus* cf. *ilex/coccifera*, fig. 5) was identified, as well as the content of small woollen bags, which proved to be a mixture of sawdust and natron, probably a part of the "embalming refuse" material. The sawdust, composed of cedar powder (80%) and unidentified vegetal remains, shows similarities to the substances found in the abdominal cavity of a mummy (726-375 BC) identified a few years ago (Asensi Amoros and Vozenin-Serra, 1998).

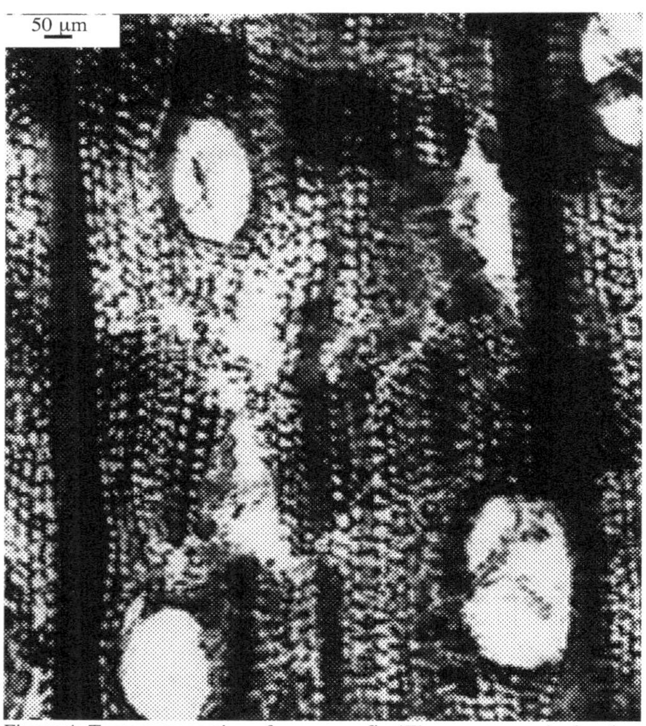

Figure 4: Transverse section of sycamore fig (*Ficus sycomorus*), showing vessels solitary or arranged by two in radial multiples. Paratracheal parenchyma in bands altering with fibres: sample from Tomb QV33-1.

Conclusion

The first study of tomb furniture from the Third Intermediate Period necropolis in the Ramesseum shows that only one fragments out of twelve was made of local acacia wood. All other samples consisted of coniferous wood that does not belong to the Egyptian flora. The Roman material found in tomb 33 in the Valley of the Queens was equally made of local wood (*Ficus sycomorus*) while the more ancient material from tomb 72 must have been imported. The wood used for different purposes seems to have been chosen according to its specific characteristics: coniferous wood was used for the sarcophagus, strong oak wood for the arrow and rotproof cedar for the embalming material. The piece of oak found in tomb 72 constitutes together with a dowel from Tutankhamun's tomb the oldest example of this species in Egypt (Vartavan and Asensi Amorós, 1997, p. 220).

The results of the identification work show the abundance of foreign wood used for the funeral material. It is commonly accepted that Egypt, lacking indigenous wood for constructions and the manufacture of precious objects, imported mainly coniferous wood from the Levant. In fact, the Egyptian vegetation is essentially composed of xerophilous savannahs, thorny steppes and, above all, desert that cover 96 % of the contry's surface (Asensi Amorós, 2000 b). Therefore, the wood trade has always held a primary place in the Egyptian economy. Since prehistoric times, remains of imported coniferous wood (*Cedrus* sp., *Cupressus* sp. and *Pinus* sp.) are attested (Vartavan and Asensi Amorós, 1997). Epigraphic, lexicographic and archaeological sources from ancient Egypt confirm the importance of imported wood (Asensi Amorós, 2000a). Already during the Old Kingdom (*c.* 3000 BC) Egypt traded with Nubia and the Levant. The south penetration and the expansion to the east continue during the Middle Kingdom (2033-1710 BC) and is particularly important during the New Kingdom (1550-1069 BC). The Ptolemaic Period (332-30 BC) shows the expansion of trade roads to the Red Sea and the Indian coast, roads that became particularly important during the following Roman period (Asensi Amorós, 2001). In fact, precious wood was imported from the whole Mediterranean area, Africa, the Near East and possibly also from India. This means that the distribution range of wood samples found in Egyptian archaeological contexts is large. For the moment, it is difficult to determi-

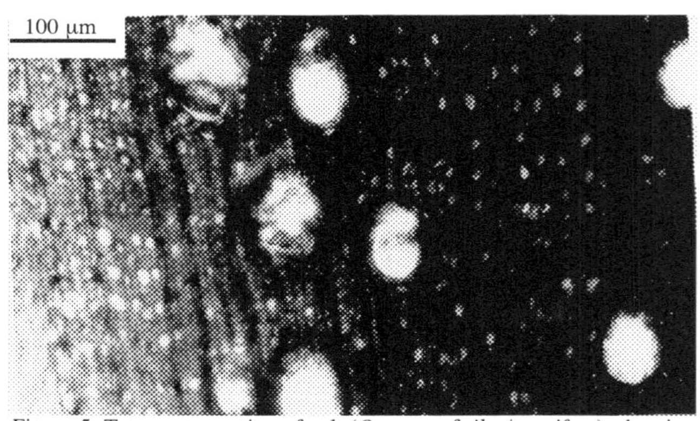

Figure 5: Transverse section of oak (*Quercus* cf. *ilex/coccifera*), showing a detail of solitary vessels. Parenchyma in narrow tangential bands, mostly one cell wide: sample (arrowpoint) from Tomb QV72.

ne the exact provenience for the wood from the Ramesseum. The presence of *Cedrus* sp., *Pinus* sp. and *Cupressus* sp. in the Ramesseum material seems to indicate an origin in the eastern Mediterranean region; in Lebanon, Syria, Cyprus...

Our knowledge about the wood used in Egypt during the Third Intermediate Period is quite poor. Even though our analyses have shown that pine wood is more represented than the traditionally imported cedar wood, we will need more material to confirm this.

The excavation campaign of 2000 at the Ramesseum temple has now ended. In 2000, the kitchens located in the economic complex have been partly excavated. The rich material found there (wood, charcoal and other vegetal macroremains) and in the newly excavated tombs will complete our knowledge of Egyptian environment, food and funeral furniture during Pharaonic times.

ACKNOWLEDGEMENTS
The author wishes to express her gratitude to Professor J. Broutin, Director of the "Laboratoire de Paléobotanique et Paléoécologie (University of Pierre and Marie Curie-Paris VI)" who kindly offered technical support as well Dr. Ch. Leblanc for having invited her to the Ramesseum excavations.

References

ASENSI AMORÓS V. 2000 (a) – *La madera en el antiguo Egipto: identifcaciones, usos y comercio*. Tesis de la Universidad de Alicante.

ASENSI AMORÓS V. 2000 (b) – *Geb, Geobotanical Bibliography of Egypt and Adjacent Countries for the Egyptologist*. London, Triade Exploration, 245 p.

ASENSI AMORÓS V. 2001 – Madera de Egipto, madera importada. I. : Contribución del estudio de la anatomía de la madera para la comprensión de la civilización egipcia. *Aula Aegyptiaca-Studia*, t. 2, p. 23-34.

ASENSI AMORÓS V., VOZENIN-SERRA C. 1998 – New Evidence for the Use of Cedar Sawdust for Embalming by Ancient Egyptians. *Journal of Egyptian Archaeology*, t. 84, p. 228-231, pl. XXIV.

BABLED Th., LEBLANC Ch. 1991 – La bibliographie du Ramesseum. *Memnonia*, t. 2, p. 65-78.

CARTER M. 1901 – Report on work done at The Ramesseum. *Annales du Service des Antiquités de l'Egypte*, t. 2, p. 193-195.

FAHN A., WERKER E., BAAS P. 1986 – *Wood anatomy and Identification of Trees and Shrubs from Israel and adjacent Regions*. Jerusalem, The Israel Academy of Sciences and Humanities, 221 p.

GREGUSS P. 1955 – *Identification of Living Gymnosperms on the basis of Xylotomy*. Budapest, Akadémiai Kiado.

GROSSER D. 1977 – *Die Hölzer Mitteleuropas*. Berlin/Heidelberg/New York, Springer Verlag.

HASSANEIN F., LECUYOT G., LOYRETTE A.M., NELSON M. 1986 – Les récentes découvertes au Ramesseum. *Bulletin de la Société Française d'Egyptologie*, t. 106, p. 7-30.

HÖLSCHER V. 1932 – *Excavations at Ancient Thebes* 1930/31. Chicago, The University of Chicago Press.

HUDSON R.H. 1960 – The Anatomy of the genus *Pinus* in relation to its classification. *Journal of the Institut of Wood Science*, t. 6, p. 26-46.

IAWA Committee 1989 – List of microscopic features for hardwood identification. *IAWA Bulletin* n. s., t. 10, p. 221-332.

KALOS M., NELSON M., LEBLANC Ch., SAINTILAN M. de, GUICHARD H., ABOUL MAGD A. M. 1996 – L'ensemble monumental dit "chapelle de la Reine blanche". *Memnonia*, t. 7, p. 69-82.

LEBLANC Ch. 1989 – *Ta Set Neferou. Une nécropole de Thèbes-Ouest et son histoire*. I. Le Caire, Printing House.

LEBLANC Ch., NELSON M. 1997 – Répertoire onomastique des propriétaires des tombes de la Troisième Période Intermédiaire du Ramesseum (I). *Memnonia*, t. 8, p. 61-91.

LECUYOT G. 1999 – The Ramesseum. The Mansion of Millions of Years of Rameses II at West Waset. KMT. *A Modern Journal of Ancient Egyp*t, t. 10, fasc. 1, p. 35-47.

LECUYOT G. 2000 – Ta Set Neferu. The Valley of the Queens: A Brief History of its Excavations. KMT. *A Modern Journal of Ancient Egypt*, t. 11, fasc. 2, p. 42-55.

LECUYOT G. 2001 – Le sanctuaire du Ramesseum. Fouille et étude. Campagnes 1997-1999, *Memnonia,* t. 10.

LEPSIUS C. R. 1900 – *Denkmäler aus Aegypten und Aethiopien*. Text III. Leipzig.

NEUMANN K., SCHOCH W., DETIENNE P., SCHWEINGRUBER F.H. 2001 – *Woods of the Sahara and the Sahel, an anatomical atlas*. Bern/Sttutgart/Wien, Paul Haupt, 465 p.

QUIBELL J.E. 1898 – *The Ramesseum*. London, B. Quaritch.

SCHIAPARELLI E. 1924 – *Relazione sui lavori della Missione archeologica italiana in Egitto*. I. Esplorazione della "Valle delle Regine" nella necropoli di Tebe. Torino.

SCHWEINGRUBER F.H. 1990 – *Anatomie of European woods*. Bern/Stuttgart, Paul Haupt, 800 p.

VARTAVAN Ch. de, ASENSI AMORÓS V. 1997 – *Codex of Ancient Egyptian Plant Remains / Codex des restes végétaux de l'Egypte ancienne*. London, Triade Exploration, 400 p.

Adaptation of colonists to Canadian woods extracts: the examples of housebuilding, shipbuilding and heating wood

Emmanuelle Miejac

8 Impasse des Herbeuses, 94200 Ivry-sur-Seine, France.
E-mail: Emiejac@aol.com

Abstract

At the beginning of the XVIIth century, Quebec was founded on the shores of the Saint Laurent. Serving as a point of entry into this new territory to conquer, the city witnessed the setting and development of the necessary infrastructures that the settlers would most need: first, habitation and housing construction, and soon after, a shipyard whose establishing had been ordered from France. The imposing and omnipresent forests of Canada served as the primary location of extraction of the necessary raw materials, but the new colonists were required to adapt to new conditions. At the same time that the people were adapting to the new needs the natural resources were evolving in conjunction with the colonists' demands. This work aims, through the study of the wood extracts used, to recognise and to understand these varied adaptations. The forest extracts chosen to build houses and for heating correspond to special criteria. The colonists quickly learned, either from their own experience or from the native population, how to choose the materials that were best suited to their needs.

Key-words: *Quebec, shipyard, fishing ground, inhabitat, wood species, adaptation.*

Résumé

Au début du XVIIe siècle, Québec est fondé au bord du Saint-Laurent. Porte d'entrée d'un vaste territoire, cette première cité vit l'installation des insfrastructures en bois nécessaires à la vie : habitat, pêcheries et chantiers de construction navale notamment. Les premiers colons devront adapter leurs techniques à ce nouveau pays : conditions climatiques et matières premières disponibles différentes. Ils utiliseront par exemple du bois franc : érable, chêne et frêne, pour le chauffage, et n'emploieront pas forcément du chêne pour la construction des navires. Cette rapide adaptation des colons est l'une des caractéristiques de la société canadienne en formation. La seconde est une certaine forme de gestion de la forêt.

Mots-clés : *Québec, construction navale, pêcherie, habitat, essence de bois, adaptation.*

At the beginning of the XVIIth century, Quebec was founded on the shores of the Saint Laurent. Serving as a point of entry into this new territory to conquer, the city witnessed the setting and development of the necessary infrastructures that the settlers would most need: first, habitation and housing construction, and soon after, a shipyard whose establishing had been ordered from France. The imposing and omnipresent forests of Canada served as the primary location of extraction of the necessary raw materials, but the new colonists were required to adapt to new conditions. At the same time that the people were adapting to the new needs the natural resources were evolving in conjunction with the colonists' demands.

Wood and house building

The first houses built in Canada reveal the initial adaptation of the settlers to their new landscape. Wood was used due to its relative adaptability and its qualities as an appropriate construction material. Indeed, wood provided effective insulation against a harsh climate with sudden and dramatic variations in temperature.

Walls

The first country houses, called billets houses, were built with wood. The walls were made with squared billets, which were piled one upon the other. The interstices were stuffed with clay. Later, mortar was used. Wall angles were finished in dovetail or wall broom joints.

A second type of wood house was also distinguished: the half timbered or Norman house. In 1663, Jacques Testard from Montreal stayed in this type of house: "une maison, de bois de charpente et coulombage, boussillée et enduitte hors doeuvre et dans œuvre de chaud, avec son comble de charpente, couverte de bardeau et planches, une cave de massonne, faite le tout a chaud et sable, trois planchers de madiers embouvetez, une eschelle louée depuis ladite cave jusqu'au grenier, quelques petis cabinets, avec ses portes, fenestres, chassis et contrevenans" (Procès-verbal d'immeubles de deffunt Jacques Testard de la Forest. Le 5 novembre 1663, notaire Bénigne Basset 297 AJM).

Houses were first built with wood, but this practice was quickly forbidden in cities (fig.1). "Défendons de bâtir aucune maison dans les villes et gros bourgs, où il se trouvera de la pierre commodément, autrement qu'en pierre, défendons de les bâtir en bois, de pièces sur pièces et de colombage, quand même ce seroit pour les recouvrir et enduire de chaux et sable." (Ordonnance portant règlement de la construction des maisons, Dupuy, 7 juin 1727). The beginning of the XVIIIth century was then marked by a development of agglomerations and repeated fires that killed settlers and devastated more and more of the new buildings.

The two main used extracts were then red spruce and white spruce. Red spruce (*Picea rubra*) could reach about 50 to 75 feet with a diameter of 1-2 feet. Its wood resembles that of white spruce but can be recognised through its more pronounced annual growth. It is also a little harder, heavier and more resistant; however, this difference is slight -sometimes so slight that it is very difficult to distinguish the two. White spruce (*Picea glauca*) reached 100 feet by 4 but was more frequently found at about 50 feet by 1.5-2 feet in diameter. The wood colour varies from white to pale yellow, without any significant difference between sapwood and "duramen". While white pine was not resistant to dry rot, it was used under favourable conditions as material for ordinary small and medium-sized buildings, stringed instruments, shop fittings, string and paddle, cooperage, stepladders and ladders, and kitchen cupboard.

First built in cities, stone houses were frequently built in the countryside beginning in the second quarter of the XVIIth century. In 1727, an order of the administrator Dupuy allowed carters and quarrymen to bring stones from the quarries of Beauport, Sillery, l'Ange Gardien, la Pointe-aux-Trembles, Château-Richer, Charlesbourg, by road and by water. For settlers, acquiring stone presented little difficulty, but there was no methodical exploitation of quarries. Moreover, the use of wood in building practices persisted in spite of administrative orders and the practical dangers mentioned above. Stone cutting techniques had not yet been mastered (perhaps due to lack of practice or lack of knowledge...). Often, fieldstones, easily collected by hand and readily available, were preferable for building stone houses, particularly in rural areas. These materials were submerged in a mortar made of sand and lime.

For the frames, the sleepers, wall plates and posts were assembled in mortise; sandpits, rafters, joists were then pinned together. Saint-André crosses were riveted between the wall plates and the rafters for added support. Beams secured principal rafters or other rafters. The frame pieces were cut from hard wood such as ash tree, oak and beech-tree. Conifer wood such as hemlock spruce (canadian *Tsuga*) and spruce was used occasionally.

Floors were made of stakes, beams and boards, in the cities as well as in the countryside (fig. 2). The stakes were usually made of oak, lime-tree, or cedar-wood. They were split into two in order to place them on the wall plates or the rafters (see picture).

In the XVIIth century, many houses were covered by thatch, but boards were more frequently used: "les unes (maisons) sont basties toutes de pierres et couvertes de planches ou aix de pin, les autres sont baties de collombages ou charpente, et masonees entre les deux, d'autres sont baties tout a fait de

Figure 1: A example of a stone and wood house.

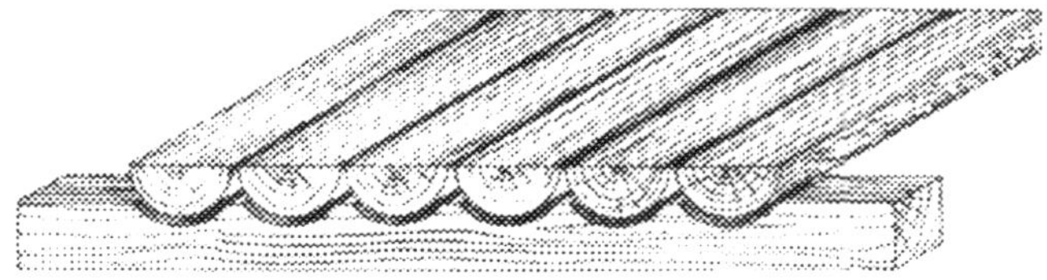

Figure 2: Construction of a floor.

bois, et toutes les dites maisons se couvrent comme dit est, de planches" (notices of Pierre Boucher 1633, 140). In August of 1687, Jean Guy Levacher du Laperle, a carpenter from Trois-Rivières, agreed to make siding from boards: "planches emboustées avec tringles" (market of August 16, 1687, notary Antoine Adhémar, n°918, AJM).

Siding

The shingle-board is dangerous because of its flammability, for Denonville it would be safer to cover roofs with wooden shingles than to line frames with sulphur-matches (ACC11A, S, fol 19).

In 1668, the upper council strictly forbade the use of shingles in cities (decisions and rules of the upper council). In the following years, an exception was made for dormer windows that were covered with shingles of oak or walnut (decisions and rules of the upper council, 1669).

On June 7, 1727, the administrator Claude Thomas Dupuis decreed that "ceux qui ont amassé du bardeau (en ville), dans le dessein d'en couvrir leurs maisons, seront tenus de s'en défaire en faveur de ceux qui bâtissent à la campagne, auxquels seulement nous permettons de couvrir en bardeau, jusqu'à ce qu'il ait été fait de la tuille, dans la colonie, suffisamment pour abolir et rejetter tout à fait une matière quasi pernicieuse que l'est le bardeau de cèdre, dont on se sert en ce pays" (administrative order and decision II, 79). He suggested furthermore the use of two boards for additional thickness, the first one from the top to the bottom and the second one as an additional lining.

Cedar was recommended for the shingles (*Thuya occidentalis*): it reaches 80 feet by 3 of diameter, often of medium size, never more than 45 feet by 1 foot. It is fairly light, weighing 21 pounds per cubic foot when it has been allowed to air dry. Soft, slightly resistant, fissile, it lasts well on the ground and under conditions that would potentially cause rotting. It is used for shingles, launches, barges, tanks, warm greenhouses, dock floats and fishing line floats.

Walnut tree (*Juglans cinerea*): 40-50 feet by 1-3 feet in diameter.

Sapwood, duramen, is darker. Light wood, soft and slightly resistant, is classified among the soft hardwoods of the lime-tree species. Oak used to be the *Quercus alba*.

In order to face the difficulties presented by using shingle boards, the administrators tried to develop tileries and slate quarries. Nevertheless, tile, of common use in France, was hardly ever used in New France. In 1685, the administrator of Meule built a small tilery where a few thousand tiles were made (Fauteux, I, 160). By request of the administrator, the minister ordered tile-makers from La Rochelle, France to come to New France "in order to improve the tile work of Sir de Meules" (Canadian Arch, part B, Vol 12, p 74). The undertaking was jeopardised, however, and the inhabitants were not able to use this material because of the extreme cold.

In 1731, an order directed the administration to study the development of the Grand-Etang slate quarry (Bill of Act of Hocquart, August 26, 1731; vol. 19, fol. 148).

Wood, while it represented an important resource in the new colony, and though it corresponded to the needs of the Native Canadians, was quickly abandoned for the construction of houses and particularly for roofing, according to the official bills. It is important to note that it was still widely used, which is illustrated by the number of orders during the XVIIIth century.

The colonists adapted quickly to the conditions of their new life and to the new climate. Once the colony was well settled, it would appear that they had mastered the frigid climate: "Il ne faut pas penser que les français habitués en Canada souffrent considérablement de ce froid, parce qu'ayant du bois en volonté, ils se chauffent tant qu'il leur plait, bien différents en cela des français de quantité de provinces de France qui manquant de bois souffrent infiniment plus de froid, quoique dans un air plus tempéré, que ne sont ceux de Canada parce qu'ils n'ont pas de quoi se chauffer et que sortant de la rue ou des champs où ils sont glacés, ils reviennent dans leurs maisons qu'ils trouvent sans feu." (RAPQ, 1920-1921, t. 24. Etat présent du Canada, dressé sur nombre de mémoires et connaissances acquises sur les lieux, par le sieur Boucault en 1754).

Fuel wood

The quantity of fuel wood is measured in "cords". The cord, according to a bill of act of May 11, 1676, measures 8 feet long by 4 feet high and 3.5 feet wide from end to end (French measure). The merchants, taking advantage of their customers, often shortened the width of the logs in order to increase their profits. A second bill of act was published on September 20, 1748, in which the administrator established the width of logs at 4 feet (French measure). Following another complaint from the settlers, Bigot enacted a new bill of act dated October 1, 1749. A cord of wood would measure 2.5 feet from end to end.

Transport

Fuel wood arrived in the cities of New France by sea from the surrounding country, secured to wooden rafts. The "cages" or "cageux" were made from big wooden boards of 25-35 feet long and often of more than 10 inches in diameter. They could be obtained with the heating wood and were then used for construction. Shipped with the help of riggings and propelled by strings of 25-30 feet long, the cages were composed of 2 or 4 smaller rafts, called "pagées", bound together by ropes. Consignment took place from the beginning of the shipping season, at the end of April, through mid-November.

The consignment locations were established little by little, so that lumber traffic did not disturb the navigation of ships and smaller barges. On June 25, 1727, a bill of act forbidding the cageux from reaching the Cul-de-Sac, a roadstead that was later transformed into a dry-dock. "il convient [...] d'empêcher l'abord des cageux de bois que l'on échoue et que l'on décharge dans le dit Cul-de-Sac[...] de confiscation des dits cageux au profit du dit hôpital (hôpital général) et du couvent des Récollets de cette ville (Québec)." (Ordonnances de Claude Thomas Dupuy du 25 juin 1727, vol. 12 fol. 125).

What kind of wood?

Heating wood was most often composed of franc wood-maple, oak and ash (Séguin, 1963, 348) or mixed wood, and sometimes of maple (tabl. 1). Franc wood was estimated at the same price, whereas mixed wood was valued at a lower price. Franc wood was the preferred heating wood, due to its higher calorific value in comparison with soft woods. Heating wood was stored in the yard, sometimes in a shed, under the coach-house, in a storeroom, in the cellar or in the granary. Consigned in cords, it had to be cut into smaller logs in order to supply the fireplace or the stove. In the XVIIIth century, an important variation in the price of heating wood was recorded. It rose from 5 pounds/cord in 1740 to 15 pounds/cord in 1758, and finally 24 pounds/cord in 1759, an increase which provoked complaints from the townspeople.

The chopping of heating wood was often determined through a contract executed by a notary, illustrated by the

Latin names	Vernacular names	Calorific value	Species used in New-France
Ulmus thomasi	Orme de Thomas	32.0	■
Carya ovata	Arbre à noix piques	30.6	
Quercus alba	Chne blanc	30.6	
Carya cordiformis	Noyer amer	29.2	
Acer saccharum	Erable blanc	29.0	■
Quercus macrocarpa	Chne à gros fruits	28.2	
	Htre à grandes feuilles	27.8	
Quercus borealis	Chne rouge	27.3	
Betula alleghaniensis	Merisier	26.2	■
Ulmus rubra	Orme gras	25.4	■
Fraxinus americana	Frne blanc	25.0	■
Ulmus americana	Orme blanc	24.5	
Acer rubrum	Plaine rouge	24.0	■
Larix laricina	Epinette rouge	24.0	
Prunus serotina	Cerisier d'automne	23.5	
Betula payrifera	Bouleau blanc	23.4	■
Fraxinus nigra	Frne gras	22.6	■
Fraxinus pennsylvanica	Frne rouge	22.1	■
Acer saccharinum	Plaine blanche	21.7	■
Pinus divaricata	Pin gris ou Cyprs	21.6	
Pinus resinosa	Pin rouge	19.7	
Acer negudo	Erable à Gigure	19.3	■
Picea mariana	Epinette noire	19.1	
Populus grandidentata	Peuplier à grandes dents	18.2	
Tsuga canadiensis	Pruche	17.9	
Populus tremulodes	Tremble	17.7	
Juglans cinerea	Arbre à noix longues	17.4	
Populus balsamifera	Peuplier	17.2	
Pinus strobus	Pin blanc	17.1	
Tilia americana	Bois blanc	17.0	■
Populus deltodes	Liard	16.8	
Thuja occidentalis	Cdre	16.3	
Abies balsamea	Sapin	15.5	

Table 1: Heating wood used in Canada.

contract of November 6, 1687 between Mr Brunet and the widow of Antoine Pichon: "quantité de 20 cordes de bon bois d'esrable, merisier, frestre, de fresne, et orme [...] a être livrer entre ... et la fin de mars dernier au devant de la porte de la dite en cette ville (Montréal) lequel il sera tenu de corder audit endroit lesquelles cordes seront de 8 pieds de long et 4 pieds d'hauteur suivant la coutume de Paris...a promis et soblige paier audit Brunet pour chacune desdites 20 cordes la comme de 5 livres a feure et a mesure quil l'es luy livrera[...] est conveneu entre lesdites parties quen cas que dans lesdites 20 cordes de bois il sen retrouve 3 de bois blanc ladite paira sera tenu de l'est recevoir comme le surplus dudit bois[...]".

At the birth of the colony, wood was abounding, but during the XVIIIth century, a shortage has been felt. The inhabitants used to cut wood on their concession but it happened sometimes that they encroached on other lands. A bill of act of 1715 forbade cutting wood on non-granted lands. Another followed in 1727, stating that the Lords were not allowed to take wood outside of their domain. An increase in the population and a lack forest management may have been the cause of this shortage. The trouble was more present in the cities than in the country. The price of wood rose constantly during the XVIIIth and XIXth centuries.

Wood in shipbuilding

Cargo ships

Cargo ships linked the colony to Europe. Throughout the XVIIth century and at the beginning of the XVIIIth century, ship traffic was at its busiest in the harbour of Quebec. From 1725-1730, a new trade route appeared with the West Indies. This link between Europe and the American colonies was essential. From 1666-1671, the first permanent shipyard was established on the Saint-Charles river in a sheltered location that was suitable for shipbuilding (Brisson, 1983, 25). From 1733-1739, military ships were built. During this period, the shipyard of the Saint Charles was transformed into a harbour and the Cul-de-Sac into a naval dockyard, subsequent to a decree from France that was carried out by the administrators. From 1733-1738, the Palace shipyard was erected (Cloutier, 1997) at the mouth of Saint Charles at the will of J. Talon and the royal establishment. A new shipyard project of was then founded on the site of Cul-de-Sac in 1746.

During the XVIIth and XVIIIth centuries, the building stages and the material used were regulated in France. The Canadian constructors first imported the most important wood pieces and tried later to find equivalents in the available resources of the colony.

The keel, an essential part of the ship, forms the basis of the entire structure (fig. 3). It must be durable and resistant, as it bears the weight of all of the other assembled elements. In France, oak (*Quercus robur*) was used because it is resistant and solid. Elm, which holds up well in humid environments, can be used as a substitute, but it is scarce in the metropolis. The difficulty finding elm and its cost often led the shipbuilders to decide against using this extract. In Canada, oak (*Quercus alba*) was also used, but it was very often of poor quality according to the notes of the administrators.

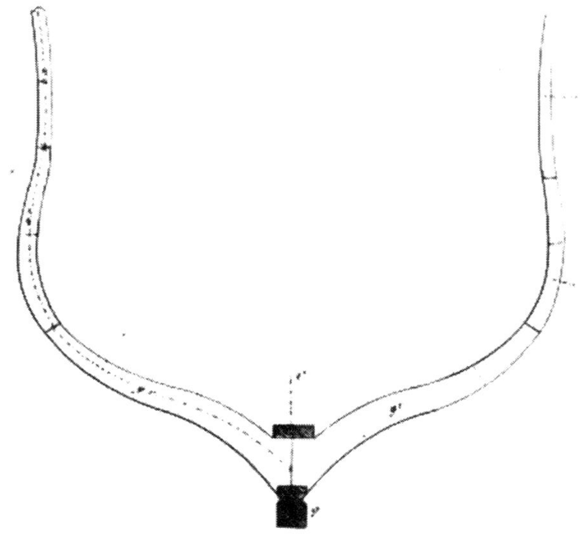

Figure 3: Vessel q.keel ; g: elbow-joint ; a: stanchions ; v: floor-timbers.

The elbow joints, floor-timbers and stanchions, pieces that make up the bottom of the vessel, form more or less the ship's frame. In France, oak pieces, with no knots that could provoke rot, were avoided, whereas big knots and twisted fibres did not keep them from using the wood; on the contrary, these variations were a sign of the quality, the strength and the durability of the material. In Canada, spruce was mainly used. A Marine carpenter used wild cherry wood in 1730 because he lacked material, but this was a very exceptional practice.

Planks complete the bottom and finish the ship's structure. In France, pine was often used because it was light and could thus reduce the weight of the vessel. Oak was also used. In Canada, the wood types used to make the planking varied from cypress, to cedar, spruce, larch or sometimes oak or chestnut.

Canadian administrators took particular care of the wood supply for shipbuilding. Exploitation was well organised: in autumn or at the beginning of winter, the areas to be logged were explored. Immediately following, a team of workers proceeded with the cutting of the billets and transported them to the river in order to build up the cajeux. During the XVIIIth century, the administrator gave precise instructions to Lanouillier de Boisclerc during his exploration of the forest: "Compter et marquer [...] tous les chesnes, examiner par les charpentiers qui l'accompagnent les racines de plusieurs de ces arbres, la tournure du tronc et des branches [...] s'il y a des chemins praticables pour conduire les bois a exploiter jusqu'au bord de la rivière [...] observer l'éloigenement des chesnes du bord de l'eau et les facilités et les difficultés qui peuvent se rencontrer dans le transport".

In the summer, wood was taken out of the water and piled up

on the shipyard for a short drying period before use.

Quality of the wood

For different administrators, wood varied accordingly:
- Levasseur: "the planking is better than that of Sweden".
- Hocquart: "Canadian wood is superior to wood from Lorraine".
- Peter Kalm: "American oak is not as good as European oak".
- Bigot: surprised by the poor quality of the wood received at the yard: "les 3/4 est tout rouli, gelif ou couronné et le cœur altéré et souvent gâté, il paraît pourtant beau à l'œil".

The quality of each piece of wood also varied. At times the wood was completely rotten. The following reasons can be evoked:
- difficulty with supply;
- scarcity of the wood, difficulty finding sufficient amounts;
- cost;
- conservation under the snow and during winter frost.

Temperature variations spoiled the quality, in particular wood would split. The cold was dangerous for the planking, and the melting of the snow could cause even more damage to the quality of the wood. Therefore, ships had to be built in one year in order to avoid these pernicious conditions. In fact, the construction of vessels that could carry a load of 500 tons was adapted to the forest features of the colony. Wood on the return (old wood) was common in Canada, whereas it did not exist in France. During this period, the French forest seemed to be well managed and renewed.

Small boats

In addition to the building of cargo ships, shipyards producing small boats were numerous in the new colony, particularly in the XVIIIth century. These boats were traditionally used by the Indians in order to travel by water. Voyageurs used them commonly, but it seems that the sedentary colonists also used them for their local travels. Some are mentioned in post-mortem inventories. These open boats were made of birch and could last up to 5-6 years. It is a little surprising to note that the Iroquois used many types of bark. According to the Jesuits who observed these practices, their boats did not last more than a month. These open boats measured approximately 20 feet long by 2 feet wide.

In regard to the "French" open boats, they were one-design boats cut from the trunk of a pine. The bottom was covered by hardwood, which was used as a keel.

Launches and barges involved more complex constructions with assembling, watches and ribs. The launch is sometimes called a "water car", since it had larger dimensions and presented settlers with the possibility of transporting passengers and goods in greater quantities. Launches were used to load or unload sea-going vessels when they arrived in the harbour. Boats can be associated with launches and barges, because of their size. An open boat built in 1683 by Jean Langlois had 3 masts, measured 37 feet of keel and 12 feet of beams and could transport a load 30 tons.

In 1698, Moïse Hilleret, master carpenter of Ville-Marie, undertook the construction of a launch with a flat bottom "40 pieds de quille portant sur terre, de 13 pieds et demy de large, 7 pieds de fond de calle [...] Laquelle barque sera construite toutte de chesne blanc à la réserve du pont quy sera de pain. Tout lequel bois sera fourni par ledit Hilleret [...]" (December 9, 1698, notary Anthoine Adhémar, 4477, AJM).

The natives would look for the biggest birch trees in order to remove the bark, which was used for making boats "Les Amérindiens recherchaient les plus gros pour lever les écorces propre à faire les canots." (Gedeon de Catalogne, 1712). White oak (*Quercus alba*) was used for building larger boats because it could resist intensive wear, could bear shock, and lasted a long time under conditions which are propitious to blight. Contact with water is then unfavourable to preservation. On the other hand, it does not shrink and can be easily riveted.

Occupation and forest management

During the XVIIth century, the Canadian colony provided a model of territory occupation that mirrored French experience, with the technical and logistical support of the metropolis. With the increase in population - births, new waves of immigration - the land that was shared in domains occupied by tenant farmers would provide the ground for an original settlement based on the exploitation of indigenous raw-material resources to emerge. At first, these practices were regulated loosely, or not at all. Step by step, governors and administrators, pushed by instructions from the royal power, implemented a regulated system. The houses were no longer made of wood, in order to avoid fire.

An official decree prohibited settlers from making fire in their courtyards, under penalty of a 100 pound fine: "ayant été informé par les officiers de police de cette ville, que plusieurs particuliers s'ingèrent de faire du feu dans leurs cours pour y faire de la bière ou pour d'autre usage, ce qui peut occasionner des incendies par les étincelles qui volent et qui peuvent s'attacher sur les écuries ou autres petits bâtiments [...] pourquoi nous faisons très expresses inhibitions et défenses à toutes personnes, de quelques qualité et condition qu'elle soient, de faire du feu dans les cours." (May 30, 1754 - Bigot, vol. 39, fol. 95).

The oak would then be reserved for shipbuilding, an important activity in Canada. A bill of act of 1670 concerning the protection of wood and shipbuilding ordered all inhabitants to give the oaks on their property to the colonial government: "Nous ordonnons à tous les habitants de ce pays ses Sujets de quelques conditsions et qualité qu'ils soient de donner par déclaration les chesnayes qui se trouveront sur leurs terres [...]" (Mr Talon; September 2, 1670, king, orders of the administrators, t. 1).

Another order prohibited owners of some regions from felling any oaks which could be suitable for the building of royal vessels: "Défense de couper aucun chêne [...] jusqu'à ce que nous en ayons fait faire la visite et que nous ayons fait marquer et retenir ceux des dits chênes qui se trouveront propres pour la construction des vaisseaux [...] à peine contre les contrevenants de confiscation des bois coupés et de dix livres d'amende, pour chaque pied d'arbre de chêne qu'ils auront ainsi indûment coupés." (Hocquart, March 20, 1740, vol. 28, fol. 23).

Permission was given to exploit 2000 cubic feet of oak on several domains for the building of a 500 ton cargo vessel for the King's service: "Il est permis d'exploiter [...] de bois de chêne, suivant les gabaris et modèles que nous avons fait remettre pour servir à la construction d 'une flûte [...] lesquels bois il fera conduire en cageux jusque dans la rivière Saint-Charles, devant le palais de la ville, pour y être reçus et visités en la manière accontumée" (Hocquart, October 15, 1731, vol. 19, fol. 151).

The implementation of harsh regulations on wood use and exploitation shows the supply problems that the carpenters faced for shipbuilding, since the wood was often "rotten or spoilt".

Eventually, the administrators passed legislation on the exploitation wood on land properties. One bill of act shows that the Lords took great precautions in the contracts that they gave to their tenant farmers, adding a clause that would make it possible for them to take all the wood that could potentially be necessary for them: "Les habitants de l'Isle de Montréal nous ayant représenté que dans leurs contrats de concessions, il y a une clause par laquelle les seigneurs de la dite isle doivent prendre tous les bois qui leur seront nécessaires dans les terres à eux concédées, laquelle clause n'étant pas bien expliquée, mettroit les dits seigneurs en état de pouvoir prendre sur chaque concession une si grande quantité de bois, que cela pourroit ruiner d'un coup toute leur habitation. [...] faire venir pardevant le sieur Cailhé afin qu'il nous explique cette clause et pour savoir de lui si l'intention des seigneurs est de prendre sur chacune habitation les bois qui leur sont nécessaires, tant de chauffage que de charpente, clôture et autres [...] que néanmoins les dits seigneurs voulant traiter favorablement les dits habitants, et leur ôter toutes les inquiétides qu 'ils pourroient avoir la-dessus" (Raudot, July 2,1706, vol. 1, fol. 55).

Conclusion

The study of building methods and of the extracts used emphasises the adaptation of the population to the available material. This adaptation is also obvious in the way natural resources were exploited. The wood chosen to build houses and for heating met specific criteria The colonists quickly learned how to choose the materials that were best suited to their needs, either from their own experience or from native Indian populations in the region.

Canadian fishermen chose wood types often used by Indians. However, for constructing large ships, which was not practiced by the Indians, the wood had to be of a very specific quality. For centuries, the Canadian administrators continued searching for the most appropriate wood. Often, they found it only after many trials. Lacking the extracts that were usually used by the French, carpenters had to adapt to new materials.

References

BERNATCHEZ L. 1991 - *Guide des poissons d'eau douce du Québec*. Ed. Broquet, La Prairie, 304 p.

CATALOGNE 1709 – Carte du gouvernement de Québec levée en l'année 1709 par le Sr Catalogne lieutenant des troupes et dressée par J. Bapt Decouagne. *Bulletin des recherches historiques*, XXI, IX (sep), 257-269; XXI, X; XXI, XI (nov) 321 p.

CHARLEVOIX F.-X. 1712 - Histoire et description générale de la Nouvelle France avec le journal historique d'un voyage fait par ordre du Roi dans l'Amérique septentrionale. *Encyclopédie ou Dictionnaire raisonné des sciences, des arts et métiers*, Lausanne, Société typographique, 1780-1782, 36 vol.

LECLERC M. 1995 - *Appropriation de l'espace et urbanisation d'un site de la Basse ville de Québec. La première campagne de l'îlot Hunt* (1991), Mem de Maitrise Univ Laval, Quebec, 168 p.

MCELHANNEY T.A. 1940 - *Les bois du Canada, leurs propriétés et leurs usages*, Imp du Roi, Ottawa, 358 p.

MOUSSETTE M. 1979 - *La pêche sur le Saint Laurent*, Boréal Express, Montréal, 212 p.

MOUSSETTE M. 1983 - *Le chauffage domestique en Nouvelle-France*.

ROY P.-G. 1919 - Inventaire des ordonnances des Intendants de Nouvelle-France conservées aux archives provinciales. Ed. *L'Eclaireur*, Beauceville, 4 vol.

SEGUIN R. 1963 - *L'habitant aux XVIIe XVIIIe siècles*, Thèse Université Laval, 3 vol.

TRUDEL M. 1963 - *Histoire de la Nouvelle-France*, Fidès, Montréal, 3 t.

www.ingramcontent.com/pod-product-compliance
Ingram Content Group UK Ltd.
Pitfield, Milton Keynes, MK11 3LW, UK
UKHW061846041026
15591UKWH00015B/928

9 781841 714431